Real-Variable Methods in Harmonic Analysis

Alberto Torchinsky

Department of Mathematics
Indiana University
Bloomington, Indiana

Dover Publications, Inc.
Mineola, New York

Copyright

Bibliographical Note

This Dover edition, first published in 2004, is an unabridged republication of the work originally published in 1986 by Academic Press, Inc., Orlando, Florida.

Library of Congress Cataloging-in-Publication Data

Torchinsky, Alberto.
Real-variable methods in harmonic analysis / Alberto Torchinsky.
p. cm.
Originally published: Orlando : Academic Press, 1986, in series: Pure and applied mathematics.
Includes bibliographical references and index.
ISBN-13: 978-0-486-43508-4 (pbk.)
ISBN-10: 0-486-43508-3 (pbk.)
1. Harmonic analysis. I. Title.

QA403.T57 2004
515'.2433—dc22

2003070050

www.doverpublications.com

To Massi

Contents

Chapter IV The Basic Principles

Chapter V The Hilbert Transform and Multipliers

Chapter VI Paley's Theorem and Fractional Integration

Chapter VII Harmonic and Subharmonic Functions

Chapter VIII Oscillation of Functions

Chapter IX A_p Weights

Chapter X More about R^n

Chapter XI Calderón-Zygmund Singular Integral Operators

Chapter XII The Littlewood-Paley Theory

Chapter XIII The Good λ Principle

Chapter XIV **Hardy Spaces of Several Real Variables**

Chapter XV **Carleson Measures**

Chapter XVI **Cauchy Integrals on Lipschitz Curves**

Chapter XVII **Boundary Value Problems on C^1-Domains**

Preface

This book is based on a set of notes from a course I gave at Indiana University during the academic year 1984–1985. My purpose in those lectures was to present some recent topics in harmonic analysis to graduate students with varied backgrounds and interests, ranging from operator theory to partial differential equations. The book is an exploration of the unity of several areas in harmonic analysis, emphasizing real-variable methods, and leading to the study of active areas of research including the Calderón–Zygmund theory of singular integral operators, the Muckenhoupt theory of A_p weights, the Fefferman–Stein theory of H^p spaces, the Burkholder–Gundy theory of good λ inequalities, and the Calderón theory of commutators.

Because I wanted this book to be essentially self-contained for those students with an elementary knowledge of the Lebesgue integral and since ideas rather than generality are stressed, the point of departure is the classical question of convergence of Fourier series of functions and distributions. Chapter I deals with pointwise convergence, Chapter II with Cesàro $(C, 1)$ convergence, Chapters III and V with norm convergence and Chapter VII with Abel convergence. Chapter IV contains the basic working principles of harmonic analysis, centered around the Calderón–Zygmund decomposition of locally integrable functions. Chapter VI discusses fractional integration, and Chapter VIII the John–Nirenberg class of *BMO* functions. A one semester course in Fourier series can easily be extracted from these first eight chapters.

From this point on our setting becomes R^n. In Chapter IX the Muckenhoupt theory of A_p weights is developed, and in Chapter X, in addition to briefly reviewing the previous results in this new context, elliptic equations in divergence form are treated. Chapter XI deals with the essentials of the Calderón–Zygmund theory of singular integral operators and Chapter XII with its vector-valued version, Littlewood–Paley theory.

Chapter XIII covers the good λ inequalities of Burkholder–Gundy, Chapter XIV the Fefferman–Stein theory of Hardy spaces of several real variables, and Chapter XV Carleson measures. Chapter XVI contains the Coifman–McIntosh–Meyer real variable approach to Calderón's commutator theorem and Chapter XVII one of its interesting applications, namely, the solution to the Dirichlet and Neumann problems on a C^1 domain by means of the layer potential methods. This second half of the book is easily adapted to a one- or two-semester topics course in harmonic analysis.

A word about where the material covered in the book fits into the existing literature: The first part of the book is essentially contained in Zygmund's treatise, where the so-called complex method is emphasized, and precedes Stein's book on singular integrals and differentiability properties of functions; the second half continues with the material discussed in Stein's book. These are the two basic sources of reference that my generation of analysts grew up with.

The notations used are standard, and we remark here only that c denotes a constant which may differ at different occurrences, even in the same chain of inequalities. "Theorem 3.2" means that the result alluded to appears as the second item in Section 3 of the same chapter, and "Theorem 3.2 in Chapter X" means that it appears as the second item in the third section of Chapter X. The same convention is used for formulas.

In order to encourage the active participation of the reader, numerous hints are provided for the problems; I hope the book will be "user friendly." It is not meant, however, to make the learning of the material effortless; many of the ideas discussed lie at the very heart of harmonic analysis and as such require some thought.

It is always a pleasure to acknowledge the contribution of those who make a project of this nature possible. A. P. Calderón, a singular analyst and teacher, has always been a source of inspiration to me; his decisive influence in contemporary harmonic analysis and its applications should be apparent to anyone browsing these pages. My colleague B. Jawerth shared with me his ideas on how results should, and should not, be presented. My largest debt, though, is to the students who attended the course and kept me honest when a simple "the proof is easy" was tempting. They are Alp Eden, Don Krug, Hung-Ju Kuo, Paul McGuire, Mohammad Rammaha, Edriss Titi, and Sung Hyun Yoon. The manuscript was cheerfully typed by Storme Day. The staff at Academic Press handled all my questions promptly and efficiently.

Real-Variable Methods in Harmonic Analysis

CHAPTER

I

Fourier Series

1. FOURIER SERIES OF FUNCTIONS

A trigonometric polynomial $p(t)$ is an expression of the form

$$p(t) = \sum_{|j| \leq n} c_j e^{ijt}, \qquad |c_n| + |c_{-n}| \neq 0. \tag{1.1}$$

n is the degree of p and the c_j's are (possibly complex) constants. Thus p is a continuous function of period 2π and is therefore determined by its values on $T = (-\pi, \pi]$, or any other interval of length 2π for that matter. On the other hand, given a trigonometric polynomial p of degree $\leq n$, we can easily compute the constants c_j by means of

$$c_j = \frac{1}{2\pi} \int_T p(t) e^{-ijt}\, dt, \qquad |j| \leq n.$$

This observation follows at once from the fact that

$$\frac{1}{2\pi} \int_T e^{ijt}\, dt = \begin{cases} 0 & \text{if } j \neq 0, \\ 1 & \text{if } j = 0. \end{cases} \tag{1.2}$$

A trigonometric series is an expression of the form

$$\sum_{j=-\infty}^{\infty} c_j e^{ijt}. \tag{1.3}$$

Since we make no assumption concerning the convergence of this series, (1.3) only formally represents a function of period 2π.

A Fourier series is a trigonometric series for which there is a periodic, Lebesgue summable function f such that

$$c_j = c_j(f) = \frac{1}{2\pi}\int_T f(t)e^{-ijt}\,dt, \qquad \text{all } j. \tag{1.4}$$

In this case we call the constants c_j the Fourier coefficients of f and denote this correspondence by

$$f \sim \sum_j c_j e^{ijt}. \tag{1.5}$$

A word about the class of functions involved in this definition. It is denoted by $L(T)$ and it consists of those periodic, Lebesgue measurable functions f with finite L^1 norm, i.e.,

$$\|f\|_1 = \frac{1}{2\pi}\int_T |f(t)|\,dt < \infty.$$

Endowed with this norm and modulo functions which coincide a.e., $L(T)$ becomes a Banach space, one in the scale of $L^p(T)$ spaces, where for $1 \leq p < \infty$

$$L^p(T) = \Big\{ f \text{ periodic, measurable: } f: T \to C, \text{ and}$$

$$\|f\|_p = \left(\frac{1}{2\pi}\int_T |f(t)|^p\,dt\right)^{1/p} < \infty \Big\}. \tag{1.6}$$

When $p = \infty$, we define the norm in $L^\infty(T)$ as the limiting expression in (1.6) as $p \to \infty$. It turns out that $L^\infty(T)$ is also a Banach space with norm $\|f\|_\infty = \text{ess sup}_T |f(t)|$. By a mild abuse of notation we also denote by $\|f\|_p$ the quantity appearing in (1.6) when $0 < p < 1$, although in this case the triangle inequality is not satisfied and elements in $L^p(T)$ are not necessarily locally integrable functions.

Still in the case of Fourier series no assumption concerning the convergence of the series (1.5) is made. More specifically, if $s_n(f, t)$ denotes the trigonometric polynomial of degree $\leq n$ corresponding to the symmetric partial sum of (1.5) of order n, i.e.,

$$s_n(f, t) = \sum_{|j| \leq n} c_j e^{ijt}, \tag{1.7}$$

then nothing is known or assumed about the existence of the $\lim_{n\to\infty} s_n(f, t)$ for any $t \in T$.

At times, and especially when dealing with examples, it is convenient to work with the so-called Fourier cosine and sine series of f; this is interchangeable with (1.5). Indeed, suppose, as we often do, that f is real. Then

$$\begin{aligned} s_n(f, t) &= c_0 + \sum_{j=1}^{n} (c_j e^{ijt} + c_{-j} e^{-ijt}) \\ &= c_0 + \sum_{j=1}^{n} (c_j + c_{-j}) \cos jt + i(c_j - c_{-j}) \sin jt \\ &= \frac{a_0}{2} + \sum_{j=1}^{n} a_j \cos jt + b_j \sin jt, \end{aligned} \tag{1.8}$$

say, where the a_j's and b_j's are real since $c_{-j} = \overline{c_j}$. Conversely, given a cosine and sine series we may recover (1.5) by letting $2c_j = a_j - ib_j$, and

$$a_j = \frac{1}{\pi} \int_T f(t) \cos jt \, dt, \qquad b_j = \frac{1}{\pi} \int_T f(t) \sin jt \, dt. \tag{1.9}$$

If the function f is even, that is $f(t) = f(-t)$, the coefficients b_j vanish and the integral defining a_j may be replaced by twice the integral over $(0, \pi)$. If f is odd, that is $f(t) = -f(-t)$, then $a_j = 0$ and the second integral above may be replaced by twice the integral over $(0, \pi)$.

Harmonic analysis studies, in a broad sense, properties of the series (1.3) and (1.5). For instance, since the sign $\sim$ in (1.5) only means that the constants c_j and the function f are connected by the formula (1.4), an important problem is to determine if, and how, the Fourier series of a function represents, or converges to, that function. We address the problem of pointwise convergence in this chapter, that of Cesàro summability in Chapter II, Abel summability in Chapter VII, and norm convergence in Chapters III and V.

We begin our discussion with some general observations concerning Fourier series. In first place note that if the partial sums of a trigonometric series (1.3) converge, in some general sense, to a function $f \in L(T)$, then actually $c_j = c_j(f)$. More precisely,

Proposition 1.1. If the symmetric partial sums $s_n(t)$ of the trigonometric series (1.3) converge in L^1 norm to $f \in L(T)$, then $c_j = c_j(f)$.

Proof. Fix an integer j and observe that the sequence $f_n(t) = (f(t) - s_n(t))e^{-ijt}$, $n = 0, 1, \ldots$ converges to 0 in $L(T)$. Moreover

$$c_j(f) = \frac{1}{2\pi} \int_T (f(t) - s_n(t)) e^{-ijt} \, dt + \frac{1}{2\pi} \int_T s_n(t) e^{-ijt} \, dt.$$

By (1.2) we readily see that the second integral above is c_j as soon as $|n| \geq j$. Therefore $|c_j(f) - c_j| \leq \|f_n\|_1 \to 0$, as $n \to \infty$, and consequently, $c_j(f) = c_j$. ■

The reader will observe that the conclusion of Proposition 1.1 also obtains from a weaker assumption, namely, the existence of a sequence $n_j \to \infty$ such that $s_{n_j}(t)$ converges to f in L^1; because such extensions are trivial we prefer to omit them unless they are clearly important. On the other hand, in the course of the above proof we have made use of the interesting fact that

$$|c_j(f)| \leq \|f\|_1, \qquad \text{all } j, \tag{1.10}$$

and there is more we can say in this direction.

Theorem 1.2 (Riemann–Lebesgue). Let $f \in L(T)$. Then $c_j \to 0$ as $|j| \to \infty$.

Proof. We invoke the well-known fact that trigonometric polynomials are dense in $L(T)$; a proof of this is given in Proposition 2.4 of Chapter II. Now, given $\varepsilon > 0$ we show that $|c_j| \leq \varepsilon$ provided $|j| > n_0$ is large enough. Let p be a trigonometric polynomial such that $\|f - p\|_1 \leq \varepsilon$, and let $n_0 =$ degree of p. Then for $|j| > n_0$ we have

$$c_j = \frac{1}{2\pi}\int_T (f(t) - p(t))e^{-ijt}\,dt$$

and consequently $|c_j| \leq \|f - p\|_1 \leq \varepsilon$. ■

Now that there is some hope that the Fourier series of $f \in L(T)$ may converge, we take a closer look at $s_n(f, x)$. It can also be written as

$$\begin{aligned} s_n(f, x) &= \sum_{|j| \leq n} \left(\frac{1}{2\pi}\int_T f(t)e^{-ijt}\,dt\right)e^{ijx} \\ &= \frac{1}{\pi}\int_T f(t)\left(\frac{1}{2}\sum_{|j|\leq n} e^{ij(x-t)}\right)dt = \frac{1}{\pi}\int_T f(t)D_n(x-t)\,dt, \end{aligned} \tag{1.11}$$

say, where we have denoted by

$$D_n(t) = \frac{1}{2}\sum_{|j|\leq n} e^{ijt}, \qquad n = 0, 1, \ldots \tag{1.12}$$

the Dirichlet kernel of order n. We list some properties of these kernels. In the first place by summing the geometric series in (1.12) we get

$$D_n(t) = \frac{1}{2}e^{-int}\frac{(e^{i(2n+1)t} - 1)}{(e^{it} - 1)}$$

$$= \frac{1}{2} \frac{e^{i(n+1)t} - e^{-int}}{e^{it/2}(e^{it/2} - e^{-it/2})}$$

$$= \frac{1}{2} \frac{e^{i(n+1/2)t} - e^{-i(n+1/2)t}}{e^{it/2} - e^{-it/2}}$$

$$= \frac{1}{2} \frac{\sin(n + 1/2)t}{\sin(t/2)}, \qquad n = 0, 1, \ldots. \tag{1.13}$$

Thus D_n is an even function, and by (1.2)

$$\frac{1}{\pi} \int_T D_n(t)\, dt = \frac{2}{\pi} \int_{[0,\pi]} D_n(t)\, dt = 1, \qquad \text{all} \quad n. \tag{1.14}$$

It is also possible to estimate $D_n(t)$. In fact by (1.12),

$$|D_n(t)| \leq \frac{1}{2} \sum_{|j| \leq n} |e^{ijt}| = \frac{2n+1}{2} = n + \frac{1}{2}, \qquad \text{all} \quad n. \tag{1.15}$$

Moreover, since as is readily seen

$$1/(2\sin(t/2)) \leq \pi/2t \qquad \text{for} \quad 0 < t < \pi, \tag{1.16}$$

by (1.13) it follows at once that

$$|D_n(t)| \leq \pi/2|t|, \qquad 0 < |t| < \pi, \qquad \text{all} \quad n. \tag{1.17}$$

This is all we need to know about this kernel.

Returning to (1.11), it is useful to replace D_n there by the symmetric expression $D_n^* = (D_{n-1} + D_n)/2$, which equals

$$D_n^*(t) = \frac{\sin((n - 1/2)t) + \sin((n + 1/2)t)}{2\sin(t/2)} = \frac{\sin nt}{2\tan(t/2)}, \qquad \text{all} \quad n. \tag{1.18}$$

Also note that since $D_n(t) - D_n^*(t) = (D_n(t) - D_{n-1}(t))/2 = \cos(nt/2)$, we can rewrite

$$s_n(f, x) = \frac{1}{\pi} \int_T f(t) D_n^*(x - t)\, dt + \frac{1}{2\pi} \int_T f(t) \cos n(x - t)\, dt$$

$$= s_n^*(f, x) + A_n, \tag{1.19}$$

say. We claim that the term A_n above is an "error term," in the sense that it tends to 0 as $n \to \infty$, uniformly in x, and may therefore be disregarded. This is easy to see since A_n equals

$$\cos(nx)\frac{1}{\pi} \int_T f(t) \cos nt\, dt + \sin(nx)\frac{1}{\pi} \int_T f(t) \sin nt\, dt,$$

and by the Riemann-Lebesgue theorem both integrals go to 0 as $n \to \infty$ and the factors in front of them are uniformly bounded by 1 for x in T.

A shorthand notation is useful to express this situation. We write $u_n = o(v_n)$ as $n \to \infty$, provided that $v_n > 0$ and $|u_n|/v_n \to 0$ as $n \to \infty$. Thus $u_n = o(1)$ means that $\lim_{n\to\infty}|u_n| = 0$. If on the other hand $|u_n|/v_n$ remains bounded as $n \to \infty$, we write $u_n = O(v_n)$. So $u_n = O(1)$ means that for some constant c, $|u_n| \leq c$, all large n. With this notation (1.19) becomes $s_n(f, x) = s_n^*(f, x) + o(1)$, uniformly for x in T.

It is also possible to introduce another expression closely related to $s_n(f, x)$. First observe that the function

$$\phi(t) = \left(\frac{1}{2\tan(t/2)} - \frac{1}{t}\right) \in L^\infty(T). \tag{1.20}$$

So if $f \in L(T)$, then $f\phi \in L(T)$, and, consequently, again by the Riemann-Lebesgue Theorem and (1.19),

$$s_n(f, x) = \frac{1}{\pi}\int_T f(t)\frac{\sin n(x-t)}{x-t}\,dt + o(1). \tag{1.21}$$

Returning to D_n^* we list some of its properties. From (1.18) it readily follows that it also is an even function, and from (1.14) that

$$\frac{1}{\pi}\int_T D_n^*(t)\,dt = \frac{2}{\pi}\int_{[0,\pi]} D_n^*(t)\,dt = 1, \qquad \text{all} \quad n. \tag{1.22}$$

Also estimates (1.15) and (1.17) have a counterpart, to wit

$$|D_n^*(t)| \leq \frac{(n-1)+\frac{1}{2}+n+\frac{1}{2}}{2} = n, \qquad \text{all} \quad n \tag{1.23}$$

and

$$|D_n^*(t)| \leq \pi/2|t|, \qquad 0 < |t| < \pi. \tag{1.24}$$

As for (1.11), since D_n^* is even we also have that $s_n^*(f, x)$ equals either

$$\frac{1}{\pi}\int_T f(t)D_n^*(x-t)\,dt \quad \text{or} \quad \frac{1}{\pi}\int_T \frac{(f(x+t)+f(x-t))}{2}D_n^*(t)\,dt. \tag{1.25}$$

Moreover since the integrand in the last integral above is an even function of t we also have that it equals

$$\frac{1}{\pi}\int_{[0,\pi]} (f(x+t)+f(x-t))D_n^*(t)\,dt. \tag{1.26}$$

We are now ready to prove our first convergence result.

Theorem 1.3 (Dini). Let $f \in L(T)$ and suppose there is a constant A such that for an x in T

$$\int_{[0,\pi]} \left| \frac{f(x+t)+f(x-t)}{2} - A \right| \frac{dt}{t} < \infty. \tag{1.27}$$

Then $\lim_{n\to\infty} s_n(f, x) = A$.

Proof. By (1.19) it suffices to prove the assertion with $s_n(f, x)$ replaced by $s_n^*(f, x)$. Moreover, by (1.26), (1.22), and (1.18), we may write

$$s_n^*(f, x) - A = \frac{2}{\pi} \int_{[0,\pi]} \left(\frac{f(x+t)+f(x-t)}{2} - A \right) \frac{\sin nt}{2\tan(t/2)}\, dt.$$

Now, assumption (1.27) is clearly equivalent to the fact that the function

$$F_x(t) = \left(\frac{f(x+t)+f(x-t)}{2} - A \right) \frac{1}{2\tan(t/2)} \in L(T),$$

since $\tan(t/2) \sim t$ near 0. Therefore $s_n^*(f, x) - A$ is nothing but the nth Fourier sine coefficient of the function $F_x(t) \in L(T)$, which by the Riemann–Lebesgue theorem tends to 0 as $n \to \infty$. ■

A word about the value of A above. If x is a point of a removable discontinuity, or a jump, of f, then A is necessarily $f(x)$ or $(f(x+0) + f(x-0))/2$. Moreover, since functions $f \in L(T)$ are only determined a.e. we may always assume that $A = f(x)$ by changing the value of f at that point if necessary. Also notice that if $f(x+t) - f(x) = O(|t|^\eta)$, $\eta > 0$, then the Fourier series of f converges to $f(x)$ at that x. In particular, this is true if $f'(x)$ exists and is finite. If any of these conditions is satisfied uniformly for x in a closed subinterval of T, then $s_n(f, x)$ converges uniformly to $f(x)$ in that interval.

To state another simple criterion, this one of a.e. nature, we need a definition. We denote by $w_1(f, x)$ the L-modulus of continuity of f, namely,

$$w_1(f, x) = \frac{1}{2\pi} \int_T |f(x+t) - f(t)|\, dt. \tag{1.28}$$

We then have

Theorem 1.4 (Marcinkiewicz). Suppose that $f \in L(T)$ and that

$$\int_{[0,\pi]} w_1(f, t) \frac{dt}{t} < \infty.$$

Then $\lim s_n(f, x) = f(x)$, a.e. in T.

Proof. Let

$$I(x) = \int_{[0,\pi]} |f(x+t) - f(x)| \frac{dt}{t} \geqslant 0.$$

By Tonelli's theorem we have

$$\frac{1}{2\pi}\int_T I(x)\,dx = \int_{[0,\pi]} \frac{1}{2\pi}\int_T |f(x+t) - f(x)|\,dx \frac{dt}{t}$$

$$= \int_{[0,\pi]} w_1(f, t)\frac{dt}{t} < \infty.$$

Consequently, $I(x) < \infty$ a.e. in T, and therefore also

$$\int_{[0,\pi]} |f(x+t) + f(x-t) - 2f(x)| \frac{dt}{t} < \infty \qquad \text{a.e. in} \quad T.$$

This implies that Dini's theorem applies with $A = f(x)$ a.e. in T. ∎

2. FOURIER SERIES OF CONTINUOUS FUNCTIONS

Although at this point we may intuitively guess that Dini's theorem suffices to assure the convergence of $s_n(f, x)$ to $f(x)$ at a point of continuity of f, nothing could be further from the truth. Indeed, the expression

$$\int_{[0,\pi]} \left| \frac{f(x+t) + f(x-t)}{2} - f(x) \right| \frac{dt}{t}$$

may diverge everywhere in T, even for a continuous function f (and even with the absolute values removed from the integral); a closely related result will be discussed in Proposition 5.1 of Chapter III. Now we turn around and guess that there may exist a continuous function whose Fourier series does not converge at a point. Statements of this nature are supported in one of two ways: either by constructing a specific function with the desired property or else by assuming that no such function exists and reaching a contradiction. Since each method has its appeal and usefulness, we present both here in our successful quest for a continuous function with a nonconvergent Fourier series at $x = 0$.

We begin by considering the so-called Lebesgue constants L_n. They are given by

$$L_n = 2\|D_n\|_1 = \frac{1}{\pi}\int_T |D_n(t)|\,dt, \qquad n \geqslant 0.$$

This is why. By (1.11) it is plain that

$$|s_n(f,0)| \leqslant \frac{1}{\pi}\int_T |f(t)||D_n(t)|\,dt \leqslant \|f\|_\infty L_n.$$

Therefore by setting $f(t) = \operatorname{sgn} D_n(t)$, $t \in T$, i.e.,

$$f_n(t) = \begin{cases} 1 & D_n(t) > 0 \\ 0 & D_n(t) = 0 \\ -1 & D_n(t) < 0 \end{cases}$$

we readily see that $\|f_n\|_\infty = 1$ and

$$\sup_{f\in L^\infty(T),\|f\|_\infty\leqslant 1} |s_n(f,0)| = L_n.$$

Since the function $f_n(t)$ is real valued and discontinuous at a finite number of points, it is easy to modify its values in small neighborhoods of those points to obtain, now, that also for continuous functions

$$\sup_{f\in C(T),\|f\|_\infty\leqslant 1} |s_n(f,0)| = L_n. \tag{2.1}$$

It becomes, then, important to study the behavior of L_n for large n.

Proposition 2.1. $L_n \sim (4/\pi^2)\ln n$, as $n \to \infty$.

Proof. Since $D_n(t)$ is even and $\sin(t/2) > 0$ for $0 < t < \pi$, we have that

$$\begin{aligned} L_n &= \frac{2}{\pi}\int_{[0,\pi]} \left|\sin\left(\left(n+\frac{1}{2}\right)t\right)\right|\left(\frac{1}{2\sin(t/2)} - \frac{1}{t}\right) dt \\ &\quad + \frac{2}{\pi}\int_{[0,\pi]} \left|\sin\left(\left(n+\frac{1}{2}\right)t\right)\right| \frac{dt}{t} = A_n + B_n, \end{aligned}$$

say. By a statement similar to (1.20) we see at once that $A_n = O(1)$. We take a look at B_n now. The change of variables $(n+\frac{1}{2})t = s$ gives

$$\begin{aligned} B_n &= \frac{2}{\pi}\int_{[0,(n+1/2)\pi]} |\sin s|\frac{ds}{s} \\ &= \frac{2}{\pi}\int_{[\pi,n\pi]} |\sin s|\frac{ds}{s} + O(1) = B_n' + O(1), \end{aligned}$$

say. Thus we will be done once we show that

$$B_n' \cong \frac{4}{\pi^2}\ln n + O(1). \tag{2.2}$$

We rewrite

$$B'_n = \frac{2}{\pi}\sum_{k=1}^{n-1}\int_{[k\pi,(k+1)\pi]}\frac{|\sin s|}{s}\,ds$$
$$= \frac{2}{\pi}\sum_{k=1}^{n-1}\int_{[0,\pi]}\frac{|\sin(k\pi+t)|}{k\pi+t}\,dt$$
$$= \frac{2}{\pi}\int_{[0,\pi]}(\sin t)\left\{\sum_{k=1}^{n-1}\frac{1}{k\pi+t}\right\}dt.$$

The expression in $\{\cdot\}$ in the above integral can be estimated below and above, uniformly for $t \in (0, \pi]$, by

$$\frac{1}{\pi}\sum_{k=1}^{n-1}\frac{1}{k+1} = \frac{1}{\pi}\sum_{k=1}^{n}\frac{1}{k} - \frac{1}{\pi} \quad \text{and} \quad \frac{1}{\pi}\sum_{k=1}^{n-1}\frac{1}{k}, \tag{2.3}$$

respectively. By (2.3) then, and since $\int_{[0,\pi]}\sin t\,dt = 2$, we finally obtain

$$\frac{4}{\pi^2}\sum_{k=1}^{n}\frac{1}{k} - O(1) \leqslant B'_n \leqslant \frac{4}{\pi^2}\sum_{k=1}^{n}\frac{1}{k}.$$

In other words (2.2) holds and we are done. ■

Corollary 2.2. If $f \in L^\infty(T)$, then $s_n(f, x) = O(\ln n)$.

We now know that for each (large) n there is a continuous function f, $|f(x)| \leqslant 1$, and

$$|s_n(f, 0)| \sim \frac{4}{\pi^2}\ln n. \tag{2.4}$$

It is natural then to search for a single continuous function f whose Fourier series has large partial sums at 0. Assuming that no such function exists we will reach a contradiction. Suppose, then, that the Fourier series of every continuous function converges at 0; in particular, the partial sums will be bounded there, i.e., $|s_n(f, 0)| \leqslant c_f < \infty$, all n, each $f \in C(T)$. By (1.11) this is equivalent to

$$\left|\int_T f(t)D_n(t)\,dt\right| \leqslant c_f < \infty, \qquad \text{all} \quad n, \qquad \text{each} \quad f \in C(T). \tag{2.5}$$

We now show that (2.5) cannot hold. Since the idea needed to do this can also be used in other settings, we prefer to cast the statement in a general context. In the application of this general result we will make use of the well known fact that $C(T)$ is a complete metric space, and that therefore any decreasing sequence of closed balls with radius approaching 0, has a nonempty intersection (consisting of a single point).

We state and prove the Uniform Boundedness Principle.

Theorem 2.3 (Banach–Steinhaus). Let X be a complete metric space and let Y be a normed linear space. Furthermore, let $\{T_\alpha\}_{\alpha\in A}$ be a family of bounded linear operators from X into Y with the property that for each $x \in X$ the family $\{T_\alpha x\}_{\alpha\in A}$ is bounded in Y, i.e., $\|T_\alpha x\|_Y \leqslant c_x < \infty$, all $\alpha \in A$. Then the family T_α is uniformly bounded, in other words there is a constant c such that

$$\sup_{\|x\|_X \leqslant 1} \|T_\alpha x\|_Y \leqslant c, \qquad \text{all} \quad \alpha \in A.$$

Proof. Suppose we can show that for some $x_0 \in X$, $\varepsilon > 0$ and a constant K we have $\|T_\alpha x\|_Y \leqslant K$ whenever $\|x_0 - x\|_X \leqslant \varepsilon$, i.e., the family $\{T_\alpha x\}_{\alpha\in A}$ is uniformly bounded at a ball $B(x_0, \varepsilon)$ about x_0. Then we are done. Indeed, for $x \neq 0$, $\|x\|_X \leqslant 1$, we put $z = \varepsilon x/\|x\|_X + x_0 \in B(x_0, \varepsilon)$. Then $\|T_\alpha z\|_X \leqslant K$ and by the triangle inequality

$$\frac{\varepsilon}{\|x\|_X}\|T_\alpha x\|_Y - \|T_\alpha x_0\|_Y \leqslant \|T_\alpha z\|_Y \leqslant K. \tag{2.6}$$

Letting $c = K + \sup_\alpha \|T_\alpha x_0\|_Y < \infty$, we may rewrite (2.6) as $\|T_\alpha x\|_Y \leqslant c/\varepsilon$, c independent of α, which is precisely what we wanted to prove. So, to complete the proof we must show that such a ball exists. We argue by contradiction and assume no such ball exists.

Fix a ball $B_0 = B(x_0, 2)$, then there exist $x_1 \in B_0$ and $\alpha_1 \in A$ such that $\|T_{\alpha_1} x_1\|_Y > 1$. Also by continuity $\|T_{\alpha_1} x\|_Y > 1$, $x \in B_1 = B(x_1, \varepsilon_1) \subseteq B_0$, $\varepsilon_1 < 1$. The family $\{T_\alpha x\}$ is still not uniformly bounded on B_1. So recursively, and after $B_0 \supseteq B_1 \supseteq \cdots \supseteq B_{k-1}$ have been chosen, with radius $\varepsilon_j < 1/j$ and centers x_j such that $\|T_{\alpha_j} x_j\|_Y > j$, $1 \leqslant j \leqslant k-1$, we then select a ball $B_k = B(x_k, \varepsilon_k)$, $B_k \subseteq B_{k-1}$, $\varepsilon_k < 1/k$, $\alpha_k \neq \alpha_j$, $j < k$, and $\|T_{\alpha_k} x\|_Y > k$ for x in B_k. Since X is complete there is a point $z \in \bar{B}\,(x_k, \varepsilon_k)$ for all k. The fact that $\|T_{\alpha_k} z\|_Y \geqslant k$, all k, contradicts the assumption that $\{T_\alpha z\}_{\alpha\in A}$ is bounded. ■

As anticipated we apply the theorem with $X = C(T)$, $Y = C$, and put

$$T_n f = s_n(f, 0) = \frac{1}{\pi}\int_T f(t) D_n(t)\, dt, \; n \geqslant 0.$$

By (2.1) and (2.4),

$$\sup_{\|f\|_{C(T)} \leqslant 1} |T_n f| \sim \frac{4}{\pi^2} \ln n. \tag{2.7}$$

Now, were (2.5) to hold, then by the Uniform Boundedness Principle, (2.7) would imply there is a constant c such that $(4/\pi^2)\ln n \leqslant c$, all n, which is impossible. Therefore there is a continuous function f such that

$\sup_n|s_n(f,0)| = \infty$. Clearly this function cannot have a convergent Fourier series at 0.

We proceed now to give a constructive proof for the existence of such a function. The argument that follows is due to Lebesgue and basically consists in taking a close look at (1.21). In first place it suffices to define f in $(0, \pi]$ and then extend it as an even function. f will be of the form

$$f(x) = \sum_{k=1}^{\infty} c_k \sin(n_k t)\chi_{I_k}(t), \qquad 0 < t < \pi,$$

where $c_k \to 0$ and χ_{I_k} = the characteristic function of the interval $[\pi/n_k, \pi/n_{k-1})$ are yet to be chosen. It is clear from (1.21) that we should expect the main contribution to $s_{n_k}(f,0)$ to come from the kth summand, namely $c_k \sin(n_k t)\chi_{I_k}(t)$; we make this statement precise. We start by letting $c_1 = 1$, $n_1 = 2$ and $I_1 = (\pi/2, \pi]$. f is defined to be $c_1 \sin(n_1 t)$ on I_1. Having chosen $n_1 < \cdots < n_{k-1}$, $c_1, \dots, c_{k-1}$ and the corresponding I_j's, we set

$$\phi(t) = \begin{cases} \sum_{j=1}^{k-1} c_j \sin(n_j t)\chi_{I_j}(t) & \text{if} \quad t \in (\pi/n_{k-1}, \pi] \\ 0 & \text{otherwise.} \end{cases}$$

Then clearly $\phi(t)/t$ is bounded, and by the Riemann–Lebesgue theorem

$$\lim_{n\to\infty} \int_{[0,\pi]} \frac{\phi(t)}{t} \sin nt \, dt = 0.$$

Pick now $n_k = n_1 \cdots n_{k-1} N_k$, $N_k \geq 2^k$, large enough so that

$$\left| \frac{2}{\pi} \int_{[0,\pi]} \frac{\phi(t)}{t} \sin(n_k t) \, dt \right| < 1. \tag{2.8}$$

With this choice of n_k we let $I_k = (\pi/n_k, \pi/n_{k-1}]$ and put $f(t) = c_k \sin(n_k t)$ in I_k; $c_k < 1$ is yet to be chosen. To estimate the partial sums $s_{n_k}(f,0)$ we consider

$$\frac{2}{\pi} \int_{(0,\pi]} f(t) \sin n_k t \frac{dt}{t} = \frac{2}{\pi}\left(\int_{(0,\pi/n_k]} + \int_{(\pi/n_k, \pi/n_{k-1}]} + \int_{(\pi/n_{k-1},\pi]} \right)$$
$$= A_k + B_k + C_k,$$

say. By (2.8) $C_k = O(1)$. Also, and independently of the choice of the c_k's,

$$|A_k| \leq \int_{(0,\pi/n_k]} |\sin n_k t| \frac{dt}{t} \leq c n_k \frac{\pi}{n_k} = O(1).$$

Finally we consider the B_k's. We have

$$B_k = c_k \int_{I_k} (\sin n_k t)^2 \frac{dt}{t} = c_k \int_{I_k} \frac{1-\cos(2n_k t)}{2t}\, dt$$
$$= c_k \frac{1}{2} \int_{I_k} \frac{dt}{t} - \frac{c_k}{2} \int_{I_k} \cos(2n_k t) \frac{dt}{t} = B'_k - B''_k,$$

say. The choice $c_k = (\ln N_k)^{-\varepsilon}$, $0 < \varepsilon < 1$, gives that $c_k \to 0$ and $B'_k = \frac{1}{2}(\ln N_k)^{1-\varepsilon} \to \infty$, as $k \to \infty$. Also the integral of the B''_k term equals

$$\frac{\sin(2n_k t)}{2n_k t}\Bigg]_{\pi/n_k}^{\pi/n_{k-1}} + \int_{I_k} \frac{\sin(2n_k t)}{2n_k} \frac{dt}{t^2}.$$

By the choice of the n_k's the integrated term is 0. As for the integral above it is dominated by $(1/2n_k)\int_{[\pi/n_k,\infty)}(dt/t^2) = O(1)$. Collecting the estimates we see that $s_{n_k}(f, 0) \sim \frac{1}{2}(\ln N_k)^{1-\varepsilon} + O(1)$, and the desired conclusion is reached.

Remark 2.4. The function $f(t)$ may also be defined in a single stroke. In fact we may set for $0 < t < \pi$, $f(t) = c_k \sin(a_k t)$ if $\pi/a_k < t \leq \pi/a_{k-1}$, where $c_k = 1/k$ and $a_k = \prod_{j=1}^k 2^j$, and $f(0) = 0$, f even.

The reader may also wish to verify an additional interesting property this function has, namely,

$$\limsup_{h\to 0^+} \int_{(h,\delta]} \frac{|f(t+h)-f(t)|}{t}\, dt > 0, \qquad \text{for every} \quad \delta > 0.$$

3. ELEMENTARY PROPERTIES OF FOURIER SERIES

In this section we discuss some of the elementary formal properties enjoyed by Fourier series; they are quite important in working with them.

In first place the correspondence between f and the sequence $\{c_j\}$, $f \sim \sum c_j e^{ijt}$, is linear, in the sense that $c_j(f + \lambda \bar{g}) = c_j(f) + \lambda \overline{c_{-j}(g)}$, all j, for f, $g \in L(T)$ and scalar λ.

Moreover if a is a real constant, then $f_a(t) = f(t+a) \in L(T)$ and its Fourier coefficients are given by

$$c_j(f_a) = \frac{1}{2\pi}\int_T f(t+a)e^{-ijt}\, dt = e^{ija}c_j(f), \qquad \text{all} \quad j. \tag{3.1}$$

In other words, $f_a \sim \sum c_j(f)e^{ij(a+t)}$, and translation by a in the argument t corresponds to multiplication by e^{ija} in the sequence space. Also, intuitively, the Fourier series of $f(t)e^{int}$ is

$$f(t)e^{int} \sim \sum_j c_{j-n}(f)e^{ijt}. \tag{3.2}$$

This can be verified by a simple computation similar to the one carried out above.

We consider now the usual operations on functions, such as differentiation, integration and convolution, and observe how they are reflected in the sequence space. First, we have

Proposition 3.1. Let $f \in L(T)$ and let $F(t)$ be its indefinite integral,

$$F(t) = c + \int_{(0,t]} f(s)\, ds, \qquad t \in T.$$

Then

$$F(t) - c_0(f)t \sim C_0 + \sum_{j \neq 0} \frac{c_j(f)}{(ij)} e^{ijt}, \qquad C_0 \text{ const.}$$

Proof. Note that

$$F(t+2\pi) - F(t) = \int_{(t,t+2\pi]} f(s)\, ds = \int_T f(s)\, ds = 2\pi c_0(f).$$

So unless $c_0(f) = 0$, F will not be periodic; therefore we consider the function $H(t) = F(t) - c_0(f)t$ instead. This function is periodic, absolutely continuous and $H'(t) = f(t) - c_0(f)$, a.e. in T. We find now the Fourier coefficients of H. If $j \neq 0$, then

$$c_j(H) = \frac{1}{2\pi} \int_T H(t) e^{-ijt}\, dt$$

$$= \frac{1}{2\pi} \frac{H(t)e^{-ijt}}{(-ij)} \Big]_{-\pi}^{\pi} + \frac{1}{2\pi(ij)} \int_T (f(t) - c_0(f)) e^{-ijt}\, dt = \frac{1}{(ij)} c_j(f).$$

Thus

$$F(t) - c_0(f)t = H(t) \sim C_0 + \sum_{j \neq 0} \frac{1}{(ij)} c_j(f) e^{ijt}. \quad \blacksquare$$

Proposition 3.2. If f is absolutely continuous, then

$$f' \sim \sum_j (ij) c_j(f) e^{ijt}.$$

Also, $c_j(f') = o(1/|j|)$.

Proof. We observe that $f(t) = f(-\pi) + \int_{(-\pi,t]} f'(s)\, ds$, and invoke Proposition 3.1. $\blacksquare$

Corollary 3.3. Let $f \in C^k(T)$, i.e., f has k continuous derivatives. Then if $D^k f$ denotes the kth derivative of f,

$$D^k f \sim \sum_j (ij)^k c_j(f) e^{ijt}.$$

Moreover, $c_j(D^k f) = o(1/|j|^k)$.

We consider now the convolution operation in $L(T)$.

Theorem 3.4. Assume $f, g \in L(T)$. Then for almost all x, $f(x-t)g(t)$, as a function of t, is integrable and if we denote by $f * g(x)$ the convolution of f and g at x, namely,

$$f * g(x) = \frac{1}{2\pi} \int_T f(x-t) g(t)\, dt,$$

we have that $f * g \in L(T)$ and $\|f * g\|_1 \leq \|f\|_1 \|g\|_1$. Moreover

$$c_j(f * g) = c_j(f) c_j(g), \qquad \text{all } j. \tag{3.3}$$

Proof. The assertions concerning $f * g$ are all easy consequences of Fubini's theorem. To prove (3.3) we note that

$$c_j(f * g) = \frac{1}{2\pi} \int_T \left(\frac{1}{2\pi} \int_T f(x-t) g(t)\, dt \right) e^{-ijx}\, dx,$$

and observe that, again by Fubini's theorem,

$$\begin{aligned} c_j(f * g) &= \frac{1}{2\pi} \int_T g(t) e^{-ijt} \left(\frac{1}{2\pi} \int_T f(x-t) e^{-ij(x-t)}\, dx \right) dt \\ &= c_j(f) c_j(g). \quad \blacksquare \end{aligned}$$

Note that with the convolution notation $s_n(f, x) = f * 2D_n(x)$. Also algebraic properties of the convolution, such as being commutative, associative, and distributive with respect to addition, are easily verified.

Corollary 3.5. If $f \in L(T)$ and $p(t) = \sum_{|j| \leq N} c_j(p) e^{ijt}$, then $f * p$ is the polynomial of degree $\leq N$,

$$f * p(x) = \sum_{|j| \leq N} c_j(f) c_j(p) e^{ijx}.$$

No new ideas are required to prove our next result either,

Proposition 3.6. Let $f, g, fg \in L(T)$. Then

$$c_j(fg) = \sum_k c_k(f) c_{j-k}(g).$$

4. FOURIER SERIES OF FUNCTIONALS

A challenging problem in harmonic analysis is to assign a natural sum to the Fourier series of an integrable function, especially when the series diverges. We open this section with an ingenious procedure which accomplishes precisely this.

Theorem 4.1 (Riemann). Given $f \in L(T)$, $f \sim \sum c_j e^{ijt}$, consider the (formal) double integral of f, i.e.,

$$\frac{c_0}{2}t^2 + At + B + \sum_{j \neq 0} \frac{c_j e^{ijt}}{(ij)^2}, \qquad A, B \text{ const.} \tag{4.1}$$

Then the series in (4.1) converges absolutely and uniformly to a continuous function H. Moreover

$$\lim_{\varepsilon \to 0^+} \frac{H(t+2\varepsilon) + H(t-2\varepsilon) - 2H(t)}{4\varepsilon^2} = f(t) \qquad \text{a.e. in} \quad T.$$

The proof of this theorem, not given here, amounts to showing that H is Riemann summable to f a.e., that is,

$$c_0 + \lim_{\varepsilon \to 0} \sum_{j \neq 0} c_j e^{ijt} \left(\frac{\sin \varepsilon j}{\varepsilon j} \right)^2 = f(t) \qquad \text{a.e.}$$

In some cases we may consider the Lebesgue method of summation corresponding to a single integral of f, namely

$$c_0 t + A + \sum_{j \neq 0} \frac{c_j e^{ijt}}{(ij)}. \tag{4.2}$$

One of the difficulties with (4.2) is that even when it converges everywhere, its sum is not necessarily continuous.

Returning to Riemann's theorem we envision the possibility of assigning a sum to a trigonometric series $\sum c_j e^{ijt}$ with coefficients of tempered growth, i.e.,

$$c_j = O(|j|^k), \qquad \text{some} \quad k \tag{4.3}$$

by simply (formally) integrating the series $(k+2)$ times and then differentiating the resulting continuous function as we did in Riemann's theorem.

Clearly sequences of tempered growth are not necessarily sequences of Fourier coefficients of summable functions, but as we shall see below they are Fourier coefficients of certain functionals on subspaces of $C(T)$.

Suppose then that X is a linear subspace of $C(T)$ and denote the norm of $u \in X$ by $\|u\|$. Examples of such spaces include $C^k(T)$, i.e., those

$u \in C(T)$ with k continuous derivatives, normed by $\|u\|_k = \sup(\|u\|_\infty, \|Du\|_\infty, \ldots, \|D^k u\|_\infty)$. Here, as usual, Du denotes the derivative of u, and the topology induced by this norm corresponds to the uniform convergence of a function and its first k derivatives.

A mapping $F: X \to C$ is a linear functional provided $F(u_1 + \lambda u_2) = F(u_1) + \lambda F(u_2)$, $u_1, u_2 \in X$, $\lambda \in C$, and is a bounded linear functional if in addition there is a constant c such that $|F(u)| \leq c\|u\|$, $u \in X$. It is readily seen that a functional is bounded in X if and only if it is continuous there, namely, $\lim_{n\to\infty} F(u_n) = F(u)$, provided $\|u_n - u\| \to 0$. This expression is equivalent to

$$\lim_{n\to\infty} F(u_n) = 0 \quad \text{provided} \quad \|u_n\| \to 0. \tag{4.4}$$

Let $C^\infty(T) = \bigcap_k C^k(T)$. It is also possible to define a notion of convergence in this space which in turn will allow us to consider a continuous linear functional on $C^\infty(T)$. We say that $u_n \to 0(C^\infty)$ if and only if $D^k u_n$ converges uniformly to 0, as $n \to \infty$, for $k = 0, 1, \ldots$. With this topology $C^\infty(T)$ becomes a complete space, in the sense that Cauchy sequences converge. It is not, however, a normed space for there is no way of choosing a norm in $C^\infty(T)$ so that convergence (C^∞) corresponds to convergence in the sense of the metric associated with a norm; we return to this point in 5.21 below.

We now consider linear functionals F in $C^\infty(T)$ which are continuous, more precisely $\lim_{n\to\infty} F(u_n) = 0$ whenever $\lim_{n\to\infty} u_n = 0(C^\infty)$.

The collection of such functionals is denoted by D' and elements $F \in D'$ are called distributions. This notion of continuity is sufficiently strong to give the following result.

Theorem 4.2. $F \in D'$ if and only if there are a constant c and an integer k such that

$$|F(u)| \leq c\|u\|_k, \qquad \text{all} \quad u \in C^\infty(T). \tag{4.5}$$

Proof. If (4.5) holds clearly F is continuous. Conversely, let $F \in D'$ and suppose that no inequality (4.5) holds. Then there is a sequence u_n of $C^\infty(T)$ functions such that $|F(u_n)| > n\|u_n\|_n$, $n = 1, 2, \ldots$. In particular, we have $u_n \neq 0$, all n. Now put $\alpha_n = \|u_n\|_n$, $v_n = u_n/n\alpha_n$. Then $v_n \in C^\infty(T)$, and $\|v_n\|_n = 1/n$. Therefore $v_n \to 0(C^\infty)$; however $|F(v_n)| > 1$, thus contradicting the fact that F is continuous. ■

This result motivates the following

Definition 4.3. We say that a distribution F is of order m if m is the smallest integer for which (4.5) holds.

It is quite simple to give examples of distribution of order m. Indeed, let $F_m(u) = D^m u(0)$, $m = 0, 1, \ldots$. Then F_m is of order m. F_0 is called the Dirac δ measure. Another example of a distribution of order 0 is given by F_f, where for $f \in L(T)$

$$F_f(u) = \frac{1}{2\pi} \int_T f(t)u(t)\,dt, \qquad u \in C^\infty(T). \tag{4.6}$$

There is also a natural notion of convergence in D'. It is given by

Definition 4.4. We say that a sequence $\{F_n\} \subset D'$ converges to 0 in D' if and only if $\lim_{n\to\infty} F_n(u) = 0$, for every $u \in C^\infty(T)$.

With this notion of convergence all usual operations become continuous in D'. For example, each $F \in D'$ is infinitely differentiable (in the sense of D') and the differentiation operator is continuous in D'. More precisely for F in D' let DF be defined by $DF(u) = F(-Du)$, $u \in C^\infty(T)$. This definition is suggested by formally integrating by parts the expression corresponding to $F_{f'}$ in (4.6). More generally, for an integer k let $D^kF(u) = F((-1)^k D^k u)$, $u \in C^\infty(T)$. It is then clear that $D^kF(u_n) \to 0$ whenever $u_n \to 0(C^\infty)$, all k, i.e., D^kF is a distribution, and that $D^kF_n(u) \to 0$ for every $u \in C^\infty(T)$ whenever $F_n \to 0$ in D', i.e., differentiation is continuous in D'.

Although the product of two distributions cannot be defined in general, we can always multiply a distribution F and a C^∞ function v. The product vF is then the distribution given by $(vF)(u) = F(uv)$, $u \in C^\infty(T)$.

It is readily seen that with this definition Leibnitz's rule holds, namely, $D(vF) = (Dv)F + vDF$.

We may also introduce the notion of Fourier series for $F \in D'$. Since for each j, $e^{ijt} \in C^\infty(T)$, the Fourier coefficients of F

$$c_j = c_j(F) = F(e^{-ijt}), \qquad \text{all} \quad j, \tag{4.7}$$

are well defined. The Fourier series of F is now

$$F \sim \sum c_j e^{ijt}. \tag{4.8}$$

An important result in the theory of distributions is

Theorem 4.5. Let F be a distribution of order m. Then $c_j = O(|j|^m)$. Moreover, if

$$s_n(F, t) = \sum_{|j| \le n} c_j e^{ijt}, \qquad c_j = c_j(F), \tag{4.9}$$

then

$$\lim_{n\to\infty} s_n(F, t) = F \qquad (\text{in } D').$$

Proof. If F is of order m, then by (4.7)

$$|c_j| = |F(e^{-ijt})| \leq c\|e^{-ijt}\|_m = c|j|^m, \qquad \text{all } j.$$

Next, for $u \in C^\infty(T)$ by (4.9) and the linearity of F

$$\begin{aligned} s_n(F)(u) &= \frac{1}{2\pi}\int_T \Big(\sum_{|j|\leq n} c_j e^{ijt}\Big) u(t)\,dt \\ &= \sum_{|j|\leq n} c_j \frac{1}{2\pi}\int_T u(t)e^{ijt}\,dt \\ &= \sum_{|j|\leq n} F(e^{-ijt})c_{-j}(u) = F\Big(\sum_{|j|\leq n} c_{-j}(u)e^{-ijt}\Big) \\ &= F(s_n(u)). \end{aligned} \tag{4.10}$$

But since $s_n(u) \to u(C^\infty)$, by (4.10) and the continuity of F it follows that $s_n(F)(u) = F(s_n(u)) \to F(u)$, all $u \in C^\infty(T)$. This means that $s_n(F) \to F$ in D'. ■

Corollary 4.6 (Uniqueness). Let $F \in D'$, $c_j(F) = 0$ for all j. Then $F = 0$.

We isolate from the above proof an important fact concerning the evaluation of F at u, namely, by (4.10)

$$F(u) = \lim_{n\to\infty} \sum_{|j|\leq n} c_j(F)c_{-j}(u), \tag{4.11}$$

where the sum is absolutely convergent since by Corollary 3.3 $c_j(u) = O(|j|^{-m})$, all m. We will have more to say about (4.11) later on. In the meantime we observe that Theorem 4.5 admits a converse,

Theorem 4.7. Let $\{c_j\}$ be a tempered sequence of order m. Then there exists a (unique) distribution F

$$F \sim \sum c_j e^{ijt}.$$

Proof. As above we observe that for each u in $C^\infty(T)$, $\sum c_j c_{-j}(u)$ is absolutely convergent. Moreover, the mapping $u \to F(u) = \sum c_j c_{-j}(u)$ is linear, but is it continuous? Well, notice that by Corollary 3.3

$$\begin{aligned} |F(u)| &\leq |c_0||c_0(u)| + c\sum_{j\neq 0} |c_j||j|^{m+2}\|u\|_{m+2} \\ &\leq C\|u\|_{m+2}. \end{aligned}$$

Consequently, F is a distribution of order $\leq m+2$. Finally, setting $u = e^{-ikt}$ we see that $F(e^{-ikt}) = \sum c_j c_{-j}(e^{-ikt}) = c_k$, all k, and since uniqueness follows at once from Corollary 4.6, we are done. ■

Theorem 4.7 is of interest in that addresses the comments following 4.1 concerning Riemann summability. In this direction we also have

Proposition 4.8. Let $F \in D'$. Then there exist a continuous function f and an integer k such that $F - c_0(f) = D^k F$, in D'.

Proof. Let m be the order of F and let $k = m+2$. Then

$$f \sim C + \sum_{j \neq 0} \frac{c_j(F)e^{ijt}}{(ij)^k}$$

will do the job. ■

Theorem 4.7 can also be applied to define convolution of distributions.

Definition 4.9. Let $F, G \in D'$. Then the sequence $\{c_j(F)c_j(G)\}$ is of tempered growth and by Theorem 4.7 there exists a unique distribution H such that $c_j(H) = c_j(F)c_j(G)$, all j. We call H the convolution of F and G and denote it by $F * G$, $F * G \sim \sum c_j(F)c_j(G)e^{ijt}$. This operation satisfies all the usual convolution properties discussed in Section 3. Moreover, if $G \in C^\infty(T)$, then also $H \in C^\infty(T)$ as is readily seen by direct inspection of the $c_j(H)$'s.

For specific computations it is often important to have functions $u \in C^\infty(T)$ at hand with arbitrarily small support. This is always possible since the function

$$u(t) = \begin{cases} ce^{-1/(1-t^2)} & |t| < 1 \\ 0 & \text{otherwise} \end{cases} \tag{4.12}$$

is $C^\infty(T)$ and is supported in $|t| \leq 1$. c is chosen so that $(1/2\pi)\int_T u(t)\,dt = 1$. Also the function $u_\varepsilon(t) = (1/\varepsilon)u(t/\varepsilon)$ has support in $|t| \leq \varepsilon$, $(1/2\pi)\int_T u_\varepsilon(t)\,dt = 1$, and is in $C^\infty(T)$.

We collect one last fact concerning distributions.

Definition 4.10. A distribution F is said to vanish in an open set $\mathcal{O} \subset T$, and we denote this by $F = 0$ in $\mathcal{O}$, if $F(u) = 0$ for every $u \in C^\infty(T)$, supp $u \subset \mathcal{O}$. It is not hard to see that if F vanishes in a collection of open sets, then it also vanishes in its union. Therefore we may define the support of F to be the complement of the largest open set in which F vanishes.

The fact we wish to single out is

Proposition 4.11. Let F be a distribution of order $\leq m$, supp $F = K$, a (compact) subset of T. Let $u \in C^\infty(T)$ be such that u and its derivatives up to order m vanish in K. Then $F(u) = 0$.

Proof. The importance of this result lies in the fact that we do not require the vanishing of u and its derivatives in a neighborhood of K, but only on K itself.

Let $K_r = \{t \in T: d(t, K) \leq r\}$, and let

$$\varepsilon(r) = \sup_{t \in K_r, 0 \leq p \leq m} |D^p u(t)|. \tag{4.13}$$

Note that $\varepsilon(r) \to 0$ as $r \to 0$. Let $\chi(t)$ denote the characteristic function of $K_{r/2}$, and finally put $\alpha_r(t) = \chi * v_{r/2}(t)$, where v is the function given in (4.12) above.

Then clearly $\alpha_r = 1$ on K, and α_r vanishes off K_r. It is also readily seen, by differentiating $v_{r/2}$ in the above convolution, that

$$|D^p \alpha_r(t)| \leq c_p r^{-p}, \qquad \text{all} \quad t, \qquad \text{all} \quad p \leq m, \tag{4.14}$$

and, moreover, this estimate is uniform in p if we replace c_p by $c = \max(c_1, \ldots, c_m)$ above.

We also want to estimate $D^p u$ for $u \in C^\infty(T)$, $u = 0$ in K. Starting out with r sufficiently small, given $\eta > 0$, by (4.13) we may assume that $|D^p u(t)| \leq \eta$, $t \in K_r$.

Therefore, since

$$D^p u(t) = \int_{[a,t]} D^{p+1} u(s)\, ds + D^p(a),$$

and since $D^p u(a) = 0$ if $a \in K_r$, it follows that

$$|D^p u(t)| \leq \eta(t - a) \leq hr. \tag{4.15}$$

In other words, from (4.15) we see that actually

$$|D^p u(t)| \leq \eta r^{m-p}, \qquad \text{all} \quad t \quad \text{in} \quad K_r, \qquad p \leq m. \tag{4.16}$$

Thus combining (4.14) and (4.15), and making use of Leibnitz's formula for the derivative of a product, we get that

$$\|D^p(\alpha_r u)\|_\infty \leq c\eta r^{m-p}, \qquad p \leq m \tag{4.17}$$

with c independent of η and r.

Moreover, since by (4.13) $\eta \to 0$ as $r \to 0$, $\{\alpha_r u\}$ is a collection of $C^\infty(T)$ functions that converges to 0 up to order m, uniformly in T, as $r \to 0$. By the continuity of F, $\lim_{r\to 0} F(\alpha_r u) = 0$, and since $\alpha_r = 1$ in a neighborhood of the support of F, and $u = u\alpha_r + u(1 - \alpha_r)$ and $u(1 - \alpha_r)$ vanishes in a neighborhood of K, actually $F(u) = \lim_{r\to 0} F(u\alpha_r) = 0$. ■

An important consequence of this result is

Theorem 4.12. Let F be a distribution supported at $t_0 \in T$, then F is a linear combination of the Dirac delta δ_{t_0} and its derivatives.

Proof. Let F be of order m. Then for $u \in C^\infty(T)$ write

$$u(t) = \sum_{p \leq m} \frac{D^p u(t_0)}{p!}(t - t_0)^p + R_m(t)$$

where the remainder $R_m(t)$ is a smooth function such that $D^p R_m(t_0) = 0$, $p \leq m$.

By Proposition 4.11, $F(R_m) = 0$, and therefore

$$F(u) = \sum_{p \leq m} \frac{F((t - t_0)^p)}{p!} D^p u(t_0)$$

$$= \sum_{p \leq m} c_p D^p \delta_{t_0}(u) = \left(\sum_{p \leq m} c_p D^p \delta_{t_0}\right)(u). \quad \blacksquare$$

5. NOTES; FURTHER RESULTS AND PROBLEMS

In the beginning there was the vibrating string. Assume that an elastic string is stretched taut along the x axis, with endpoints at $-\pi$ and π. If this string is displaced and released it vibrates in such a way that the ordinate of a point in the string is a function $y = y(x, t)$ of the time t and the x coordinate of the point. As early as 1747 d'Alembert knew that this function satisfies the differential equation $\partial^2 y/\partial t^2 = a^2(\partial^2 y/\partial x^2)$. In the mid-1750's, d'Alembert himself, Euler, and Bernoulli showed that the solution of this equation involved representing the initial position of the string at the time of release, i.e., $y(x, 0) = f(x)$, by a trigonometric series of the form (1.8), namely

$$\frac{1}{2}a_0 + \sum_{j=1}^{\infty} a_j \cos jt + b_j \sin jt.$$

This posed two natural questions:

(i) If f could be so represented, how could the sequence of coefficients be determined?

(ii) Is it reasonable to expect that a single expression of the form (1.8) could represent a straight-line on a part of the interval T, while at the same time representing a sine curve, say, on another part of T?

It was Bernoulli who suggested, in 1755, that the motion $y(x, t)$ is expressible in the form

$$a_j = \frac{1}{\pi}\int_T f(t)\cos jt\,dt, \qquad b_j = \frac{1}{\pi}\int_T f(t)\sin jt\,dt.$$

Fourier, while considering in 1867 the transfer of heat in a conducting medium, stated and used the fact that any summable function defined in T is representable there by a trigonometric series whose coefficients are determined by the above formula; of course summable meant something else in those days.

But it was only in 1829 that Dirichlet established the following result: if the function f defined on T has only a finite number of simple discontinuities and only a finite number of maxima and minima there, then the partial sums $s_n(f, x)$ of the Fourier series of $f(x)$ tend to $f(x)$ at each point of continuity of f and to $(f(x+0)+f(x-0))/2$ at a jump.

In 1876 du Bois-Reymond constructed a continuous function with a nonconvergent Fourier series at a single point; this led to an example where the convergence fails at each point of an everywhere dense set of points in T.

In 1926 Kolmogorov produced an integrable function whose Fourier series diverges everywhere in T. Until 1966 it was not known whether or not there exists a continuous function with this property. In that year Carleson [1966] proved that "Lusin's conjecture" holds, namely, the Fourier series of an $L^2(T)$ function converges a.e. Shortly after Hunt [1968] generalized Carleson's argument to prove that the Fourier series of L^p functions converge a.e. in T, provided $1 < p$. A different proof of this result was later given by C. Fefferman [1973]. We shall return to discuss Kolmogorov's example in the next chapter; unfortunately, the proof of Carleson's result is beyond the scope of this book.

Concerning arbitrary trigonometric series, Menshov showed in 1916 that there exists a trigonometric series that converges a.e. to a function $f \in L(T)$; yet this series is not the Fourier series of f.

Further Results and Problems

5.1 (Jensen's Inequality) Let f, p be two functions defined in $I = [a, b]$ such that $\alpha < f(t) < \beta$ for t in I, $p \geq 0$, nonidentically 0. Let $\phi(u)$ be a convex function defined for $\alpha < u < \beta$. Then

$$\phi\left(\int_I f(t)p(t)\,dt \Big/ \int_I p(t)\,dt\right) \leq \int_I \phi(f(t))p(t)\,dt \Big/ \int_I p(t)\,dt.$$

5.2 (Young's Inequality) Let $\phi(t)$, $\psi(t)$, $t > 0$, be two continuous functions, vanishing at 0, strictly increasing, tending to ∞ at ∞, and inverse to each other. Then for $a, b > 0$, we have $ab \leqslant \Phi(a) + \Psi(b)$, where $\Phi(a) = \int_{[0,a]} \phi(t)\,dt$ and $\Psi(b) = \int_{[0,b]} \psi(t)\,dt$.

5.3 (Hölder's Inequality) If $1/p + 1/p' = 1$, $1 \leqslant p, p' \leqslant \infty$, then $\|fg\| \leqslant \|f\|_p \|g\|_{p'}$. Discuss a converse. (*Hint*: Consider $\phi(t) = t^{p-1}$ in (5.2).)

5.4 (Minkowski's Inequality) If $f, g \in L^p(T)$, $1 < p < \infty$, then

$$\|f + g\|_p \leqslant \|f\|_p + \|g\|_p.$$

(*Hint*: Apply to Hölder's inequality to $|f + g|^{p-1}(|f| + |g|)$.)

5.5 (Yet Another Minkowski's Inequality) Prove the following integral version of Minkowski's inequality; namely, for $1 \leqslant p < \infty$,

$$\frac{1}{2\pi}\left(\int_T\left(\int_T |f(x, t)|\,dx\right)^p dt\right)^{1/p} \leqslant \int_T\left(\frac{1}{2\pi}\int_T |f(x, t)|^p\,dt\right)^{1/p} dx$$

or

$$\left\|\int_T f(\cdot, t)\,dt\right\|_p \leqslant \int_T \|f(\cdot, t)\|_p\,dt.$$

With this notation the result holds for $p = \infty$ also.

5.6 (Hardy's Inequality) Let $r > 1$, $s < r - 1$, $f(t) > 0$ and $F(x) = \int_{[0,x]} f(t)\,dt$. If $f(x)^r x^s$ is integrable over $I = [0, 1]$, then so is

$$\left(\frac{F(x)}{x}\right)^r x^s \quad \text{and} \quad \int_I (F(x)/x)^r x^s\,dx \leqslant \frac{r}{r - s - 1}\int_I f(x)^r x^s\,dx.$$

(*Hint*: Rewrite $F(x)/x = \int_{[0,1]} f(xt)\,dt$ and apply (5.5).)

5.7 (Abel's Summation Formula) If $u_0, u_1, \ldots, v_0, v_1, \ldots$ are (complex) numbers and $V_n = v_0 + \cdots + v_n$, $V_{-1} = 0$, then for $n > m > 0$,

$$\sum_{k=m}^{n} u_k v_k = \sum_{k=m}^{n-1} (u_k - u_{k-1}) V_k + u_n V_n - u_m V_{m-1}.$$

5.8 If a real sequence a_k decreases to 0, then the trigonometric series $(a_0/2) + \sum_{k=1}^{\infty} a_k \cos kt$ converges everywhere except, perhaps, at $x = 0$. Moreover, for any $\delta > 0$ the series converges uniformly in $\delta < |x| < \pi$. (*Hint*: Put $u_k = a_k$, $v_k = \cos kt$ in (5.7), and use estimate (1.17).)

5.9 Let f be a real-valued, summable function in T. Show that if all of the Fourier coefficients of f vanish, then $f = 0$, a.e. (*Hint*: First assume f is continuous and not identically 0. Then there is a point x_0 and two numbers δ and h such that $f(x) > h$ for $x \in I = (x_0 - \delta, x_0 + \delta)$. If we can produce a sequence $\{T_n(x)\}$ of trigonometric polynomials such that $T_n(x) > 1$ in I, $T_n(x)$ tends uniformly to ∞ in every proper subinterval of I and $T_n(x)$ is uniformly bounded in $T\backslash I$, then we are done. $T_n(x) = t(x)^n$, with $t(x) = 1 + \cos(x - x_0) - \cos\delta$, does the job. In the general case use Proposition 3.1.

This proof is due to Lebesgue. We shall return to this "completeness" result in the next chapter.)

5.10 If

$$f(x) - f(x_0) = O\left(\ln\left(\ln\left(\frac{1}{|x - x_0|}\right)\right)^{-(1+h)} \ln\left(\frac{1}{|x - x_0|}\right)\right)$$

for some $h > 0$, in a neighborhood of x_0 in T, then

$$\lim s_n(f, x_0) = f(x_0).$$

5.11 If f, g are integrable functions and for a given x the function $(f(t) - g(t))/(x - t)$ is integrable in a neighborhood of x, then

$$\lim_{n\to\infty} (s_n(f, x) - s_n(g, x)) = 0.$$

5.12 (Hardy-Littlewood) A necessary and sufficient condition that f should be a.e. equal to a function of bounded variation in T is that $w_1(f, h) = O(|h|)$ as $h \to 0$. (*Hint*: That the condition is necessary follows at once upon writing f as a difference of nondecreasing functions. Conversely, if the condition is satisfied, put $F_n(x) = n \int_{(x,x+1/n)} f(t)\, dt$. Then $\int_T |F_n(x + h) - F_n(x)|\, dx = O(h)$, and if $(x_j, x_j + h_j)$ is any set of nonoverlapping intervals, $\sum_j |F_n(x_j + h_j) - F_n(x_j)| = O(1)$. But $F_n(x) \to f(x)$ a.e. The result follows without much difficulty from this.)

5.13 Let $f \in L(T)$, then $w_1(f, h) = o(|h|)$ if and only if $f = c$ a.e.

5.14 Let $f \in L(T)$, then $|c_j(f)| \leq (1/2) w_1(f, \pi/|j|)$, $j \neq 0$. (*Hint*: $c_j(f) = (1/2\pi) \int_T f(t + \pi/j) e^{-ijt}\, dt$.)

5.15 (Jordan's Test) Let f be a function of bounded variation in T. Then the Fourier series of f converges to $(f(x + 0) + f(x - 0))/2$ at any point interior to T. If in addition f is continuous in $(a, b) \subset T$, then the convergence is uniform in every interval $(a + h, b - h)$, $h > 0$. This test and Dini's test are not comparable. Indeed, let $f(x)$, $g(x)$ be even and let $f(x) = 1/|\log(x/2\pi)|$, $g(x) = x^\eta \sin(1/x)$, $0 < \eta < 1$, $0 < x < \pi$. At the point 0, f satisfies Jordan's condition but not Dini's, and conversely g satisfies Dini's condition but not Jordan's.

5.16 For $a_k = 2\pi k/|j|$, $0 < k < |j|$, $|j| \neq 0$, let

$$g(t) = \sum_{k=1}^{|j|} c_k \chi_{(a_{k-1}, a_k)}(t),$$

where the c_k's are constants and $\chi_{(a_{k-1}, a_k)}(t)$ denotes the characteristic function of (a_{k-1}, a_k). Show that $c_j(g) = 0$ and obtain as a corollary of this that, if f is a function of bounded variation, then $c_j(f) = O(1/|j|)$. (*Hint*: Use a judicious choice of c_k's.)

5.17 Let $f(t) = t$ in T, and have period 2π. Show that

$$f \sim \sum_{j=1}^{\infty} (-1)^{j+1} (\sin jt)/j$$

and obtain by Jordan's test and the change of variables $t \to \pi - t$ that

$$\frac{\pi - t}{2} = \sum_{j=1}^{\infty} \frac{\sin jt}{j}, \qquad 0 < t < 2\pi.$$

Furthermore show that there is a constant c so that the partial sums $s_n(f, t) = \sum_{j=1}^{n} (\sin jt)/j$ are uniformly bounded, $|s_n(f, t)| \leq c$, all n, t in $(0, 2\pi)$. Show that $s_n(f, t) = -t/2 + \int_{[0, nt]} (\sin s)/s \, ds + O(1)$, and observe that, although $s_n(f, t)$ converges to $(\pi - t)/2$, all t in $0 < t < 2\pi$, by letting $t = \pi/n, 2\pi/n, \ldots$ the functions $s_n(f, t)$ concentrate around the interval $(0, L)$, where $L = \int_{[0, \pi]} (\sin t)/t \, dt$, to the right of the origin. Conclude that around the point $t = 0$ the values of these functions oscillate not between $-\pi/2$ and $\pi/2$ as expected, but around $[-L, L]$, $L \sim 1.8519 \cdots > \pi/2$. This is known as Gibbs' phenomenon. (*Hint*: $s_n(f, t) = \int_{[0, t]} (D_n(s) - \frac{1}{2}) \, ds$. Also use relation (1.21).)

5.18 Show that Gibbs' phenomenon holds for every function f of bounded variation about its points of discontinuity, as long as they are isolated. (*Hint*: Subtract from f another function, such as the one in (5.17), with the same jump at a point in question.)

5.19 (Poisson Summation Formula) Let g be a function defined for $-\infty < t < \infty$, tending to 0 as $|t| \to \infty$, and integrable over any finite interval. Suppose that the series $\sum_{k=-\infty}^{\infty} g(t + 2\pi k) = G(t)$ converges uniformly for $t \in T$. (By this we actually mean that the sequence of symmetric partial sums $G_n(t) = \sum_{|k| \leq n} g(t + 2\pi k)$ converges uniformly in T.) Then the sum G is a periodic function of period 2π and its Fourier coefficients are given by

$$c_j(G) = \lim_{n \to \infty} \frac{1}{2\pi} \int_T G_n(t) e^{-ijt} \, dt = \frac{1}{2\pi} \int_{(-\infty, \infty)} g(t) e^{-ijt} \, dt.$$

Suppose that at $t = 0$, G satisfies one of the conditions insuring the convergence of the Fourier series of G to $G(0)$ and obtain conditions under which

$$\sum_{j=-\infty}^{\infty} g(2\pi k) = \frac{1}{2\pi} \sum_{j=-\infty}^{\infty} \int_{(-\infty, \infty)} g(t) e^{-ijt} \, dt.$$

Illustrate the result obtained when $g(t) = e^{-|t|}$ to show that

$$\frac{1}{2} + \frac{1}{e^{2\pi} - 1} = \frac{1}{2\pi} + \frac{1}{\pi} \sum_{j=1}^{\infty} \frac{1}{1 + j^2}.$$

5.20 Let $f(t) = f(t + a)$, $a \neq \pi/k$, all k, for t in T. Show that $f = c$ a.e. in T.

5.21 Show that there is no norm $|\cdot|$ in $C^\infty(T)$ such that the sequence $u_n \to 0$ (C^∞) if and only if $|u_n| \to 0$. (*Hint*: Suppose such a norm exists, and show that $|u| \leq M\|u\|_n$, some n. This leads to a contradiction.)
5.22 Show that $F * \delta = F$ for each $F \in D'$.
5.23 Show that $D^k(F * G) = D^kF * G = F * D^kG$, for $F, G \in D'$.
5.24 Let $F_n \to F(D')$, $G_n \to G(D')$ and show that $F_n * G_n \to F * G(D')$.
5.25 Find necessary and sufficient conditions on $A \in D'$ so that the equation $A * F = B$ is solvable for each $B \in D'$. (*Hint*: It suffices to consider $B = \delta$, for if $A * F = \delta$, then $A * (F * B) = \delta * B = B$. Solve now $c_j(A)c_j(F) = 1$, all j; since F is of finite order a necessary condition on A follows at once.)

CHAPTER

II

Cesàro Summability

1. $(C, 1)$ SUMMABILITY

In Chapter I we saw that the notion of pointwise convergence is not the ideal one in dealing with Fourier series of summable functions; in fact in 5.20 below we outline Kolmogorov's construction of a function $f \in L(T)$ with an everywhere divergent Fourier series. In this chapter we address the convergence question from the more general point of view of the arithmetic means. We begin by defining the notion of Cesàro $(C, 1)$ summability.

Definition 1.1. Given a sequence of complex numbers $\{c_j\}$, we say that it converges to L in the Cesàro $(C, 1)$ sense, and we write

$$\lim_{j\to\infty} c_j = L(C, 1)$$

provided that $\lim_{j\to\infty} C_j = L$, where C_j is the average $(c_1 + \cdots + c_j)/j$. It would be reassuring to know that convergent sequences also converge $(C, 1)$ to the same limit.

Proposition 1.2. Let $\lim_{j\to\infty} c_j = L$; then $\lim_{j\to\infty} c_j = L(C, 1)$.

Proof. We may assume that $L = 0$ by replacing c_j by $c_j - L$ if necessary. Observe that the c_j's have the following properties

(i) $|c_j| \leqslant K$, all j
(ii) Given $\varepsilon > 0$, there is a j_0 such that $|c_j| \leqslant \varepsilon$ provided $j \geqslant j_0$.

It is now a simple matter to estimate the C_j's. Indeed

$$|C_j| \leqslant \frac{|c_1| + \cdots + |c_{j_0-1}| + |c_{j_0}| + \cdots + |c_j|}{j}$$

$$\leqslant \frac{j_0 K}{j} + \frac{\varepsilon(j - j_0)}{j} \leqslant \frac{j_0 K}{j} + \varepsilon, \qquad j \geqslant j_0.$$

Therefore, by first picking ε, thus fixing j_0, and then letting $j \to \infty$, we see that $|C_j|$ can be made arbitrarily small for j large. ■

Observe that the oscillating sequence $c_j = 1 + (-1)^j$ has limit 1 in the $(C, 1)$ sense.

In a similar vein we define Cesàro summability for series.

Definition 1.3. Let $\{c_j\}$ be a sequence of complex numbers and put

$$s_n = \sum_{j=1}^{n} c_j, \qquad \sigma_n = \frac{1}{n} \sum_{j=1}^{n} s_j.$$

If $\lim_{n\to\infty} \sigma_n = s$, then we say that the series $\sum c_j$ is C_1-summable to s and we write $\sum c_j = s(C, 1)$.

By Proposition 1.2 it follows that, if $\lim_{n\to\infty} s_n = s$, then $\sum c_j = s(C, 1)$. On the other hand, if $c_j = z^j$, $z \neq 1$ a complex number with $|z| = 1$, then $\sum z^j$ does not converge, yet

$$\sum_{j=0}^{\infty} z^j = \frac{1}{1 - z}(C, 1).$$

2. FEJÉR'S KERNEL

We explore now how the notion of Cesàro summability applies to Fourier series. As in Chapter I we begin by determining the integral, or convolution, representation of the Cesàro means $\sigma_n(f, x)$ corresponding to (the Fourier series of) f. More precisely, given $f \sim \sum c_j e^{ijx}$, let

$$s_n(f, x) = \sum_{|j| \leqslant n} c_j e^{ijx}$$

and (2.1)

$$\sigma_n(f, x) = \frac{s_0(f, x) + \cdots + s_n(f, x)}{n + 1}, \qquad n \geqslant 0.$$

An explicit expression of $\sigma_n(f, x)$ is fairly easy to obtain. Indeed, observe that the numerator of (2.1) is the sum of

$$\begin{array}{c} c_0 \\ c_{-1}e^{-ix} + c_0 + c_1 e^{ix} \\ \vdots \\ c_{-n}e^{-inx} + \cdots + c_0 + \cdots + c_n e^{inx} \end{array}$$

which equals $c_{-n}e^{-inx} + 2c_{-n+1}e^{-i(n-1)x} + \cdots + (n+1)c_0 + \cdots + c_n e^{inx}$. Thus, by dividing by $(n+1)$ we get that

$$\sigma_n(f, x) = \sum_{|j| \le n} \left(1 - \frac{|j|}{n+1}\right) c_j e^{ijx}. \tag{2.2}$$

As for the integral representation alluded to above, by formula (1.11) of Chapter I we readily see that

$$\begin{aligned} \sigma_n(f, x) &= \frac{f * 2D_0(x) + \cdots + f * 2D_n(x)}{n+1} \\ &= \frac{f * (2\sum_{j=0}^{n} D_j)(x)}{n+1} = f * 2K_n(x), \end{aligned} \tag{2.3}$$

say, where

$$K_n(t) = \frac{\sum_{j=0}^{n} D_j(t)}{(n+1)} \tag{2.4}$$

is the Fejér kernel of order n. It is not hard to compute $K_n(t)$. In the first place observe that the numerator in (2.4) equals

$$\sum_{j=0}^{n} \frac{\sin((j+1/2)t)}{2\sin(t/2)} = \frac{1}{2\sin(t/2)} \operatorname{Im}\left(\sum_{j=0}^{n} e^{i(j+1/2)t}\right).$$

By summing the geometric series we see that the imaginary part of the above sum equals

$$\begin{aligned} &\operatorname{Im}\left(e^{it/2}\frac{(1 - e^{i(n+1)t})}{(1 - e^{it})}\right) \\ &= \operatorname{Im}\left(e^{i(n+1)t/2}\frac{(e^{-i(n+1)t/2} - e^{i(n+1)t/2})}{(e^{-it/2} - e^{it/2})}\right) \\ &= \frac{\sin((n+1)t/2)}{\sin(t/2)} \operatorname{Im}(e^{i(n+1)t/2}) = \frac{\sin^2((n+1)t/2)}{\sin(t/2)}. \end{aligned}$$

Thus

$$K_n(t) = \frac{1}{2(n+1)} \left(\frac{\sin((n+1)t/2)}{\sin(t/2)}\right)^2. \tag{2.5}$$

We emphasize that on account of (2.4) and (2.3), also

$$K_n(t) = \frac{1}{2} \sum_{|j| \le n} \left(1 - \frac{|j|}{n+1}\right) e^{ijt}. \tag{2.6}$$

The following properties of $K_n(t)$ are readily verified. In the first place it is a positive even function, and by (2.6) above

$$\frac{1}{\pi} \int_T K_n(t)\, dt = \frac{2}{\pi} \int_{[0,\pi]} K_n(t)\, dt = 1. \tag{2.7}$$

It is also immediate to estimate $K_n(t)$. By (1.15) of Chapter I,

$$\begin{aligned} K_n(t) &\le \frac{1}{n+1} \sum_{j=0}^{n+1} |D_j(t)| \le \frac{1}{n+1} \sum_{j=0}^{n} (j + \tfrac{1}{2}) \\ &= \frac{1}{(n+1)} \left(\frac{n(n+1)}{2} + \frac{(n+1)}{2} \right) = \frac{n+1}{2}. \end{aligned} \tag{2.8}$$

Similarly, by (1.16) of Chapter I,

$$K_n(t) \le \frac{2\pi}{2(n+1)t^2}, \qquad 0 < |t| < \pi. \tag{2.9}$$

This is all we need to know about this kernel.

Concerning the representation of $\sigma_n(f, x)$, since K_n is even, we may rewrite (2.3) as either

$$\frac{1}{\pi} \int_T \left(\frac{f(x+t) + f(x-t)}{2} \right) K_n(t)\, dt$$

or (2.10)

$$\frac{1}{\pi} \int_{[0,\pi]} (f(x+t) + f(x-t)) K_n(t)\, dt.$$

We can now prove

Theorem 2.1 (Fejér). Let $f \in L(T), f \sim \sum c_j e^{ijt}$. If the limits $f(x \pm 0)$ exist, then

$$\sum c_j e^{ijx} = \left(\frac{f(x+0) + f(x-0)}{2} \right) (C, 1). \tag{2.11}$$

In particular, if f is continuous at every point of an interval $I \subset T$, then the convergence is uniform over I.

Proof. We may, and do, assume that $f(x) = (f(x+0) + f(x-0))/2$. On account of (2.10) and (2.7) we have

$$\sigma_n(f, x) - f(x) = \frac{2}{\pi} \int_{[0,\pi]} \left(\frac{f(x+t) + f(x-t)}{2} - f(x) \right) K_n(t)\, dt. \tag{2.12}$$

Therefore, for $0 < \eta < \pi$, we see that

$$\begin{aligned} &|\sigma_n(f, x) - f(x)| \\ &\quad \leqslant \frac{2}{\pi} \left(\int_{[0,\eta)} + \int_{[\eta,\pi)} \right) \left| \frac{f(x+t) + f(x-t)}{2} - f(x) \right| K_n(t)\, dt \\ &\quad = I_1 + I_2, \end{aligned} \tag{2.13}$$

say. We estimate I_1 first. By assumption, given $\varepsilon > 0$, there exists $\delta > 0$ such that

$$\left| \frac{f(x+t) + f(x-t)}{2} - f(x) \right| \leqslant \varepsilon$$

provided $0 \leqslant t < \delta$. We set $\eta = \delta$ in (2.13) and observe that

$$I_1 \leqslant \varepsilon \frac{2}{\pi} \int_{[0,\delta]} K_n(t)\, dt \leqslant \varepsilon.$$

To bound I_2, we introduce the quantity $M_n(\delta) = \max_{\delta < t < \pi} K_n(t)$. Then, by (2.9), $M_n(\delta) \leqslant c/(N+1)\delta^2$. Thus $\lim_{n \to \infty} M_n(\delta) = 0$, for each $\delta > 0$, and

$$I_2 \leqslant M_n(\delta) \frac{2}{\pi} \int_{[0,\pi]} (|f(x+t)| + |f(x-t)| + 2|f(x)|)\, dt = o(1) \tag{2.14}$$

as $n \to \infty$. This completes the proof of the first assertion. If now f is continuous in a closed interval $I \subset T$, then (2.14) holds uniformly over I, and the theorem is completely proved. ∎

Remark 2.2. A closer look at the preceding proof shows that the following also holds: if $m \leqslant f(x) \leqslant M$ in (a, b), then for any $\delta > 0$ there exists $n_0 = n_0(\delta)$ such that for $n \leqslant n_0$

$$m - \delta \leqslant \sigma_n(f, x) \leqslant M + \delta, \qquad x \in [a + \delta, b - \delta].$$

Also Theorem 2.1 remains valid if the limit is $\pm\infty$.

Corollary 2.3. (Weierstrass). Let $f \in C(T)$. Then given $\varepsilon > 0$ there is a trigonometric polynomial p such that $|f(x) - p(x)| \leqslant \varepsilon$, all t in T.

Proof. The natural candidate for p is $\sigma_n(f, x)$, with n sufficiently large. By Theorem 2.1 this choice works. ■

In order to complete the (self-contained) proof of Theorem 1.2 of Chapter I, we now prove

Proposition 2.4. Let $f \in L(T)$. Then given $\varepsilon > 0$ there is a trigonometric polynomial p such that $\|f - p\|_1 \leqslant \varepsilon$.

Proof. We may assume that f is a real-valued nonnegative function. Since f is finite a.e., it is the pointwise a.e. and norm limit of an increasing sequence of simple functions, i.e., of linear combinations $\sum \lambda_j \chi_{E_j}$ where the λ_j are positive constants and the χ_{E_j} are the characteristic functions of Lebesgue measurable sets. By the regularity of the Lebesgue measure these characteristic functions can in turn be approximated pointwise a.e. and in norm by characteristic functions of disjoint unions of finitely many open intervals. It is a simple matter now to construct an approximating sequence of continuous functions. By invoking Corollary 2.3 the proof is complete. ■

Remark 2.5. Clearly a similar result, with similar proof, holds for $L^p(T)$, $1 < p < \infty$, as well. The limiting case $p = \infty$ is Corollary 2.3, i.e., the result for $C(T)$.

Corollary 2.6. $L^p(T)$, $1 \leqslant p < \infty$, functions are norm-continuous, namely,

$$\lim_{h \to 0} \frac{1}{2\pi} \int_T |f(t+h) - f(t)|^p \, dt = 0.$$

Proof. The statement is clearly true for continuous functions f, which are dense in $L^p(T)$. ■

Theorem 2.7. Let $1 \leqslant p < \infty$. Then

$$\lim_{n \to \infty} \|\sigma_n(f) - f\|_p = 0, \quad \text{all} \quad f \in L^p(T).$$

Proof. Let

$$F(t) = \left(\frac{1}{2\pi} \int_T |f(x+t) - f(x)|^p \, dx \right)^{1/p};$$

by Corollary 2.6 $F(t)$ is continuous at $t = 0$. Thus Fejér's theorem (Theorem 2.1) gives $\lim_{n \to \infty} \sigma_n(F, 0) = F(0) = 0$. Also by (2.3) and (2.7) we have

$$\sigma_n(f, x) - f(x) = \frac{1}{\pi} \int_T (f(x+t) - f(x)) K_n(t) \, dt,$$

and by Minkowski's integral inequality (see 5.5 of Chapter I),

$$\|\sigma_n(f) - f\|_p \leq \frac{1}{\pi}\int_T \|f(\cdot + t) - f(\cdot)\|_p K_n(t)\,dt$$

$$= \frac{1}{\pi}\int_T F(t)K_n(t)\,dt = \sigma_n(F, 0).$$

Whence $\lim_{n\to\infty}\|\sigma_n(f) - f\|_p \leq \lim_{n\to\infty}\sigma_n(F, 0) = 0.$ ■

Theorem 2.7 makes Proposition 2.4 more precise and has important applications. For instance, we have

Corollary 2.8 (Uniqueness). Let $f \in L(T)$, $f \sim \sum c_j e^{ijt}$. If $c_j = 0$ for all j, then $f = 0$ a.e.

Proof. Observe that $\sigma_n(f, x) = 0$, all x, and apply Theorem 2.7. ■

3. CHARACTERIZATION OF FOURIER SERIES OF FUNCTIONS AND MEASURES

Fejér's theorem asserts that for continuous functions f, $\sigma_n(f, x)$ converges uniformly to f; it is natural to consider whether this statement admits a converse. More generally, is it possible to state a simple criteria that will enable us to identify among all trigonometric series those which correspond to Fourier series of continuous functions? or $L^p(T)$ functions, or even measures for that matter?

So, let

$$\sum c_j e^{ijt} \tag{3.1}$$

be a trigonometric series, and let

$$\sigma_n(t) = \sum_{|j|\leq n}\left(1 - \frac{|j|}{n+1}\right)c_j e^{ijt} \tag{3.2}$$

denote the sequence of Cesàro means of the series (3.1). We can then prove

Theorem 3.1. The trigonometric series (3.1) is the Fourier series of a function $f \in C(T)$ if and only if the sequence $\{\sigma_n(t)\}$ converges uniformly in T.

Proof. By Theorem 2.1 (Fejér's) it is sufficient to show that if $\sigma_n(t)$ converges uniformly in T, then there is a continuous function f such that $c_j(f) = c_j$, all j. Let $f \in C(T)$ be the (uniform) limit of the σ_n's, and observe that for a fixed index j and all $n \geqslant |j|$,

$$\left(1 - \frac{|j|}{n+1}\right) c_j = \frac{1}{2\pi} \int_T \sigma_n(t)\, e^{-ijt}\, dt. \tag{3.3}$$

We let now $n \to \infty$. The left-hand side of (3.3) converges to c_j, whereas the right-hand side converges, on account of the uniform convergence of the σ_n's, to $c_j(f)$. ■

To consider the $L^p(T)$ criteria, $1 \leqslant p \leqslant \infty$, we begin by observing that, from (2.3) and Minkowski's inequality, we obtain that

$$\|\sigma_n(f)\|_p \leqslant \frac{1}{\pi} \int_T \|f(\cdot - t)\|_p K_n(t)\, dt \leqslant \|f\|_p. \tag{3.4}$$

In other words, the sequence $\sigma_n(f)$ is bounded in $L^p(T)$ (by the L^p norm of f). To determine whether this condition is also sufficient to insure that (3.1) be the Fourier series of an L^p function requires some knowledge about L^p-bounded sequences of functions. What we need is contained in the "weak compactness" statement that follows, namely,

Proposition 3.2. Let $\{f_n\}$ be a bounded sequence in $L^p(T)$, $1 < p \leqslant \infty$, with bound M. Then there exist a subsequence n_k and an L^p function f, $\|f\|_p \leqslant M$, such that

$$\lim_{n_k \to \infty} \frac{1}{2\pi} \int_T f_{n_k}(t) g(t)\, dt = \frac{1}{2\pi} \int_T f(t) g(t)\, dt \tag{3.5}$$

for each $g \in L^{p'}(T)$, $1/p + 1/p' = 1$ (here $p' = 1$ when $p = \infty$). When (3.5) holds, we say that f_{n_k} converges weakly to f in $L^p(T)$.

Proof. We divide the proof into three steps:

(i) There is a subsequence n_k such that for all trigonometric polynomials g with rational coefficients

$$\lim_{n_k \to \infty} \frac{1}{2\pi} \int_T f_{n_k}(t) g(t)\, dt \qquad \text{exists.} \tag{3.6}$$

(ii) If for trigonometric polynomials g as in (i) above we set

$$L(g) = \lim_{n_k \to \infty} \frac{1}{2\pi} \int_T f_{n_k}(t) g(t)\, dt,$$

then L is a linear functional on these polynomials which can be extended to a bounded linear functional on $L^{p'}(T)$; moreover $\|L(h)\|_{p'} \leqslant M \|h\|_{p'}$, all $h \in L^{p'}(T)$.

(iii) (Riesz Representation theorem). Each bounded linear functional on $L^{p'}(T)$, as in (ii), can be represented as

$$L(h) = \frac{1}{2\pi} \int_T h(t) f(t) \, dt,$$

where $f \in L^p(T)$ and $\|f\|_p = \sup(|L(h)|/\|h\|_{p'}) \leqslant M$.

Proof of (i). Enumerate the polynomials $g_1, g_2, \ldots$ and let

$$c_{n,m} = \frac{1}{2\pi} \int_T f_n(t) g_m(t) \, dt, \quad \text{all} \quad n, m \geqslant 1.$$

Then by Hölder's inequality

$$|c_{n,m}| \leqslant \|f_n\|_p \|g_m\|_{p'} \leqslant M \|g_m\|_{p'}, \qquad \text{all } n, m. \tag{3.7}$$

Fix $m = 1$ now. By (3.7) $\{c_{n,1}\}$ is a bounded sequence and there is a subsequence $n_1 \to \infty$ such that $\lim n_{1\to\infty} c_{n_1,1}$ exists. Repeating this argument with $\{c_{n_1,2}\}$ in place of $\{c_{n,1}\}$ above we obtain a new subsequence $n_2 \to \infty$, say, such that $\lim_{n_2\to\infty}, c_{n_2,j}$ exists, $j = 1, 2$. These are the first steps of the so-called Cantor diagonal process which ensures the existence of a subsequence $n_k \to \infty$ so that $\lim_{n_k\to\infty} c_{n_k,j}$ exists, $j = 1, \ldots, k$. (i) follows by letting $k \to \infty$.

Proof of (ii). L is decidedly linear over these polynomials, i.e., $L(g_1 + \lambda g_2) = L(g_1) + \lambda L(g_2)$ for each scalar λ and is also bounded since by (3.7) $|L(g)| \leqslant M\|g\|_{p'}$; next we extend L linearly and continuously to all of $L^{p'}(T)$. By Proposition 2.4 and Remark 2.5 this class of polynomials is dense in $L^{p'}(T)$, therefore to each h in $L^{p'}(T)$ there corresponds a sequence of polynomials g_n such that $\lim_{n\to\infty}\|h - g_n\|_{p'} = 0$, $\limsup_{n\to\infty}\|g_n\|_{p'} \leqslant \|h\|_{p'}$. Moreover, since, as is readily verified, for these g_n's the sequence of scalars $\{L(g_n)\}$ is Cauchy, it is also convergent. Putting $L(h) = \lim_{n\to\infty} L(g_n)$, L turns out to be a well-defined linear functional in $L^{p'}(T)$. It only remains to show it is bounded. Well,

$$|L(h)| \leqslant \limsup_{n\to\infty} |L(h) - L(g_n)| + \limsup_{n\to\infty} |L(g_n)| \leqslant M\|h\|_{p'}.$$

Proof of (iii). We begin by considering the action of L on the simplest function that comes to mind, namely $\chi_{(-\pi,x]}$, the characteristic function of the interval $(-\pi, x]$, for each x in T. Let then

$$F(x) = L(\chi_{(-\pi,x]}). \tag{3.8}$$

We first show that F is absolutely continuous in T. For this purpose let $\{I_j\}_{j=1}^n$ be a collection of disjoint intervals, $I_j = (a_j, b_j)$ say, of T. We must show that given $\varepsilon > 0$ there is $\delta > 0$ so that

$$\sum_{j=1}^n |F(b_j) - F(a_j)| < \varepsilon, \qquad \text{provided } \sum_{j=1}^n |I_j| < \delta. \tag{3.9}$$

But this is pretty straightforward. Let

$$\phi(t) = \sum_{j=1}^{n} \operatorname{sgn}(F(b_j) - F(a_j))\chi_{I_j}(t).$$

Since the I_j's are disjoint we readily see that

$$\|\phi\|_{p'}^{p'} = \sum_{j=1}^{n} |I_j|. \tag{3.10}$$

Furthermore

$$L(\phi) = \sum_{j=1}^{n} \operatorname{sgn}(F(b_j) - F(a_j))L(\chi_{I_j}) = \sum_{j=1}^{n} |F(b_j) - F(a_j)|. \tag{3.11}$$

Whence, by combining (3.11) and (3.10), the continuity of L gives

$$\sum_{j=1}^{n} |F(b_j) - F(a_j)| \leq M\left(\sum_{j=1}^{n} |I_j|\right)^{1/p'}$$

and (3.9) holds with the choice $\delta = (\varepsilon/M)^{p'}$.

Let $f \in L(T)$ be such that

$$F(x) = \frac{1}{2\pi}\int_{(-\pi,x]} f(t)\, dt. \tag{3.12}$$

On account of (3.8) we we may rewrite (3.12) as

$$L(\chi_{(-\pi,x]}) = \frac{1}{2\pi}\int_T \chi_{(-\pi,x]}(t) f(t)\, dt.$$

Moreover, since every step function ϕ in T is (equal a.e. to) a finite linear combination of the $\chi_{(-\pi,x]}$'s we also have

$$L(\phi) = \frac{1}{2\pi}\int_T \phi(t) f(t)\, dt \tag{3.13}$$

by the linearity of L and of the integral. Let now h be a real trigonometric polynomial and let ϕ_n be a sequence of step functions converging a.e. and in $L^{p'}$ norm to h. By the boundedness of L

$$L(h) = \lim_{n\to\infty} L(\phi_n) \tag{3.14}$$

and by the Lebesgue dominated convergence theorem

$$\frac{1}{2\pi}\int_T h(t) f(t)\, dt = \lim_{n\to\infty} \frac{1}{\pi}\int_T \phi_n(t) f(t)\, dt. \tag{3.15}$$

Thus combining (3.13), (3.14), and (3.15) we see that (3.13) also holds with ϕ replaced by h there. Furthermore, by applying (3.13) to the real and imaginary parts of an arbitrary trigonometric polynomial, we see that (3.13) holds for all trigonometric polynomials.

Since by the boundedness of L we have

$$\left|\frac{1}{2\pi}\int_T h(t)f(t)\,dt\right| \leq M\|h\|_{p'},$$

by the converse to Hölder's inequality it follows that $\|f\|_p \leq M$, $1/p+1/p'=1$. Thus it only remains to show that (3.13) holds for an arbitrary $h \in L^{p'}(T)$. But this is easy since, for a sequence of trigonometric polynomials h_n converging to h in $L^{p'}$, we have on one hand that $\lim_{n\to\infty} L(h_n) = L(h)$ and on the other hand, by Hölder's inequality, that

$$\lim_{n\to\infty}\frac{1}{2\pi}\int_T (h(t)-h_n(t))f(t)\,dt = 0. \quad ■$$

We are now in a position to characterize those trigonometric series which are the Fourier series of $L^p(T)$ functions, $1 < p \leq \infty$.

Theorem 3.3 (Fejér). A necessary and sufficient condition for the trigonometric series (3.1) to be the Fourier series of an L^p function f, $1 < p \leq \infty$, is that the sequence (3.2) of its Cesàro means be bounded in L^p, i.e., $\|\sigma_n\|_p \leq K$. In this case $\|f\|_p \leq K$.

Proof. On account of the estimate (3.4) it is enough to show the sufficiency of the statement. By Proposition 3.2, there exist a sequence $n_k \to \infty$ and a function f in $L^p(T)$ such that σ_{n_k} converges weakly to f, as $n_k \to \infty$. But since $e^{-ijt} \in L^\infty(T)$ we see that

$$\lim_{n_k\to\infty}\frac{1}{2\pi}\int_T \sigma_{n_k}(t)e^{-ijt}\,dt = \lim_{n_k\to\infty}\left(1-\frac{|j|}{n_k+1}\right)c_j$$

$$= c_j = \frac{1}{2\pi}\int_T f(t)e^{-ijt}\,dt = c_j(f). \quad ■$$

Theorem 3.3 does not hold for $p=1$ since $L(T)$ is not a dual space, but the space of finite measures is; the question we consider is, then, under what conditions (3.1) is the Fourier series of a finite measure μ. In first place observe that, if (3.1) is the Fourier series of a finite measure μ, then

$$\sigma_n(\mu, x) = \frac{1}{\pi}\int_T K_n(x-t)\,d\mu(t)$$

and

$$\|\sigma_n(\mu)\|_1 \leq \frac{1}{2\pi}\int_T |d\mu(t)|, \qquad \text{all } n. \tag{3.16}$$

So the natural question is, then, what can we say about $L(T)$-bounded sequences $\{f_n\}$. In the first place, by setting

$$F_n(x) = \int_{(-\pi,x]} f_n(t)\,dt,$$

we readily see that the sequence $\{F_n\}$ is of uniformly bounded variation in T. These functions satisfy the following weak-compactness criteria.

Lemma 3.4. (Helly). Given a sequence $\{F_n\}$ of uniformly bounded variation in T, either there exists a uniformly bounded subsequence $\{F_{n_k}(x)\}$ converging everywhere to a function $F(x)$ of bounded variation in T, or else $\{|F_n(x)|\}$ diverges uniformly to ∞ in T, as $n \to \infty$.

Proof. Suppose first that each F_n is nonnegative, nondecreasing, and bounded above by K, say. As in the proof of step (i) of Proposition 3.2, by Cantor's diagonal process we find a subsequence $n_{k\to\infty}$ such that $\lim_{n_k\to\infty} F_{n_k}(r) = F(r)$ exists for each rational number r in T. Wherever the limit $F(r)$ exists, it represents a nondecreasing, nonnegative, bounded function. For $x \in T$, x irrational, put $d(x) = \lim_{r\to x^+} F(r) - \lim_{r\to x^-} F(r)$, where r runs through the rational numbers in T. As is readily seen for any finite collection $x_1, \dots, x_n$ in T we have $d(x_1) + \cdots + d(x_n) \le K$; therefore the set $N = \{x \in T : d(x) > 0\}$ is at most countable. We claim that for $x \in T\backslash N$, $\lim_{n_k\to\infty} F_{n_k}(x)$ exists. In fact, given $\varepsilon > 0$ and x in $T\backslash N$, $x \ne -\pi$, π, there are rational numbers $r_1 < x < r_2$ such that $0 \le F(r_2) - F(r_1) < \varepsilon$. Since $F_{n_k}(r_1) \le F_{n_k}(x) \le F_{n_k}(r_2)$, and the extreme terms of this inequality tend to $F(r_1)$ and $F(r_2)$, respectively, as $n_k \to \infty$, we see that the oscillation of $F_{n_k}(x)$ does not exceed ε, and the sequence converges at x. Let $D = \{x \in T : F_{n_k}(x) \text{ diverges}\}$; D is at most countable. Repeating with D the argument we gave above for the rational numbers in T, we find a subsequence of n_k, which we denote n_k again for convenience, such that $F_{n_k}(x)$ converges in D. Clearly, this sequence converges everywhere in T.

In the general case, and with no loss of generality, assume that the F_n's are real and put $F_n(x) = F_n(-\pi) + P_n(x) - N_n(x)$, where $P_n(x)$ and $N_n(x)$ denote the positive and negative variations of $F_n(x) - F_n(-\pi)$ respectively. If there is a subsequence n_k such that $F_{n_k}(-\pi)$ converges to a finite limit as $n_k \to \infty$, then from n_k first select a subsequence n'_k such that $P_{n_{k'}}(x)$ converges and from $n_{k'}$ another subsequence $n_{k''}$, say, so that $N_{n_{k''}}(x)$ converges. Then also $\lim_{n_{k''}\to\infty} F_{n_{k''}}(x) = F(x)$ exists, and it is of bounded variation since

$$F(x) = F(-\pi) + \lim_{n_{k''}\to\infty} P_{n_{k''}}(x) - \lim_{n_{k''}\to\infty} N_{n_{k''}}(x),$$

where the last two terms are nondecreasing, bounded functions of x. If, on the other hand, our assumption concerning $\{F_n(-\pi)\}$ does not hold, then $\lim_{n\to\infty}|F_n(-\pi)| = \infty$. Since the oscillations of the functions $F_n(x)$ are uniformly bounded, it is readily seen that $\{|F_n(x)|\}$ diverges uniformly to ∞ as $n \to \infty$. ■

We can now prove

Theorem 3.5. A necessary and sufficient condition for the trigonometric series (3.1) to be the Fourier series of a finite measure μ, or a Fourier–Stieltjes series, is that the sequence (3.2) of its Cesàro means be uniformly in $L^1(T)$, i.e., $\|\sigma_n\|_1 \leq M$. In this case $(1/2\pi)\int_T |d\mu(t)| \leq M$.

Proof. By inequality (3.16), it is enough to show the sufficiency of the statement. Let $F_n(x) = \int_{(-\pi,x]} \sigma_n(t)\,dt$; the functions $F_n(x)$ are of uniformly bounded variation over T. Since $F_n(-\pi) = 0$ for each n, $\{|F_n(x)|\}$ cannot diverge uniformly to ∞ in T, and by Helly's Lemma 3.4 there exist a subsequence $n_k \to \infty$ and a function $F(x)$ of bounded variation over T such that

$$\lim_{n_k\to\infty} \int_{(-\pi,x]} \sigma_{n_k}(t)\,dt = F(x), \qquad x \in T.$$

Let now j be fixed. Then for $n_k > |j|$ we have (cf. Proposition 3.2 of Chapter I)

$$\begin{aligned}\left(1 - \frac{|j|}{n_k+1}\right)c_j &= \frac{1}{2\pi}\int_T \sigma_{n_k}(t)e^{-ijt}\,dt \\ &= \frac{1}{2\pi}F_{n_k}(\pi) + (ij)\frac{1}{2\pi}\int_T F_{n_k}(t)e^{-ijt}\,dt.\end{aligned}$$

Whence, by letting $n_k \to \infty$ we get

$$c_j = \frac{1}{2\pi}F(\pi) + (ij)\frac{1}{2\pi}\int_T F(t)e^{-ijt}\,dt = \frac{1}{2\pi}\int_T e^{-ijt}\,dF(t),$$

and our conclusion obtains with $d\mu = dF$. ■

Only the case of (3.1) being a Fourier series remains to be discussed. But this is an easy matter now, as it suffices to check under what conditions the measure μ of Theorem 3.5 or, equivalently, the function F there is absolutely continuous.

Theorem 3.6. A necessary and sufficient condition for the trigonometric seris (3.1) to be a Fourier series is that the sequence (3.2) of its Cesàro means be Cauchy in $L(T)$, i.e.,

$$\lim_{m,n\to\infty} \|\sigma_m - \sigma_n\|_1 = 0.$$

Proof. Let $f \in L(T)$, then, by Theorem 2.7, $\{\sigma_n(f)\}$ is Cauchy in $L(T)$. Conversely, since Cauchy sequences are bounded, by Theorem 3.5 there is a function F of bounded variation so that

$$dF \sim \sum c_j e^{ijt}. \tag{3.17}$$

On the other hand, since $L(T)$ is complete, there is a function $f \in L(T)$ such that $\lim_{n\to\infty}\|\sigma_n - f\|_1 = 0$, and by Proposition 1.1 of Chapter I,

$$c_j(f) = \lim_{n\to\infty} c_j(\sigma_n) = c_j, \qquad \text{all} \quad j. \tag{3.18}$$

Whence, by combining (3.17) and (3.18), we see that $c_j = c_j(dF) = c_j(f)$, all j, and, by the uniqueness principle 4.6 of Chapter I, $dF(t) = f(t)\,dt$. ∎

4. A.E. CONVERGENCE OF $(C, 1)$ MEANS OF SUMMABLE FUNCTIONS

Since for $f \in L(T)$, $\sigma_n(f) \to f$ in L^1-norm, there is a subsequence $n_k \to \infty$ so that $\sigma_{n_k}(f, x) \to f(x)$ a.e., as $n_k \to \infty$. The sequence n_k depends on f, though. Is the stronger statement, namely, $\sigma_n(f, x) \to f(x)$ a.e., true? Some positive evidence in this direction is provided by the Lebesgue differentiation theorem. This result, a particular case of Theorem 2.2 of Chapter IV, states that if $\chi_\varepsilon(x) = (1/2\varepsilon)\chi_I(x/\varepsilon)$, where χ_I denotes the characteristic function of the interval $(-1, 1)$, then

$$\lim_{\varepsilon\to 0} f * \chi_\varepsilon(x) = \lim_{\varepsilon\to 0}\frac{1}{2\varepsilon}\int_{(-\varepsilon,\varepsilon)} f(x+t)\,dt = f(x) \text{ a.e.} \tag{4.1}$$

Now $\sigma_n(f, x) = f * 2K_n(x)$ looks rather like the expression in (4.1) in the sense that it also is a convolution of f with a positive kernel of integral 1. In other words, Fejér's kernel seems to be a good candidate for an "approximate identity," i.e., a kernel for which a conclusion such as (4.1) obtains. This statement will be made precise in Chapter IV; in the meantime we give a positive answer to the question concerning the a.e. convergence of the Cesàro means of a summable function.

To carry out this proof we need a strengthened version of (4.1), one which nevertheless can be deduced from it, namely,

Proposition 4.1. Let $f \in L(T)$, then

$$\int_{[0,\varepsilon]} |f(x+t) - f(x)|\,dt = o(\varepsilon) \qquad \text{a.e.} \quad x \text{ in } T. \tag{4.2}$$

Proof. The points x in T where (4.2) holds are called the Lebesgue points of f; our assertion, then, is that almost all points of T are in the Lebesgue set of $f \in L(T)$. The proof itself is quite straightforward. For each rational number r let

$$G_r = \left\{x \in T: \lim_{\varepsilon \to 0^+} \int_{[0,\varepsilon]} |f(x+t) - r|\, dt = |f(x) - r|\right\},$$

and put $B_r = T \backslash G_r$. By (4.1) applied to $|f(t) - r|$, we readily see that each (bad) set B_r is of measure 0 as is the union B of all the B_r's. B will be discarded. In addition we may disregard $N = \{x \in T: |f(x)| = \infty\}$, which is also of measure 0. We now show that points x in the good set $G = T\backslash(B \cup N)$ are Lebesgue points of f. For $x \in G$ and a rational number r we have

$$\frac{1}{\eta}\int_{[0,\eta]} |f(x+t) - f(x)|\, dt$$
$$\leqslant \frac{1}{\eta}\int_{[0,\eta]} |f(x+t) - r|\, dt + |r - f(x)| = I + J,$$

say. Let $\varepsilon > 0$ be given. Since $x \notin B$, we have that $I \leqslant \varepsilon/2$ provided η is sufficiently small. Also since $x \notin N$ there is a rational number r so that $I \leqslant \varepsilon/2$. Thus

$$\frac{1}{\eta}\int_{[0,\eta]} |f(x+t) - f(x)|\, dt \leqslant \varepsilon, \qquad \eta \text{ sufficiently small.} \quad \blacksquare$$

We are now in a position to prove

Theorem 4.2 (Lebesgue). Suppose $f \in L(T)$, then

$$\lim \sigma_n(f, x) = f(x) \tag{4.3}$$

at each Lebesgue point x of f. Thus (4.3) holds a.e. in T.

Proof. We proceed as in the proof of Theorem 2.1, but in a more deliberate fashion. By (2.12) of that theorem

$$|\sigma_n(f, x) - f(x)| \leqslant \frac{2}{\pi}\left(\int_{[0,\eta)} + \int_{[\eta,\pi]}\right)\left|\frac{f(x+t) + f(x-t)}{2} - f(x)\right| K_n(t)\, dt$$
$$= I_1 + I_2,$$

say. Moreover, by (2.9)

$$I_2 \leqslant \frac{c}{n+1}\int_{[\eta,\pi]}\left|\frac{f(x+t) + f(x-t)}{2} - f(x)\right|\frac{dt}{t^2} \leqslant \frac{c}{(n+1)\eta^2} c_f,$$

where c_f depends, of course, on f. So this term can be made small, but there must be a balance between η and n. Choosing $\eta = 1/n^{1/4}$, for instance, we see at once that I_2 can be made arbitrarily small for sufficiently large n. To bound I_1 is a more delicate pursuit. We split I_1 into two integrals,

$$\frac{2}{\pi}\left(\int_{[0,1/n)} + \int_{[1/n,1/n^{1/4}]}\right) = I_3 + I_4,$$

say. Now, by (2.8) it readily follows that

$$I_3 \leqslant cn \int_{[0,1/n)} \left| \frac{f(x+t)+f(x-t)}{2} - f(x) \right| dt = o(1),$$

at every Lebesgue point x of f by virtue of Proposition 4.1. It only remains to estimate I_4. Again by (2.9) it is dominated by

$$\frac{c}{n} \int_{[1/n,1/n^{1/4}]} \left| \frac{f(x+t)+f(x-t)}{2} - f(x) \right| \frac{dt}{t^2}. \tag{4.4}$$

Let now

$$F_x(t) = F(t) = \int_{[0,t)} \left| \frac{f(x+s)+f(x-s)}{2} - f(x) \right| ds.$$

Since F is absolutely continuous we may rewrite (4.4) as

$$\frac{c}{n} \int_{[1/n,1/n^{1/4}]} F'(t) \frac{dt}{t^2},$$

whence integrating by parts we see that

$$I_4 \leqslant \frac{c}{n} \frac{F(t)}{t^2}\Big]_{1/n}^{1/n^{1/4}} + \frac{2c}{n} \int_{[1/n,1/n^{1/4}]} F(t) \frac{dt}{t^3}. \tag{4.5}$$

Since $F(t)/t = o(1)$ at each Lebesgue point x of f, the integrated term in (4.5) is $o(1)$ as $n \to \infty$. The same is true for the integral in (4.5), since given $\varepsilon > 0$ we may first choose n large enough so that $F(t)/t < \varepsilon$ in $[1/n, 1/n^{1/4}]$. Therefore the integral does not exceed

$$\frac{c\varepsilon}{n} \int_{[1/n,\infty)} \frac{dt}{t^2} = c\varepsilon. \quad \blacksquare$$

5. NOTES; FURTHER RESULTS AND PROBLEMS

It was not always the case that the concepts of convergence and divergence were well understood; in fact, it was not until the time of Cauchy that the definitions were explictly formulated. Although the first mathematicians to use series were not interested in working with "divergent" series, it soon became apparent that some results obtained by formal manipulation of series led to correct results which could be verified independently. The important fact was to interpret these results properly. The concept of $(C, 1)$ convergence can be traced as far back as 1771 in the work of D. Bernoulli, and, applied to special series, to 1713 in the work of Leibnitz. The modern definition was given by Cesàro in 1890 in a paper dealing with multiplication of series. This gap in time is explained, in part, by the powerful opposition that mathematicians such as d'Alembert, Laplace, and even Lagrange expressed to these methods. Any result which asserts that a summation method sums every convergent series to its ordinary sum is called an Abelian theorem, and the method itself is called regular. Proposition 1.2 is such an example. In order to have an interesting summation method the converse to an Abelian result should be false. However, if by the addition of a hypothesis such a converse is true, we call this result a Tauberian theorem; 5.5 is an example of this. Lebesgue's Theorem 4.2, proved in 1904 and often called the Fejér–Lebesgue theorem, is one of the first applications of the new concept of Lebesgue integration, introduced in 1902.

Further Results and Problems

5.1 If $\sum c_j$ is $(C, 1)$ summable to a finite sum, then

$$\lim_{j\to\infty} \frac{c_j}{j} = \lim_{j\to\infty} \frac{s_j}{j} = 0.$$

5.2 Show that $\sum_{j=0}^{\infty} \cos jt = \frac{1}{2}(C, 1)$ and

$$\sum_{j=1}^{\infty} \sin jt = \frac{\cos(t/2)}{2\sin(t/2)}(C, 1).$$

5.3 If a series with positive terms is $(C, 1)$ summable to s, $0 \leqslant s \leqslant \infty$, then it also converges, or diverges, to s. (*Hint*: The sequence of partial sums is increasing.)

5.4 If $\sum c_j = s(C, 1)$ and $c_j = o(1/j)$, then $\sum c_j = s$. Show that the growth condition on the c_j's can be replaced by $\sum j^p |c_j|^{p+1} < \infty$, $0 < p < \infty$. (*Hint*: $s_n - \sigma_n = (1/n)\sum_{j=1}^{n} jc_j$.)

5.5 (Hardy) If $\sum c_j = s(C, 1)$ and $c_j = O(1/j)$, then $\sum c_j = s$. Can the growth condition on the c_j's be weakened to read: for every $\varepsilon > 0$ there exists a $\lambda > 1$ such that $\lim_{n\to\infty} \sum_{n<j<\lambda n} |c_j| < \varepsilon$? (*Hint*: Assume $s = 0$ and

$c_j \leqslant 1/j$, j large. If $m > n$ the following identity holds,

$$s_m = \frac{m\sigma_m - n\sigma_n}{m-n} - \frac{\sum_{j=n+2}^{m}(n-j+1)c_j}{m-n}.$$

Given $0 < \varepsilon < 1$ choose n large enough so that $|\sigma_n|, |\sigma_m| < \varepsilon$. Then

$$|s_n| \leqslant \varepsilon\frac{m+n}{m-n} + \frac{m-n}{2n} = \frac{\varepsilon((m/n)+1)}{(m/n)-1} + \frac{1}{2}\left(\frac{m}{n} - 1\right).$$

Minimizing this expression in terms of m/n, we see that the choice $m/(1+\sqrt{\varepsilon}) \leqslant n \leqslant m/(1+2\sqrt{\varepsilon})$ will do.)

5.6 The Fourier series of a function f of bounded variation converges to $(f(x+0)+f(x-0))/2$ a.e. (Dirichlet). Also, if $f \sim \sum c_j e^{ijt}$ and $c_j = O(1/|j|)$, then $s_n(f, x) \to f(x)$ a.e. (*Hint*: Combine 5.16 of Chapter I with 5.5 above to obtain Dirichlet's result; the second statement follows from Theorem 4.2.)

5.7 If for some $p \geqslant 1$, $\|\sigma_n(f) - f\|_p = o(1/n)$, then $f = c$. (*Hint*: Observe that $c_j(\sigma_n - f) = |j|c_j(f)/(n+1)$, $|j| \leqslant n$, and invoke estimate (1.10) of Chapter I.)

5.8 Let $f \in L(T)$ and $g \in L^\infty(T)$. The Fourier coefficients of the function $\chi_x(t) = f(x+t)g(t)$, which depend on x, tend to 0 uniformly for $x \in T$. (*Hint*: By 5.14 of Chapter I it suffices to show that $w_1(\chi_x, h) = o(1)$ as $h \to 0$, uniformly for x in T. Use the density of trigonometric polynomials in $L(T)$ to show this.)

5.9 (Riemann Localizaton Principle) Let $f, g \in L(T)$ coincide in some open interval $I \subset T$. Then $\lim_{n\to\infty}|s_n(f, x) - s_n(g, x)| = 0$ for all $x \in I$. Moreover, this convergence is uniform on every closed subinterval of I. (*Hint*: The statement, without the uniformity stressed, was proved in 5.11 of Chapter I. Here one must use 5.8.)

5.10 If the Cesàro means σ_n of the trigonometric series (3.1) satisfy $\sigma_n \geqslant 0$, then there is a positive measure μ such that $c_j = c_j(\mu)$. Is the statement true if we assume only that $\sigma_{n_k} \geqslant 0$ for a subsequence $n_k \to \infty$?

5.11 If the partial sums s_n of the trigonometric series (3.1) satisfy $s_n(x) \geqslant f(x)$, $x \in T$, $f \in L(T)$, then there is a finite measure μ such that $c_j = c_j(\mu)$. (*Hint*: $\sigma_n(x) - f(x) \geqslant 0$.)

5.12 $F \in D'$ is said to be positive of $F(u) \geqq 0$ whenever $u \geqslant 0$, $u \in C^\infty(T)$. Show that a positive distribution is necessarily a measure.

5.13 A complex-valued sequence $\{c_n\}$ is said to be positive definite if $\sum_{m,n} c_{m-n}\bar{z}_m z_n \geqslant 0$ for each finite sequence $\{z_n\}$ of complex numbers. Show that if μ is a positive measure then $\{c_n(\mu)\}$ is positive definite. Prove Herglotz' theorem, namely, given a positive definite sequence $\{c_n\}$ there is a positive measure μ such that $c_n = c_n(\mu)$. (*Hint*: Show that for all $m \geqslant 0$

$\sigma_m(t) \geqslant 0$, where $\sigma_m(t)$ are the Cesàro means of the series (3.1). This is readily done by choosing $z_n = e^{-int}$ in the definition).

5.14 The Cesàro means σ_n of the trigonometric series (3.1) satisfy $\|\sigma_n'\|_1 \leqslant K$ if and only if $c_j = c_j(f)$, f of bounded variation.

5.15 Using 5.14 above prove Hardy and Littlewood's result concerning functions of bounded variation stated in 5.12 of Chapter I. (*Hint*: If $w_1(f, h) = O(|h|)$, then by inequality (3.4) $\|\sigma_n(f, \cdot + h) - \sigma_n(f, \cdot)\|_1 \leqslant c|h|$ and consequently $\|\sigma_n'(f)\|_1 \leqslant c$.)

5.16 If $\{\phi_n\} \subset L(T)$ and $\lim_{n\to\infty} \|f - f * \phi_n\|_1 = 0$ for each $f \in L(T)$, then $\lim_{n\to\infty} c_j(\phi_n) = 1$, all j. Is some kind of converse true?

5.17 Let $f \in L(T)$, $f \sim \sum c_j e^{ijt}$. Then the Fourier series of f may be integrated termwise over any measurable subset E of T, and the resulting series is $(C, 1)$ summable to the integral of f. More precisely,

$$\sum_j c_j \int_E e^{ijt}\, dt = \int_E f(t)\, dt (C, 1).$$

(*Hint*: Write $\int_E (f(t) - \sigma_n(f, t))\, dt + \int_E \sigma_n(f, t)\, dt$ and use Theorem 2.7.)

5.18 Assume $f \in L^p(T)$, $g \in L^{p'}(T)$, $1/p + 1/p' = 1$, $1 < p, p' < \infty$. If $f \sim \sum c_j e^{ijt}$ and $g \sim \sum d_j e^{ijt}$, then

$$\sum c_j d_{-j} = \frac{1}{2\pi} \int_T f(t) g(t)\, dt (C, 1).$$

5.19 Show that the conclusion of 5.18 still holds if either $f \in L^\infty(T)$ and $g \in L(T)$ or $f \in C(T)$ and g is replaced by dG, a finite measure.

5.20 (Kolmogorov's Example of a Function $f \in L(T)$ with an Everywhere Divergent Fourier Series) We actually do the a.e. case here only. Suppose we can find a sequence $\{\phi_n\}$ of trigonometric polynomials satisfying the following properties. (i) $\phi_n \geqslant 0$; (ii) $(1/\pi) \int_T \phi_n(t)\, dt = 1$; and (iii) to each ϕ_n there correspond (a) a scalar $M_n \to \infty$, (b) a set E_n whose measure $\to 2\pi$, and (c) an integer q_n such that $|s_{p_n}(\phi_n, x)| > M_n$ for everywhere $x \in E_n$ and integer $p_n = p_n(x) < q_n$.

Then it is easy to construct f with the desired property. Since $M_n \to \infty$ we can choose a sequence n_k so that $\sum M_{n_k}^{-1/2} < \infty$. Write ϕ_{n_k} as a sum of exponentials and put

$$F_{n_k}(x) = \frac{e^{ir_k x}}{M_{n_k}^{1/2}} \phi_{n_k}(x) = \frac{e^{ir_k x}}{M_{n_k}^{1/2}} \sum_{|j| \leqslant m_{n_k}} c_j e^{ijx}.$$

The indices r_k are chosen so that there is no overlapping between the F_{n_k}'s. Moreover, since

$$\int_T \sum |F_{n_k}(x)|\, dx \leqslant \sum M_{n_k}^{-1/2} \int_T \phi_{n_k}(x)\, dx < \infty,$$

the series $\sum F_{n_k}(x)$ converges a.e. to a function $f \in L(T)$. It is readily verified that actually $f \sim \sum F_{n_k}$. On the other hand, if $x \in E_{n_k}$, then $F_{n_k}(x)$ contains a block of successive terms whose sum is at least $M_{n_k}^{1/2}$. Hence the series diverges whenever x belongs to infinitely many E_{n_k}'s, and this happens a.e. as $|E_{n_k}| \to 2\pi$. Thus it only remains to construct the polynomials ϕ_n; this is achieved as follows: write $A_j = (4j\pi/(2n+1))$, $0 \le j \le n$, suppose that $m_0, m_1, \ldots, m_n$ are integers with the properties (a) $m_0 \ge n^4$, (b) $m_{l+1} > 2m_l$ and (c) $2m_l + 1$ is divisible by $2n+1$ and put

$$\phi_n(x) = \frac{1}{n+1}\{K_{m_0}(x - A_0) + \cdots + K_{m_n}(x - A_n)\},$$

where K_m is Fejérs kernel of order m. These polynomials will do the job; this clearly is an outline of the construction, the verification of the actual details is quite involved.

5.21 (Bernstein's Inequality) Let t be a trigonometric polynomial of degree $\le N$. Then $\|t'\|_p \le 2N\|t\|_p$, $1 \le p \le \infty$. (*Hint*: Let $\phi_N(y) = K_{N-1}(y)(e^{iNy} + e^{-iNy}) = \sum_{|j| \le N}(1 - |j|/N)(e^{-(j+N)y} + e^{-i(j-N)y})$, $\|\phi_N\|_1 \le 2$. Then for $k \le N$, $\phi_N * e^{ikx} = (e^{ikx})'/(iN)$. The constant $2N$ is not best possible, N is.)

CHAPTER

III

Norm Convergence of Fourier Series

1. THE CASE $L^2(T)$; HILBERT SPACE

Because of the simplicity and completeness of the convergence results, the greatest success in representing functions by their Fourier series occurs in the L^2 setting. The reason is that $L^2(T)$ carries the geometric structure of a Hilbert space, its inner product given by

$$(f, g) = \frac{1}{2\pi}\int_T f(t)\overline{g(t)}\,dt. \tag{1.1}$$

In this Hilbert space the exponentials $\{e^{ijt}\}_{-\infty}^{\infty}$ form a complete orthonormal system. We start this section with a brief discussion of the basic properties of Fourier series in Hilbert space and conclude with the corresponding statements for $L^2(T)$.

Definition 1.1. A complex vector space H is called an inner product space if to each ordered pair x, y in H there is associated a complex number (x, y) called the inner product of x and y, with the following properties

(i) $(x, y) = \overline{(y, x)}$,
(ii) $(x + y, z) = (x, z) + (y, z)$,
(iii) $(\lambda x, y) = \lambda(x, y)$, and
(iv) $(x, x) \geqslant 0$ and $= 0$ only if $x = 0$.

By property (iv) we may define the norm $\|x\|$ of x as the number $\|x\| = (x, x)^{1/2}$, $x \in H$.

This quantity satisfies the usual properties of norms, including the triangle inequality. Therefore, we may also define the distance $d(x, y)$ of x to y as

$d(x, y) = \|x - y\|$ and this function satisfies all the conditions of a metric. If H endowed with this metric is complete, then it is called a Hilbert Space. All spaces we consider are Hilbert.

The concept of inner product also allows for the notion of orthogonality. We say that x and y are orthogonal if $(x, y) = 0$. If E is a subset of H we say that x is orthogonal to E provided $(x, y) = 0$ for every y in E. A subset E of H is called orthogonal if any two vectors in E are orthogonal to each other. Finally, a subset E of H is said to be an orthonormal system, and we abbreviate this by ONS, if E is orthogonal and the norm of each vector in E is 1, i.e., $(x, y) = 0$ and $(x, x) = 1$ for all $x \neq y$ in E.

Definition 1.2. If $\{x_\alpha\}_{\alpha \in A}$ is an ONS, we associate with each $x \in H$ a complex function on the index set A by means of

$$x(\alpha) = (x, x_\alpha), \qquad \alpha \in A. \tag{1.2}$$

These are the Fourier coefficients of x with respect to the ONS $\{x_\alpha\}$.

Lemma 1.3. If $\{x_j\}_{j=1}^n$ is an ONS and $x = \sum_{j=1}^n c_j x_j$, then $c_j = (x, x_j)$, $1 \leq j \leq n$, and $\|x\|^2 = \sum_{j=1}^n |c_j|^2$.

Proof. Just apply the relations given in Definition 1.1. ∎

Corollary 1.4. If $\{x_j\}_{j=1}^\infty$ is an ONS and $\{c_j\}$ is a complex sequence with $\sum |c_j|^2 < \infty$, then $\sum c_j x$ converges in H. If we denote the sum by x, then $c_n = x(n)$, all n.

Proof. To assure the convergence of the sum it suffices to show that the sequence $s_N = \sum_{j=1}^N c_j x_j$ is Cauchy in H. But by Lemma 1.3

$$\|s_N - s_M\|^2 = \sum_{j=N+1}^{M} |c_j|^2, \qquad N < M,$$

and this expression, being the tail of a convergent series, tends to 0 as N, $M \to \infty$. Moreover, $x(n) = \lim_{N\to\infty}(s_N, x) = c_n$. ∎

Next we seek to minimize $\|x - \sum_{j=1}^n \lambda_j x_j\|$, where $\{x_j\}_{j=1}^n$ is an ONS and the λ_j's are complex scalars.

Theorem 1.5 (Best Approximation). Let $\{x_j\}_{j=1}^n$ be an ONS, and let $x \in H$. Then

$$\left\| x - \sum_{j=1}^{n} x(j) x_j \right\| \leq \left\| x - \sum_{j=1}^{n} \lambda_j x_j \right\|, \tag{1.3}$$

for all scalars $\lambda_1, \ldots, \lambda_n$. Equality holds in (1.3) if and only if $\lambda_j = x(j)$, $1 \leqslant j \leqslant n$, and in this case

$$0 \leqslant \left\| x - \sum_{j=1}^{n} x(j)x_j \right\|^2 = \|x\|^2 - \sum_{j=1}^{n} |x(j)|^2. \tag{1.4}$$

Proof. Observe that

$$\left\| x - \sum_{j=1}^{n} \lambda_j x_j \right\|^2 = \|x\|^2 - \sum_{j=1}^{n} \bar{\lambda}_j x(j) - \sum_{j=1}^{n} \lambda_j \overline{x(j)} + \sum_{j=1}^{n} |\lambda_j|^2.$$

Let $\lambda_j = \alpha_j + i\beta_j$ and $x(j) = a_j + ib_j$; then the last three sums equal

$$-\sum_{j=1}^{n} (\alpha_j(2a_j - \alpha_j) + \beta_j(2b_j - \beta_j)),$$

and since the function $t(a - t)$ assumes its maximum when $t = a/2$ we see that the minimum of the above expression is assumed when $\alpha_j = a_j$, $\beta_j = b_j$, i.e.,

$$\lambda_j = x(j), \qquad 1 \leqslant j \leqslant n. \tag{1.5}$$

Moreover, equality can hold only if equality holds in (1.5) for all $j \leqslant n$. (1.3) then holds and (1.4) follows from Lemma 1.3 at once. ■

Corollary 1.6 (Bessel's Inequality). If $\{x_\alpha\}_{\alpha \in A}$ is an ONS, then $\sum_\alpha |x(\alpha)|^2 \leqslant \|x\|^2$.

Proof. We must give a meaning to the sum above; the conclusion asserts that, in the first place, the set of $x(\alpha) \neq 0$ is at most countable and, moreover, that

$$\sup\left(\sum_{\text{finitely many } \alpha\text{'s}} |x(\alpha)|^2 \right) \leqslant \|x\|^2.$$

The second assertion follows from inequality (1.4) and the first follows readily from this. ■

Proposition 1.7 (Riesz-Fischer). Let $\{x_\alpha\}_{\alpha \in A}$ be an ONS, and suppose that $\sum_\alpha |\lambda_\alpha|^2 < \infty$. Then there is $x \in H$ such that $x = \sum_\alpha \lambda_\alpha x_\alpha$.

Proof. Since $\sum_\alpha |\lambda_\alpha|^2 < \infty$, $\lambda_\alpha \neq 0$ only for countably many α's. To this countable set apply Corollary 1.4. ■

The basic properties of ONS, in addition to the ones discussed above, are given by

Theorem 1.8. Let $\{x_\alpha\}_{\alpha \in A}$ be an ONS in H. Each of the following conditions implies the others:

(i) (Completeness)$\{x_\alpha\}_{\alpha \in A}$ is a maximal orthonormal set in H.

(ii) $S = \{\sum_{\text{finitely many } \alpha\text{'s}} \lambda_\alpha x_\alpha\}$ is dense in H.
(iii) For every x in H, $\|x\|^2 = \sum_\alpha |x(\alpha)|^2$.
(iv) (Parseval's Identity) For x, y in $H(x, y) = \sum_\alpha x(\alpha)\overline{y(\alpha)}$.

Proof. (i) *implies* (ii). Let M denote the closure of S; M is also a subspace of H. If $M \neq H$, then by well-known properties of Hilbert spaces there is a nonzero vector x orthogonal to M. This contradicts the fact that $\{x_\alpha\}_{\alpha \in A}$ is a maximal orthogonal set.

(ii) *implies* (iii). If x is a finite linear combination of the x_α's, then (iii) follows from Lemma 1.3. As for the general case, take limits in the identity (1.4).

(iii) *implies* (iv). We polarize (iii). More precisely we apply (ii) to $x + \lambda y$ to get $(x + \lambda y, x + \lambda y) = \sum_\alpha |x(\alpha) + \lambda y(\alpha)|^2$. We then take $\lambda = 1$ and $\lambda = i$ to obtain the desired conclusion.

(iv) *implies* (i). If (i) does not hold, then there is $y \neq 0$ such that $(y, x_\alpha) = 0$ for all α. By (iv), with $x = y$ there, we get that $\|y\|^2 = 0$, or $y = 0$, which is a contradiction. ∎

How do these results apply to $L^2(T)$? Corollary 2.8 of Chapter II states that the ONS $\{e^{ijt}\}_{-\infty}^{\infty}$ is complete, i.e., (i) of theorem 1.8 above holds. Therefore, (ii)–(iv) of that theorem, as well as the rest of the results in this section also hold. In particular, by the identity (1.4) in theorem 1.5

$$\lim_{n\to\infty} \|f - s_n(f)\|_2 = 0, \qquad f \in L^2(T), \tag{1.6}$$

and

$$\frac{1}{2\pi}\int_T |f(t)|^2\, dt = \sum_{j=-\infty}^{\infty} |c_j(f)|^2. \tag{1.7}$$

2. NORM CONVERGENCE IN $L^p(T)$, $1 \leq p \leq \infty$

We start out in an upbeat note because of the positive result, i.e., statement (1.6), for the case $p = 2$. On the other hand, the case $p = \infty$ was ruled out, even with $L^\infty(T)$ replaced by $C(T)$, by the results in Section 2 of Chapter I. Moreover, since for $1 \leq p < \infty$ the $L^p(T)$ spaces are Hilbert only when $p = 2$, a different approach is needed here.

Our first result, although elementary, is of interest since it reduces the question of norm convergence to one of establishing the continuity of a convolution operator in $L^p(T)$; this task seems more accessible.

Proposition 2.1. Let $1 \leq p < \infty$. Then the following are equivalent:

(i) For every $f \in L^p(T)$, $\lim_{n\to\infty} \|s_n(f) - f\|_p = 0$.
(ii) $\|s_n(f)\|_p \leq c_p \|f\|_p$, for some constant c_p independent of n and f.

Proof. (i) *implies* (ii). If (i) holds, then in particular $|s_n(f, x)| \leqslant c_f < \infty$ a.e. in T, for every f in $L^p(T)$. Putting $X = Y = L^p(T)$ in the uniform boundedness principle, Theorem 2.3 of Chapter I, this estimate gives

$$\sup_{\|f\|_p \leqslant 1} \|s_n(f)\|_p \leqslant c, \quad \text{all} \quad n.$$

(ii) *implies* (i). Let $f \in L^p(T)$. Then given $\varepsilon > 0$, let g denote a trigonometric polynomial such that $\|f - g\|_p \leqslant \varepsilon$. Clearly, $s_n(g, x) \to g(x)$ as $n \to \infty$; in fact, both expressions coincide if $n \geqslant$ degree g. Therefore by Minkowski's inequality

$$\begin{aligned}\|f - s_n(f)\|_p &\leqslant \|f - g\|_p + \|g - s_n(g)\|_p + \|s_n(g) - s_n(f)\|_p \\ &= I + J + K,\end{aligned}$$

say. As we pointed out, $J = 0$, provided n is sufficiently large. Also $K = \|s_n(f - g)\|_p \leqslant c_p\|f - g\|_p \leqslant c_p\varepsilon$. Therefore $\|f - s_n(f)\|_p \leqslant \varepsilon(1 + c_p)$, ε arbitrary, provided n is sufficiently large. ■

Proposition 2.1 suffices to rule out norm convergence for $p = 1$. Indeed, since $\|2K_N\|_1 = 1$, all N, we have that

$$I = \sup_{\|f\|_1 \leqslant 1} \|s_n(f)\|_1 \geqslant \|s_n(2K_N)\|_1, \quad \text{all} \quad N. \tag{2.1}$$

Now we take a closer look at $s_n(2K_N, x)$; it equals $2K_N * 2D_n(x) = \sigma_N(2D_n, x)$, $x \in T$. Moreover, since by Lebesgue's Theorem 4.2 of Chapter II, $\sigma_N(2D_n, x) \to 2D_n(x)$ a.e. in T, from (2.1) and Proposition 2.1 in Chapter I it follows that $I \geqslant L_n \sim (4/\pi^2) \ln n$, n large. Therefore (ii) of Proposition (2.1) cannot hold and convergence in norm cannot occur.

The next reduction in making the problem of norm convergence tractable is to consider an analytic expression which roughly corresponds to "one-sided Fourier series." This is motivated by noting that

$$s_n(f, x) = \sum_{k=-\infty}^{\infty} \chi_{[-n,n]}(k) c_k e^{ikx}$$

and is achieved by studying a new object, the conjugate operator.

3. THE CONJUGATE MAPPING

To each $f \in L(T)$, $f \sim \sum c_j e^{ijt}$, we associate the trigonometric series

$$\sum (-i)(\operatorname{sgn} j) c_j e^{ijt} \tag{3.1}$$

where

$$\operatorname{sgn} j = \begin{cases} 1 & \text{if} \quad j > 0, \\ 0 & \text{if} \quad j = 0, \\ -1 & \text{if} \quad j < 0. \end{cases}$$

If there is an integrable function with Fourier series (3.1), we call this function the conjugate to f and denote it by $\tilde{f}$.

For instance, if $f \in L^2(T)$ then also $\tilde{f} \in L^2(T)$ and $\|\tilde{f}\|_2 \leqslant \|f\|_2$. This is quite simple, for in this case

$$\sum |(-i)(\operatorname{sgn} j)c_j|^2 = \|f\|_2^2 - |c_0|^2 < \infty$$

and by Riesz-Fischer there is a unique L^2 function g such that $c_j(g) = (-i)(\operatorname{sgn} j)c_j$; this function is $\tilde{f}$. In general we say that $L^p(T)$ admits conjugation if to every $f \in L^p(T)$, $f \sim \sum c_j e^{ijt}$, there corresponds a function $g \in L^p(T)$ with Fourier coefficients $c_j(g) = (-i)(\operatorname{sgn} j)c_j$ and $\|g\|_p \leqslant c_p \|f\|_p$, where c_p is a constant independent of f; in this case $g = \tilde{f}$.

Closely related to the concept of conjugation is that of projection. Given $f \in L(T)$, $f \sim \sum c_j e^{ijt}$, we define its projection Pf to be the trigonometric series $\sum_{j=0}^{\infty} c_j e^{ijt}$. We then say that $L^p(T)$ admits projections if, for every $f \in L^p(T)$, Pf is the Fourier series of an $L^p(T)$ function g and $\|g\|_p \leqslant c_p \|f\|_p$, where c_p is a constant independent of f.

The relation between these concepts is given by

Proposition 3.1. The following statements are equivalent

(i) $L^p(T)$ admits conjugation.
(ii) $L^p(T)$ admits projections.

Proof. (i) implies (ii). Let $f \in L^p(T)$, $f \sim \sum c_j e^{ijt}$, and put $g = c_0/2 + (f + i\tilde{f})/2$. Since L^p admits conjugation, $g \in L^p(T)$, and $\|g\|_p \leqslant c_p \|f\|_p$. By inspection, we observe that $c_j(g) = c_j(f)$, $j \geqslant 0$, and $c_j(g) = 0$ if $j < 0$; thus $g = Pf$ and we are done.

(ii) implies (i). Conversely, now let

$$g = \frac{2}{i}\left(Pf - \frac{c_0}{2} - \frac{f}{2}\right).$$

Since L^p admits projections, $g \in L^p(T)$ and $\|g\|_p \leqslant c_p \|f\|_p$. Again by inspection we notice that $c_j(g) = (-i)(\operatorname{sgn} j)c_j(f)$, and therefore $g = \tilde{f}$. ■

The main reason for the study of the conjugate operator is given by

Theorem 3.2. The following statements are equivalent

(i) For every $f \in L^p(T)$, $\lim_{n\to\infty} \|s_n(f) - f\|_p = 0$.
(ii) $L^p(T)$ admits conjugation.

Proof. By combining Propositions 2.1 and 3.1, it suffices to prove that the following pair of statements are equivalent:

(i′) $\|s_n(f)\|_p \leqslant c_p \|f\|_p$, c_p independent of n and f.
(ii′) $L^p(T)$ admits projections.

(i′) implies (ii′). let $f \in L^p(T)$, $f \sim \sum c_j e^{ijt}$. Since by identity (3.2) of Chapter I $c_j(e^{-int}f) = c_{n+j}$, a simple computation gives

$$e^{int}s_n(e^{-int}f, t) = \sum_{j=0}^{2n} c_j e^{ijt}. \tag{3.2}$$

We denote the polynomial on the right-hand side of (3.2) above by $P_n f$, the nth partial projection of f. Therefore

$$\|P_n f\|_p = \|s_n(e^{-int}f)\|_p \leqslant c_p \|f\|_p, \qquad \text{all} \quad n. \tag{3.3}$$

Next we show that the sequence $P_n f$ is actually Cauchy in $L^p(T)$. Indeed, given $\varepsilon > 0$, let g be a trigonometric polynomial of degree N such that $\|f - g\|_p \leqslant \varepsilon$. Then by (3.3) $\|P_n(f) - P_n(g)\|_p = \|P_n(f - g)\|_p \leqslant c_p\varepsilon$. Now, if $n, m > N/2$, $P_n(g) = P_m(g)$ and consequently

$$\|P_n(f) - P_m(f)\|_p \leqslant \|P_n(f) - P_n(g)\|_p + \|P_m(f) - P_m(g)\|_p \leqslant 2c_p\varepsilon,$$

and the sequence is Cauchy. Since $L^p(T)$ is complete, there is a function $h \in L^p(T)$ such that $\lim_{n\to\infty}\|P_n(f) - h\|_p = 0$. Moreover, by (3.3), $\|h\|_p \leqslant c_p\|f\|_p$. It is clear that $\lim_{n\to\infty} c_j(P_n f) = c_j$, $j \geqslant 0$. Also by Theorem 1.1 of Chapter I, $c_j(h) = \lim_{n\to\infty} c_j(P_n f), j \geqslant 0$. Consequently, $h = Pf$, and we are done.

(ii′) implies (i′). We rewrite (3.2) as

$$e^{int}s_n(e^{-int}f, t) = Pf(t) - e^{i(2n+1)t}P(e^{-i(2n+1)t}f, t). \tag{3.4}$$

Consequently, multiplying through by e^{-int} and replacing f by $e^{int}f$ (3.4) becomes

$$s_n(f, t) = e^{-int}P(e^{int}f, t) - e^{i(n+1)t}P(e^{-i(n+1)t}f, t). \tag{3.5}$$

Whence it follows that $\|s_n(f)\|_p \leqslant \|P(e^{int}f)\|_p + \|P(e^{-i(n+1)t}f)\|_p \leqslant 2c_p\|f\|_p$, with c_p independent of n and f. ■

Corollary 3.3. $L(T)$ does not admit conjugation.

Proof. Since, by the example following Proposition 2.1, (i) of Theorem 3.2 does not hold, (ii) cannot hold either. ■

4. MORE ON INTEGRABLE FUNCTIONS

We already know that $L(T)$ does not admit conjugation; in this section we show that, in fact, the situation is as hopeless as it could be.

Example 4.1. There exists an integrable function $f, f \sim \sum c_j e^{ijt}$, such that the trigonometric series $\sum(-i)(\operatorname{sgn} j)c_j e^{ijt}$

(a) converges everywhere in T and
(b) is not a Fourier series.

The construction of this example is achieved through a sequence of results of independent interest, which are, therefore, stated separately. As a byproduct of the construction we are able to show that although the Fourier coefficients of an integrable function tend to 0, they do so arbitrarily slowly.

We begin introducing a particular kind of sequence, one which is relatively simple to handle.

Definition 4.2. We say that a real sequence $\{a_n\}$, $n \geq 0$ is convex if

$$a_{n-1} + a_{n+1} - 2a_n \geq 0, \qquad \text{all} \quad n \geq 1. \tag{4.1}$$

Examples of convex sequences abound; indeed, it suffices to put $a_n = \phi(n)$, where ϕ is a convex, real-valued function. A basic property of these sequences is

Proposition 4.3. If $\{a_n\}_{n=0}^{\infty}$ is positive, convex, and bounded, then it is decreasing, $\lim_{n\to\infty} n(a_{n-1} - a_n) = 0$, and

$$\sum_{j=1}^{\infty} j(a_{j-1} + a_{j+1} - 2a_j) = a_0 - \lim_{n\to\infty} a_n.$$

Proof. Convexity is clearly equivalent to

$$a_{j-1} - a_j \geq a_j - a_{j+1}, \text{ all } j. \tag{4.2}$$

We begin by showing that

$$a_{n-1} - a_n \geq 0, \text{ for all } n. \tag{4.3}$$

Suppose that (4.3) does not hold. Then there exists a value of n such that $a_{n-1} - a_n < 0$, and by (4.2) we also have that $|a_{j-1} - a_j| \geq |a_{n-1} - a_n|$, all $j \geq n$. Therefore, since for $m > n$

$$a_m - a_n = \sum_{j=n}^{m-1} |a_{j-1} - a_j| \geq (m-n)|a_{n-1} - a_n|,$$

it follows that $\lim_{m\to\infty} a_m = \infty$, which is not true. Consequently, (4.3) holds and the sequence is decreasing.

Let $a = \lim_{n\to\infty} a_n$. Then

$$a_0 - a = (a_0 - a_1) + \cdots + (a_{n-1} - a_n) + \cdots,$$

where the series on the right has monotone decreasing terms and converges. It is therefore well known and readily seen that $\lim_{n\to\infty} n(a_{n-1} - a_n) = 0$. Finally, since

$$s_n = \sum_{j=1}^{n} j(a_{j-1} + a_{j+1} - 2a_j) = a_0 - a_n - n(a_n - a_{n-1}),$$

$$\lim_{n\to\infty} s_n = a_0 - \lim_{n\to\infty} a_n. \quad \blacksquare$$

We would like to show that given a positive, convex sequence $\{a_n\}_{n=0}^{\infty}$, $\lim_{n\to\infty} a_n = 0$, there exists a unique $f \in L(T)$ such that

$$f \sim \sum a_{|j|} e^{ijt}. \tag{4.4}$$

One of the problems we are faced with now is that if even we had = in (4.4) it would be nontrivial to show that $f \in L(T)$. On the other hand, this difficulty is rather easily overcome if we can also write the relation (4.4) as

$$f(t) = \sum_{n=1}^{\infty} \lambda_n K_{n-1}(t), \tag{4.5}$$

where $\lambda_n \geq 0$ and the K_{n-1}'s are the Fejér kernels of order $n - 1$, because then the Fourier coefficients of f are essentially known (once the λ_n's are fixed) and the convergence and integrability of the series can be readily established by the fact that we are only dealing with positive quantities. More precisely, f will be integrable provided that $\sum \lambda_n < \infty$ since by the identity (2.7) in Chapter II

$$\|f\|_1 = \sum_{n=1}^{\infty} \lambda_n \frac{1}{2\pi} \int_T K_{n-1}(t)\, dt = \frac{1}{2} \sum_{n=1}^{\infty} \lambda_n.$$

Moreover, by the formula (2.2) in Chapter II, it also follows from (4.5) at once that

$$\begin{aligned} c_j(f) &= \frac{1}{2\pi} \int_T \left(\sum_{n=1}^{\infty} \lambda_n K_{n-1}(t) \right) e^{-ijt}\, dt \\ &= \sum_{n=1}^{\infty} \lambda_n c_j(K_{n-1}) = \sum_{n=j+1}^{\infty} \lambda_n \left(1 - \frac{|j|}{n}\right). \end{aligned} \tag{4.6}$$

The λ_n's are still unknown, whereas the c_j's are precisely the $a_{|j|}$'s; therefore, in order to complete our proof we must be able to show that the linear system with infinitely many unknowns given by (4.6) above, i.e.,

$$a_j = \sum_{n=j+1}^{\infty} \lambda_n\left(1-\frac{j}{n}\right), \qquad j=0,1,\ldots \tag{4.7}$$

can be solved when the a_j's are convex and tend to 0. Written in matrix form, (4.7) becomes

$$\begin{bmatrix} a_0 \\ a_1 \\ a_2 \\ \vdots \end{bmatrix} = \begin{bmatrix} 1 & 1 & 1 & 1 & \cdots \\ 0 & \frac{1}{2} & \frac{2}{3} & \frac{3}{4} & \cdots \\ 0 & 0 & \frac{1}{3} & \frac{2}{4} & \cdots \\ \vdots & \vdots & \vdots & \vdots & \cdots \end{bmatrix} \begin{bmatrix} \lambda_1 \\ \lambda_2 \\ \lambda_3 \\ \vdots \end{bmatrix}. \tag{4.8}$$

Thus by "inverting" the matrix in (4.8) by means of elementary operations, we readily see that under our assumptions this system of equations is equivalent to

$$\begin{bmatrix} 1 & -2 & 1 & 0 & 0 & \cdots \\ 0 & 1 & -2 & 0 & 0 & \cdots \\ 0 & 0 & 1 & 1 & 0 & \cdots \\ \vdots & \vdots & \vdots & \vdots & \vdots & \vdots \end{bmatrix} \begin{bmatrix} a_0 \\ a_1 \\ a_2 \\ \vdots \end{bmatrix} = \begin{bmatrix} 1 & 0 & 0 & \cdots \\ 0 & \frac{1}{2} & 0 & \cdots \\ 0 & 0 & \frac{1}{3} & \cdots \\ \vdots & \vdots & \vdots & \vdots \end{bmatrix} \begin{bmatrix} \lambda_1 \\ \lambda_2 \\ \lambda_3 \\ \vdots \end{bmatrix}$$

In other words, the solution is given by

$$\lambda_n = n(a_{n-1} + a_{n+1} - 2a_n), \qquad n \geqslant 1. \tag{4.9}$$

Also by Proposition (4.3), $\sum \lambda_n < \infty$, which is one of the required conditions. Summing up, we have

Proposition 4.4. Given a convex, positive sequence of real numbers $\{c_j\}_{j=0}^{\infty}$ which tends to zero, there is a positive, integrable function f such that $f \sim \sum c_{|j|} e^{ijt}$.

Proof. Put $f(t) = \sum_{n=1}^{\infty} n(c_{n-1} + c_{n+1} - 2c_n) K_{n-1}(t)$. ■

We collect one more observation, this one dealing with Fourier series with odd coefficients.

Proposition 4.5. Let $f \in L(T)$, $f \sim \sum c_n e^{int}$, and assume that $0 \leqslant c_n = -c_{-n}$, all $n \geqslant 0$. Then $\sum_{n=1}^{\infty} c_n/n < \infty$.

Proof. Since $c_0 = 0$, by Proposition 3.1 of Chapter I, $F(t) = \int_{(-\pi, t]} f(s)\,ds$ is a periodic, continuous function and $c_n(F) = (1/in)c_n$, $|n| \geqslant 1$. In particular, since F is continuous at 0, by Fejér's theorem in Chapter II, $\lim_{n\to\infty} \sigma_n(F, 0) = F(0)$.

Now, by formula (2.2) in Chapter II

$$\begin{aligned}\sigma_N(F,0) &= \sum_{|n|\leq N}\left(1-\frac{|n|}{N+1}\right)c_n(F)\\ &= c_0(F)+\sum_{n=1}^{N}\left(1-\frac{n}{N+1}\right)\left(\frac{c_n}{in}+\frac{c_{-n}}{(-in)}\right)\\ &= c_0(F)+\frac{2}{i}\sum_{n=1}^{N}\left(1-\frac{n}{N+1}\right)\frac{c_n}{n},\end{aligned} \tag{4.10}$$

and this expression converges to $F(0)$ as $N\to\infty$. Therefore

$$\sum_{n=1}^{N}\left(1-\frac{n}{N+1}\right)\frac{c_n}{n}=\sum_{n=1}^{N}\frac{c_n}{n}-\frac{1}{N+1}\sum_{n=1}^{N}c_n=A_N+B_N,$$

say, also converges as $N\to\infty$. But since $f\in L(T)$, $c_n\to 0$ as $n\to\infty$, and by Proposition 1.2 of Chapter II also $c_n\to 0(C,1)$, i.e., $B_N\to 0$ as $N\to\infty$. Thus $\lim_{N\to\infty}A_N$ exists. ■

Construction of Example 4.1. Let $c_j=(1/(2\ln|j|))$, $|j|\geq 2$. By Proposition 4.4 there is a positive, integrable function f such that $f\sim\sum c_je^{ijt}$. Since the sequence c_j is even we also have $f\sim\sum_{j=2}^{\infty}\cos(jt)/\ln(j)$. The formal, or distributional, conjugate of f is given by the trigonometric series $\sum_{|j|\geq 2}(-i)\times(\operatorname{sgn} j)(1/2\ln|j|))e^{ijt}$ or $\sum_{j=2}^{\infty}(\sin jt)/(\ln j)$. This is an everywhere convergent trigonometric series (cf. 7.23 below), which by Proposition 4.5 is not a Fourier series since $\sum_{j=2}^{\infty}1/j\ln j$ diverges.

Finally, observe that if $\{c_n\}_{n=0}^{\infty}$ is any complex sequence which converges to 0 as $n\to\infty$, then there is a positive, real sequence $\{a_n\}_{n=0}^{\infty}$ which is convex, decreases to zero, and $a_n\geq|c_n|$, all n. The integrable function f of Proposition 4.4 corresponding to this sequence of a_n's, then, has Fourier coefficients which tend to 0 at the rate of the c_n's.

There is yet a different way to state this property.

Proposition 4.6. Let $f\in L^p(T)$, $1\leq p<\infty$. Then we can find a positive, even sequence $\{\lambda_j\}$, $\lambda_{|j|}$ increasing to ∞, and a function $g\in L^p(T)$ such that $c_j(g)=\lambda_{|j|}c_j(f)$, all j.

Proof. For each n, let m_n be an integer such that $\|f-\sigma_{m_n}(f)\|_p\leq 2^{-n}$. This choice is possible on account of Theorem 2.7 of Chapter II. We may assume that m_n increases to ∞. Then put

$$g(x)=f(x)+\lim_{k\to\infty}\sum_{n=1}^{k}(f(x)-\sigma_{m_n}(f,x)),$$

where the limit is taken in the L^p sense. By Proposition 1.1 in Chapter I

$$c_j(g) = \left(1 + \sum_{n=1}^{\infty} \min\left(1, \frac{|j|}{m_n + 1}\right)\right) c_j(f). \quad \blacksquare$$

The above proof clearly holds for $C(T)$ as well.

5. INTEGRAL REPRESENTATION OF THE CONJUGATE OPERATOR

A key ingredient in the consideration of the partial sums or Cesàro means of an integrable function is the integral (actually a convolution) representation of the linear operator determined by these mappings. We now attempt a similar approach for the conjugate mapping given for $f \sim \sum c_j e^{ijt}$ by

$$f \to \sum_j (-i)\,(\operatorname{sgn} j) c_j e^{ijt}. \tag{5.1}$$

The first step in this direction is to study the behavior of the partial sums and Cesàro means of the distribution F with Fourier series

$$F \sim \sum (-i)(\operatorname{sgn} j) e^{ijt}. \tag{5.2}$$

These are denoted by $2\tilde{D}_n(t)$ and $2\tilde{K}_n(t)$ as they correspond to the conjugate Dirichlet and Fejér kernels respectively. It is not hard to obtain the explicit expression for these kernels. Indeed,

$$2\tilde{D}_n(t) = \sum_{|j| \le n} (-i)(\operatorname{sgn} j) e^{ijt} = (-i) \sum_{j=1}^{n} (e^{ijt} - e^{-ijt}).$$

To evaluate this, observe that the geometric sum

$$\sum_{j=1}^{n} e^{ijt} = e^{it} \frac{(1 - e^{int})}{(1 - e^{it})} = \frac{e^{it} e^{int/2} (e^{-int/2} - e^{int/2})}{e^{it/2} (e^{-it/2} - e^{it/2})}$$

$$= e^{i(n+1)t/2} \frac{\sin(nt/2)}{\sin(t/2)}.$$

By combining this with the expression obtained by replacing t by $-t$ above, we see that

$$\tilde{D}_n(t) = \frac{\sin(nt/2)}{2\sin(t/2)} (-i)(e^{i(n+1)t/2} - e^{-i(n+1)t/2})$$

$$= 2\sin(n/2) \frac{\sin((n+1)t/2)}{2\sin(t/2)}$$

$$= \frac{\cos(t/2)}{2\sin(t/2)} - \frac{\cos((n+1/2)t)}{2\sin(t/2)}, \qquad n = 0, 1, \ldots. \tag{5.3}$$

Thus $\tilde{D}_n(t)$ is an odd function bounded by

$$\frac{1}{2}\left|\sum_{|j|\leqslant n}(-i)(\operatorname{sgn} j)e^{ijt}\right| \leqslant \frac{2n}{2} = n. \tag{5.4}$$

Also as in estimate (1.17) of Chapter I it follows that

$$|\tilde{D}_n(t)| \leqslant \pi/|t|, \qquad 0<|t|<\pi, \quad \text{all} \quad n. \tag{5.5}$$

As the for conjugate Fejér kernel $\tilde{K}_n(t)$ we have

$$\begin{aligned}\tilde{K}_n(t) &= \frac{\tilde{D}_0(t)+\cdots+\tilde{D}_n(t)}{n+1}\\ &= \frac{\cos(t/2)}{2\sin(t/2)} - \frac{1}{(n+1)}\sum_{j=0}^{n}\frac{\cos((j+1/2)t)}{2\sin(t/2)}.\end{aligned} \tag{5.6}$$

The last sum in (5.6) equals

$$\begin{aligned}\operatorname{Re}\left(\sum_{j=0}^{n} e^{i(j+1/2)t}\right) &= \operatorname{Re}\left(e^{it/2}\sum_{j=0}^{n} e^{ijt}\right)\\ &= \operatorname{Re}\left(e^{i(n+1)t/2}\frac{\sin((n+1)t/2)}{\sin(t/2)}\right)\\ &= \frac{2\cos((n+1)t/2)\sin((n+1)t/2)}{2\sin(t/2)}\\ &= \frac{\sin((n+1)t)}{2\sin(t/2)}.\end{aligned}$$

Consequently,

$$\tilde{K}_n(t) = \frac{\cos(t/2)}{2\sin(t/2)} - \frac{1}{(n+1)}\frac{\sin((n+1)t)}{(2\sin(t/2))^2}, \qquad n=0,1,\ldots. \tag{5.7}$$

Thus $\tilde{K}_n(t)$ is also an odd function bounded by

$$\frac{1}{(n+1)}\sum_{j=0}^{n}|\tilde{D}_j(t)| \leqslant \frac{1}{(n+1)}\sum_{j=0}^{n} j = \frac{n}{2}, \qquad n=0,1,\ldots. \tag{5.8}$$

Moreover, by estimate 1.16 of Chapter I it also readily follows that

$$\left|\tilde{K}_n(t) - \frac{\cos(t/2)}{2\sin(t/2)}\right| \leqslant \frac{\pi^2}{4(n+1)t^2}. \tag{5.9}$$

This is all we need to know about these kernels.

Now, one way to insure that the mapping (5.1) be bounded in $L^p(T)$ is to show that the distribution F with Fourier series (5.2) is actually the

Fourier series of a finite measure μ on T. By Theorem 3.5 of Chapter II, this will be the case if (and only if)

$$\|\tilde{K}_n\|_1 \leq c, \qquad \text{all} \quad n. \tag{5.10}$$

Rewrite

$$\tilde{K}_n(t) = \frac{\cos(t/2)}{2\sin(t/2)}\left(1 - \frac{\sin((n+1)t)}{(n+1)\sin t}\right).$$

Since as it is readily seen

$$\left|\frac{\sin((n+1)t)}{(n+1)\sin t}\right| \leq \frac{\pi}{2t(n+1)} \qquad \text{for} \quad 0 < t < \pi/2,$$

we have

$$\left|1 - \frac{\sin((n+1)t)}{(n+1)\sin t}\right| \geq \frac{1}{2} \qquad \text{provided} \quad \frac{\pi}{n+1} < t < \frac{\pi}{2}.$$

Consequently,

$$\begin{aligned}\|\tilde{K}_n\|_1 &\geq \frac{1}{2}\int_{[\pi/(n+1),\pi/2]} \frac{\cos(t/2)}{2\sin(t/2)}\,dt = \tfrac{1}{2}\ln(\sin(t/2))]_{\pi/n+1}^{\pi/2}\\ &= \tfrac{1}{2}\ln(\sqrt{2}/2 - \ln(\sin(\pi/(n+1)))) \to \infty\end{aligned}$$

as $n \to \infty$. This makes it impossible for (5.10) to hold.

Let us return, then, to our original program of obtaining an integral representation for (5.1). Assume first that $f = \sum_{|j|\leq N} c_j e^{ijt}$ is a trigonometric polynomial of degree $\leq N$ and let

$$\tilde{s}_n(f, t) = \sum_{|j|\leq n} (-i)(\operatorname{sgn} j)c_j e^{ijt}. \tag{5.11}$$

By the convolution formula (3.3) of Chapter I we readily see that $\tilde{s}_n(f, x) =$

$$2\tilde{D}_n * f(x) = \frac{1}{\pi}\int_T f(x-t)\tilde{D}_n(t)\,dt. \tag{5.12}$$

Since $\tilde{D}_n(t)$ is odd (5.12) can be rewritten as either

$$\frac{-1}{\pi}\int_T f(x+t)\tilde{D}_n(t)\,dt \ \text{ or } \ \frac{1}{\pi}\int_T\left(\frac{f(x-t)-f(x+t)}{2}\right)\tilde{D}_n(t)\,dt. \tag{5.13}$$

Moreover, since the integrand in the last integral of (5.13) is even we also have

$$\tilde{s}_n(f, x) = \frac{-1}{\pi}\int_{[0,\pi]} (f(x+t) - f(x-t))\left(\frac{1}{2\tan(t/2)} - \frac{\cos((n+1/2)t)}{2\sin(t/2)}\right)dt. \tag{5.14}$$

Since f is a trigonometric polynomial, $\tilde{s}_n(f, x)$ actually coincides with $\tilde{f}(x)$ for $n \geq N$. In particular, $\lim_{n\to\infty} \tilde{s}_n(f, x) = \tilde{f}(x)$ everywhere. On the other hand, since

$$f(x+t) - f(x-t) = \sum_{|j|\leq N} c_j(e^{ij(x+t)} - e^{ij(x-t)})$$
$$= (-2i) \sum_{1\leq|j|\leq N} c_j e^{ijx}\sin(jt)$$

it is clear that $(f(x+t) - f(x-t))/(\sin(t/2))$ is bounded, and hence integrable, as a function of t in $[0, \pi]$. Therefore, by the Riemann–Lebesgue theorem,

$$\int_{[0,\pi]} \left(\frac{f(x+t) - f(x-t)}{\sin(t/2)}\right) \cos\left(\left(n + \frac{1}{2}\right)t\right) dt \to 0$$

as $n \to \infty$, since this integral can be expressed as the nth Fourier sine and cosine coefficients of integrable functions. Whence by letting $n \to \infty$ in (5.14) we see that $\tilde{f}(x)$ is given by the absolutely convergent integral

$$\tilde{f}(x) = \frac{-1}{\pi} \int_{[0,\pi]} \frac{f(x+t) - f(x-t)}{2\tan(t/2)}\, dt. \tag{5.15}$$

It now becomes apparent that the problem we set to solve may be a difficult one for it is not clear that the integral in (5.15) converges for arbitrary integrable, or even p-integrable, functions f. Indeed, as we shall see below, there is a continuous, periodic function f such that

$$\int_{[0,\pi]} \frac{|f(x+t) - f(x-t)|}{t}\, dt = +\infty, \qquad \text{every } x \text{ in } T. \tag{5.16}$$

Thus the existence of the integral in (5.15) is due not to the smallness of the $f(x+t) - f(x-t)$ for small t, but rather to the cancellation of positive and negative values.

Proposition 5.1 (Lusin). (5.16) holds.

Proof. Following Kaczmarz, we construct first a continuous function g with period 1 satisfying the following three properties, to wit:

(i) $|g(x)| \leq 1$;
(ii) $|g(x+t) - g(x)| \leq A|t|$;
(iii) $\int_{[1/n,1]} |g(nx+nt) - g(nx-nt)|/t\, dt \sim \ln n$.

Clearly condition (iii) above is the only one which is nontrivial. Nevertheless, let g be chosen so that $g(x+t)-g(x-t)$ is not identically equal to 0 as a function of t, for each x in $[0,1]$, while at the same time it verifies conditions (i) and (ii) above. For example, the function g with $g(0)=g(1)=0$ and $g, (\frac{1}{4})=1$ which is interpolated linearly in $(0,\frac{1}{4})$ and $(\frac{1}{4},1)$ will do.

It is clear that for the periodic extension of this function with period 1, which we also call g, we have

$$\int_{[0,1]} |g(x+t)-g(x-t)|\, dt \geqslant c > 0, \qquad x \in [0,1], \tag{5.17}$$

since the integral, as a continuous periodic function of x now, does not assume the value 0 for any x and has therefore a positive lower bound. Also by (i)

$$\int_{[0,1]} |g(x+t)-g(x-t)|\, dt \leqslant 2. \tag{5.18}$$

Let us replace x and t by nx and nt respectively and consider

$$\begin{aligned} I_n &= \int_{[1/n,1]} \left| \frac{g(nx+nt)-g(nx-nt)}{t} \right| dt \\ &= \int_{[1,n]} \left| \frac{g(nx+y)-g(nx-y)}{y} \right| dy \\ &= \int_{[0,1]} |g(nx+t)-g(nx-t)| \left\{ \frac{1}{t+1} + \cdots + \frac{1}{t+n-1} \right\} dt. \end{aligned} \tag{5.19}$$

Since the term $\{\cdot\}$ in (5.19) is of order $\ln n$, we see that property (iii) holds for g on account of (5.17) and (5.18).

By (ii) it also follows that

$$\int_{[0,1/n]} \left| \frac{g(nx+nt)-g(nx-nt)}{t} \right| dt \leqslant 2An \cdot \frac{1}{n} = 2A. \tag{5.20}$$

Consequently, combining (iii) and (5.20), we see at once that

$$\int_{[0,1]} \left| \frac{g(nx+nt)-g(nx-nt)}{t} \right| dt \leqslant c \ln n. \tag{5.21}$$

We now turn to construct f. The idea is to set $f(x)=\sum_{k=1}^{\infty} \varepsilon_k g(n_k x)$, where $\varepsilon_k > 0, \sum \varepsilon_k < \infty$, and n_k are to be chosen in such a way that there is an appropriate interplay between the ε_k's and the I_{n_k}'s. To show that (5.16) holds for f, note that

$$\begin{aligned}
J_k &= \int_{[1/n_k,1]} \left| \frac{f(x+t)-f(x-t)}{t} \right| dt \\
&\geqslant \varepsilon_k \int_{[1/n_k,1]} \left| \frac{g(n_k x+n_k t)-g(n_k x-n_k t)}{t} \right| dt \\
&\quad - \sum_{j=1, j\neq k}^{\infty} \varepsilon_j \int_{[1/n_k,1]} \left| \frac{g(n_j x+n_j t)-g(n_j x-n_j t)}{t} \right| dt \\
&\geqslant \varepsilon_k I_{n_k} - c\sum_{j=1}^{k-1} \varepsilon_j \ln n_j - 2\sum_{j=k+1}^{\infty} \varepsilon_j \int_{[1/n_k,1]} \frac{dt}{t} \\
&\geqslant c\left(\varepsilon_k \ln n_k - \sum_{j=1}^{k-1} \varepsilon_j \ln n_j \right) - 2\ln(n_k) \sum_{j=k+1}^{\infty} \varepsilon_j. \qquad (5.22)
\end{aligned}$$

It only remains now to choose the ε_k's and n_k's so that the right-hand side of (5.22) goes to infinity with k; the reader can verify that

$$\varepsilon_k = 1/k!, \qquad n_k = 2^{(k!)^2}$$

will do. Therefore $J_k \to \infty$ as $k \to \infty$ and (5.16) holds for a function f of period 1. The desired result follows without difficulty from this. ■

We need one more notation before we go on. By the convolution formula (3.3) of Chapter I we readily see that if

$$\tilde{\sigma}_n(f,t) = \sum_{|j|\leqslant n} \left(1 - \frac{|j|}{n+1}\right)(-i)(\operatorname{sgn} j)c_j e^{ijt}, \qquad (5.23)$$

then also

$$\tilde{\sigma}_n(f,x) = 2\tilde{K}_n * f(x) = \frac{1}{\pi}\int_T f(x-t)\tilde{K}_n(t)\,dt. \qquad (5.24)$$

Moreover, since $\tilde{K}_n(t)$ is odd (5.24) can be rewritten as either

$$\frac{-1}{\pi}\int_T f(x+t)\tilde{K}_n(t)\,dt \qquad \text{or} \qquad \frac{1}{\pi}\int_T \frac{(f(x-t)-f(x+t))}{2}\tilde{K}_n(t)\,dt. \qquad (5.25)$$

Furthermore, since the integrand in the last integral of (5.25) is even, we also have

$$\begin{aligned}
\tilde{\sigma}_n(f,x) &= \frac{-1}{\pi}\int_{[0,\pi]} (f(x+t)-f(x-t)) \\
&\quad \times \left(\frac{\cos(t/2)}{2\sin(t/2)} - \frac{1}{(n+1)}\frac{\sin((n+1)t)}{(2\sin(t/2))^2} \right) dt. \qquad (5.26)
\end{aligned}$$

6. THE TRUNCATED HILBERT TRANSFORM

Proposition 5.1 motivates the consideration of a new integral operator, namely, the truncated Hilbert transform $H_\varepsilon f$, which is given by an absolutely convergent integral and which we expect will converge, in some sense, to $\tilde{f}$.

Definition 6.1. For $f \in L(T)$ and $0 < \varepsilon < \pi$, let

$$H_\varepsilon f(x) = \frac{-1}{\pi}\int_{(\varepsilon,\pi]} \frac{f(x+t) - f(x-t)}{2\tan(t/2)}\,dt$$

$$= \frac{-1}{\pi}\int_{\varepsilon<|t|<\pi} \frac{f(x+t)}{2\tan(t/2)}\,dt. \tag{6.1}$$

As usual, we do the L^2 case first.

Proposition 6.2. Suppose $f \in L^2(T)$. Then $\lim_{\varepsilon\to 0^+}\|H_\varepsilon f - \tilde{f}\|_2 = 0$.

Proof. We begin by showing that

$$\|H_\varepsilon f\|_2 \leqslant c\|f\|_2 \tag{6.2}$$

with c independent of ε and f.

In first place observe that

$$H_\varepsilon f(x) = \frac{1}{\pi}\int_{\varepsilon<|t|\leqslant\pi} f(x-t)\left(\frac{1}{2\tan(t/2)} - \frac{1}{t}\right) dt$$

$$+\frac{1}{\pi}\int_{\varepsilon<|t|\leqslant\pi} f(x-t)\frac{dt}{t} = I + J$$

say. It will therefore suffice to show that (6.2) holds with I and J in place of $H_\varepsilon f$ there. Now, I and J represent convolution operators with integrable functions, and as such they are bounded in $L^2(T)$. It only remains to check that the norms are independent of ε. For this purpose put

$$\phi_1(t) = \chi_{[-\pi,-\varepsilon)\cup(\varepsilon,\pi]}(t)\left(\frac{1}{2\tan(t/2)} - \frac{1}{t}\right)$$

and

$$\phi_2(t) = \chi_{[-\pi,-\varepsilon)\cup(\varepsilon,\pi]}(t)(1/t).$$

If we can show that the sequences $\{c_j(\phi_1)\}$ and $\{c_j(\phi_2)\}$ of the Fourier coefficients of ϕ_1 and ϕ_2 are uniformly bounded, by A say, then by relation (1.7) of Chapter II,

$$\|I\|_2 = \|f * 2\phi_1\|_2 = 2\left(\sum |c_j(f)|^2|c_j(\phi_1)|^2\right)^{1/2} \leqslant 2A\|f\|_2,$$

and similarly for J. The bound for $|c_j(\phi_1)|$ is readily obtained since

$$|c_j(\phi_1)| \leq \|\phi_1\|_1 \leq \frac{1}{\pi}\int_{[0,\pi]}\left|\frac{1}{2\tan(t/2)} - \frac{1}{t}\right| dt = A < \infty.$$

As for the $c_j(\phi_2)$'s, to fix ideas suppose that $j > 0$ and consider

$$\begin{aligned} c_j(\phi_2) &= \frac{1}{2\pi}\int_{[-\pi,-\varepsilon)} e^{-ijt}\frac{dt}{t} + \frac{1}{2\pi}\int_{(\varepsilon,\pi]} e^{-ijt}\frac{dt}{t} \\ &= \frac{(-2i)}{2\pi}\int_{(\varepsilon,\pi]}\frac{\sin(jt)\,dt}{t} = \frac{(-i)}{\pi}\int_{(\varepsilon/j,\pi/j]}\frac{\sin t}{t}\,dt. \end{aligned}$$

Now we invoke the well-known and readily seen fact that

$$\left|\int_{(a,b]}\frac{\sin t}{t}\,dt\right| \leq A,$$

A independent of a, b. Therefore, also in this case $|c_j(\phi_2)| \leq A$, and (6.2) follows.

Next, we show that the conclusion holds for polynomials p. Indeed, by (5.14) and (6.1) we readily see that

$$\lim_{\varepsilon\to 0} H_\varepsilon p(x) = \tilde{p}(x) \qquad \text{a.e. in} \quad T. \tag{6.3}$$

Since clearly by (5.14) and (6.1) again, $|H_\varepsilon p(x)|$, $|\tilde{p}(x)| \leq K$ for x in T, by the Lebesgue dominated convergence theorem we get that

$$\|H_\varepsilon p - \tilde{p}\|_2 \to 0, \qquad \text{as} \quad \varepsilon \to 0^+. \tag{6.4}$$

Finally, suppose that $f \in L^2(T)$ and let p be a trigonometric polynomial. Then

$$\begin{aligned} \|H_\varepsilon f - \tilde{f}\|_2 &\leq \|H_\varepsilon f - H_\varepsilon p\|_2 + \|H_\varepsilon p - \tilde{p}\|_2 + \|\tilde{p} - \tilde{f}\|_2 \\ &= \|H_\varepsilon(f - p)\|_2 + \|H_\varepsilon p - \tilde{p}\|_2 + \|(p - f)^{\sim}\|_2 \\ &= I_1 + I_2 + I_3, \end{aligned}$$

say. By (6.2) we have that $I_1 \leq c\|f - p\|_2$, and as observed in the comments following (3.1), $I_3 \leq \|p - f\|_2$ as well. These observations suffice to show that $\|H_\varepsilon f - \tilde{f}\|_2$ can be made small provided ε is sufficiently small. Indeed, first choose p so that $\|f - p\|_2 < \eta$, η arbitrary; this is possible by Remark 2.5 in Chapter II. Once p has been chosen, let ε be small enough so that $I_2 \leq \eta$; this is possible by (6.4). Thus adding the estimates for the I_j's we see that for ε sufficiently small $\|H_\varepsilon f - \tilde{f}\|_2 \leq c\eta + \eta + \eta = (c+2)\eta$. ■

The study of the behavior of $H_\varepsilon f$ for f merely integrable is more subtle, as is the study of the existence of norm and pointwise limits. In this direction we begin by proving a qualitative result relating $H_\varepsilon f(x)$ to $\tilde{\sigma}_n(f, x)$; the quantitative relation will be discussed later.

Theorem 6.3 (Lebesgue). Let $f \in L(T)$. Then

$$\lim_{n\to\infty} \tilde{\sigma}_n(f, x) - H_{1/n}f(x) = 0 \qquad \text{a.e.}$$

Moreover if $\tilde{f}(x)$ exists, then $\tilde{\sigma}_n(f, x) = \sigma_n(\tilde{f}, x)$.

Proof. It is quite similar to the proof of Theorem 4.2 of Chapter II. In first place from the definitions it follows that

$$\tilde{\sigma}_n(f, x) - H_{1/n}f(x) = \frac{-1}{\pi}\int_{[0,1/n]} (f(x+t) - f(x-t))\tilde{K}_n(t)\, dt$$

$$-\frac{1}{\pi}\int_{(1/n,\pi]} (f(x+t) - f(x-t))\left(\tilde{K}_n(t) - \frac{1}{2\tan(t/2)}\right) dt = I + J, \quad (6.5)$$

say. Now, on account of (5.7)

$$|I| \leqslant cn \int_{[0,1/n]} |f(x+t) - f(x-t)|\, dt, \quad (6.6)$$

and this expression is $o(1)$ as $n \to \infty$ at each Lebesgue point x of f, that is a.e. in T. Also from (5.8) it readily follows that

$$|J| \leqslant \frac{c}{n}\int_{(1/n,\pi]} |f(x+t) - f(x-t)|\frac{dt}{t^2}, \quad (6.7)$$

and also this expression is $o(1)$ as $n \to \infty$ a.e. in T, as we have already shown in the proof of the Lebesgue theorem 4.2 of Chapter II. The remark concerning $\tilde{\sigma}_n(f, x)$ follows at once from the relation $\tilde{\sigma}_n(f, x) = \tilde{f} * 2K_n(x)$, which is readily seen to hold by a direct examination of the Fourier series of $\tilde{f}$. ∎

Corollary 6.4. Let $f \in L(T)$. Then $\lim_{\varepsilon\to 0^+} H_\varepsilon f(x)$ exists, a.e. in T, if and only if $\lim_{n\to\infty} \tilde{\sigma}_n(f, x)$ exists, a.e. in T. Moreover, these limits coincide whenever both exist.

Proof. Only the sufficiency has to be proven. Given $\varepsilon > 0$ let n be the integer such that $1/(n+1) \leqslant \varepsilon < 1/n$. Then

$$H_\varepsilon f(x) - H_{1/n}f(x) = \frac{-1}{\pi}\int_{(\varepsilon,1/n]} \frac{f(x+t) - f(x-t)}{2\tan(t/2)}\, dt,$$

and

$$|H_\varepsilon f(x) - H_{1/n}f(x)| \leq \frac{1}{\pi}\int_{[1/n+1,1/n]}\left|\frac{f(x+t)-f(x-t)}{2\tan(t/2)}\right| dt$$

$$\leq cn\int_{[0,1/n]} |f(x+t)-f(x-t)|\, dt. \tag{6.8}$$

Since the last integral tends to 0 at each Lebesgue point x of f, that is a.e. in T, it follows that, as $\varepsilon \to 0$ and as $n \to \infty$, $H_\varepsilon f(x)$ and $H_{1/n}f(x)$ are equiconvergent. By Theorem 6.3 also $\tilde{\sigma}_n(f, x)$ and $H_{1/n}f(x)$ are equiconvergent as $n \to \infty$. ■

Corollary 6.5. Let $f^2(T)$. Then $\lim_{\varepsilon\to 0^+} H_\varepsilon f(x) = \tilde{f}(x)$ a.e.

Proof. Since $f \in L^2(T)$, $\tilde{f}$ is also in $L^2(T)$. Therefore by Theorem 6.3 and Theorem 4.2 of Chapter II, $\tilde{\sigma}_n(f, x) = \sigma_n(\tilde{f}, x) \to \tilde{f}(x)$, as $n \to \infty$, a.e. in T. Moreover, by Corollary 6.4, $H_\varepsilon f(x)$ is equiconvergent with $\tilde{\sigma}_n(f, x)$. ■

To deal with the truncated Hilbert transform in the general case we need a new idea, namely, the Calderón–Zygmund decomposition, which is developed in the next chapter.

7. NOTES; FURTHER RESULTS AND PROBLEMS

The study of some important integral equations, similar to the ones we consider in Chapter XVII, can be reduced to that of a system of linear equations in an infinite number of unknowns by the introduction of a set S verifying the assumption (ii) of Theorem 1.8. Thus the simplest space which we introduce consists of those complex sequences $\{c_n\}$ such that $\sum|c_n|^2 < \infty$, this is called the elementary "Hilbert coordinate space," or $l^2(Z)$. Parseval's identity, or relation (iv) of Theorem 1.8, leads to the recognition of the fact that actually $L^2(T)$ and $l^2(Z)$ have the same linear and metric structure, and that they may be therefore considered as realizations of an "abstract" space which is precisely defined in terms of these common properties, namely, linearity, existence of the notions of inner product and norm, and, very importantly, the Riesz–Fischer property of completeness (Proposition 1.7). These are the Hilbert spaces H of interest to us. By dimension of H we meant the smallest cardinal number equivalent to that of a set S which verifies (ii) of Theorem 1.8, or, more precisely, that of a set of elements in H whose finite linear combinations are dense in H. The dimension of a separable Hilbert space is at most countable and it is quite elementary to construct ONS there by means of the Gram–Schmidt process. Both geometric and analytic methods are available in dealing with Hilbert spaces, and this makes for a rich and powerful theory. Some purely analytic methods, such as the one used to deal with the concept of "almost orthogonality" given in 7.14 below, are quite important in harmonic analysis.

Further Results and Problems

7.1 Suppose $\{x_j\}_{j=1}^{\infty} \subset H$ satisfy (i) of Theorem 1.8, i.e. it is a maximal ONS, and suppose that $\{y_j\}_{j=1}^{\infty} \subset H$ is such that $\sum_j \|x_j - y_j\|^2 < 1$. Then $\{y_j\}$ is complete. In other words, if y in H is such that $(y, y_j) = 0$ for all j, then $y = 0$.

7.2 Let $\{x_j\}_{j=1}^{\infty} \subset H$ satisfy (i) of Theorem 1.8, and let $y_j = x_j - x_{j+1}, j \geqslant 1$. Then $\{y_j\}$ is complete.

7.3 (Paley-Wiener) Let $\{x_j\}_{j=1}^{\infty} \subset H$ satisfy (iv) of Theorem 1.8, and suppose that $\{y_j\}$ is a sequence of elements in H such that

$$\left\|\sum_{j=1}^{n} \lambda_j(x_j - y_j)\right\| \leqslant A\left\|\sum_{j=1}^{n} \lambda_j x_j\right\|, \qquad 0 \leqslant A < 1,$$

for all scalars $\lambda_j, j \geqslant 1$. Then each $x \in H$ can be written as a linear combination of the y_j's, i.e., $x = \sum_{j=1}^{\infty} c_j y_j$, c_j scalar, where the sum converges in H. (*Hint*: Since the sum $\sum_{j=1}^{\infty} c_j(x_j - y_j)$ converges in H whenever $\sum_{j=1}^{\infty} c_j x_j$ converges, the mapping $T(\sum_j c_j x_j) = \sum_j c_j(x_j - y_j)$ is well defined, has norm $\|T\| \leqslant A < 1$, and consequently $I - T$ is invertible.)

When $H = L^2(T)$ the Paley-Wiener result seems to indicate that only under small perturbations of the integers, i.e., only when $|\lambda_n - n|$ is small, does the system $\{e^{i\lambda_n t}\}_{n=-\infty}^{\infty}$ span $L^2(T)$. This is because the condition

$$\left\|\sum_n c_n(e^{int} - e^{i\lambda_n t})\right\|_2 < A \leqslant 1 \tag{7.1}$$

whenever $\sum_n |c_n|^2 \leqslant 1$ suffices to insure that $f \in L^2(T)$ can be written as $f(t) = \sum_n c_n e^{i\lambda_n t}$, in L^2. Surprisingly, the following is true. Kadec's $\frac{1}{4}$ theorem: If $\{\lambda_n\}$ is a real sequence and $|\lambda_n - n| \leqslant L < \frac{1}{4}$, all n, then (7.1) holds. The proof is elementary in the sense that it is entirely computational and sharp. The reader may consult Young's book [1980] for this and other interesting results in nonharmonic Fourier series.

7.4 Let A denote the Banach algebra of continuous functions f with absolutely convergent Fourier series normed by $\|f\|_A = \sum_j |c_j(f)| < \infty$. Show that A is a proper subset of $C(T)$ and that $A = L^2 * L^2$, i.e., each f in A can be written as $f = g * h$, $g, h \in L^2(T)$.

7.5 Suppose that f is absolutely continuous and that $Df \in L^2(T)$. Show that $f \in A$ and $\|f\|_A \leqslant |c_0(f)| + c\|Df\|_2$, c an absolute constant. (*Hint*: $c_j(f) = (ij)c_j(f)/(ij)$, $j \neq 0$.)
7.6 Suppose $f \in L(T)$ and $\sum_{n=1}^{\infty} w_1(f, \pi/n)^2 < \infty$. Then $f \in L^2(T)$.
7.7 $\sum c_j(f)c_j(g)$ is not necessarily convergent for all $f \in L(T)$ and $g \in L^2(T)$ or even $g \in C(T)$; is the same true for $\sum(|c_j(f)|\,|c_j(g)|)^{2-\varepsilon}$, $\varepsilon > 0$?
7.8 Let $f \sim \sum_{n=1}^{\infty} b_n \sin nx$, $f \in L^2(T)$. Then

$$\sum_{n=1}^{\infty}(b_n/n) = (1/2\pi)\int_T (\pi - x)f(x)\,dx.$$

7.9 For $\phi \in L^\infty(T)$ and square-summable sequences $x = \{x_n\}$, $y = \{y_n\}$, let $\lambda_n = c_n(\phi)$ and put $A_N(x, y) = \sum_{n,m=0}^{N} \lambda_{n+m}x_n y_m$. Then

$$|A_N(x, y)| \leqslant \|\phi\|_\infty\left(\sum_{n=0}^{N}|x_n|^2\right)^{1/2}\left(\sum_{n=0}^{N}|y_n|^2\right)^{1/2}.$$

(*Hint*: If $p(t) = \sum_{n=0}^{N} x_n e^{-int}$, then $A_N(x, x) = (1/2\pi)\int_T p(t)^2\phi(t)\,dt$ and $|A_N(x, x)| \leqslant \|\phi\|_\infty \sum_{n=0}^{N}|x_n|^2$. The result now follows with not much difficulty from the identity $A_N(x, y) = (A_N(x + y, x + y) - A_N(x - y, x - y))/4$.)

7.10 (Hilbert)

$$\left|\sum_{n,m=0}^{\infty} \frac{x_n y_m}{n + m + 1}\right| \leqslant \pi\left(\sum_{n=0}^{\infty}|x_n|^2\right)^{1/2}\left(\sum_{n=0}^{\infty}|y_n|^2\right)^{1/2}.$$

(*Hint*: For $\phi(t) = i(\pi - t)e^{-it}$, $c_n(\phi) = 1/(n + 1)$ and $\|\phi\|_\infty = \pi$; for further results in this direction see Duren [1970].)
7.11 (Wiener) Let μ be a measure in T with discrete part $\sum a_j\delta_{x_j}$, where the x_j's are distinct points in T and the a_j's are complex scalars. Then $\lim_{N\to\infty}(1/(2N + 1))\sum_{|n|\leqslant N}|c_j(\mu)|^2 = \sum_j|a_j|^2$. In particular, a necessary and sufficient condition for the continuity of μ is that the above limit be 0. (*Hint*: Let $\tilde{\mu} = \sum \bar{a}_j\delta_{-x_j}$; by the convolution formula, $\mu * \tilde{\mu}(0) = \sum|a_j|^2$. On the other hand,

$$\begin{aligned}\sum_{|n|\leqslant N}|c_j(\mu)|^2 &= \sum_{|n|\leqslant N} c_j(\mu * \tilde{\mu})\\ &= \frac{1}{\pi}\left(\int_{|t|<\eta} + \int_{\eta<|t|\leqslant\pi}\right)D_N(t)\,d(\mu * \tilde{\mu})(t);\end{aligned}$$

finally show that

$$\left|\frac{1}{2N + 1}\frac{1}{\pi}\int_{|t|<\eta} - \mu * \tilde{\mu}(0)\right| \quad \text{and} \quad \left|\frac{1}{N}\int_{\eta<|t|\leqslant\pi}\right|$$

go to 0 as $N \to \infty$.)

Next we discuss a couple of results of the so-called Littlewood–Paley type. We shall return to them in Chapter XII.

7.12 Given $f \in L^2(T)$, Put $F(x) = (\sum_{n=1}^{\infty} |s_n(f,x) - \sigma_n(f,x)|^2/n)^{1/2}$. Then $\|F\|_2 \leqslant \|f\|_2$; in particular, $F(x) < \infty$ a.e. (*Hint*: Find an appropriate expression for $s_n(f,x) - \sigma_n(f,x)$ and use relation (1.7).)

7.13 (Kaczmarz-Zygmund) Given $f \in L^2(T)$ put

$$K(f,x) = \left(\sum_{n=2}^{\infty} n|\sigma_n(f,x) - \sigma_{n-1}(f,x)|^2 \right)^{1/2}.$$

Then $\|K(f)\|_2 \leqslant c\|f\|_2$; in particular $K(f,x) < \infty$ a.e.

7.14 (Almost Orthogonality) Let $\{f_k\}_{k=1}^{\infty}$ be a sequence of 2π periodic, real-valued functions verifying the following properties uniformly in k: (i) $|f_k(t)| \leqslant 1$; (ii) $\int_T f_k(t)\,dt = 0$; (iii) there exist a number η, $0 < \eta \leqslant 1$, and a constant A such that $|f_k(x) - f_k(t)| \leqslant A|x-t|^{\eta}$. Furthermore let $\{\lambda_k\}$ be a Hadamard sequence, i.e., $\lambda_{k+1}/\lambda_k \geqslant \theta > 1$, all k. Under these conditions, there is a constant $c = c(\eta, \theta)$ such that for all sequences $\{r_k\}_{k=1}^{\infty}$,

$$\int_{[0,2\pi]} \left| \sum_k r_k f_k(\lambda_k t) \right|^2 dt \leqslant c \sum_k |r_k|^2.$$

(*Hint*: Assume first that the sequence $\{r_k\}$ is finite; the infinite sum is understood in the L^2 sense. If $I_{j,k} = \int_{[0,2\pi]} f_j(\lambda_j t) f_k(\lambda_k t)\,dt$, it suffices to show that $|I_{j,k}| \leqslant c2^{-\varepsilon|j-k|}$, for some constant c and $0 < \varepsilon \leqslant 1$, since then the integral in question is bounded by $\sum_{j,k} |I_{j,k}|\,|r_j \bar{r}_k|$ and we are done. It is also enough to do the case $j < k$, for the case $k > j$ is symmetric with this and the case $j = k$ is trivial. For each k fixed, then, and $1 \leqslant j < k$, write $I_{j,k}$ as a sum of integrals J_q over the intervals $[2\pi q/\lambda_k, 2\pi(q+1)/\lambda_k]$, $q \geqslant 0$; there will also be an error term, possibly, corresponding to $[2\pi q_1/\lambda_k, 2\pi]$ where q_1 is the largest integer such that $q_1/\lambda_k < 1$. The contribution of the error term is readily seen to be $\leqslant 2\pi/\lambda_k$. Moreover, since

$$|J_q| \leqslant \frac{1}{\lambda_k} \int_{[2\pi q, 2\pi(q+1)]} \left| f_j\left(\frac{\lambda_j 2\pi q}{\lambda_k}\right) - f_j\left(\frac{\lambda_j t}{\lambda_k}\right) \right| |f_k(t)|\,dt$$

the desired conclusion follows by summation. This proof, which is the prototype of many making use of properties of Hadamard, or lacunary, sequences, is from Meyer [1979], and we shall return to it later on.)

7.15 Assume that $\sum c_j$ is $(C,1)$ summable to a finite limit and that its partial sums s_n verify $s_n = o(\mu_n)$, where $\{1/\mu_n\}$ is convex and tends to 0. Then $\sum_{j=0}^{\infty} c_j/\mu_j$ converges. Consequently, if $f \in L^{\infty}(T)$, $f \sim \sum c_n e^{int}$, then $\sum (c_n/(1+|n|)^{\eta}) e^{int}$ converges a.e., whenever $\eta > 0$. (*Hint*: Sum by parts twice and invoke Corollary 2.2 of Chapter I.)

7.16 $L^p(T) = L(T) * L^p(T)$, $1 \leqslant p < \infty$. (*Hint*: the proof of Proposition 4.6 requires the following adjustment: If $\alpha_n = 1/(\ln(n+1))^2$, then $\sum \alpha_n = \infty$, but $\sum \alpha_n/n < \infty$; now let $\{m_n\}$ be as in the proposition and put $g(x) = f(x) + \lim_{k\to\infty} \sum_{n=1}^{k} \alpha_n (f(x) - \sigma_{m_n}(f,x))$, where the limit is taken in the L^p sense.

Next note that the sequence $\{1/\sum_n \alpha_n \min(1, |j|(m_n + 1))\}$ is even, convex, and tends to 0, and invoke Proposition 4.4. This idea, in the case $p = 1$, is Rudin's [1958] and for general p, Edwards's [1967].)

7.17 Does the above result hold for $C^k(T)$, $k \geqslant 0$? How about D'?

7.18 If $\tilde{D}_n$ is the conjugate Dirichlet kernel, then

$$\lim_{n\to\infty}[1/(\ln n)] \int_{[0,\pi]} \tilde{D}_n(t)\, dt = 1.$$

7.19 Let $f \in L(T)$ have a jump discontinuity $f(x+0) - f(x-0) = d$, $x \in T$. Then $\lim_{n\to\infty}(\tilde{s}_n(f, x))/(\ln n) = -d/\pi$. Obtain as a corollary that at any point x where f has a jump, $\tilde{s}_n(f, x)$ diverges and that if $\tilde{s}_n(f, x) = O(\ln n)$, then f cannot have a jump discontinuity at x. (*Hint*: Use (5.12).)

7.20 Let $f \in L(T)$, then $f + i\tilde{f}$ cannot have a simple discontinuity at any point x in T. (*Hint*: If x is a simple discontinuity of $f + i\tilde{f}$, then by Fejér's Theorem 2.1 of Chapter II $\sigma_n(f + i\tilde{f}, x)$ remains bounded. On the other hand, if $f(x+0) - f(x-0) = d > 0$, say, then by 7.19 $\tilde{s}_n(f, x) \to -\infty$, and again by Fejér's theorem $\sigma_n(\tilde{f}, x) = \tilde{\sigma}_n(f, x) \to -\infty$.)

7.21 (Localization) If f vanishes in an interval $I \subset T$, then $\tilde{s}_n(f, x)$ converges uniformly in any subinterval of the interior of I. (*Hint*: It is a straightforward variant of 5.8 and 5.9 of Chapter II.)

7.22 The distribution p.v. $(\frac{1}{2}\tan(t/2))$ has Fourier coefficients $c_k = (-i)\operatorname{sgn} k$. (*Hint*: Apply (5.14) to $f(x) = e^{ikx}$.)

7.23 If a real sequence b_k decreases to 0, then the trigonometric series $\sum_{k=1}^{\infty} b_k \sin kt$ converges everywhere. Moreover, for any $\delta > 0$ the series converges uniformly in $\delta < |x| < \pi$. (*Hint*: As in 5.8 of Chapter I use Abel's summation formula and estimate (5.5) this time. Since all sine terms vanish when $t = 0$ the series converges now everywhere.)

The question of the uniform convergence of the above series must be decided in a different manner.

7.24 If a real sequence b_k decreases to 0, then for the uniform convergence of $\sum b_k \sin kt$ in T it is necessary and sufficient that $kb_k \to 0$ as $k \to \infty$. (*Hint*: The necessity follows at once upon observing that $|\sum_{k=m+1}^{2m} b_k \sin kx|$ can be made uniformly small and letting $x = \pi/4m$ there. As for the sufficiency, on account of 7.23 it is enough to show that, if $r_n(t) = \sum_{k=n}^{\infty} b_k \sin kt$ and $\varepsilon_n = \max_{k\geqslant n} kb_k$, then $|r_n(t)| \leqslant c\varepsilon_n$ for t in $[0, \pi/4]$. Since it is possible to find an integer N such that $1/N < t \leqslant 1/(N-1)$, then write

$$r_n(t) = \left(\sum_{k=n}^{N-1} + \sum_{k=N}^{\infty}\right) b_k \sin(kt) = r_{n,1}(t) + r_{n,2}(t),$$

say, where $r_{n,1}(t) = 0$ if $N \leqslant n$. To estimate $r_{n,1}(t)$, use that $|\sin kt| \leqslant k|t|$; to estimate $r_{n,2}(t)$, distinguish two cases.)

7.25 There exists a trigonometric series that converges uniformly but not absolutely in T. (*Hint*: $\sum_{k=2}^{\infty}(\sin(kt))/(k \ln k)$.)

7.26 If b_k is a real sequence decreasing to 0, and kb_k is bounded, then the partial sums of the trigonometric series $\sum_{k=1}^{\infty} b_k \sin(kt)$ are bounded. In particular $|\sum_{k=1}^{n}(\sin(kt)/k)| \leqslant C$, all t.

7.27 If the real sequence b_k decreases to 0, the condition $kb_k \to 0$ is necessary and sufficient for $\sum b_k \sin(kt)$ to be the Fourier series of a continuous function. (Hint: It is enough to prove the necessity of the condition; if the $(C, 1)$ means $\sigma_n(t)$ of the above series converge uniformly, then $\sigma_n(\pi/2n) \to 0$ and

$$\sum_{k=1}^{n} b_k\left(1 - \frac{k}{n+1}\right)\frac{2}{\pi}\left(\frac{\pi k}{2n}\right) \to 0.$$

The result follows without much difficulty from this.)

CHAPTER

IV

The Basic Principles

1. THE CALDERÓN–ZYGMUND INTERVAL DECOMPOSITION

In this section we discuss one of the most important topics of harmonic analysis, namely, the Calderón–Zygmund decomposition of an integrable function. This is a principle which roughly states that an integrable function f can be written as a sum of two functions, $f = g + b$ say, where the "good part" g is essentially bounded and the "bad part" b has a cancellation property and bounded averages over a particular collection of open, disjoint subintervals of T.

Let us be more precise. In first place we assume that f is a nonnegative function, for otherwise we can apply the result to $|f|$ to include arbitrary real- or complex-valued functions as well. Let then $\lambda > 0$ be a number exceeding f_T, the average of f over T, i.e.,

$$\frac{1}{2\pi}\int_T f(y)\,dy < \lambda.$$

We will work with the family of open, dyadic subintervals of T, i.e., those intervals obtained by successively subdividing T into equal, open subintervals. The first subdivision of T thus consists of $T_1 = (-\pi, 0)$ and $T_2 = (0, \pi)$. Observe that since

$$\frac{1}{2}(f_{T_1} + f_{T_2}) = \frac{1}{2}\left(\frac{1}{\pi}\int_{T_1} f(y)\,dy + \frac{1}{\pi}\int_{T_2} f(y)\,dy\right) = f_T < \lambda,$$

then at least one of the averages f_{T_1} or f_{T_2} does not exceed λ. Also,

$$f_{T_i} \leqslant \frac{2}{2\pi}\int_T f(y)\,dy < 2\lambda, \qquad i = 1, 2,$$

so that even if an average exceeds λ, it remains bounded by 2λ. We can now get the selection process for intervals going. In case the average of f over a subinterval does not exceed λ, then we continue subdividing that interval. If, on the other hand, the average of f over an interval does exceed λ, then we separate that interval and rename it I_1. Clearly,

$$\lambda < \frac{1}{|I_1|}\int_{I_1} f(y)\,dy = f_{I_1} < 2\lambda.$$

Suppose, then, that after k steps $I_1, I_2, \ldots, I_n$ have been separated. The intervals $\{I\}$ which have not been renamed and separated at that step have the property that $f_I < \lambda$. We therefore let each I in this collection play the role of T above and subdivide it into two equal, open subintervals. The average of f over either of these intervals does not exceed 2λ, and over at least one of them it does not exceed λ. Those subintervals of $\{I\}$ for which the average of f exceeds λ and is bounded by 2λ are renamed $I_{n+1}, \ldots, I_m$, say, and separated. These are the intervals obtained in the $(k+1)$-step. This selection procedure thus produces a collection of open, disjoint, dyadic subintervals $\{I_j\}$ of T with the following properties:

$$\lambda < \frac{1}{|I_j|}\int_{I_j} f(y)\,dy < 2\lambda, \qquad \text{all } j \tag{1.1}$$

and

$$\sum_j |I_j| \leqslant \frac{1}{\lambda}\sum_j \int_{I_j} f(y)\,dy \leqslant \frac{1}{\lambda}\int_T f(y)\,dy. \tag{1.2}$$

Moreover, if we put $\Omega = \bigcup I$, then the following assertion holds: for almost all x in $T\backslash\Omega$ there is a sequence $\mathscr{I} = \{I\}$ consisting of dyadic subintervals of T which converges to x and

$$\frac{1}{|I|}\int_I f(y)\,dy < \lambda, \qquad \text{all} \quad I \in \mathscr{I}. \tag{1.3}$$

Clearly, (1.3) holds only a.e. in $T\backslash\Omega$ as it does not necessarily hold for those points, such as 0, which are the endpoints of the dyadic subintervals of T.

We call the collection $\{I_j\}$ the Calderón–Zygmund interval decomposition at level λ. Referring to the functions g and b alluded to above we set

$$g(x) = f(x)\chi_{T\backslash\Omega}(x) + \sum_j f_{I_j}\chi_{I_j}(x) \tag{1.4}$$

and $b(x) = f(x) - g(x)$. It is clear that $b(x) = 0$ off Ω, that

$$\int_{I_j} b(x)\, dx = 0, \qquad \text{all} \quad j, \tag{1.5}$$

and that

$$\frac{1}{|I_j|}\int_{I_j} |b(x)|\, dx = \frac{1}{|I_j|}\int_{I_j} |f(x) - f_{I_j}|\, dx \leqslant 4\lambda, \qquad \text{all} \quad j. \tag{1.6}$$

As for g, we readily see that

$$g(x) \leqslant 2\lambda, \qquad x \text{ in } \Omega. \tag{1.7}$$

By (1.3) above it is clear that to estimate g in $T\backslash\Omega$, we need to consider the behavior of the averages of f. This we do with the study of our next topic, the Hardy-Littlewood maximal function.

2. THE HARDY-LITTLEWOOD MAXIMAL FUNCTION

Suppose that f is an integrable, periodic function, defined in T, and put

$$Mf(x) = \sup_I \frac{1}{|I|}\int_I |f(y)|\, dy,$$

where the supremum is taken over the collection $\mathcal{I} = \{I\}$ of those open intervals I centered at x, of length $\leqslant 2\pi$.

It is readily seen that $MF(x)$ is lower semicontinuous, i.e., $\mathcal{O}_\lambda = \{x \in T\colon Mf(x) > \lambda\} = \{Mf > \lambda\}$, is open for every $\lambda > 0$. Indeed, let $\{x_n\}$ be a sequence of points in $T\backslash\mathcal{O}_\lambda$ which converges to x; we show that $x \in T\backslash\mathcal{O}_\lambda$ as well. It clearly suffices to show that for every interval $I = (x - \eta, x + \eta)$, $0 < \eta \leqslant \pi$, we have

$$\frac{1}{|I|}\int_I |f(y)|\, dy \leqslant \lambda.$$

Let $I_n = (x_n - \eta,\ x_n + \eta)$ and put $f_n(y) = f(y)\chi_{I_n \Delta I}(y)$, where $I_n \Delta I = (I_n\backslash I) \cup (I\backslash I_n)$ is the symmetric difference of I_n and I. Since $|f_n(y)| \leqslant |f(y)|$ and $f_n(y) \to 0$ as $n \to \infty$, by the Lebesgue dominated convergence theorem it follows that

$$\frac{1}{|I|}\int_I |f_n(y)|\, dy \to 0, \qquad \text{as} \quad n \to \infty. \tag{2.1}$$

Moreover, since $x_n \in T\backslash\mathcal{O}_\lambda$

$$\frac{1}{|I|}\int_{I_n} |f(y)|\, dy = \frac{1}{|I_n|}\int_{I_n} |f(y)|\, dy \leqslant \lambda, \qquad \text{all} \quad n. \tag{2.2}$$

Now, by (2.2),

$$\frac{1}{|I|}\int_I |f(y)|\,dy \leqslant \frac{1}{|I|}\int_{I_n \Delta I} |f(y)|\,dy + \frac{1}{|I|}\int_{I_n} |f(y)|\,dy,$$
$$\leqslant \frac{1}{|I|}\int_I |f_n(y)|\,dy + \lambda.$$

Thus letting $n \to \infty$ above, and on account of (2.1), we obtain that $(1/|I|)\int_I |f(y)|\,dy \leqslant \lambda$, which is precisely what we wanted to show.

So $Mf(x)$ is a positive, measurable function, but is it integrable? A simple example shows that this is not necessarily the case. Indeed, let $f(y) = \chi_{(0,1/2)}(y)(d/dy)(1/\ln(1/y))$. Then $f(y) \geqslant 0$ and $f \in L(T)$. Now for x in $(-\frac{1}{2}, 0)$ we have

$$Mf(x) \geqslant \frac{1}{2\eta}\int_{(x-\eta,x+\eta)} f(y)\,dy, \qquad \text{all} \quad \eta > 0.$$

In particular setting $\eta = 2|x|$, we get

$$Mf(x) \geqslant \frac{1}{4|x|}\int_{(0,|x|)} f(y)\,dy = \frac{1}{4|x|}\frac{1}{\ln(1/|x|)},$$

and this function is not integrable in a neighborhood of 0.

To deal with this inconvenience we introduce the weak-L class of Marcinkiewicz. We say that a measurable function $f(x)$ is in wk-$L(T)$ provided there is a constant c such that

$$\lambda|\{x \in T: |f(x)| > \lambda\}| = \lambda|\{|f| > \lambda\}| \leqslant c, \qquad \text{all} \quad \lambda > 0. \tag{2.3}$$

The infimum of the constants c appearing in (2.3) is called the "wk-L norm" of f, although this quantity does not satisfy the usual requirement of a norm. By Chebychev's inequality we see that integrable functions f are in wk-$L(T)$, with norm not exceeding $\|f\|_1$. On the other hand the function $f(x) = (1/x)(\chi_{(0,1)}(x))$ is in wk-$L(T)$, although it is not even locally integrable in a neighborhood of 0.

We can now prove the Hardy–Littlewood maximal theorem, namely,

Theorem 2.1. Suppose $f \in L(T)$. Then $Mf \in$ wk-$L(T)$ and the wk-L norm of Mf does not exceed $72\pi\|f\|_1$. More explicitly

$$\lambda|\{Mf > \lambda\}| \leqslant 36\int_T |f(y)|\,dy, \qquad \text{all} \quad \lambda > 0. \tag{2.4}$$

Proof. Since $Mf(x) = M(|f|)(x)$, we may assume that f is a real-valued, nonnegative function. Moreover, since $\lambda|\{Mf > \lambda\}| \leqslant \lambda 2\pi$, all λ, we may also assume that $(1/2\pi)\int_T f(y)\,dy < \lambda/18$, for otherwise there is nothing to prove.

Let then $f_T < \lambda/18$, and invoke the Calderón–Zygmund interval decomposition at level $\lambda/18$. We then have the collection $\{I_j\}$ of open, disjoint, dyadic subintervals of T as well as the functions g and $b, f = g + b$, verifying properties (1.1)–(1.7). Now, it is clear that M is a sublinear mapping, i.e., $M(f_1 + f_2)(x) \leqslant Mf_1(x) + Mf_2(x)$, and, consequently,

$$Mf(x) \leqslant Mg(x) + Mb(x), \qquad \text{all} \quad x \in T. \tag{2.5}$$

For each of the intervals I_j, let $2I_j$ denote the concentric interval with I_j with measure (or length) twice that of I_j. Finally, put $\Omega^* = \bigcup_j 2I_j$. With the aid of (2.5) we will be able to show that

$$Mf(x) \leqslant \lambda, \qquad x \in T\backslash\Omega^*. \tag{2.6}$$

If this is the case, then $\{Mf > \lambda\} \subset \Omega^*$ and by (1.2),

$$\begin{aligned}\lambda|(Mf > \lambda\}| &\leqslant \lambda \sum_j |2I_j| = 2\lambda \sum_j |I_j| \\ &\leqslant 2\lambda\frac{18}{\lambda}\int_T f(y)\,dy = 36\int_T f(y)\,dy,\end{aligned}$$

and we are done.

It thus remains to prove (2.6). Let I be an interval centered at $x \in T\backslash\Omega^*$; we begin by showing that

$$\frac{1}{|I|}\int_I |b(y)|\,dy \leqslant \frac{16\lambda}{18}. \tag{2.7}$$

This is not hard. Since $b = 0$ off Ω, we have that

$$\int_I |b(y)|\,dy = \sum_j \int_{I\cap I_j} |b(y)|\,dy,$$

where the sum extends only over those j's with $I \cap I_j \neq \varnothing$. We divide these I_j's into two mutually exclusive families, namely,

(i) $|I \cap I_j| \geqslant |I_j|/2$, we call these intervals $I^{(1)}$'s, and
(ii) $|I \cap I_j| < |I_j|/2$, we call these intervals $I^{(2)}$'s.

For the $I^{(1)}$'s we immediately see that

$$\begin{aligned}\frac{|I\cap I_j|}{|I\cap I_j|}\int_{I\cap I_j}|b(y)|\,dy &\leqslant 2\frac{|I\cap I_j|}{|I_j|}\int_{I_j}|b(y)|\,dy \\ &\leqslant 4\frac{|I\cap I_j|}{|I_j|}\int_{I_j}|f(y)|\,dy \leqslant 8\left(\frac{\lambda}{18}\right)|I\cap I_j|.\end{aligned} \tag{2.8}$$

For the $I^{(2)}$'s we invoke the fact that $x \in T\backslash\Omega^*$. Since in particular $x \notin 2I_j$, all j, we have $|I \cap I_j| \geq |I_j|/2$. Thus we see that

$$\begin{aligned}\int_{I\cap I_j} |b(y)|\,dy &\leq \frac{|2I \cap I_j|}{|2I \cap I_j|}\int_{2I\cap I_j} |b(y)|\,dy \\ &\leq \frac{2|2I \cap I_j|}{|I_j|}\int_{I_j} |b(y)|\,dy \leq 8\frac{|I \cap I_j|}{|I_j|}\int_{I_j} f(y)\,dy \\ &\leq 16(\lambda/18)|I \cap I_j|. \end{aligned} \tag{2.9}$$

Whence collecting estimates (2.8) and (2.9) we obtain

$$\begin{aligned}\frac{1}{|I|}\int_I |b(y)|\,dy &= \sum_j \frac{1}{|I|}\int_{I\cap I_j} |b(y)|\,dy \\ &\leq 16\left(\frac{\lambda}{18}\right)\sum_j \frac{|I \cap I_j|}{|I|} \leq \frac{16\lambda}{18},\end{aligned}$$

which is precisely (2.7). Since I is arbitrary in (2.7) this means that

$$Mb(x) \leq 16\lambda/18, \qquad x \in T\backslash\Omega^*. \tag{2.10}$$

Again for an interval I centered at x, we estimate now $(1/|I|)\int_I g(y)\,dy$. By the definition (1.4) of g we readily see that

$$\int_I g(y)\,dy = \sum_{j, I\cap I_j \neq \varnothing} \frac{|I \cap I_j|}{|I_j|}\int_{I_j} f(y)\,dy + \int_{I\backslash\Omega} f(y)\,dy = A_1 + A_2,$$

say. Also from (1.1) it readily follows that

$$A_1 \leq 2\left(\frac{\lambda}{18}\right)\sum_j |I \cap I_j|. \tag{2.11}$$

To estimate A_2 we again resort to the geometry of the situation. In first place it will suffice to consider the integral extended to the open set $I\backslash\bigcup_j \bar{I}_j \subset I$. This set can be written as a countable, disjoint union of open intervals J_k, say. Each of the J_k's such that $\bar{J}_k$ does not contain an endpoint of I (and there may be only two of these at that) is, in turn an a.e. union of disjoint, dyadic subintervals $J_{k,n}$, say, of T which were subdivided in the Calderón–Zygmund decomposition process. Thus by (1.1),

$$\frac{1}{|J_{k,n}|}\int_{J_{k,n}} f(y)\,dy < \frac{\lambda}{18}, \qquad \text{all } j, n. \tag{2.12}$$

On the other hand, if $\bar{J}_{k,n}$ contains an endpoint of I, let n be the integer such that $(2\pi/2^{n+1}) \leq |J_k| < (2\pi/2^n)$. Then $J_k \subset J_{k,1} \cup J_{k,2}$, where the $J_{k,i}$'s are disjoint, dyadic subintervals of T, each of measure $\leq 2\pi/(2n+1)$, which were not separated in the Calderón–Zygmund decomposition process. Thus we have

$$\int_{J_k} f(y)\,dy \leq \int_{J_{k,1}} f(y)\,dy + \int_{J_{k,2}} f(y)\,dy$$
$$\leq (\lambda/18)(|J_{k,1}| + |J_{k,2}|) < (\lambda/18)2|J_k|. \tag{2.13}$$

Combining the estimates (2.12) and (2.13), we readily see that

$$A_2 \leq 2(\lambda/18)\left|I \setminus \bigcup_j I_j\right|. \tag{2.14}$$

Thus adding (2.11) and (2.14) we get

$$\int_I g(y)\,dy \leq 2\left(\frac{\lambda}{18}\right)\left(\sum_j |I \cap I_j| + |I \setminus \bigcup I_j|\right)$$
$$\leq 2(\lambda/18)|I|.$$

Since I is arbitrary from this estimate it follows that

$$Mg(x) \leq 2(\lambda/18), \qquad \text{all} \quad x \in T. \tag{2.15}$$

Finally, by estimating the right-hand side of (2.5) by (2.10) and (2.15), we obtain that

$$Mf(x) \leq 16\lambda/18 + 2\lambda/18 = \lambda, \qquad x \in T \setminus \Omega^*,$$

which is precisely (2.6). ■

An interesting corollary to the maximal theorem is a version of the Lebesgue differentiation theorem, which we hope will be useful in estimating the right-hand side of (1.3). Observe, however, that what we really need is a "noncentered" version of our results. Let us do this then. Let $\tilde{M}f(x)$ denote the noncentered maximal function defined by

$$\tilde{M}f(x) = \sup \frac{1}{|I|} \int_I |f(y)|\,dy$$

where the I's are open intervals of length $\leq 2\pi$ containing x. It is trivial to see that $\tilde{M}f$ is lower-semicontinuous and therefore measurable. Moreover, the Hardy–Littlewood maximal theorem also gives that $\tilde{M}f \in$ wk-$L(T)$ with norm $\leq 216\pi\|f\|_1$. This is easy to see since for each open interval I containing x the interval $I_x = (x - \frac{3}{2}|I|, x + \frac{3}{2}|I|)$ contains I. Therefore

$$\frac{1}{|I|}\int_I |f(y)|\,dy \leq \frac{|I_x|}{|I|}\frac{1}{|I_x|}\int_{I_x} |f(y)|\,dy \leq 3Mf(x),$$

and, consequently, $\tilde{M}f(x) \leqslant 3Mf(x)$. Thus $\{\tilde{M}f > \lambda\} \subset \{Mf > \lambda/3\}$ and the desired estimate follows at once from (2.4). We can now state, and prove

Theorem 2.2 (Lebesgue Differentiation Theorem). Suppose that $f \in L(T)$. For each $x \in T$ let $\mathscr{I}_x = \{I_x\}$ be a family of open intervals containing x, converging to x as $|I_x| \to 0$. Then

$$\lim_{|I_x|\to 0} \frac{1}{|I_x|} \int_{I_x} f(y)\, dy = f(x) \qquad \text{a.e. in} \quad T.$$

Proof. With no loss of generality we may assume that f is real valued. It is also obvious that the desired conclusion holds true for a trigonometric polynomial, or even a continuous function f. Such functions are dense in $L(T)$, and we may assume that they are also real-valued when f is (cf. Remark 2.5 of Chapter II). Let

$$\phi(f, x) = \limsup_{|I_x|\to 0} \frac{1}{|I_x|} \int_{I_x} f(y)\, dy - \liminf_{|I_x|\to 0} \frac{1}{|I_x|} \int_{I_x} f(y)\, dy \geqslant 0.$$

Clearly, $\phi(f, x) = \phi(f - p, x)$ for every real-valued trigonometric polynomial p. Moreover, since $\phi(g, x) \leqslant 2\tilde{M}g(x)$, we also have that $\phi(f, x) \leqslant 2\tilde{M}(f - p)(x)$, all x in T. Thus for each $\lambda > 0$, $\{\phi(f) > \lambda\} \subset \{\tilde{M}(f - p) > \lambda/2\}$, and by Theorem 2.1

$$|\{\phi(f) > \lambda\}| \leqslant (c/\lambda)\|f - p\|_1. \tag{2.16}$$

But since the right-hand side of (2.16) can be made arbitrarily small it follows that $|\{\phi(f) > \lambda\}| = 0$, each $\lambda > 0$. This in turn implies that for a.e. x in T

$$\limsup_{|I_x|\to 0} \frac{1}{|I_x|} \int_{I_x} f(y)\, dy = \liminf_{|I_x|\to 0} \frac{1}{|I_x|} \int_{I_x} f(y)\, dy,$$

and, consequently, $\lim_{|I_x|\to 0}(1/|I_x|) \int_{I_x} f(y)\, dy$ exists for almost all $x \in T$. It is now immediate to see that this limit is actually $f(x)$ a.e. Indeed, for $\lambda > 0$ put

$$\mathscr{U}_f(\lambda) = \left\{ x \in T\colon \left| \lim_{|I_x|\to 0} \frac{1}{|I_x|} \int_{I_x} f(y)\, dy - f(x) \right| > \lambda \right\}.$$

Now, again for an arbitrary trigomometric polynomial p, we have

$$\begin{aligned} &\left| \lim_{|I_x|\to 0} \frac{1}{|I_x|} \int_{I_x} f(y)\, dy - f(x) \right| \\ &\qquad \leqslant \lim_{|I_x|\to 0} \frac{1}{|I_x|} \int_{I_x} |f(y) - p(y)|\, dy + |p(x) - f(x)| \\ &\qquad \leqslant \tilde{M}(f - p)(x) + |p(x) - f(x)|. \end{aligned}$$

Therefore

$$\mathcal{U}_f(\lambda) \subset \{\tilde{M}(f-p) > \lambda/2\} \cup \{|p-f| > \lambda/2\}$$
$$= \mathcal{U}_{f,1} + \mathcal{U}_{f,2},$$

say. By choosing p appropriately, as was done above, we see at once that $|\mathcal{U}_{f,1}|$ can be made arbitrarily small. But a similar conclusion holds for $\mathcal{U}_{f,2}$. since by Chebychev's inequality

$$|\mathcal{U}_{f,2}|\frac{\lambda}{2} \leq \int_T |f(x) - p(x)|\, dx = 2\pi\|f-p\|_1.$$

Whence $|\mathcal{U}_f(\lambda)| = 0$ for each $\lambda > 0$ and

$$\lim_{|I_x|\to 0} \frac{1}{|I_x|}\int_{I_x} f(y)\, dy = f(x), \qquad \text{a.e. in } T. \quad \blacksquare$$

Theorem 2.2, in one of its simplest formulations can be restated as folows. Let $\chi(t)$ denote the characteristic function of the interval $(-\frac{1}{2}, \frac{1}{2})$ and let $\chi_\varepsilon(t) = (2\pi/\varepsilon)\chi(t/\varepsilon)$. Then

$$f * \chi_\varepsilon(x) = \frac{1}{2\pi}\int_T f(x-t)\chi_\varepsilon(t)\, dt$$
$$= \frac{1}{\varepsilon}\int_{|t|<\varepsilon/2} f(x-t)\, dt \xrightarrow[\varepsilon\to 0]{} f(x) \qquad \text{a.e.} \tag{2.17}$$

In particular, if f is nonnegative (and in the general case also by replacing f by $|f|$), $\sup_{0<\varepsilon<2\pi}|f * \chi_\varepsilon)| \leq \tilde{M}f(x)$. Thus the convolution with the kernal $\chi_\varepsilon(t)$, which roughly represents a "bump" at the origin of width ε, height $2\pi/\varepsilon$, and total mass 1, is controlled by the Hardy-Littlewood maximal operator, and the differentiation of the integral, namely, statement (2.17) holds a.e. in T. The natural questions then is, under what general conditions on an arbitrary kernel will (2.17) remain valid. The particular examples we have in mind are those kernels which look like the Fejér K_n kernels; also note that the story is totally different for the Dirichlet kernels D_n since they have variable sign and more importantly $\|D_n\|_1 \to \infty$ as $n \to \infty$. More precisely, then, for an integrable function ϕ on the line R, under that conditions does

$$\lim_{n\to\infty}\int_T f(x-t)n\phi(nt)\, dt = \left(\int_R \phi(t)\, dt\right) f(x) \tag{2.18}$$

for x a.e. in T? A closer look at the proof of Theorem 2.2 indicates that a positive answer to (2.18) depends upon the following two properties of ϕ:

(i) $\sup_n|\int_T f(x-t)n\phi(nt)\, dt| \in$ wk-$L(T)$ whenever $f \in L(T)$;

(ii) $\lim_{n\to\infty}\int_T f(x-t)n\phi(nt)\, dt = (\int_R \phi(t)\, dt)f(x)$, for a dense class of smooth functions $f \in L(T)$.

(ii) requires no further assumption on ϕ if we restrict ourselves to trigonometric polynomials. In fact, by linearity, we may assume that $f(t) = e^{ijt}$, and in this case

$$\int_{(-\pi,\pi]} e^{ij(x-t)} n\phi(nt)\,dt = e^{ijx} \int_R \chi_{(-n\pi,n\pi]}(t) e^{-ijt/n}\phi(t)\,dt.$$

If we denote by $\phi_n(t)$ the integrand of the above integral it readily follows that $|\phi_n(t)| \leq |\phi(t)|$ and $\lim_{n\to\infty} \phi_n(t) = \phi(t)$ a.e.; consequently, by the Lebesgue dominated convergence theorem (ii) holds.

To prove (i) it actually suffices to show that the sup in question is majorized by a wk-L function; a constant multiple of $Mf(x)$ will do. This requires the following argument.

Proposition 2.3. Let ϕ be an integrable function in R which admits a nonincreasing, even, integrable majorant η; more precisely,

$$|\phi(t)| \leq \eta(t), \qquad t \in R,$$

where η is even, $\int_R \eta(t)\,dt < \infty$ and $\eta(s) \geq \eta(t)$, $0 < s < t$. Then

$$\sup_n \left| \int_T f(x-t) n\phi(nt)\,dt \right| \leq \left(4 \int_R \eta(t)\,dt\right) Mf(x).$$

Proof. Clearly,

$$\begin{aligned} \left| \int_T f(x-t) n\phi(nt)\,dt \right| &\leq \int_T |f(x-t)| n\eta(nt)\,dt \\ &= \sum_{k=0}^{\infty} n \int_{\pi/2^{k+1}<|t|\leq\pi/2^k} |f(x-t)|\eta(nt)\,dt \\ &\leq \sum_{k=0}^{\infty} n\eta(n\pi/2^{k+1}) \int_{|t|\leq\pi/2^k} |f(x-t)|\,dt = \sum_{k=0}^{\infty} A_k, \end{aligned} \tag{2.19}$$

say. It is also readily seen that the integral in A_k is bounded by

$$\frac{2\pi}{2^k}\frac{2^k}{2\pi} \int_{|t|\leq\pi/2^k} |f(x-t)|\,dt \leq \frac{2\pi}{2^k} Mf(x)$$

and, consequently,

$$A_k \leq (2\pi/2^k) n\eta(n\pi/2^{k+1}) Mf(x), \qquad \text{all} \quad k, n, x.$$

Thus summing over k we get

$$\sum_{k=0}^{\infty} A_k \leq 2\left(\sum_{k=0}^{\infty} \frac{n\pi}{2^k}\eta(n\pi/2^{k+1})\right) Mf(x). \tag{2.20}$$

Moreover, since

$$t\eta(t) \leq 2\int_{(t/2,t]} \eta(s)\,ds, \qquad t \geq 0,$$

the sum on the right-hand side of (2.20) is bounded by

$$8\sum_{k=0}^{\infty}\int_{(n\pi/2^{k+1},n\pi/2^k]} \eta(s)\,ds \leq 4\int_R \eta(s)\,ds, \tag{2.21}$$

independently of n. Our conclusion follows combining (2.19)-(2.21). ■

Corollary 2.4. Under the assumptions of Proposition 2.3, for $f \in L(T)$,

$$\lim_{n\to\infty}\int_T f(x-t)n\phi(nt)\,dt = \left(\int_R \phi(t)\,dt\right)f(x), \qquad x \text{ a.e. in } \ T.$$

Corollary 2.4 gives, in particular, a new proof of the result covered by Fejér's theorem, Theorem 4.2 in Chapter II, but it does not enable us to identify the set of x's where convergence does occur. To see that this is the case, it suffices to note that an account of the estimates (2.8) and (2.9) of Chapter II, $K_n(t) \leq cn\phi(nt)$, where $\phi(t) = \chi_{[0,1)}(-t) + t^{-2}\chi_{[1,\infty)}(t)$, ϕ even.

Because of the importance of this particular application, we state it separately as

Proposition 2.5. Suppose $f \in L(T)$ and let $\sigma^*(f,x) = \sup_n|\sigma_n(f,x)|$. Then $\sigma^*(f) \in \text{wk-}L(T)$ and

$$\lambda|\{\sigma^*(f) > \lambda\}| \leq c\|f\|_1, \qquad \text{all} \quad \lambda > 0.$$

Proof. Follows at once from Proposition 2.3. ■

3. THE CALDERÓN–ZYGMUND DECOMPOSITION

We may now return to the Calderón–Zygmund decomposition of an integrable function f and express it in its most useful form.

Theorem 3.1 (The Calderón–Zygmund Decomposition at Level λ). Let $f \in L(G)$ and let $\lambda > (1/2\pi)\int_T|f(t)|\,dt$. Then there exists a sequence $\{I_j\}$ of open, disjoint, dyadic subintervals of T such that

$$|f(x)| \leq \lambda, \qquad x \ \text{ a.e. in } \ T\setminus\bigcup I_j \tag{3.1}$$

$$\lambda \leq \frac{1}{|I_j|}\int_{I_j}|f(y)|\,dy \leq 2\lambda, \qquad \text{all } j \tag{3.2}$$

and if $\mathcal{O}_\lambda = \Omega = \bigcup I_j$, then

$$|\Omega| \leqslant \frac{1}{\lambda} \int_\Omega |f(y)|\, dy. \tag{3.3}$$

Moreover, if we set

$$g(x) = f(x)\chi_{T\setminus\Omega}(x) + \sum_j \left(\frac{1}{|I_j|} \int_{I_j} f(y)\, dy \right) \chi_{I_j}(x) \tag{3.4}$$

and

$$b(x) = \sum_j \left(f(x) - \frac{1}{|I_j|} \int_{I_j} f(y)\, dy \right) \chi_{I_j}(x), \tag{3.5}$$

then $f(x) = g(x) + b(x)$; the "good" function g and the "bad" function b are called the Calderón–Zygmund decomposition of f at level λ and have the following properties

$$|g(x)| \leqslant 2\lambda, \qquad x \text{ a.e. in } T, \tag{3.6}$$

$$\|g\|_p \leqslant (2\lambda)^{p-1} \|f\|_1, \qquad 1 \leqslant p < \infty, \tag{3.7}$$

and

$$\int_{I_j} b(y)\, dy = 0, \qquad \frac{1}{|I_j|} \int_{I_j} |b(y)|\, dy \leqslant \frac{2}{|I_j|} \int_{I_j} |f(y)|\, dy; \tag{3.8}$$

$$\|b\|_1 \leqslant 2\|f\|_1. \tag{3.9}$$

Proof. In view of the results discussed in the first section of this chapter it only remains to prove (3.6) and (3.7), which follows at once from (3.6). By the definition of g, $|g(x)| \leqslant 2\lambda$ in Ω so we must show that this estimate holds a.e. in $T\setminus\Omega$ as well. But this is an immediate consequence of the Lebesgue differentiation Theorem 2.2 since (a.e.) to x in $T\setminus\Omega$ there corresponds a sequence $\{I_x\}$ of dyadic, open intervals containing x, which converges to x as $|I_x| \to 0$ and such that

$$\frac{1}{|I_x|} \int_{I_x} |f(y)|\, dy < \lambda, \qquad \text{all } I_x.$$

Thus $|f(x)| \leqslant \lambda$ a.e. in $T\setminus\Omega$, and since $f(x) = g(x)$ there, our conclusion follows.

4. THE MARCINKIEWICZ INTERPOLATION THEOREM

The Hardy–Littlewood maximal operator Mf maps $L(T)$ into wk-$L(T)$ and it also maps $L^\infty(T)$ into itself, with norm 1, since

$$\frac{1}{|I|}\int_I |f(y)|\,dy \leqslant \|f\|_\infty, \qquad \text{every} \quad I \subset T.$$

What can we say about the behavior of Mf for L^p functions f, $1 < p < \infty$? Since these functions are in particular integrable, Mf is a well defined, wk-L function, but can we say something more precise about this function? It is convenient to cast this question in a general setting; first we present some definitions.

An operation $f \to Tf$ is said to be sublinear if $T(f_1 + f_2)$ and $T(cf_1)$ are well defined whenever Tf_1 and Tf_2 are defined, c is a scalar, and in this case

$$|T(f_1+f_2)(t)| \leqslant |Tf_1(t)| + |Tf_2(t)| \qquad \text{a.e.} \tag{4.1}$$

and

$$|T(cf_1)(t) \leqslant |c|\,|Tf_1(t)| \qquad \text{a.e.} \tag{4.2}$$

Such an operation T is said to be of (strong-) type (p, p), $1 \leqslant p \leqslant \infty$, if the mapping is defined in $L^p(T)$ and for some constant c

$$\|Tf\|_p \leqslant c\|f\|_p \tag{4.3}$$

with c independent of f; the smallest constant in (4.3) above is the norm of T as a bounded mapping in L^p. Similarly, an operation T is said to be of weak-type (p, p), $1 \leqslant p < \infty$, if the mapping is defined in L^p and for a constant c

$$\lambda^p|\{|Tf| > \lambda\}| \leqslant c^p\|f\|_p^p, \qquad \text{all} \quad \lambda > 0 \tag{4.4}$$

with c independent of f; the smallest constant in (4.4) is called the weak-type norm of T as an operator of weak-type (p, p).

Since we will consider operations T naturally defined in a couple of spaces, it is convenient to introduce the sum and intersection of the Lebesgue spaces. More specifically, let $L^{p_0}(T) + L^{p_1}(T) = \{$measurable, complex-valued functions f: $f = f_0 + f_1, f_0 \in L^{p_0}(T), f_1 \in L^{p_1}(T)\}$, and introduce there the norm

$$\|f\|_{L^{p_0}+L^{p_1}} = \inf\{\|f_0\|_{p_0} + \|f_1\|_{p_1}\}, \tag{4.5}$$

where the infimum is taken over all pairs $f_0 \in L^{p_0}(T)$, $f_1 \in L^{p_1}(T)$ such that $f = f_0 + f_1$; then $L^{p_0} + L^{p_1}$ also becomes a Banach space. Similarly, in $L^{p_0}(T) \cap L^{p_1}(T)$ we introduce the norm

$$\|f\|_{L^{p_0}\cap L^{p_1}} = \max(\|f\|_{p_0}, \|f\|_{p_1}) \tag{4.6}$$

and also $L^{p_0} \cap L^{p_1}$ becomes a Banach space. Because only the linear structure of these spaces will be used at this time, we postpone discussing further properties until 7.3.

We begin by proving a mixed weak-type–strong-type version of the Marcinkiewicz interpolation theorem; this is precisely what we need to handle the maximal function.

Theorem 4.1. Assume that a sublinear operator T is defined in $L^{p_0} + L^{p_1}$ and is simultaneously of weak-type (p_0, p_0) with norm $\leqslant c_0$ and of type (p_1, p_1) with norml $\leqslant c_1$, for $1 \leqslant p_0 < p_1 \leqslant \infty$. If now $p_0 < p < p_1$ and $1/p = (1-\eta)/p_0 + \eta/p_1$, $0 < \eta < 1$, then T is also of type (p, p) with norm $\leqslant c(1/(p-p_0))^{(1-\eta)/p_0} c_0^{1-\eta} c_1^{\eta}$. Here, c is an absolute constant $\leqslant 8e^{1/e}$ independent of the mapping T.

Proof. To compute the L^p norm of a function g we use the expression

$$2\pi \|g\|_p^p = p \int_{[0,\infty)} \lambda^{p-1} |\{|g| > \lambda\}| \, d\lambda, \tag{4.7}$$

which is readily obtained from the identity

$$\int_T |g(t)|^p \, dt = \int_T \int_{[0,|g(t)|]} d\lambda^p \, dt$$

by Tonelli's theorem.

In the first place, T is well defined in L^p, since, as is readily seen by considering large and small values of functions in L^p separately, $L^p \subseteq L^{p_0} + L^{p_1}$. This consideration is not strictly necessary in our particular case as the underlying space is of finite measure and, consequently, $L^p \subseteq L^{p_0}$; nevertheless, we prefer to give an argument that extends to the infinite measure case, such as the line R, with no modifications. Thus if $f \in L^p$, we also have $f = f_0 + f_1$, $f_i \in L^{p_i}$, $i = 0, 1$, and, by the sublinearity of T,

$$|Tf(t)| \leqslant |Tf_0(t)| + |Tf_1(t)| \qquad \text{a.e.} \tag{4.8}$$

We assume $p_1 = \infty$ first as this is the easiest case. From (4.8) it follows at once that for $\lambda > 0$,

$$\{|Tf| > \lambda\} \subseteq \{|Tf_0| > \lambda/2\} \cup \{|Tf_1| > \lambda/2\}. \tag{4.9}$$

Moreover, since $\|Tf_1\|_\infty \leqslant c_1 \|f_1\|_\infty$, the second set in the right-hand side of (4.9) is empty provided that

$$\|f_1\|_\infty \leqslant \lambda/2c_1. \tag{4.10}$$

This suggests that we actually consider a family of decompositions $f = f_{0,\lambda} + f_{1,\lambda}$ parametrized by λ. On account of (4.10) it is natural to set

$$f_{1,\lambda}(t) = \begin{cases} f(t) & \text{if } |f(t)| \leqslant \lambda/2c_1, \\ (\lambda/2c_1)\operatorname{sgn} f & \text{otherwise,} \end{cases} \tag{4.11}$$

and $f_{0,\lambda}(t) = f(t) - f_{1,\lambda}(t)$. By (4.9), (4.11) and the weak-type estimate, it follows that

$$
\begin{aligned}
|\{|Tf| > \lambda\}| &\leqslant |\{|Tf_{0,\lambda}| > \lambda/2\}| \leqslant (2/\lambda)^{p_0} c_0^{p_0} \|f_{0,\lambda}\|_{p_0}^{p_0} \\
&\leqslant \left(\frac{2}{\lambda}\right)^{p_0} c_0^{p_0} \left(\int_{\{|f| > \lambda/2c_1\}} |f(t)|^{p_0}\, dt + \left(\frac{\lambda}{2c_1}\right)^{p_0} \left| \left\{ |f| > \frac{\lambda}{2c_1} \right\} \right| \right) \\
&\leqslant (2c_0)^{p_0} \lambda^{-p_0} \int_{\{|f| > \lambda/2c_1\}} |f(t)|^{p_0}\, dt + \left(\frac{c_0}{c_1}\right)^{p_0} \left| \left\{ |f| > \frac{\lambda}{2c_1} \right\} \right|.
\end{aligned} \tag{4.12}
$$

We combine (4.7) and (4.12) and invoke Tonelli's theorem to obtain

$$
\begin{aligned}
2\pi \|Tf\|_p^p &= p \int_{[0,\infty)} \lambda^{p-1} |\{|Tf| > \lambda\}|\, d\lambda \\
&\leqslant p(2c_0)^{p_0} \int_{[0,\infty)} \lambda^{p-p_0-1} \left(\int_{\{|f| > \lambda/2c_1\}} |f(t)|^{p_0}\, dt \right) d\lambda \\
&\quad + \left(\frac{c_0}{c_1}\right)^{p_0} p \int_{[0,\infty)} \lambda^{p-1} |\{|f| > \lambda/2c_1\}|\, d\lambda \\
&= p(2c_0)^{p_0} \int_T |f(t)|^{p_0} \int_{[0, 2c_1|f(t)|)} \lambda^{p-p_0-1}\, d\lambda\, dt \\
&\quad + 2^p \left(\frac{c_1}{c_0}\right)^p \int_T |f(t)|^p\, dt \\
&\leqslant \frac{2p2^p}{p - p_0} c_0^{p_0} c_1^{p-p_0} \int_T |f(t)|^p\, dt.
\end{aligned} \tag{4.13}
$$

Whence, upon dividing by 2π and taking pth roots, (4.13) reads

$$
\|Tf\|_p \leqslant 4p^{1/p} \frac{1}{(p - p_0)^{1/p}} c_0^{p_0/p} c_1^{1-p_0/p} \|f\|_p. \tag{4.14}
$$

But since $p_0/p = 1 - \eta$, $1 - p_0/p = \eta$ and $p^{1/p} \leqslant e^{1/e}$, $1 < p < \infty$, we can rewrite (4.14) as

$$
\|Tf\|_p \leqslant c \frac{1}{(p - p_0)^{(1-\eta)/p_0}} c_0^{1-\eta} c_1^{\eta} \|f\|_p, \tag{4.15}
$$

$c \leqslant 4e^{1/e}$ independent of f and T, and we are done in this case.

Next, suppose that $p_1 < \infty$. As usual we put $f = f_{0,\lambda} + f_{1,\lambda}$, but rather than using (4.11) above we set

$$
f_{1,\lambda}(t) = \begin{cases} f(t) & |f(t)| \leqslant \varepsilon\lambda, \\ \varepsilon\lambda \operatorname{sgn} f & \text{otherwise}, \end{cases} \tag{4.16}
$$

where ε is a parameter yet to be chosen, $f_{0,\lambda} = f - f_{1,\lambda}$. From (4.8) we see at once that

$$2\pi\|Tf\|_p^p \le p\int_{[0,\infty)} \lambda^{p-1}|\{|Tf_{0,\lambda}| > \lambda/2\}|\, d\lambda$$
$$+ p\int_{[0,\infty)} \lambda^{p-1}|\{|Tf_{1,\lambda}| > \lambda/2\}|\, d\lambda$$
$$= I_0 + I_1, \tag{4.17}$$

say. The estimate for I_0 is carried out as before and it now reads

$$I_0 \le 2^p \frac{p2^{p_0}}{p - p_0} c_0^{p_0} \varepsilon^{p_0 - p} 2\pi\|f\|_p^p. \tag{4.18}$$

To estimate I_1 we introduce the function

$$\phi(s, \lambda) = \int_{[0,s]} u^{p_1 - 1}\left|\left\{|Tf_{1,\lambda}| > \frac{u}{2}\right\}\right| du. \tag{4.19}$$

Clearly, for $t > 0$

$$\left.\frac{\partial\phi(s, \lambda)}{\partial s}\right]_{s=t} = t^{p_1-1}\left|\left\{|Tf_{1,\lambda}| > \frac{t}{2}\right\}\right|.$$

With this notation we have

$$I_1 = p\int_{[0,\infty)} \lambda^{p-p_1}\left.\frac{\partial\phi(s, \lambda)}{\partial s}\right]_{s=\lambda} d\lambda,$$

whence, by integrating by parts, we readily see that

$$I_1 = p\lambda^{p-p_1}\phi(\lambda, \lambda)\Big]_0^\infty - p\int_{[0,\infty)} \phi(\lambda, \lambda)\, d\lambda^{p-p_1} = J + K, \tag{4.20}$$

say.

We estimate each term separately and do J first. For this purpose let $J(\lambda) = p\lambda^{p-p_1}\phi(\lambda, \lambda)$. Since $J(0) \ge 0$ the lower limit in J can be disregarded. As for $J(\infty)$, we begin by observing that

$$\phi(\lambda, \lambda) = \int_{[0,\lambda)} u^{p_1-1}\left|\left\{|Tf_{1,\lambda}| > \frac{u}{2}\right\}\right| du$$
$$= 2\pi(2^{p_1}/p_1)\|Tf_{1,\lambda}\|_{p_1}^{p_1} \le 2\pi(2^{p_1}/p_1)c_1^{p_1}\|f_{1,\lambda}\|_{p_1}^{p_1}. \tag{4.21}$$

Thus

$$J(\lambda) \le 2\pi p(2c_1)^{p_1}\int_{\{|f| \le \varepsilon\lambda\}} |f(t)|^{p_1}\, dt \Big/ p_1\lambda^{p_1-p} \tag{4.22}$$

and consequently

$$J \leqslant c \limsup_{\lambda \to \infty} \int_{\{|f| \leqslant \varepsilon\lambda\}} |f(t)|^{p_1} \, dt \Big/ \lambda^{p_1 - p}. \tag{4.23}$$

We now show that the right-hand side of (4.23) is 0. There are two cases, depending upon whether f is in L^{p_1} or not. In the former case the numerator of (4.23) is a bounded function of λ, whereas the denominator goes to ∞ with λ, and the limit is 0. On the other hand, if f is not in L^{p_1}, then the expression in (4.23) is indeterminate of the form ∞/∞ as $\lambda \to \infty$; this calls for L'Hopital's rule. The numerator of (4.23) can be written as

$$c \int_{[0, \varepsilon\lambda]} u^{p_1 - 1} |\{|f| > u\}| \, du$$

and the limit in question actually equals

$$c \lim_{\lambda \to \infty} \frac{(\varepsilon\lambda)^{p_1 - 1} |\{|f| > \varepsilon\lambda\}| \varepsilon}{\lambda^{p_1 - p - 1}} = c \lim_{\lambda \to \infty} (\varepsilon\lambda)^p |\{|f| > \varepsilon\lambda\}|.$$

But this expression is readily seen to be 0 since by Chebychev's inequality

$$(\varepsilon\lambda)^p |\{|f| > \varepsilon\lambda\}| \leqslant \int_{\{|f| > \varepsilon\lambda\}} |f(t)|^p \, dt \tag{4.24}$$

and the right-hand side of (4.24) goes to 0 as $\lambda \to \infty$ whenever $f \in L^p$.

It remains to estimate K. On account of (4.21)

$$\begin{aligned} K &\leqslant -2\pi p \left(\frac{2c_1}{p_1} \right)^{p_1} \int_{[0,\infty)} \int_{\{|f| \leqslant \varepsilon\lambda\}} |f(t)|^{p_1} \, dt \, d\lambda^{p - p_1} \\ &= -2\pi p \left(\frac{2c_1}{p_1} \right)^{p_1} \int_T |f(t)|^{p_1} \int_{[|f(t)|/\varepsilon, \infty)} d\lambda^{p - p_1} \, dt \\ &= 2\pi p \left(\frac{2c_1}{p_1} \right)^{p_1} \int_T |f(t)|^p \left(\frac{|f(t)|}{\varepsilon} \right)^{p - p_1} dt \\ &= p \left(\frac{2c_1}{p_1} \right)^{p_1} \varepsilon^{p_1 - p} 2\pi \|f\|_p^p. \end{aligned} \tag{4.25}$$

Thus combining (4.17), (4.18), (4.20), and (4.25) we finally obtain

$$\|Tf\|_p^p \leqslant p \left(2^p \left(\frac{2c_0}{p - p_0} \right)^{p_0} \varepsilon^{p_0 - p} + \left(\frac{2c_1}{p_1} \right)^{p_1} \varepsilon^{p_1 - p} \right) \|f\|_p^p. \tag{4.26}$$

The time has come to select ε, a prudent choice being that which makes both summands in (4.26) equal. This gives

$$\varepsilon = 2^{p/(p_1 - p_0)} \tfrac{1}{2} (c_0^{p_0} p_1 / c_1^{p_1} (p - p_0))^{1/(p_1 - p_0)}$$

and consequently the constant in the right-hand side above is

$$cp^{\frac{1}{p}} \frac{1}{(p-p_0)^{\frac{1}{p}\frac{p_1-p}{p_1-p_0}} p_1^{\frac{1}{p}\frac{p-p_0}{p_1-p_0}}} c_0^{\frac{p_0}{p}\frac{p_1-p}{p_1-p_0}} c_1^{\frac{p_1}{p}\frac{p-p_0}{p_1-p_0}}.$$

It is now a simple matter to verify that the constant c is as it should be. ■

Because of its importance we emphasize the following particular case of the interpolation theorem.

Proposition 4.2. Assume that ϕ is an integrable function which satisfies the conditions of Proposition 2.3, and let $F(x) = \sup_n |\int_T f(x-t)n\phi(nt)\,dt|$. Then

$$\|F\|_p \leqslant \frac{c}{(p-1)^{1/p}} \|f\|_p, \qquad 1 < p < \infty.$$

Proof. Apply Theorem 4.1 with $p_0 = 1$, $p_1 = \infty$. ■

Next we consider the Marcinkiewicz interpolation theorem in its original version, namely,

Theorem 4.3. Assume that a sublinear operator T is defined $L^{p_0} + L^{p_1}$, and is simultaneously of weak-types (p_0, p_0) and (p_1, p_1) with norm $\leqslant c_0$, c_1, respectively. If $p_0 < p < p_1$ and $1/p = (1-\eta)/p_0 + \eta/p_1$, $0 < \eta < 1$, then T is of type (p, p) with norm

$$\leqslant c \frac{1}{(p-p_0)^{(1-\eta)/p_0}} \frac{1}{(p_1-p)^{\eta/p_1}} c_0^{1-\eta} c_1^{\eta},$$

where c is an absolute constant.

Proof. The proof is similar to, yet simpler than, that of Theorem 4.3, since to estimate the term corresponding to $Tf_{1,\lambda}$ we simply use the fact that

$$|\{|Tf_{1,\lambda}| > \lambda/2\}| \leqslant \left(\frac{2}{\lambda}\right)^{p_1} c_1^{p_1} \int_{\{|f| < \varepsilon\lambda\}} |f(t)|^{p_1}\,dt. \quad ■$$

5. EXTRAPOLATION AND THE ZYGMUND $L \ln L$ CLASS

It is an easy pursuit to characterize those sublinear operations T which are simultaneously of weak-type $(1, 1)$ and of type (∞, ∞). Indeed, we have

Proposition 5.1. A sublinear mapping T defined in $L + L^\infty$ is simultaneously of weak-type $(1, 1)$ and of type (∞, ∞) if and only if there are constants

c_1, c_2 such that for all $f \in L + L^\infty$ and $\lambda > 0$

$$|\{|Tf| > \lambda\}| \leqslant \frac{c_1}{\lambda} \int_{[\lambda/c_2,\infty)} |\{|f| > t\}| \, dt. \tag{5.1}$$

In this case c_1 and c_2 bound the norm of T as a weak-type $(1, 1)$ and type (∞, ∞) mapping, respectively.

Proof. The necessity of the condition follows at once from (4.12). Conversely, if (5.1) holds and $f \in L(T)$, then the weak-type assertion obtains by replacing λ/c_2 by 0 in the integral above. Also, if $f \in L^\infty(T)$, then $|\{|f| > t\}| = 0$ as soon as $t \geqslant \|f\|_\infty$, and the integral vanishes if $\lambda \geqslant c_2\|f\|_\infty$; consequently, the same is true for $|\{|Tf| > \lambda\}|$ and $\|Tf\|_\infty \leqslant c_2\|f\|_\infty$. ■

Although the estimate (5.1) is equivalent to the weak-type $(1, 1)$ and type (∞, ∞) statements, it contains additional information concerning the integrability of Tf. This information, which is actually "extrapolated" from (5.1), allows us to decide, for instance, under what conditions Tf is integrable. Clearly $Tf \in L(T)$ whenever $f \in \bigcup_{p>1} L^p(T)$, but is this the optimal result? The answer is no, and to make a precise statement we introduce the Zygmund class $L \ln L$.

Definition 5.2. We say that a measurable function f is in $L \ln L(T)$ provided that

$$\int_T |f(t)| \ln^+|f(t)| \, dt = \int_{[0,\infty)} |\{|f| > \lambda\}| \frac{d(\lambda \ln^+ \lambda)}{d\lambda} < \infty,$$

where

$$\ln^+ t = \begin{cases} \ln t & t \geqslant 1, \\ 0 & \text{otherwise.} \end{cases}$$

We can now prove

Theorem 5.3. Assume that T is a sublinear operation defined in $L + L^\infty$ which is simultaneously of weak-type $(1, 1)$ and of type (∞, ∞). Then T maps $L \ln L$ into L and

$$\|Tf\|_1 \leqslant c + c \int_T |f(t)| \ln^+|f(t)| \, dt \tag{5.2}$$

where c is an absolute constant independent of f.

Proof. Since $|\{|Tf| \leqslant 1\}| \leqslant 2\pi$ it actually suffices to show that for some constant c

$$I = \int_{\{|Tf|>1\}} |Tf(t)| \, dt \leqslant c \int_T |f(t)| \ln^+|f(t)| \, dt. \tag{5.3}$$

But this is immediate since by the estimate (5.1) and Tonelli's theorem

$$I = \int_{[1,\infty)} |\{|Tf| > \lambda\}| \, d\lambda \leq c \int_{[1,\infty)} \frac{1}{\lambda} \int_{[c\lambda,\infty)} |\{|f| > s\}| \, ds \, d\lambda$$

$$= c_1 \int_{[c,\infty)} |\{|f| > s\}| \int_{[1,s/c]} \frac{d\lambda}{\lambda} \, ds$$

$$= c_1 \int_{[c,\infty)} |\{|f| > s\}| \ln^+\left(\frac{s}{c}\right) ds. \quad \blacksquare$$

That Theorem 5.3 is sharp is readily seen by observing that for the Hardy–Littlewood maximal operator the reverse inequality to (5.2) holds, namely,

Theorem 5.4 (Stein). Suppose $f \in L(T)$ is such that $Mf \in L(T)$. Then $f \in L \ln L(T)$.

Proof. We begin by showing that for $\lambda > |f|_T$

$$\frac{1}{2\lambda} \int_{\{|f| > \lambda\}} |f(t)| \, dt \leq |\{Mf > \lambda\}|. \tag{5.4}$$

Indeed, let $\{I_j\}$ be the Calderón–Zygmund interval decomposition at level λ; recall that in particular

$$\lambda < \frac{1}{|I_j|} \int_{I_j} |f(t)| \, dt \leq 2\lambda, \qquad \text{all } j, \tag{5.5}$$

and

$$|f(t)| \leq \lambda \qquad \text{a.e. in } T \setminus \bigcup_j I_j. \tag{5.6}$$

By the first inequality in (5.5), we see that $\bigcup_j I_j \subseteq \{Mf > \lambda\}$, and, by the second one, that

$$\int_{\bigcup I_j} |f(t)| \, dt \leq 2\lambda \left|\bigcup I_j\right| \leq 2\lambda |\{Mf > \lambda\}|. \tag{5.7}$$

Moreover, since by (5.6) $\{|f| > \lambda\} \subseteq \bigcup I_j$ the estimate (5.4) follows at once from (5.7). Now integrating (5.4) we obtain

$$\int_{[|f|_T,\infty)} \frac{1}{\lambda} \int_{\{|f| > \lambda\}} |f(t)| \, dt \, d\lambda$$

$$= \int_{\{|f| > |f|_T\}} |f(t)| \int_{[|f|_T, |f(t)|)} \frac{d\lambda}{\lambda} \, dt$$

$$\leq 2 \int_{[0,\infty)} |\{Mf > \lambda\}| \, d\lambda$$

and the desired conclusion follows from this with no difficulty. $\blacksquare$

6. THE BANACH CONTINUITY PRINCIPLE AND a.e. CONVERGENCE

In this section we discuss some of the basic ingredients in the proof of an a.e. convergence result.

Let B denote either a Lebesgue $L^p(T)$ space, $1 \leq p \leq \infty$ or $C(T)$. We consider a sequence of operators, much like the partial sums of a Fourier series, defined in B and continuous in measure. More precisely, each of the operators T in the sequence verifies the following properties

$$|Tf(x)| < \infty \qquad \text{a.e. for each} \quad f \in B \tag{6.1}$$

and

$$\text{if } f_n, f \text{ are in } B \text{ and } \|f_n - f\|_B \to 0, \text{ then } Tf_n \to Tf \text{ in measure,}$$

i.e., for each $\varepsilon > 0$

$$|\{|Tf_n - Tf| > \varepsilon\}| \to 0 \qquad \text{as} \quad n \to \infty. \tag{6.2}$$

For a sequence $\{T_n\}$ of such operators we set

$$T_N^* f(x) = \max_{1 \leq n \leq N} |T_n f(x)|, \qquad f \in B \tag{6.3}$$

and

$$T^* f(x) = \sup_N T_N^* f(x), \qquad f \in B. \tag{6.4}$$

In case

$$T_n f(x) \text{ converges, as } n \to \infty, \text{ a.e. for each } f \text{ in } B, \tag{6.5}$$

then also

$$T^* f(x) < \infty \qquad \text{a.e. for each} \quad f \in B. \tag{6.6}$$

We are interested in showing that property (6.6) by itself gives the continuity in measure for T^*f and that this in turn may be used to obtain (6.5). We begin by considering the continuity result

Theorem 6.1 (Banach Principle). Assume $\{T_n\}$ is a sequence of linear mappings in B which verifies property (6.6) above. Then there is a positive, decreasing function $C(\lambda)$, $\lambda > 0$, which tends to 0 as $\lambda \to \infty$ and such that for all f in B

$$|\{T^* f > \lambda \|f\|_B\}| \leq C(\lambda), \qquad \text{all} \quad \lambda > 0. \tag{6.7}$$

Proof. It suffices to establish the conclusion for f with $\|f\|_B = 1$. Fix $\varepsilon > 0$. Then by (6.6) to each f in B there corresponds an integer $n = n(f)$ such that $|\{T^*f > n\}| \leq \varepsilon$. In other words, we have

$$B = \bigcup_{n=1}^{\infty} \{f: |\{T^*f > n\}| \leq \varepsilon\} = \bigcup_{n=1}^{\infty} B_n, \tag{6.8}$$

say. Observe that each B_n is closed in B. Indeed, since

$$B_n = \bigcap_{N=1}^{\infty} \{f: |\{T_N^*f > n\}| \leq \varepsilon\}$$

and since the operators T_N^* are also continuous in measure, each of the sets in the above intersection is closed, and so is B_n. Statement (6.8) expresses then the fact that B is a countable union of closed sets and by the Baire category theorem (cf. 7.32 below) one of the B_n's contains a ball. In other words, there are an integer n, $f_0 \in B$ and $\eta > 0$ such that

$$|\{T^*f > n\}| \leq \varepsilon \qquad \text{whenever} \quad \|f_0 - f\|_B \leq \eta. \tag{6.9}$$

By translating this ball so that it is centered at the origin, (6.9) can be restated as

$$|\{T^*(f_0 + \eta f) > n\}| \leq \varepsilon \qquad \text{when} \quad \|f\|_B \leq \eta. \tag{6.10}$$

Moreover, since T^* is sublinear, it follows that $T^*f(x) \leq T^*(f_0 + \eta f)(x)/\eta + T^*f_0(x)/\eta$, and, consequently, by (6.10) we see that for $\|f\|_B \leq 1$

$$|\{T^*f > 2n/\eta\}| \leq |\{T^*(f_0 + \eta f) > n\}| + |\{T^*f_0 > n\}| \leq 2\varepsilon. \tag{6.11}$$

But $\varepsilon > 0$ above is arbitrary, whence (6.11) implies that, if $C(\lambda) = \sup_{\|f\|_B \leq 1} |\{T^*f > \lambda\}|$, then $\lim_{\lambda \to \infty} C(\lambda) = 0$, which is precisely what we wanted to show. ■

In most applications, the a.e. convergence of $\{T_nf(x)\}$ can be established for f in a dense class of functions in B. In this case the Banach principle gives

Proposition 6.2. Suppose $\{T_n\}$ is a sequence of operators in B which verifies (6.6). Then the set $B_0 = \{f \in B$: $T_nf(x)$ is a.e. convergent$\}$ is closed in B. In particular if B_0 is dense, then $\{T_nf(x)\}$ is a.e. convergent for every $f \in B$.

Proof. Let f be in the norm closure of B_0. We must show that $f \in B_0$ as well. Let $\tau f(x) = \limsup_{m,n \to \infty} |T_nf(x) - T_mf(x)|$. We will be done if we can show that $\tau f(x) = 0$ a.e. Since $\tau f(x) \leq 2T^*f(x)$, (6.7) gives

$$|\{\tau f > \lambda \|f\|_B\}| \leq C(\lambda/2). \tag{6.12}$$

Now for $f \in B$ and $g \in B_0$, $\tau f(x) = \tau(f-g)(x)$ for almost all x, and by (6.12) $|\{\tau f > \lambda \|f-g\|_B\}| \leq C(\lambda/2)$. Upon choosing $\lambda = 1/\varepsilon$ and g so that $\|f-g\|_B \leq \varepsilon^2$, we get $|\{\tau f > \varepsilon\}| \leq C(1/2\varepsilon)$. Since $C(1/2\varepsilon) \to 0$ with ε, we see at once that $\tau f(x) = 0$ a.e. ■

Before we are in a position to apply the Banach principle to the particular case of Fourier series, we need to have some control on the growth of the partial sums $s_n(f, x)$ of L^2 functions f. We will assume we know that

$$s_n(f, x) = 0(\lambda_n) \qquad \text{a.e., all} \quad f \in L^2(T), \tag{6.13}$$

where λ_n is a fixed, nondecreasing sequence tending to ∞ with n and will then show that for functions $f \sim \sum c_j e^{ijx}$ satisfying

$$\sum |c_j|^2 \lambda_{|j|}^2 < \infty \tag{6.14}$$

the Fourier series of f converges a.e. in T. Clearly, (6.14) is more effective the slower is the growth of the λ_n's. The best result we give here is the Kolmogorov–Seliverstov Theorem in 7.27; the order of the λ_n's there is $(\ln n)^{1/2}$ as $n \to \infty$.

Proposition 6.3. Suppose that (6.13) holds; if $f \in L^2(T)$ verifies the additional assumption (6.14), then $\lim_{n\to\infty} s_n(f, x)$ exists a.e.

Proof. For $h \in L^2(T)$ put $T_n h(x) = s_n(h, x)/\lambda_n$, $T^*h(x) = \sup_n |T_n h(x)|$. On account of Theorem 6.1, there is a function $C(\lambda) \to 0$ as $\lambda \to \infty$ such that

$$|\{T^*h > \lambda \|h\|_2\}| \leq C(\lambda). \tag{6.15}$$

We proceed now along the lines of the proof of Proposition 6.2 and introduce

$$\tau h(x) = \limsup_{n\to\infty} |T_n h(x)|. \tag{6.16}$$

Since $\tau h(x) \leq T^*h(x)$, (6.15) holds with τ in place of T^*, i.e., for all $h \in L^2(T)$ and $\lambda > 0$, $|\{\tau h > \lambda \|h\|_2\}| \leq C(\lambda)$. But $\tau h(x)$ is not affected if we replace h by $h - s_n(h)$, so we also have $|\{\tau h > \lambda \|h - s_n(h)\|_2\}| \leq C(\lambda)$ and consequently

$$|\{\tau h > \lambda\}| \leq C(\lambda / \|h - s_n(h)\|_2). \tag{6.17}$$

Invoking (1.6) of Chapter III it follows that the right-hand side of (6.17) goes to 0 as $n \to \infty$ and, therefore, $|\{\tau h > \lambda\}| = 0$ for each $\lambda > 0$; thus

$$\lim_{n\to\infty} (s_n(h, x)/\lambda_n) = 0 \qquad \text{a.e. in} \quad T, \quad h \in L^2(T). \tag{6.18}$$

It is now an easy matter to complete the proof. In first place observe that by the Riesz–Fischer theorem there is an L^2 function $g \sim \sum c_j \lambda_{|j|} e^{ijx}$.

Moreover, as is readily seen

$$s_n(f, x) = \sum_{j=0}^{n-1} \left(\frac{1}{\lambda_j} - \frac{1}{\lambda_{j+1}}\right) s_j(g, x) + \frac{s_n(g, x)}{\lambda_n}$$
$$= I_n + J_n,$$

say. By (6.18) $J_n \to 0$ as $n \to \infty$, a.e. in T. Furthermore since $\|s_j(g)\|_1 \leq \|s_j(g)\|_2 \leq c\|g\|_2$, it also follows that

$$\sum_{j=0}^{n-1} \left(\frac{1}{\lambda_j} - \frac{1}{\lambda_{j+1}}\right) \|s_j(g)\|_1 \leq \frac{1}{\lambda_0}\|g\|_2 < \infty,$$

and the series I_n converges (absolutely) a.e. in T. This gives the a.e. convergence of $s_n(f, x)$. ■

It is also quite simple to show that $\lim_{n\to\infty} s_n(f, x) = f(x)$ a.e. The proof we just gave is interesting in that the convergence result follows from a growth estimate; it is not in the spirit of the proof of the a.e. convergence for $f * \phi_n(x)$, which follows from a maximal estimate instead. The Banach principle itself, on the other hand, is a result in this direction but a rather incomplete one because of the unknown nature of the function $C(\lambda)$. To address this situation we begin by proving an interesting result of independent interest.

Lemma 6.4 (Calderón). Suppose that $\{E_k\}$ is a sequence of subsets in T such that $\sum |E_k| = \infty$. Then there is a sequence $\{x_k\} \subset T$ such that almost every x in T is in infinitely many of the sets $(x_k + E_k)$.

Proof. The assumption on the $|E_k|$'s is equivalent to the fact that the product $\prod_{k=1}^{\infty}(1 - (|E_k|/2\pi))$ diverges to 0. With the x_k's yet to be chosen, the subset of those x's in T which belong to infinitely many of the $(x_k + E_k)$'s is

$$\bigcap_{n=1}^{\infty}\left(\bigcup_{k\geq n}(x_k + E_k)\right). \tag{6.19}$$

We now show that for an appropriate choice of the x_k's the complement of the set in (6.19) has measure 0. With the notation $F_k = T\backslash(x_k + E_k)$ this complement is $\bigcup_{n=1}^{\infty}(\bigcap_{k\geq n} F_k)$. So we will be done once we prove that we can choose the x_k's so that

$$\left|\bigcap_{k\geq n} F_k\right| = 0, \qquad n = 1, 2, \ldots. \tag{6.20}$$

We consider the case $n = 1$ first and take a look at $F = \bigcap_{k=1}^{m} F_k$. Let χ_j denote the characteristic function of F_j, $1 \leq j \leq m$. Then the characteristic function χ of F is given by

$$\chi(t) = \chi_1(t - x_1) \cdots \chi_m(t - x_m) \tag{6.21}$$

and the measure $|F|$ is

$$\int_T \chi(t)\, dt = \int_T \chi_1(t - x_1) \cdots \chi_m(t - x_m)\, dt. \tag{6.22}$$

Interpreting χ as a function of the $m+1$ variables $t, x_1, \ldots, x_m$ now, and integrating $(1/(2\pi)^m)\chi(t)$ over all variables, each one ranging over T, we note that

$$\frac{1}{(2\pi)^m}\int \chi(t)\, dx_1 \cdots dx_m\, dt$$

$$= \int_T \left(\frac{1}{2\pi}\int_T \chi_1(t - x_1)\, dx_1\right) \cdots \left(\frac{1}{2\pi}\int_T \chi_m(t - x_m)\, dx_m\right) dt$$

$$= 2\pi \textstyle\sum_{k=1}^m \left(1 - \dfrac{|E_k|}{2\pi}\right). \tag{6.23}$$

The expression on the right-hand side of (6.23) can be made arbitrarily small, $<\frac{1}{2}$ say, provided m is sufficiently large on account of the divergence of the infinite product. Now that m has been fixed it is easy to choose $x_1, \ldots, x_m$ so that $|F| < \frac{1}{2}$. Indeed, suppose that for every $x_1, \ldots, x_m$, each one ranging over T, the integral in (6.22) exceeds $\frac{1}{2}$. Then dividing by $(2\pi)^m$ and integrating the expression in (6.22) we readily see that the integral in (6.23) exceeds $\frac{1}{2}$, and this is not the case. Thus there are points $x_1^{(1)}, \ldots, x_{m_1}^{(1)}$ in T so that $|\bigcap_{k=1}^{m_1} T\backslash(x_k^{(1)} + E_k)| < \frac{1}{2}$; more generally, it is clear that we can choose points $x_1, \ldots, x_{m_j}^{(j)}$ in T such that

$$\left|\bigcap_{k=m_{(j-1)}+1}^{m_j} (T\backslash(x_k^{(j)} + E_k))\right| < \frac{1}{2^j},$$

$j = 1, 2, \ldots$, $m_0 = 0$, m_j increasing to ∞. Consequently, for each n, and as soon as $m_{j-1} > n$,

$$\left|\bigcap_{k\geqslant n} F_k\right| \leqslant \left|\bigcap_{k=m_{j-1}+1}^{m_j} F_k\right| < \frac{1}{2^j}, \qquad \text{all } j \text{ large},$$

(6.20) holds, and we are done. ■

We can now prove

Theorem 6.5 (Calderón). The Fourier series of each $f \in L^2(T)$ converges a.e. in T if and only if the mapping $f \to s^*(f) = \sup_n |s_n(f)|$ is of weak-type (2, 2).

Proof. That the condition is sufficient follows by an argument analogous to that of Proposition 6.2, and we say no more. Conversely, if the Fourier series of each $f \in L^2(T)$ converges a.e. in T, then $s^*(f, x) < \infty$ and by the Banach continuity principle, there is a function $C(\lambda) \to 0$ as $\lambda \to \infty$ and $|\{s^*(f) > \lambda \|f\|_2\}| \leq C(\lambda)$; it only remains to show that $C(\lambda) = O(\lambda^{-2})$. Suppose this is not the case, i.e., corresponding to each integer $N \geq 0$ there is an L^2 function f_N with $\|f_N\|_2 = 1$ and a scalar $\lambda_N > 0$ such that

$$|E_N| = |\{s^*(f_N) > \lambda_N\}| \geq 2^N/\lambda_N^2. \tag{6.24}$$

By the density of the trigonometric polynomials in L^2 we may take f_N to be a polynomial p_N of norm 1. Also note that since by (6.24)

$$2^N/\lambda_N^2 \leq 2\pi, \qquad \text{all} \quad N, \tag{6.25}$$

$\lambda_N \to \infty$ with N. Out of the λ_N's we construct a new sequence $\{\mu_j\}$ as follows: if $k_0 = 1$ and for $N \geq 1$, k_N denotes the smallest integer such that

$$1 \leq 2^N k_N/\lambda_N^2 \leq 2\pi, \tag{6.26}$$

set

$$\mu_j = \lambda_N \chi_{[\sum_{n=1}^N k_{n-1}, \sum_{n=1}^N k_n]}(j), \qquad j, N > 1. \tag{6.27}$$

This is a complicated way of saying that the μ_j's consist of successive blocks of λ_N's, the first k_1 terms being equal to λ_1, the next k_2 terms equal to λ_2, and so on. We also introduce $\{v_j\}$, replacing λ_N by 2^N in (6.27). By (6.25) and (6.26), we readily verify the following two properties of the sequence $\{\mu_j\}$:

$$\sum_j \frac{1}{\mu_j^2} = \sum_N \frac{k_N}{\lambda_N^2} \leq 2\pi \sum_N 2^{-N} < \infty \tag{6.28}$$

and

$$\sum_j \frac{v_j}{\mu_j^2} = \sum_N \frac{k_N 2^N}{\lambda_N^2} \geq \sum_N 1 = \infty. \tag{6.29}$$

Note that (6.24) can also be written in an obvious fashion as

$$|E_j| = |\{s^*(p_j) > \mu_j\}| \geq v_j/\mu_j^2, \tag{6.30}$$

where the p_j's are the polynomials of norm 1 corresponding to the μ_j's (which are, after all, the λ_N's renamed). By (6.29), $|E_j| = \infty$ and, consequently by Lemma 6.4, there are points $\{x_1, \ldots, x_j, \ldots\}$ in T so that almost every x in T belongs to infinitely many of the sets $(x_j + E_j)$. Choose now a sequence m_j increasing to ∞ so rapidly that the polynomials $e^{im_j x} p_j(x)$ have no common terms, and, finally, consider

$$\sum_{j=1}^{\infty} \frac{e^{im_j x} p_j(x - x_j)}{\mu_j}. \tag{6.31}$$

Next, we show that the trigonometric series (6.31) converges in L^2 to a function f, of which it necessarily is the Fourier series, and that $s_n(f, x)$

does not converge for almost every x in T; this will clearly produce a contradiction. In first place

$$\left\| \sum_{j=N}^{M} \frac{e^{im_j x} p_j(x - x_j)}{\mu_j} \right\|_2^2 \leqslant \sum_{j=N}^{M} \frac{\|p_j\|_2^2}{\mu_j^2} = \sum_{j=N}^{M} \frac{1}{\mu_j^2} \to 0$$

as $N, M \to \infty$, and (6.31) is the Fourier series of an L^2 function f. Moreover, on account of the definition of E_j, $s^*(p_j, x - x_j) > \mu_j$ any time $x \in (x_j + E_j)$; this means in particular that some complete block arbitrarily advanced in the tail of the Fourier series of f is in modulus $\geqslant 1$. Since almost every x in T belongs to infinitely many of the $(x_j + E_j)$'s, the Fourier series of f cannot converge on a subset of T of full measure. ■

7. NOTES; FURTHER RESULTS AND PROBLEMS

Regardless of the form it assumes, the Calderón–Zygmund decomposition is an indespensable tool in the study of the continuity of operators which arise in the course of normal events in harmonic analysis. The idea of breaking a function into pieces and then handling each resulting term by an appropriate technique lies at the heart of every important result in this book.

The development of the different topics in this chapter is closely intertwined. The continuity of the conjugate operator in $L^p(T)$, $1 < p < \infty$, was established by M. Riesz in 1927; his proof relied exclusively on "complex variable methods." We will touch briefly on this subject in Section 3 of Chapter V and in Chapter VII; Chapter VII in Zygmund's treatise [1968] contains the full scope of this technique. Very shortly after this, Kolmogorov and Zygmund produced substitute results for the $L(T)$ case, Kolmogorov [1925] gave a noncomputational proof of the fact that the conjugate of an integrable function is in the class wk-$L(T)$ (this is the starting point for the material discussed in Section 6), and Zygmund [1929] introduced the class $L \ln L(T)$ in order to discuss a sharp condition that insures that the conjugate of a function in that class is actually integrable.

At about the same time Hardy and Littlewood were interested in maximal inequalities to solve a question in the theory of functions; this problem stated in a simple form is the following: suppose $\lambda > 0$, $f(z)$ is regular in $|z| < 1$ and $F(x) = \sup_{0<r<1} |f(re^{ix})|$ is the maximum modulus of f in the direction of $x \in T$. Is it true, then, that for some constant c_λ, depending only on λ, it follows that $\int_T F(x)^\lambda \, dx \leqslant c_\lambda \int_T |f(e^{ix})|^\lambda \, dx$? The answer turns out to be yes and the applications, including the boundedness of the Hardy–Littlewood maximal operator in $L^p(T)$, $1 < p < \infty$, are extremely important. Concerning the method of proof of this result, which relies on averaging of sequences, this is what the authors had to say: "The problem is most easily grasped when stated in the language of cricket, or any other game in which a player compiles a series of scores of which an average is recorded," [1930]. From this point on things get even better as the authors discuss, among other things, the maximal theorem for $1 < p < \infty$, the estimate $\sigma^*(f, x) \leq cMf(x)$, and Theorem 5.3, again for the maximal operator.

The weak-type (1, 1) estimate, which seems to be the most natural setting for these results was proved by F. Riesz [1932] a couple of years later; his proof makes use of the rising sun lemma given in 7.1 below. This lemma is the one-dimensional version of the Calderón–Zygmund decomposition, which first appeared in an article dedicated to M. Riesz on the occasion of his 65th birthday [1952].

The Marcinkiewicz interpolation Theorem 4.3 seems to be the last paper announced by the author. This was in 1939 at the age of 29; at about this time Marcinkiewicz was mobilized as a reserve officer in the Polish army, was taken prisoner on the Eastern front, and vanished without trace. The proof of the theorem was communicated by letter to Zygmund and the complete proof, as well as important extensions, first appeared in Zygmund's article [1956].

The results of Kolmogorov and Calderón, expressing the relation between the finiteness a.e. and the type of maximal operators were cast in definite form by Stein in the early 1960's. Stein's result roughly states that for a sequence of linear operators $\{T_k\}$ defined in $L^p(X)$, $1 \leq p \leq 2$, where X is a finite measure space, the existence of a family of mixing, measure-preserving mappings on X which commute with the T_k's implies that the following conditions are equivalent: (i) for each $f \in L^p(X)$, $T^*f(x) < \infty$ a.e., and (ii) $f \to T^*f$ is of weak-type (p, p). The technique of proof is quite interesting and uses the Rademacher functions. We present in 7.30 and 7.31 a similar result, due to Sawyer [1966], for positive operators; the conclusion holds now, however, for $1 < p < \infty$.

Further Results and Problems

7.1 (F. Riesz, Rising Sun Lemma) Given $f \in C(T)$, let $H = \{x \in T\colon$ there is a point y in T, $y < x$, such that $f(y) < f(x)\}$. Then H can be written as a union of countably many disjoint intervals I_k with endpoints a_k and b_k such that $f(a_k) < f(b_k)$. All the I_k's are open except possibly that with

$b_k = \pi$. (*Hint*: That H is open, except possibly for π, is immediate; thus H is of the desired form. Now assume that $f(a_k) > f(b_k)$ and reach a contradiction; we can actually say more, namely $f(a_k) = f(b_k)$ unless $b_k = \pi$.)

7.2 For $f \in L^p(T)$, $1 < p < \infty$, and $\lambda > \|f\|_1$ let

$$b(x) = \sum_j \left(f(x) - \frac{1}{|I_j|} \int_{I_j} f(y)\, dy \right) \chi_{I_j}(x) = \sum_j b_j(x)$$

denote the "bad" function in the Calderón–Zygmund decomposition of f at level λ. Show that $\lim_{N\to\infty} \|\sum_{j=1}^N b_j - b\|_p = 0$. (*Hint*: It suffices to show there is an L^p function ϕ such that $|\sum_{j=1}^N b_j(x)|, |b(x)| \leq \phi(x)$. The function

$$\phi(x) = \sum_j (|f(x)| + 2\lambda)\chi_{I_j}(x) + |f(x)| + |g(x)|,$$

where g is the "good" part, does the job.)

7.3 Let $1 < p < q < \infty$ and show that $L^p(T) \cap L^q(T)$ and $L^p(T) + L^q(T)$ are Banach spaces. (*Hint*: All the norm properties for the intersection are obvious except for the completeness. If $\{f_n\}$ is Cauchy in $L^p \cap L^q$, then it is also Cauchy in L^p and L^q and $\lim f_n(L^p) = \lim f_n(L^q) = \lim f_n(L^p \cap L^q)$. For the sum there are two nonobvious properties; (i) that the norm of $f = 0$ implies $f = 0$ a.e. and (ii) completeness. To show (i) observe that there are sequences $\{g_n\} \subseteq L^p$ and $\{h_n\} \subseteq L^q$ such that $g_n + h_n = f$, all n, and $\lim \|g_n\|_p = \lim \|h_n\|_q = 0$. Then $g_n + h_n = g_1 + h_1$ and $g_n - g_1 = h_1 - h_n$ converges to $-g_1$ (in L^p) and to h_1 (in L^q, and thus also in L^p); this means that $f = g_1 + h_1 = 0$. To show (ii) just prove that if $\sum \|f_n\|_{L^p + L^q} < \infty$, then $\lim_{N\to\infty} \sum_{n=1}^N f_n$ exists in $L^p + L^q$.)

7.4 Let $1/p + 1/p' = 1$, $1/q + 1/q' = 1$, $1 < p, q < \infty$. Consider the relation between the space of bounded linear functionals on $L^p \cap L^q$ or the space dual to $L^p \cap L^q$ and $L^{p'} + L^{q'}$; do the same for the functionals on $L^p + L^q$ and $L^{p'} \cap L^{q'}$. (*Hint*: It may be helpful to think about the relationship between the norm in $L^p \cap L^q$ and $\int_{[0,\infty)} |\{|f| > \lambda\}|\, d \max(\lambda^p, \lambda^q)$ and the norm in $L^p + L^q$ and $\int_{[0,\infty)} |\{|f| > \lambda\}|\, d \min(\lambda^p, \lambda^q)$.)

7.5 For $0 < \alpha < \infty$, $0 < \beta < 1$, the following statements are equivalent:

(i) For every measurable set $E \subseteq T$, $\int_E |f(x)|^\alpha\, dx \leq c|E|^\beta$.
(ii) $f \in \text{wk-}L^{\alpha/(1-\beta)}(T)$.

How is the constant c in (i) related to the weak norm of f? (*Hint*: (i) implies (ii) Put $E = \{|f| > \lambda\}$. (ii) *implies* (i). Use

$$\int_E |f(x)|^\alpha\, dx \leq \int_{[0,\infty)} \min\{|E|, |\{x \in E : |f(x)| > \lambda\}|\}\, d\lambda^\alpha.)$$

7.6 Show by means of an example that $L \ln L$ strictly contains $\bigcup_{p>1} L^p$.

7.7 Show that condition (5.1) is actually equivalent to the single condition $|\{|Tf| > 1\}| \leq c_1 \int_{[1/c_2,\infty)} |\{|f| > t\}\, dt$.

7.8 Let $a(t)$ be a nondecreasing function such that $\int_{[1,s)} a(t)\,dt/t = O(b(s))$, and let $A(t) = \int_{[0,t)} a(s)\,ds$, $B(t) = \int_{[0,t)} b(s)\,ds$. If T is a sublinear mapping simultaneously of weak-type $(1, 1)$ and type (∞, ∞) then $\int_T A(|Tf(x)|)\,dx \leq c + c\int_T B(|f(x)|)\,dx$. (*Hint*: Theorem 5.3 corresponds to the case $a(t) = 1$, $b(t) = \ln^+ t$.)

7.9 Let $\{g_n\}$ be a sequence of wk-L functions with norm (uniformly) ≤ 1 and furthermore let $\{c_n\}$ be a sequence of positive numbers, $\sum c_n = 1$, and $\sum c_n|\ln c_n| = K < \infty$. Then $\lambda|\{\sum c_n g_n > \lambda\}| \leq 2(K+2)$, all $\lambda > 0$. (*Hint*: For each $n > 1$, put $u_n(x) = g_n(x)$ if $g_n(x) < \lambda/2$ and 0 otherwise, $v_n(x) = g_n(x)$ if $g_n(x) > \lambda/2c_n$ and 0 otherwise, and $w_n(x) = g_n(x) - (u_n(x) + v_n(x))$. Put $u(x) = \sum c_n u_n(x)$ and similarly for $v(x)$ and $w(x)$. The following properties are readily verified: $u(x) < \lambda/2$, $|\{v \neq 0\}| \leq \sum_n |\{g_n > \lambda/2c_n\}| < 2/\lambda$ and

$$\int_T w(x)\,dx = -\sum c_n \int_{[\lambda/2,\lambda/2cn]} s\,d|\{g_n > s\}| \leq \sum c_n|\ln c_n| + 1.$$

Finally estimate

$$|\{\sum c_n g_n > \lambda\}| \leq \left|\left\{u > \frac{\lambda}{2}\right\}\right| + |\{v \neq 0\}| + \left|\left\{w > \frac{\lambda}{2}\right\}\right|.$$

This result is from Stein–N. Weiss [1969].)

7.10 Let $\{T_n\}$ be a sequence of sublinear operators defined in $L(T)$ and assume that the weak-type $(1, 1)$ norm of these operators is (uniformly) ≤ 1. If $\{c_n\}$ is a sequence of positive real numbers so that $\sum c_n|\ln c_n| < \infty$, then $T = \sum c_n T_n$ is of weak-type $(1, 1)$.

7.11 A sublinear operator T defined in $L^p + L^\infty$ is simultaneously of weak-type (p, p) and of type (∞, ∞) if and only if there are constants c_1, c_2 such that for all $f \in L^p + L^\infty$ and $\lambda > 0$

$$|\{|Tf| > \lambda\}| \leq \frac{c_1}{\lambda^p}\int_{[c_2\lambda,\infty)} |\{|f| > s\}|s^{p-1}\,ds.$$

Extrapolate and compare with 7.7.

7.12 A sublinear operator T defined in $L^p + L^\infty$ is simultaneously of types (p, p) and (∞, ∞) if and only if there are constants c_1, c_2 such that for all $f \in L^p + L^\infty$ and $\lambda > 0$

$$\int_{[\lambda,\infty)} |\{|Tf| > s\}|s^{p-1}\,ds \leq c_1 \int_{[c_2\lambda,\infty)} |\{|f| > s\}|s^{p-1}\,ds.$$

7.13 Show that for t large,

$$\sup_{\|f\|_p \leq 1} \int_{[t,\infty)} |\{|f| > s\}|\,ds \sim \frac{1}{t^{p-1}}.$$

7.14 Extend the Marcinkiewicz theorem to the case where the operator T in question is simultaneously of types (p_0, p_0) and (p_1, p_1). What can you say about the norm of T, as a bounded mapping in L^p, $p_0 < p < p_1$? What if T is of type (p_0, p_0), weak-type (p_1, p_1) $1 < p_0 < p_1 < \infty$?

7.15 State and prove the "abstract" version of the Marcinkiewicz interpolation theorem, i.e., T acts now on $L^{p_0}(X) + L^{p_1}(X)$, where $(X, d\mu)$ is an arbitrary measure space. Show that the requirement $p_0, p_1, p > 1$ is not necessary, i.e., we may actually assume $p_0, p_1, p > 0$.

7.16 There are ways to sharpen the conclusion of the Marcinkiewicz interpolation theorem by considering, for instance, a finer scale of intermediate spaces between the Lebesgue L^p classes. We may consider the Lorentz spaces $L^{p,q}(T)$ consisting of those measurable f's such that

$$\|f\|_{p,q} = \left(p \int_{[0,\infty)} |\{|f| > \lambda\}|^{q/p} \lambda^{q-1}\, d\lambda\right)^{1/q} < \infty.$$

State and prove such a result. The work of O'Neil [1968] is relevant here.

7.17 Can we use $L \ln L$ as an endpoint in an interpolation theorem rather than as an intermediate space? Compare with 7.7 above. The results of Torchinsky [1976] are also of interest in this question.

7.18 A linear operator T defined in $L + L^{p_1}$, $1 < p_1 < \infty$, is said to be of pseudo-type $(1, 1)$ if to each $f \in L + L^{p_1}$ and $\lambda > 0$ there correspond a measurable set $G \subseteq T$ and a function $g \in L(T)$ such that $|G| \leq c\|f\|_1/\lambda$, $|g(x)| \leq c\lambda$, $\|g\|_1 \leq c\|f\|_1$ and $\int_{T\setminus G} |T(f - g)(x)|\, dx \leq c\|f\|_1$; all constants c above are independent of f and λ. Prove that if T is of pseudo-type $(1, 1)$ and of weak-type (p_1, p_1), then T is of weak-type $(1, 1)$ and consequently of type (p, p) for $1 < p < p_1$. (*Hint*: Estimate $|\{|Tf| > 2\lambda\}| \leq |\{|T(f - g)| > \lambda\}| + |\{|Tg| > \lambda\}| = |I| + |J|$, say. $|J|$ can be easily handled on account of the weak-type (p_1, p_1) assumption. Also $|I| \leq |G| + |(T\setminus G) \cap \{|T(f - g)| > \lambda\}|$. This result is of interest in that it brings the Calderón-Zygmund decomposition into play. We will have more to say about this later on; the proof is Cotlar's (see Cotlar and Cignoli [1974]).)

7.19 A sublinear operator T is of weak-type (p, p) if and only if for each measurable subset $E \subseteq T$ and $0 < r < p$ Kolmogorov's inequality holds, namely, for all $f \in L^p(T)$

$$\int_E |Tf(x)|^r\, dx \leq c^r \frac{p}{p - r} |E|^{1-r/p} \|f\|_p^r. \tag{7.1}$$

(*Hint*: Compare with 7.5.)

7.20 Assume that T and S are sublinear operators and that T is majorized by S in the following sense: if $C(x, r) = \{y \in T: r \leq |x - y| \leq 2r\}$, to each x in T and $f \in L(T)$ there corresponds $0 < \bar{r} < \pi$ such that $|Tf(x)| \leq \inf_{y \in C(x,\bar{r})} |Sf(y)|$. Then, if S is of weak-type (p, p) for some $p > 0$, so is T.

(*Hint*: Let $0 < q < p$. Then $|Tf(x)|^q \le \inf_{y \in C(x,\bar r)} |Sf(y)|^q$ and averaging over $C(x, \bar r)$

$$|Tf(x)|^q \le \frac{1}{|C(x,\bar r)|} \int_{C(x,\bar r)} |Sf(y)|^q \, dy$$
$$\le \frac{|I(x,\bar r)|}{|C(x,\bar r)|} \frac{1}{|I(x,\bar r)|} \int_{I(x,\bar r)} |Sf(y)|^q \, dy,$$

where $I(x, \bar r)$ denotes the interval centered at x of length $4\bar r$. The above estimate then gives $|Tf(x)|^q \le cM(|SF|^q)(x)$, with c independent of x and f now; therefore, it will suffice to show that $M(|Sf|^q)(x)^{1/q}$ is of weak-type (p, p). But this is quite easy since by Proposition 5.1

$$|\{M(|Sf|^q > \lambda^q\}| \le c\lambda^{-q} \int_{[c\lambda^q,\infty)} |\{Sf| > t^{1/q}\}| \, dt$$
$$\le c\lambda^{-q} \int_{[c\lambda^q,\infty)} \|f\|_p^p t^{-p/q} \, dt = c\lambda^{-p} \|f\|_p^p;$$

the result is also Cotlar's.)

7.21 Assume that T and S are sublinear operators and that T is majorized by S in the following sense: to each x in T and $f \in L(T)$ there corresponds $0 < \bar r < \pi$ such that if $I(x, \bar r)$ denotes the interval centered at x of length $2\bar r$, then

$$|Tf(x)| \le \inf_{y \in I(x,\bar r)} |S(f\chi_{I(x,2\bar r)})(y)|.$$

Then, if S is of weak-type (p, p) for some $p > 0$, so is T. (*Hint*: Use (7.1) now, this also is Cotlar's observation.)

7.22 With the notation of Section 6, we say that $E \subseteq T$ is a set of divergence for B if there exists $f \in B$ whose Fourier series diverges at every point of E. Prove that E is a set of divergence for B if and only if there is $f \in B$ such that $s^*(f, x) = \infty$ for $x \in E$. (*Hint*: The condition is clearly sufficient; to show that it is necessary let $g \in B$ have a divergent Fourier series at each point of E and let $f \in B$ and $\{\lambda_j\}$ be the function and sequence corresponding to g constructed in Proposition 4.6. Then, as in the proof of Proposition 6.3, we see that $|s_n(g, x) - s_m(g, x)| \le 2s^*(f, x)\lambda_{m+1}^{-1}$ and the Fourier series of g converges whenever $s^*(f, x) < \infty$. This observation is Katznelson's [1968].)

7.23 E is a set of divergence for B if and only if there exists a sequence of trigonometric polynomials $\{p_n\}$ such that

$$\|p_n\|_B < \infty \quad \text{and} \quad \sup_n s^*(p_n, x) = \infty \quad \text{for} \quad x \in E.$$

(*Hint*: The proof is very much like that of Theorem 6.5. To show the necessity assume that the polynomials p_n of degree N_n exist and with integers k_n such that $k_n > k_{n-1} + N_{n-1} + N_n$, put $f(x) = \sum_n e^{ik_n x} p_n(x)$; then

$$s_{m_n+m}(f, x) - s_{m_n-m-1}(f, x) = e^{im_n x} s_m(p_n, x)$$

whenever $m < N_n$ and the partial sums of f diverge on E. Conversely, if E is a set of divergence for B there are a monotonic sequence $\mu_n \to \infty$ and $f \in B$ such that $|s_n(f, x)| > \mu_n$ infinitely often, for every x in E. Let now $\{\lambda_n\}$ be a sequence of integers such that $\|f - \sigma_{\lambda_n}(f)\|_B < 2^{-n}$ and choose integers η_n such that $\mu_{\eta_n} > 2 \sup_x s^*(\sigma_{\lambda_n}, x)$. If we set now $p_n = (2K_{2\eta_n+2} - K_{\eta_n+1}) * (f - \sigma_{\lambda_n})$, then $\sum \|p_n\|_B < \infty$, and if $x \in E$ and n is an integer such that $|s_n(f, x)| > \mu_n$, then for some j,

$$\eta_j < n < \eta_{j+1} \quad \text{and} \quad |s_n(p_j, x)| = |s_n(f, x) - s_n(\sigma_{\lambda_j}(f), x)| > \mu_n/2.$$

This proof is also from Katznelson's book [1968].)

7.24 Show that if the E_j's are sets of divergence for B, then so is $E = \bigcup_{j=1}^{\infty} E_j$.

7.25 (Kolmogorov) Let $\{\eta_k\}$ be an Hadamard sequence, i.e., $n_{k+1}/n_k \geq \theta > 1$, and assume that $f \in L(T)$ has a lacunary series, i.e., $f \sim \sum c_{n_k} e^{in_k x}$. Then show that $s^*(f, x) \leq c\sigma^*(f, x)$, where c depends only on θ, and $\lim_{k\to\infty} s_{n_k}(f, x) = f(x)$ a.e. (*Hint*: Observe in the first place that, if $\sum_k c_{n_k} = s(C, 1)$, then also $\sum_k c_{n_k} = s$. This is not hard to see since we may suppose that $s = 0$, and in that case the partial sums s_{n_k} of the series in question verify the following relation: if $j > k$, then $(n_j - n_k)s_{nk} \doteq n_j\sigma_{n_j-1} - n_k\sigma_{n_k-1}$. Thus $(n_j - n_k)s_{n_k} = o(n_j) + o(n_k) = o(n_j - n_k)$, and $s_{n_k} = o(1)$. Also $|s_{n_k}| \leq ((n_j + n_k)/(n_j - n_k))\sigma^* \leq ((\theta + 1)/(\theta - 1))\sigma^*$.)

7.26 (Kolmogorov) Let $\{n_k\}$ be a fixed Hadanard sequence and $f \in L^2(T)$. Then $\lim_{k\to\infty} s_{n_k}(f, x) = f(x)$ a.e., and, if $\Omega^*(f, x) = \sup_k |s_{n_k}(f, x)|$, then $\|\Omega^*(f)\|_2 \leq c\|f\|_2$ where c depends only on θ. (*Hint*: Both statements follow from the estimate

$$\sum_k \int_T |\sigma_{n_k}(f, x) - s_{n_k}(f, x)|^2 \, dx \leq c\|f\|_2^2,$$

where c depends only on θ; but this is easy to obtain from an appropriate expression for $s_n(f, x) - \sigma_n(f, x)$ (cf. 5.4 of Chapter II). The second assertion also requires the observation that

$$\sup_k |s_{n_k}(x)|^2 \leq \sum_k |s_{n_k}(x) - \sigma_{n_k}(x)|^2 + \sigma^*(f, x).)$$

7.27 (Kolmogorov-Seliverstov) If $f \in L^2(T)$, then $s_n(f, x) = o((\ln n)^{1/2})$ a.e. Furthermore, the function $T^*f(x) = \sup_{n\geq 2}(|s_n(f, x)|/(\ln n)^{1/2})$ is in $L^2(T)$ and $\|T^*f\|_2 \leq c\|f\|_2$, with c independent of f. (*Hint*: It suffices to prove the norm estimate for $T_N^* f(x) = \sup_{n \le N}(|s_n(f, x)|/(\ln n)^{1/2})$, with a constant independent of N. Let $n(x)$ be any step function taking integer values $2 \le n(x) \le N$, and put $\lambda(x) = 1/\ln n(x)$; it is enough to show that

$$I = \int_T \lambda(x) s_{n(x)}^2(f, x)\, dx \le c\|f\|_2^2$$

since $T_N^* f(x) = \lambda(x)^{1/2} s_{n(x)}(f, x)$. By the estimates (1.15) and (1.17) in Chapter I, $\int_T |D_{n(x)}(x)|\, dx \le c \ln N$. Now I equals

$$\frac{1}{2\pi}\int_T \lambda^{1/2}(x) s_{n(x)}(f, x)\phi(x)\, dx, \qquad \text{with} \quad \|\phi\|_2 = 1.$$

Put $\psi(x) = \lambda^{1/2}(x)\phi(x)$. Then

$$I = c\int_T f(t)\left(\int_T \psi(x) D_{n(x)}(x - t)\, dx\right) dt$$

$$\le c\left\|\int_T \psi(x) D_{n(x)}(x - \cdot)\, dx\right\|_2 \|f\|_2 = A\|f\|_2,$$

say. Clearly

$$A^2 \le c\int_T\int_T \psi(x)\psi(t) D_{\min(n(x),n(t))}(x - t)\, dx\, dt$$

$$\le c\int_T\int_T \psi^2(x)|D_{\min(n(x),n(t))}(x - t)|\, dx\, dt$$

$$+ c\int_T\int_T \psi^2(t)|D_{\min(n(x),n(t))}(x - t)|\, dx\, dt,$$

and the estimate obtains using the above bound for $\int_T |D_{n(x)}(x)|\, dx$. This shows that $s_n(f, x) = O((\ln n)^{1/2})$ a.e. To refine O to o we pass to lim sup instead.)

7.28 (Plessner) If $f \in L(T)$ and

$$F(x) = \left(\int_{[0,\pi]}\left|\frac{f(x+t) - f(x-t)}{t}\right|^2 dt\right)^{1/2}$$

is in $L^2(T)$, then the Fourier series of f converges a.e. (*Hint*: Show that the assumption is actually equivalent to $\sum_{|n|>1}|c_n(f)|^2 \ln n < \infty$.)

7.29 With the same notation and assumptions of 7.14 of Chapter III, show that if

$$s^*(x) = \sup_{N \ge 0}|s_N(x)| = \sup_{N \ge 0}\left|\sum_{k=0}^{N} r_k f_k(t_k x)\right|,$$

then there is a constant c, depending only on η and θ such that

$$s^*(x) \leqslant Mf(x) + c\left(\sum_{k=0}^{\infty}|r_k|^2\right)^{1/2}.$$

Furthermore $\|s^*\|_2 \leqslant c\|f\|_2$ and $s_N(x) \to f(x)$ a.e. (*Hint*: The proof is an interesting application of an "averaging" method. Let $I = [x, x + 2\pi/t_N]$. If $y \in I$, then

$$|s_N(x) - s_N(y)| \leqslant \sum_{k=0}^{N}|r_k||f_k(t_kx) - f_k(t_ky)|$$

$$\leqslant (2\pi)^\eta \sum_{k=0}^{N}|r_k|\left(\frac{t_k}{t_N}\right)^\eta \leqslant c\left(\sum|r_k|^2\right)^{1/2}$$

and consequently $|s_N(x)| \leqslant |s_N(y)| + c(\sum|r_k|^2)^{1/2}$. By averaging now over I, it follows that

$$|s_N(x)| \leqslant \frac{1}{|I|}\int_I |f(y)|\, dy + \frac{1}{|I|}\int_I |R_{N+1}(y)|\, dy + c\left(\sum|r_k|^2\right)^{1/2} = A + B + C,$$

say, where $R_{N+1}(y)$ in B is the "tail" of the series. Clearly, C is of the right order and $A \leqslant cMf(x)$. To do B change variables $y = x + s/t_N$, $0 \leqslant s \leqslant 2\pi$, and apply 7.14 of Chapter III to $R_{N+1}(x + s/t_N) = \sum_{k=N+1}^{\infty} r_k f_k(x + s/t_N)$. Thus the first assertion is proved; that $\|s^*\|_2 \leqslant c(\sum|r_k|^2)^{1/2}$ follows at once. To prove the a.e. convergence apply the norm result to $R_N^*(x) = \sup_{j \geqslant k \geqslant N}|s_j(x) - s_k(x)|$; since we have $r_0 = \cdots = r_N = 0$, $\|R_N^*\|_2 \to 0$ as $N \to \infty$. Finally, note that since $R_1^* \geqslant \cdots \geqslant R_N^* \geqslant \cdots \geqslant 0$ and $\int_T R_N^*(x)\, dx \to 0$ we also have $\lim_{N\to\infty} R_N^*(x) = 0$ a.e. This result is also from Meyer's work [1979].)

7.30 Suppose $\{S_\alpha\}_{\alpha\in I}$ is a family of transformations from T into T verifying the following properties: (i) each S_α is measure preserving, i.e., if E is a measurable subset of T, then $|S_\alpha^{-1}(E)| = |E|$, and (ii) the family is mixing, i.e., if E_1 and E_2 are measurable subsets of T and $r > 1$, then there is an S_α in the family such that $|E_1 \cap S_\alpha^{-1}(E_2)| \leqslant r|E_1||E_2|$. (Observe that if the S_α's were such that $|E_1 \cap S_\alpha^{-1}(E_2)| = |E_1||E_2|$, then E_1 and $S_\alpha^{-1}(E_2)$ would be probabilistically independent; our assumption is not so restrictive.) Prove that if $\{E_k\}$ is a sequence of measurable subsets of T such that $\sum|E_k| = \infty$, then there exists a sequence $\{S_k\}$ of S_α's such that almost every x in T is in infinitely many of the sets $S_k^{-1}(E_k)$.

7.31 (Sawyer) Suppose that $\{T_k\}$ is a sequence of linear operators defined in $L^p(T)$ for some $1 \leqslant p \leqslant \infty$ which verifies the following assumptions: (i) each T_k is continuous in measure, (ii) each T_k is positive, i.e., if $f \geqslant 0$, then $T_kf \geqslant 0$, (iii) there is a family $\{S_\alpha\}_{\alpha\in I}$ verifying the assumptions of 7.30 so that the T_k's and S_α's commute in the following sense: if $f \in L^p(T)$ and

$(S_\alpha f)(x) = f(S_\alpha x)$, then for each T_k, $T_k S_\alpha = S_\alpha T_k$; more precisely, for each $x \in T$, $T_k(S_\alpha f)(x) = (S_\alpha(T_k f)(x)$. Prove that for such a sequence of operators the following conditions are equivalent: (a) $\sup_k|T_k f(x)| = T^*f(x)$ is of weak-type (p, p) and (b) for each $f \in L^p(T)$, $T^*f(x) < \infty$ a.e. (*Hint*: As in the proof of Calderón's theorem, assume that T^* is not of weak-type (p, p) and reach a contradiction; 7.30 is an important step in the proof which follows de Guzmán's presentation [1981]. Sawyer's principle has wider applications than it may seen on the surface; in fact, although the T_k's themselves may not be positive, they may be of the form $T_k = P_k T$, where the P_k's are positive and T is some fixed, bounded, linear operator in L^p. Then the estimate for T^* will automatically follow from the one for P^*.)

7.32 (Baire Category Theorem) A metric space is said to be of first category if it can be written as a countable union of sets that are nowhere dense. If a metric space is not of first category, then we say it is of second category. The rational numbers on the line with the usual metric are of first category; on the other hand, the real line is of second category. This last assertion is a special case of the Baire category theorem: A complete metric space X is of second category. (*Hint*: Suppose $X = \bigcup_{j=1}^{\infty} X_j$, where the X_j are nowhere dense, that is, the sets $\bar{X}_j$ have no interior points. Fix a ball $B(x_0, 1)$; since $\bar{X}_1$ does not contain it, there is a point $x_1 \in B(x_0, 1)\backslash\bar{X}_1$; from this point on the proof is identical to that of the uniform boundedness principle, Theorem 2.3 of Chapter I.)

CHAPTER

V

The Hilbert Transform and Multipliers

1. EXISTENCE OF THE HILBERT TRANSFORM OF INTEGRABLE FUNCTIONS

This section covers one of the basic results in harmonic analysis, namely, the (a.e.) existence of the principal value integral defining the Hilbert transform of a summable function. This limit will not, in general, be a locally integrable function, though. The proof we present here is purely real variable and it serves as an inspiration for the extension of these results to the Euclidean n-dimensional case.

Theorem 1.1. Suppose that $f \in L(T)$. Then

$$\lim_{\varepsilon \to 0} H_\varepsilon f(x) = \lim_{\varepsilon \to 0} -\frac{1}{\pi} \int_{\varepsilon<|t|<\pi} \frac{f(x-t)}{2\tan(t/2)\, dt} \tag{1.1}$$

exists for almost every x in T. We call this limit $Hf(x)$, the Hilbert transform of f, and the relation (1.1) is denoted by

$$Hf(x) = \text{p.v.} -\frac{1}{\pi} \int_T \frac{f(x-t)}{2\tan(t/2)}\, dt.$$

The principal value notation (p.v.) above emphasizes that a symmetric neighborhood about the origin is deleted before the limit is taken; the origin corresponds to the singularity of the convolution kernel here.

Proof. The idea of the proof is to show that the p.v. integral converges a.e. in the complement of a sequence of sets with measure approaching 0; once this is achieved, clearly the p.v. integral exists also in the union of these sets, namely, in a subset of T of full measure.

In first place we may, and do, assume that f is nonnegative. Let λ be a large constant; $\lambda > |f|_T$, the average of $|f|$ over T, will do. We then invoke the Calderón–Zygmund decomposition of f at level λ, Theorem 3.1 of Chapter IV, and obtain a sequence of open, disjoint subintervals $\{I_j\}$ of T and a decomposition $f = g + b$, induced by these intervals. Clearly, for each $\varepsilon > 0$ $H_\varepsilon f(x) = H_\varepsilon g(x) + H_\varepsilon b(x)$, x in T. Moreover, since $g \in L^2(T)$, by Corollary 6.5 in Chapter III, $\lim_{\varepsilon\to 0} H_\varepsilon g(x) = \tilde{g}(x)$ exists a.e. in T. Let, as usual, $\Omega^* = \bigcup 2I_j$, $|\Omega^*| \leq (2/\lambda)\int_T |f(y)|\,dy$. Suppose we can show that

$$\lim_{\varepsilon\to 0} H_\varepsilon b(x) \qquad \text{exists a.e. in} \quad T\backslash\Omega^*. \tag{1.2}$$

Then by choosing a sequence $\lambda_k \to \infty$, and if $\{\Omega_k^*\}$ denotes the corresponding sequence of open sets, the p.v. integral in (1.2) exists, for the corresponding b's of course, a.e. in the complement of Ω_k^* and the same is true for the p.v. integral corresponding to f. Thus the Hilbert transform of f exists a.e. in $\bigcup_k (T\backslash\Omega_k^*)$, which is a subset of T of full measure.

Returning to the proof of (1.2), then, it suffices to show that $\{H_\varepsilon b(x)\}$ is Cauchy a.e. in $T\backslash\Omega^*$ as $\varepsilon \to 0$. In other words, we must show that

$$\lim_{\varepsilon,\eta\to 0} |H_\varepsilon b(x) - H_\eta b(x)| = 0, \qquad \text{a.e. for } x \text{ in } \quad T\backslash\Omega^*. \tag{1.3}$$

We break the proof of (1.3) into two steps. In the first place, we show that

$$\limsup_{\varepsilon,\eta\to 0} |H_\varepsilon b(x) - H_\eta b(x)| < \infty, \qquad \text{a.e. for} \quad x \quad \text{in} \quad T\backslash\Omega^*, \tag{1.4}$$

and once we know that this lim sup is finite we show, in the second step, that it is 0.

For $\varepsilon > \eta > 0$ fixed, note that $H_\varepsilon b(x) - H_\eta b(x)$ equals

$$-\frac{1}{\pi}\left(\int_{[x-\varepsilon,x-\eta)} + \int_{(x+\eta,x+\varepsilon]}\right)\frac{b(t)}{2\tan((x-t)/2)}\,dt. \tag{1.5}$$

Moreover, since $b(t) = \sum_j b(t)\chi_{I_j}(t)$ and $x \notin I_j$ for any j, we can divide the I_j's into 2 families, namely, those intersecting $[x-\varepsilon, x-\eta)$ and those intersecting $(x+\eta, x+\varepsilon]$; all other intervals contribute nothing to the integral in (1.5) and may be disregarded. As the argument for each family is identical we only consider those intervals I_j intersecting $(x+\eta, x+\varepsilon]$. These intervals can in turn be separated into three subfamilies, to wit

(i) Possibly an I_j containing $x+\eta$, call it I_j^η.
(ii) Possibly an I_j containing $x+\varepsilon$, call it I_j^ε.
(iii) Those I's totally contained in $(x+\eta, x+\varepsilon]$, call them still I_j.

The main contribution to (1.5) comes from the integrals extended over the I_j's; the other terms represent basically "error" terms. We make this statement precise.

Let L_j^η = length of I_j^η, x_j^η = center of I_j^η and put

$$A_j^\eta = \int_{(x+\eta,x+\varepsilon]\cap I_j^\eta} \frac{b(t)}{2\tan((x-t)/2)}\,dt.$$

The following two geometric observations are clear:

(a) Since $x \notin 2I_j^\eta$, then $\eta > L_j^\eta/2$.
(b) $(x+\eta, x+\varepsilon] \cap I_j^\eta \subseteq (x+\eta, x+\eta+L_j^\eta) \subseteq (x+\eta, x+3\eta)$.

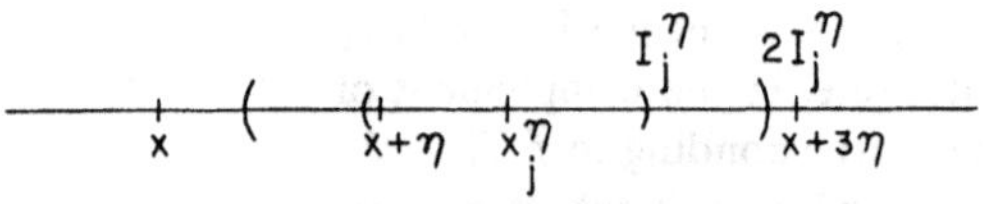

Thus

$$|A_j^\eta| \leqslant \int_{(x+\eta,x+3\eta)} \frac{|b(t)|}{2|\tan((x-t)/2)|}\,dt. \tag{1.6}$$

Moreover since $\tan u \sim \eta$ when $u \sim \eta$ and η is small, the denominator of the integrand in (1.6) is of order c/η and it follows that

$$|A_j^\eta| \leqslant \frac{c}{\eta}\int_{(x+\eta,x+3\eta)} |b(t)|\,dt = \frac{c}{\eta}\int_{(\eta,3\eta)} |b(x+t)|\,dt$$

$$\leqslant \frac{c}{\eta}\int_{(0,3\eta)} |b(x+t)|\,dt. \tag{1.7}$$

Now $b(x) = 0$ since $x \in T\backslash\Omega^*$ and (1.7) can finally be rewritten as

$$|A_j^\eta| \leqslant \frac{c}{3\eta}\int_{(0,3\eta)} |b(x+t)-b(x)|\,dt. \tag{1.8}$$

The expression appearing in the right-hand side of (1.8) is a familiar one, as we have previously encountered it in Proposition 4.1 of Chapter II; it actually is $o(1)$ as $\eta \to 0$ at every Lebesgue point x of b that is a.e. in T. Similarly,

$$|A_j^\varepsilon| \leqslant \frac{c}{3\varepsilon}\int_{(0,3\varepsilon)} |b(x+t)-b(x)|\,dt = o(1) \quad \text{as} \quad \varepsilon \to 0 \tag{1.9}$$

also a.e. in T. So these are indeed error terms.

To estimate the main contribution we invoke the cancellation property $\int_{I_j} b(t)\,dt = 0$ and rewrite with x_j = center of I_j

$$A_j = \int_{I_j} \frac{b(t)}{2\tan((x-t)/2)}\,dt$$
$$= \frac{1}{2}\int_{I_j} b(t)\left(\frac{1}{\tan((x-t)/2)} - \frac{1}{\tan((x-x_j)/2)}\right) dt$$
$$= \frac{1}{2}\int_{I_j} b(t)k(x, t, x_j)\,dt, \tag{1.10}$$

say. Furthermore, a simple computation making use of elementary trigonometric identities gives

$$k(x, t, x_j) = \frac{\sin((t-x_j)/2)}{\sin((x-t)/2)\sin((x-x_j)/2)} \tag{1.11}$$

and to estimate $k(x, t, x_j)$ we refer once again to the geometry of the situation. In the first place, notice that if $t, x_j \in I_j$ and $x \notin 2I_j$, then

$$1/3 \leqslant \frac{|x-t|}{|x-x_j|} \leqslant 3. \tag{1.12}$$

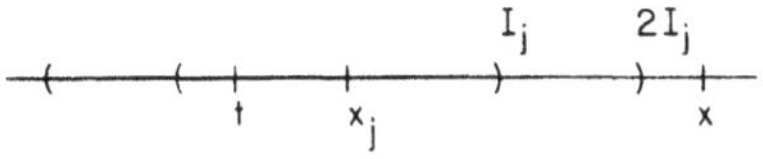

Indeed, since with L_j = length of I_j we have $|x - x_j| > L_j/2$, then we see at once that $|x-t| \leqslant |x-x_j| + |x_j - t| \leqslant |x - x_j| + L_j \leqslant 3|x-x_j|$. Exchanging the role of x_j and t we also note that $|x - x_j| \leqslant 3|x - t|$, and combining these two estimates (1.12) follows. Moreover, since we are dealing with small values in the above expressions, we may replace the sines there by their arguments and on account of (1.12) (and estimate (1.16) of Chapter I) we observe that

$$|k(x, t, x_j)| \leqslant c\frac{|t-x_j|}{|x-t|\,|x-x_j|} \leqslant c\frac{L_j}{|x-x_j|^2}. \tag{1.13}$$

This is all we need to complete the estimate of the A_j's. Indeed, by (1.10) and (1.13) it follows that

$$|A_j| \leqslant c\frac{L_j}{|x-x_j|^2}\int_{I_j}|b(t)|\,dt \leqslant c\frac{L_j}{|x-x_j|^2}\int_{I_j}|f(t)|\,dt,$$

with c independent of x and j. Combining these bounds with the ones obtained for the interval $[x-\varepsilon, x-\eta)$ we finally see that for almost every x in $T\backslash\Omega^*$,

$$\begin{aligned}|H_\varepsilon b(x) - H_\eta b(x)| &\\ &\le c \sum_{j, I_j \cap([x-\varepsilon, x-\eta)\cup(x+\eta, x+\varepsilon))\neq\phi} \frac{L_j}{|x-x_j|^2}\int_{I_j}|f(t)|\,dt + o(\varepsilon)+o(\eta)\\ &= L_{\varepsilon,\eta}(x),\end{aligned} \tag{1.14}$$

say. Thus

$$\limsup_{\varepsilon,\eta\to 0}|H_\varepsilon b(x) - H_\eta b(x)| \le \limsup_{\varepsilon,\eta\to 0} L_{\varepsilon,\eta}(x) = L(x), \tag{1.15}$$

say. In view of (1.15) estimate (1.4) will hold once we show that $L(x) < \infty$ a.e. For this purpose we intoduce a majorant of $L(x)$, the Marcinkiewicz function $\Delta(f, x)$ associated to $\Omega = \bigcup I_j$ and given by

$$\Delta(f, x) = \sum_j \frac{L_j}{|x-x_j|^2}\int_{I_j}|f(t)|\,dt, \qquad x \notin \Omega. \tag{1.16}$$

We claim that

$$\Delta(f, x) < \infty \qquad \text{a.e. in} \quad T\backslash\Omega^*. \tag{1.17}$$

The usual procedure for this is to note that the stronger statement

$$I = \int_{T\backslash\Omega^*}\Delta(f, x)\,dx < \infty \tag{1.18}$$

actually holds.

First, observe that

$$\begin{aligned}L_j\int_{T\backslash\Omega^*}\frac{1}{|x-x_j|^2}\,dx &\le L_j\int_{T\backslash 2I_j}\frac{1}{|x-x_j|^2}\,dx\\ &\le L_j\int_{[L_j/2,\infty)}\frac{ds}{s^2} = c,\end{aligned}$$

where c is independent of j. Whence by Tonelli's theorem we get that

$$I = \sum_j\left(\int_{I_j}|f(t)|\,dt\right)L_j\int_{T\backslash\Omega^*}\frac{1}{|x-x_j|^2}\,dx \le c\int_\Omega|f(t)|\,dt, \tag{1.19}$$

thus (1.18) obtains and the first step of our proof is complete.

By the way, Chebychev's inequality applied to (1.19) gives at once that

$$\lambda|\{t\in T\backslash\Omega^*: \Delta(f, t) > \lambda\}| \le c\int_\Omega|f(t)|\,dt. \tag{1.20}$$

To see that (1.3) also holds is now straightforward. Note that we actually have

$$L(x) \leq c \sum_{j=N+1}^{\infty} \frac{L_j}{|x - x_j|^2} \int_{I_j} |f(t)|\, dt, \qquad \text{each} \quad N \geq 0, \tag{1.21}$$

since by the fact that $x \in T \setminus \bigcup_{j=1}^{N} I_j$ for each $N > 0$, we may always miss the first N intervals in the family $\{I_j\}$ as soon as ε, $\eta > 0$ are small enough in (1.14). Furthermore, since the right-hand side of (1.21) is the tail of a convergent series for almost every x in $T \setminus \Omega^*$ we obtain that $L(x) = 0$ for those x's by letting $N \to \infty$ in (1.21); this means that (1.3) holds. ■

Remark 1.2. We isolate two facts from the proof above. First, by putting $\varepsilon = \pi$ in (1.14) we see that for almost every x in $T \setminus \Omega^*$

$$\begin{aligned} |H_\eta b(x)| &\leq c\Delta(f, x) + \frac{c}{\eta} \int_{(x+\eta, x+3\eta)} |b(t)|\, dt + \frac{c}{\eta} \int_{(x-3\eta, x-\eta)} |b(t)|\, dt \\ &\leq c(\Delta(f, x) + Mb(x)), \qquad \eta > 0 \end{aligned} \tag{1.22}$$

with c independent of η. Moreover, once again by (1.14) but taking now limits as $\eta \to 0$, we note that for almost every x in $T \setminus \Omega^*$

$$|Hb(x)| \leq c\, \Delta(f, x). \tag{1.23}$$

2. THE HILBERT TRANSFORM IN $L^p(T)$, $1 \leq p < \infty$

Although the Hilbert transform Hf is well defined for any $f \in L(T)$, it is not in general integrable. The following simple example illustrates this point: let $f \in L(T)$ be a positive function, vanishing off $[0, \pi/2)$. Then for $x \in [-\pi/2, 0)$,

$$Hf(x) = \lim_{\varepsilon \to 0} H_\varepsilon f(x) = -\frac{1}{\pi} \int_{[0,\pi/2)} \frac{f(t)}{2\tan((x-t)/2)}\, dt$$

Since $x < 0$ and $t > 0$ above, $\tan((x-t)/2) = -\tan((|x| + t)/2)$ and consequently

$$\begin{aligned} Hf(x) &= \frac{1}{\pi} \int_{[0,\pi/2)} \frac{f(t)}{2\tan((|x|+t)/2)}\, dt \\ &\geq \frac{1}{\pi} \int_{[0,|x|]} \frac{f(t)}{2\tan(|x|/2)}\, dt \\ &\geq \frac{c}{|x|} \int_{[0,|x|]} f(t)\, dt. \end{aligned} \tag{2.1}$$

But as we have already seen in Section 2 of Chapter IV, to make this integral large it suffices to put $f(t) = d(1/\ln(1/t))/dt \geq 0$; then $f \in L(T)$ but Hf is not integrable. Nevertheless, as in the case of the Hardy–Littlewood maximal function, we have the following substitute weak-type result:

Theorem 2.1. Let $f \in L(T)$. Then for each $0 < \varepsilon < \pi$ there is a constant c, independent of ε and f, such that for all $\lambda > 0$

$$\lambda|\{|H_\varepsilon f| > \lambda\}| \leq c\|f\|_1. \tag{2.2}$$

Proof. We may assume that $\lambda > \|f\|_1$; we then invoke the Calderón–Zygmund decomposition at level λ and write, in the notation of Theorem 1.1, $f = g + b$. Also for $x \in T\backslash\Omega^*$,

$$H_\varepsilon f(x) = H_\varepsilon g(x) + H_\varepsilon b(x). \tag{2.3}$$

Moreover, since

$$\{|H_\varepsilon f(x)| > \lambda\} \subseteq \Omega^* \cup \{x \in T\backslash\Omega^* : |H_\varepsilon f(x)| > \lambda\}, \tag{2.4}$$

it suffices to estimate the measure of each set on the right-hand side of (2.4). In first place

$$|\Omega^*| \leq 2\sum |I_j| \leq 4\pi\|f\|_1/\lambda \tag{2.5}$$

and this bound is of the right order. Also by (2.3)

$$\begin{aligned} &\{x \in T\backslash\Omega^*: |H_\varepsilon f(x)| > \lambda\} \\ &\quad \subseteq \{x \in T\backslash\Omega^*: |H_\varepsilon g(x)| > \lambda/2\} \cup \{x \in T\backslash\Omega^*: |H_\varepsilon b(x)| > \lambda/2\} \\ &\quad = I \cup J, \end{aligned}$$

say. To estimate I we make use of the L^2 result for $H_\varepsilon g$. From Chebychev's inequality, estimate (6.2) in Chapter III and estimate (3.7) in Chapter IV,

$$\left(\frac{\lambda}{2}\right)^2 |I| \leq \int_T |H_\varepsilon g(x)|^2\, dx \leq c\|g\|_2^2 \leq c\lambda\,\|f\|_1,$$

which in turn implies that

$$|I| \leq c\|f\|_1/\lambda, \tag{2.6}$$

and this term is also of the right order. As for the set J, by estimate (1.22) of Remark 1.2, we note that it is contained in

$$\begin{aligned} &\{x \in T\backslash\Omega^*: \Delta(f, x) > \lambda/4\} \cup \{x \in T\backslash\Omega^*: Mb(x) > \lambda/4\} \\ &\quad = J_1 \cup J_2, \end{aligned}$$

say. By estimate (1.20),

$$|J_1| \leq c\|f\|_1/\lambda. \tag{2.7}$$

Also by the maximal theorem and estimate (3.9) of Chapter IV we obtain

$$|J_2| \leq c\|b\|_1/\lambda \leq c\|f\|_1/\lambda. \tag{2.8}$$

Thus, by combining (2.5)–(2.8), and since all constants are independent of ε and f, the desired conclusion follows. ■

Observe that actually $H_\varepsilon f(x)$, being the convolution of integrable functions, is also integrable for each $\varepsilon > 0$; nevertheless, $\lim_{\varepsilon\to 0}\|H_\varepsilon f\|_1 = \infty$ in general.

A reasoning essentially analogous to that of Theorem 2.1 obtains

Theorem 2.2 (Kolmogorov). Let $f \in L(T)$. Then there is a constant c, independent of f, such that for all $\lambda > 0$

$$\lambda|\{|Hf| > \lambda\}| \leq c\|f\|_1. \tag{2.9}$$

Proof. We use estimate (1.23) of Remark 1.2 in this case; the proof is simpler than that of Theorem 2.1 since the term J_2 does not appear. ■

It is now an easy task to complete the consideration of the L^p result as well. We do the case $1 < p < 2$ first, which in view of the L^2 and weak-type (1, 1) estimates we expect to be rather straightforward.

Theorem 2.3. Suppose $f \in L^p(T)$, $1 < p < 2$. Then there is a constant $c_p = O(1/(p-1))$, independent of ε and f, such that

$$\|H_\varepsilon f\|_p \leq c_p\|f\|_p. \tag{2.10}$$

Theorem 2.4 (M. Riesz). Suppose $f \in L^p(T)$, $1 < p < 2$. Then there is a constant $c_p = O(1/(p-1))$, independent of f, such that

$$\|Hf\|_p \leq c_p\|f\|_p. \tag{2.11}$$

Moreover, $Hf(x) = \tilde{f}(x)$ a.e. and consequently $L^p(T)$ admits conjugation.

***Proof of Theorem* 2.4.** Let $1/p = (1-\eta) + \eta/2$, $0 < \eta < 1$; then by Theorem 4.1 in Chapter IV (with $p_0 = 1$, $p_1 = 2$ there) H is an L^p bounded mapping with norm $O(1/(p-1)^{(1-\eta)})$. Furthermore, since $(p-1)^\eta \leq 1$ and actually tends to 1 as $p \to 1$, $c_p = O(1/(p-1))$ as anticipated.

In order to show that $Hf(x)$ coincides with $\tilde{f}(x)$ a.e., we consider $H(f - \sigma_n(f))(x)$, which equals $Hf(x) - \tilde{\sigma}_n(f, x)$ a.e. since $\sigma_n(f) \in L^2(T)$. Now by (2.11)

$$\|Hf - \tilde{\sigma}_n(f)\|_p \leq c_p\|f - \sigma_n(f)\|_p, \tag{2.12}$$

with c_p independent of n. But the right-hand side of (2.12) goes to 0 as $n \to \infty$ by Fejér's Theorem 2.7 of Chapter II, and so does the left-hand side. Thus, on account of Proposition 1.1 of Chapter I,

$$c_j(Hf) = \lim_{n\to\infty} c_j(\tilde{\sigma}_n(f)), \quad \text{all } j.$$

Moreover, since $c_j(\tilde{\sigma}_n(f)) = (1 - |j|/(n+1))(-i)(\operatorname{sgn} j)c_j(f)$ when $|j| \leq n$ and 0 otherwise, we see at once that the above limit is $(-i)(\operatorname{sgn} j)c_j(f)$; this implies that $\tilde{f} = Hf \in L^p(T)$. ■

The case $2 < p \leq \infty$ remains to be discussed. When $p = \infty$ we may, and will, give more than one answer; this is because $f \in L^\infty(T)$ does not, in general, imply that $Hf \in L^\infty(T)$, as the simple example in Remark 2.7 below shows. Before discussing this negative result, we consider the positive results for $2 < p < \infty$. They are readily obtained by a "duality" argument.

Theorem 2.5. Suppose $f \in L^p(T)$, $2 < p < \infty$. Then there is a constant $c_p = O(p)$, independent of ε and f, such that

$$\|H_\varepsilon f\|_p \leq c_p \|f\|_p. \tag{2.13}$$

Theorem 2.6. Suppose $f \in L^p(T)$, $2 < p < \infty$. Then there is a constant $c_p = O(p)$, independent of f, such that

$$\|Hf\|_p \leq c_p \|f\|_p. \tag{2.14}$$

Moreover, $Hf(x) = \tilde{f}(x)$ a.e. and consequently $L^p(T)$ admits conjugation.

***Proof of Theorem* 2.6.** Since $L^p \subset L^2$, we already have $Hf = \tilde{f} \in L^2(T)$ by the results in Section 6 of Chapter III; it only remains to prove that $Hf \in L^p(T)$ as well. By Theorem 3.3 of Chapter II it suffices to show that

$$\|\sigma_n(\tilde{f})\|_p \leq c_p \|f\|_p \tag{2.15}$$

with $c_p = O(p)$ independent of n and f; this is not hard to do. Indeed, let g be a trigonometric polynomial with $\|g\|_{p'} \leq 1$, $1/p + 1/p' = 1$. Then by Parseval's identity (iv) of Theorem 1.8 in Chapter III it follows that

$$I = \frac{1}{2\pi}\int_T \sigma_n(\tilde{f}, t)\overline{g(t)}\, dt = \sum_{|j| \leq n}\left(1 - \frac{|j|}{n+1}\right)(-i)(\operatorname{sgn} j)c_j(f)\overline{c_j(g)}$$
$$= -\sum_{|j| \leq n} c_j(f)\overline{\left(1 - \frac{|j|}{n+1}\right)(-i)(\operatorname{sgn} j)c_j(g)} = \frac{1}{2\pi}\int_T f(t)\overline{\sigma_n(\tilde{g}, t)}\, dt.$$

Whence by Hölder's inequality, estimate (3.4) of Chapter II, and Theorem 2.4 we obtain

$$|I| \leq \|\sigma_n(\tilde{g})\|_{p'}\|f\|_p \leq c_{p'}\|f\|_p.$$

(2.15) follows now by the converse to Hölder's inequality but with $c_{p'}$ instead of c_p there. This is only a minor inconvenience because since $p = p'/(p'-1)$, p' near 1 and $c_{p'} = O(1/(p'-1))$, then also $c_{p'} = c_p = O(p)$. ■

Remark 2.7. The order of the constant c_p in Theorem 2.7 is sharp. Indeed, let $f(t) = \chi_{[0,1]}(t)$. Then clearly $\|f\|_p = 1$, all $p \geqslant 1$, $Hf(x) = c(\ln|\sin(x/2)| - \ln|\sin((1-x)/2)|)$, and

$$\|Hf\|_p \geqslant c\left(\int_{[0,1]} |\ln x|^p \, dx\right)^{1/p}$$

$$= c\left(\int_{[0,\infty]} s^p e^{-s} \, ds\right)^{1/p} \geqslant c\Gamma(p+1)^{1/p}. \tag{2.16}$$

Here $\Gamma(p+1)$ denotes the gamma function of order p and by known estimates from (2.16) we see at once that $\liminf_{p\to\infty} \|Hf\|_p / p \geqslant c > 0$. Note in passing that, although Hf is not bounded, $Hf \in \bigcap_{p<\infty} L^p$, Hf is exponentially integrable, and $|Hf(x)|$ has logarithmic growth. These properties remain true for the Hilbert transform of every L^∞ function, as we will prove shortly (however, the last property will have to wait until Chapter VIII).

Since $L^p(T)$ admits conjugation for $1 < p < \infty$, $\|S_n(f) - f\|_p \to 0$ as $n \to \infty$. We also have in this case

Proposition 2.8. Assume that $f \in L^p(T)$, $1 < p < \infty$. Then $\|\tilde{s}_n(f) - \tilde{f}\|_p \to 0$ as $n \to \infty$.

Proof. Since $\tilde{s}_n(f) - \tilde{f} = (s_n(f) - f)^{\sim}$, $\|\tilde{s}_n(f) - \tilde{f}\|_p \leqslant c_p \|s_n(f) - f\|_p \to 0$ as $n \to \infty$. ■

To complete the discussion of the case $p = 1$ we show

Proposition 2.9. Assume that $f \in L(T)$ and $0 < p < 1$. Then $\|Hf\|_p \leqslant c_p \|f\|_1$, $c_p = O(1/1-p)$ independent of f. A similar estimate holds for H_ε with c_p also independent of ε.

Proof. Since $Hf \in wk\text{-}L(T)$ we are in a position to use 7.5 of Chapter IV; we put $\alpha = p$ and $\beta = 1 - p$ there and note that the constant is of the right order. ■

One of the first results we discovered was that $\|H_\varepsilon f - Hf\|_2 \to 0$ as $\varepsilon \to 0$. It is reasonable to expect that a similar result will hold for L^p norms, $1 < p < \infty$, as well as substitute results for $p = 1$; we prove this next.

Proposition 2.10. Assume that $f \in L^p(T)$, $1 < p < \infty$. Then

$$\lim_{\varepsilon\to 0} \|H_\varepsilon f - Hf\|_p = 0. \tag{2.17}$$

Proof. For an arbitrary $f \in L^p(T)$ let g be a trigonometric polynomial so that $\|f - g\|_p \leqslant \eta$ is arbitrarily small. Then by Minkowski's inequality we have

$$\begin{aligned}\|H_\varepsilon f - Hf\|_p &\leqslant \|H_\varepsilon f - H_\varepsilon g\|_p + \|H_\varepsilon g - Hg\|_p + \|Hg - Hf\|_p \\ &\leqslant c_p\|f-g\|_p + \|H_\varepsilon g - Hg\|_p + c_p\|f-g\|_p \\ &\leqslant 2c\eta + \|H_\varepsilon g - Hg\|_p.\end{aligned}$$

Consequently, it suffices to show that $\lim_{\varepsilon\to 0}\|H_\varepsilon g - Hg\|_p = 0$. But since g is a trigonometric polynomial, the proof for estimate (6.4) of Proposition 6.2 in Chapter III also works in this case. ■

Proposition 2.11. Assume that $f \in L(T)$. Then

$$\lim_{\varepsilon\to 0} H_\varepsilon f = Hf, \qquad \text{in measure,} \tag{2.18}$$

and

$$\lim_{\varepsilon\to 0}\|H_\varepsilon f - Hf\|_p = 0, \qquad 0 < p < 1.$$

Proof. The convergence in the L^p metric follows as above and (2.18) is an immediate consequence of this. ■

The truncated Hilbert transforms, as well as the Hilbert transform itself, are uniformly controlled by the "maximal Hilbert transform H^*" given by

$$H^*f(x) = \sup_{0<\varepsilon<\pi} |H_\varepsilon f(x)|. \tag{2.19}$$

The behavior of H^* is similar to that of H itself. The next two results make this precise.

Theorem 2.12. Assume that $f \in L^p(T)$, $1 < p < \infty$. Then there is a constant c independent of f such that

$$H^*f(x) \leqslant c(Mf(x) + M(Hf)(x)). \tag{2.20}$$

Thus $\|H^*f\|_p \leqslant c_p\|f\|_p$ with $c_p = O(1/(p-1)^2)$ as $p \to 1^+$ and $c_p = O(p)$ as $p \to \infty$.

Proof. $H_\varepsilon f$ can be readily estimated when $\varepsilon \geqslant 1$ since

$$\begin{aligned}|H_\varepsilon f(x)| &\leqslant \frac{1}{\pi}\int_{1<|x-t|<\pi} \frac{|f(t)|}{|2\tan((x-t)/2)|}\,dt \\ &\leqslant c\int_T |f(t)|\,dt \leqslant cMf(x).\end{aligned}$$

Now if $0 < \varepsilon < 1$ and n is the integer such that $1/(n+1) \leqslant \varepsilon < 1/n$, write

$$\begin{aligned}H_\varepsilon f(x) &= (H_\varepsilon f(x) - H_{1/n} f(x)) + (H_{1/n} f(x) - \tilde{\sigma}_n(f, x)) \\ &\quad + \tilde{\sigma}_n(f, x) = I + J + K,\end{aligned}$$

say. Thus $|H_\varepsilon f(x)| \leq |I| + |J| + |K|$, and it suffices to estimate each term separately; we do $|I|$ first. By taking sup over n in estimate (6.8) in Corollary 6.4 of Chapter III, it follows that $|I| \leq cMf(x)$. The same is true for $|J|$ if we consider estimates (6.6) and (6.7) in Theorem 6.3 of Chapter III instead; the former gives $cMf(x)$ directly whereas the latter must be combined with Proposition 2.3 of Chapter IV first (with $\phi(t) = t^{-2}\chi_{[1,\infty)}(t)$ there). To bound $|K|$ we invoke Proposition 2.3 of Chapter IV. This gives $|K| \leq cM(Hf)(x)$ and (2.20) holds; the norm estimate with the growth order of the c_p's obtains from (2.20) at once. ■

We also have the weak-type estimate when f is merely integrable

Theorem 2.13. Assume $f \in L(T)$, then $H^*f \in wk - L(T)$ and there is a constant c, independent of f, such that

$$\lambda|\{H^*f > \lambda\}| \leq c\|f\|_1, \qquad \text{all} \quad \lambda > 0. \tag{2.21}$$

Proof. As in the proof of the weak type for Hf itself we invoke the Calderón–Zygmund decomposition at level λ sufficiently large, and write $f = g + b$, $\Omega^* = \bigcup 2I_j$. Then

$$H^*f(x) \leq H^*g(x) + H^*b(x) \tag{2.22}$$

and it suffices to show that

$$\lambda|\{H^*g > \lambda/2\}| \leq c\|f\|_1 \tag{2.23}$$

and

$$\lambda|\{x \in T\backslash\Omega^*: H^*b(x) > \lambda/2\}| \leq c\|f\|_1. \tag{2.24}$$

This is straightforward. Indeed, by Theorem 2.13 $\|H^*g\|_2 \leq c\|g\|_2 \leq c\lambda\|f\|_1$ and (2.23) follows from Chebychev's inequality. Moreover, by estimate (1.22) in Remark 1.2 we note that $H^*b(x) \leq c(\Delta f(x) + Mb(x))$, $x \in T\backslash\Omega^*$ and (2.24) follows at once on account of estimate (1.2), the fact that the maximal operator is of weak-type (1, 1), and $\|b\|_1 \leq c\|f\|_1$. ■

Remark 2.14. We may now combine Theorem 2.13 with the Marcinkiewicz interpolation theorem to find that, in Theorem 2.12, $c_p = O(1/(p-1))$; thus the analogy with the Hilbert transform itself is complete.

3. LIMITING RESULTS

Assume $f \in L(T)$, $f \sim \sum c_j e^{ijx}$. Still the question remains as to when

$$\sum(-i)(\operatorname{sgn} j)c_j e^{ijx} \tag{3.1}$$

is a Fourier series. Suppose $Hf \in L(T)$. Then since $Hf = f * \text{p.v.}(1/(\tan(t/2)))$ (in D') by 7.22 in Chapter III we get that $c_j(Hf) = (-i)(\operatorname{sgn} j)c_j$ and (3.1) is a Fourier series in this case. To decide when Hf is integrable we proceed as we did for the maximal operator and search for an extrapolation result first.

Proposition 3.1. A sublinear mapping T defined in $L + L^p$, $1 < p < \infty$, is simultaneously of weak-types $(1, 1)$ and (p, p) if and only if there are constants c_1, c_2, c_3 such that for all $f \in L + L^p$ and $\lambda > 0$

$$|\{|Tf| > \lambda\}| \leq c_1\left(\frac{1}{\lambda^p}\int_{[0,c_2\lambda)} s^{p-1}|\{|f| > s\}|\, ds + \frac{1}{\lambda}\int_{[c_3\lambda,\infty)} |\{|f| > s\}|\, ds\right). \tag{3.2}$$

Proof. The necessity follows from the decomposition (4.14) of Chapter IV with $\varepsilon = 1$ there. As for the sufficiency, if (3.2) holds and $f \in L^p$ say, then in the first integral we just replace $c_2\lambda$ by ∞ in the limit of integration, and this term is of the right order. It is also readily seen that the second integral does not exceed

$$\frac{1}{\lambda}\frac{1}{(c_3\lambda)^{p-1}}\int_{[0,\infty)} s^{p-1}|\{|f| > s\}|\, ds,$$

which is also of the right order. The weak-type $(1, 1)$ case is handled similarly. Finally note that (3.2) is equivalent to the single condition obtained by setting $\lambda = 1$ there. ∎

We can now extrapolate.

Proposition 3.2. Assume T is a sublinear operation defined in $L + L^p$, $1 < p < \infty$, which is simultaneously of weak-types $(1, 1)$ and (p, p). Then T maps $L \ln L$ into L and there are constants A, B independent of f such that

$$\|Tf\|_1 \leq A + B\int_T |f(t)|\ln^+|f(t)|\, dt. \tag{3.3}$$

Proof. It suffices to show that $\int_{\{|Tf|>1\}}|Tf(t)|\, dt$ is bounded by the right-hand side of (3.3). But this is immediate since by (3.2) it suffices to show that the same is true for

$$\int_{[1,\infty)}\frac{1}{\lambda^p}\int_{[0,c_2\lambda)} s^{p-1}|\{|f| > s\}|\, ds\, d\lambda \tag{3.4}$$

and

$$\int_{[1,\infty)} \frac{1}{\lambda} \int_{[c_3\lambda,\infty)} |\{|f| > s\}| \, ds \, d\lambda. \tag{3.5}$$

That the expression in (3.5) is of the right order was already proved in Theorem 5.3 of Chapter IV; the expression in (3.4) is quite easy to handle. Indeed, suppose as we may, that $c_2 \leqslant 1$ and observe that (3.4) is dominated by

$$\begin{aligned} &\int_{[1,\infty)} \frac{1}{\lambda^p} 2\pi \int_{[0,c_2)} s^{p-1} \, ds \, d\lambda \\ &\quad + \int_{[1,\infty)} \frac{1}{\lambda^p} \int_{[c_2,c_2\lambda)} s^{p-1} |\{|f| > s\}| \, ds \, d\lambda \\ &\leqslant A + \int_{[c_2,\infty)} s^{p-1} |\{|f| > s\}| \int_{[s/c_2,\infty)} \frac{1}{\lambda^p} \, d\lambda \, ds \\ &\leqslant A + B \int_{[c_2,\infty)} |\{|f| > s\}| \, ds. \quad \blacksquare \end{aligned}$$

Proposition 3.2 clearly applies to H_ε, H, and H^*. There also is some evidence, furnished by estimate (2.1), that, as in the case of the maximal operator, some kind of converse result may be true. Indeed, for positive, integrable functions f vanishing off $[0, \pi/2)$ with $Hf \in L(T)$, we have that

$$\begin{aligned} \infty > \int_{[-\pi/2,0)} |Hf(x)| \, dx &\geqslant c \int_{(0,\pi/2]} \frac{1}{x} \int_{[0,x]} f(t) \, dt \, dx \\ &= c \int_{[0,\pi/2]} f(t) \int_{[t,\pi/2]} \frac{1}{x} \, dx \, dt \\ &= c \int_{[0,\pi/2]} f(t) \ln(\pi/2t) \, dt, \end{aligned}$$

which is readily seen to imply the integrability of $f(t) \ln^+ f(t)$ in $[0, \pi/2)$. The positivity of f is important here. We shall return to this question in Chapter VII and remove the restriction on the support of f then.

There is yet another extrapolation result and it corresponds to the spaces near L^∞, since, unlike the maximal operator, the Hilbert transform is not bounded in $L^\infty(T)$. We may search for this result by a duality argument as follows: for trigonometric polynomials f and g observe that

$$\begin{aligned} \left| \int_T \tilde{f}(t) g(t) \, dt \right| &= \left| \int_T f(t) \tilde{g}(t) \, dt \right| \\ &\leqslant \left(A + B \int_T |g(t)| \ln^+ |g(t)| \, dt \right) \|f\|_\infty. \end{aligned}$$

On account of Young's inequality (see 5.2 of Chapter I), we expect now some kind of exponential integrability for $\tilde{f}$ when $f \in L^\infty$. The precise statement for general operators T that behave like the Hilbert transform is

Theorem 3.3. Assume T is a sublinear operator bounded in L^p, $1 < p_0 \leqslant p < \infty$, with norm $O(p)$ as $p \to \infty$. Then there exist constants c and A independent of f, such that

$$\int_T e^{A|Tf(x)|}\, dx \leqslant c, \qquad \|f\|_\infty \leqslant 1. \tag{3.6}$$

Proof. We estimate the integral in (3.6) by

$$\sum_{k=0}^{\infty} \frac{1}{k!} A^k \int_T |Tf(x)|^k\, dx$$
$$\leqslant c + c \sum_{k \geqslant p_0} \frac{1}{k!} A^k (k+1)^k \|f\|_\infty^k,$$

which is readily seen to be dominated by an absolute constant provided A is sufficiently small, since $(k/e)^k \leqslant k!$. ■

This general theorem indeed applies to H_ε, H, and H^*, but is there a more precise result for the particular case of the Hilbert transform? In other words, can we be more explicit about the value of A in the conclusion of Theorem 3.3? In order to do this and to illustrate the simplicity and power of the so-called complex method, we prove the following result

Proposition 3.4 (Zygmund). Assume f is a real-valued function, $|f(x)| \leqslant \pi/2$. Then

$$\frac{1}{2\pi} \int_T e^{|\tilde{f}(x)|} \cos(f(x))\, dx \leqslant 2 \cos\left(\frac{1}{2\pi} \int_T f(t)\, dt \right) \leqslant 2. \tag{3.7}$$

Proof. Suppose first that f is a trigonometric polynomial, $f(x) = \sum_{|j| \leqslant N} c_j e^{ijx}$ and put $F(x) = i(f(x) - i\tilde{f}(x))$. $F(x)$ has the following properties

(i) It equals $i(c_0 + 2\sum_{j=1}^N c_j e^{ijx})$.

(ii) If $F(z) = i(c_0 + 2\sum_{j=1}^N c_j z^j)$, then $F(z)$ is a harmonic function in $|z| \leqslant 1$ (actually it is analytic there).

(iii) $F(0) = ic_0 = i((1/2\pi)\int_T f(t)\, dt)$, $|c_0| \leqslant \pi/2$.

(iv) $e^{F(z)}$ is also harmonic and analytic in $|z| \leqslant 1$.

By the mean value property of harmonic functions, to be proved in Chapter VII, we have

$$\frac{1}{2\pi} \int_{|z| \leqslant 1} \frac{e^{F(z)}}{z}\, dz = \frac{1}{2\pi} \int_T e^{F(e^{ix})}\, dx = e^{F(0)}. \tag{3.8}$$

Whence replacing $F(e^{ix})$ by its explicit form and taking real parts in (3.8) we see that

$$\frac{1}{2\pi}\int_T e^{\tilde{f}(x)}\cos(f(x))\,dx = \mathrm{Re}\left((1/2\pi)\int_T e^{\tilde{f}(x)}e^{if(x)}\,dx\right)$$
$$= \mathrm{Re}(e^{F(0)}) = \cos(c_0). \qquad (3.9)$$

Also, replacing f by $-f$ in (3.9), we see that

$$\frac{1}{2\pi}\int_T e^{-\tilde{f}(x)}\cos(f(x))\,dx = \cos(c_0) \qquad (3.10)$$

and adding (3.9) and (3.10) we get

$$\frac{1}{2\pi}\int_T e^{|\tilde{f}(x)|}\cos(f(x))\,dx \leq 2\cos(c_0). \qquad (3.11)$$

Next we would like to show that (3.11) still holds for an arbitrary, real-valued function f with $|f(x)| \leq \pi/2$. To see this, consider the Fejér polynomials $\sigma_n(f, x) = f * 2K_n(x)$; $\sigma_n(f, x)$ is real valued, $\|\sigma_n(f)\|_\infty \leq (\pi/2)\|2K_n\|_1 = \pi/2$, and $c_0(\sigma_n(f)) = c_0$, all n. Furthermore, since $\|\sigma_n(f) - f\|_2 \to 0$, then also $\|\tilde{\sigma}_n(f) - \tilde{f}\|_2 \to 0$ and there is a subsequence $n_k \to \infty$ such that $\sigma_{n_k}(f, x) \to f(x)$ and $\tilde{\sigma}_{n_k}(f, x) \to \tilde{f}(x)$ a.e. in T. Since (3.11) holds with f replaced by $\sigma_{n_k}(f)$ there and the right-hand side is independent of n_k, by Fatou's lemma we see that (3.11) holds for f as well. ■

Corollary 3.5 (Zygmund). Suppose f is real valued and $\|f\|_\infty < \pi/2$. Then $\int_T e^{|\tilde{f}(x)|}\,dx < \infty$.

Proof. Let $\|f\|_\infty = L < \pi/2$. By (3.7)

$$\int_T e^{|\tilde{f}(x)|}\,dx \leq \frac{4\pi}{\cos L} < \infty. \quad ■$$

That this result is sharp follows at once by looking at $f(x) = \frac{1}{2}\pi(2\chi_{(-\eta,\eta)}(x) - 1)$.

Corollary 3.6. Let $\|f\|_\infty < \pi/2$, then

$$|\{|\tilde{f}| > \lambda\}| \leq \frac{8\pi}{\cos(\|f\|_\infty)} e^{-\lambda/\sqrt{2}}. \qquad (3.12)$$

Proof. Write $f = f_1 + if_2, f_1, f_2$ real valued; then $\tilde{f} = \tilde{f}_1 + i\tilde{f}_2$ and $\{|\tilde{f}| < \lambda\} \subseteq \{|\tilde{f}_1| > \lambda/\sqrt{2}\} \cup \{|\tilde{f}_2| > \lambda/\sqrt{2}\}$. By (3.7),

$$|\{|\tilde{f}_j| > \lambda/\sqrt{2}\}| \leq \frac{4\pi}{\cos(\|f\|_\infty)} e^{-\lambda/\sqrt{2}}, \qquad j = 1, 2,$$

4. MULTIPLIERS

The fact that H is a bounded mapping in L^2 constitutes an important step in establishing the continuity of the Hilbert transform in L^p, $1 < p < \infty$, as well. The proof of the L^2 result depends on the following 2 facts: for $f \sim \sum c_j e^{ijx}$, $Hf \sim \sum (-i)(\operatorname{sgn} j) c_j e^{ijx}$ and $\{(-i) \operatorname{sgn} j\} \in l^\infty(Z) = l^\infty$, the (Banach) space of bounded sequences. In fact there is nothing special about $\{(-i) \operatorname{sgn} j\}$ as any l^∞ sequence will do. More precisely, suppose that $\{\lambda_j\} \in l^\infty$ and let T be the (linear) mapping defined by

$$f \sim \sum c_j e^{ijx} \to Tf \sim \sum \lambda_j c_j e^{ijx}. \tag{4.1}$$

These T's are called "multiplier operators," since they are obtained by pointwise, or coordinatewise, multiplication of the Fourier coefficients of f by the "multiplier sequence" or, plainly, "multiplier" $\{\lambda_j\}$. To show that T is bounded in L^2 is quite easy, since by the Riesz–Fischer theorem $Tf \in L^2$ and

$$\| Tf \|_2^2 = \sum |\lambda_j|^2 |c_j|^2 \leqslant \left(\sup_j |\lambda_j| \right)^2 \| f \|_2^2. \tag{4.2}$$

Furthermore, if $\lim_{n \to \infty} |\lambda_{j_n}| = \sup_j |\lambda_j|$, by putting $f(x) = e^{ij_n x}$ in (4.2) we readily see that the norm $\| T \|$ of T as a mapping in L^2 is

$$\| T \| = \sup_j |\lambda_j|. \tag{4.3}$$

It is also natural to consider the possibility of obtaining L^p continuity results for multipliers for values of p other than 2. It is convenient to introduce the notation M_p for the collection of bounded L^p multipliers, $1 \leqslant p \leqslant \infty$, i.e., a sequence $\{\lambda_j\} \in M_p$ if and only if for trigonometric polynomials f,

$$\left\| \sum \lambda_j c_j(f) e^{ijx} \right\|_p \leqslant c \| f \|_p \tag{4.4}$$

with c independent of f; the infimum over the c's in (4.4) is called the "norm" of the multiplier.

Short of characterizing M_p, which still remains an open question for $1 < p < \infty$, $p \neq 2$, we are often satisfied to decide whether a given individual sequence $\{\lambda_j\}$ belongs to M_p or not. To illustrate this we give this typical, although not sharp, result.

Proposition 4.1 (Hirschman). Let $\{\lambda_j\}$ be a bounded sequence so that $|\lambda_j| = O(|j|^{-\alpha})$, $0 < \alpha < 1$. Then $\{\lambda_j\} \in M_p$, $(1-\alpha)/2 < 1/p < (1+\alpha)/2$.

Proof. Our assumption implies that

$$\sup_{2^k \leqslant |j| < 2^{k+1}} |\lambda_j| = O(2^{-\alpha k}), \qquad k \geqslant 0. \tag{4.5}$$

For trigonometric polynomials $f \sim \sum c_j e^{ijx}$ and $k \geqslant 0$ put

$$T_k f(x) = \sum_{2^k \leqslant |j| < 2^{k+1}} \lambda_j c_j e^{ijx}. \tag{4.6}$$

Clearly, T_k is of type (p, p) for all $p \geqslant 1$, but we need a good estimate on its norm; this can be easily done for $p = 1, 2$. In first place, observe that

$$\| T_k f\|_1 \leqslant \sum_{2^k \leqslant |j| < 2^{k+1}} |\lambda_j|\,|c_j| \leqslant c 2^k 2^{-\alpha k} \|f\|_1 \tag{4.7}$$

and

$$\| T_k f\|_2^2 = \sum_{2^k \leqslant |j| < 2^{k+1}} |\lambda_j|^2 |c_j|^2 \leqslant c 2^{-2\alpha k} \|f\|_2^2. \tag{4.8}$$

(4.7) and (4.8) give that the norm of T_k in L is of order $2^{(1-\alpha)k}$ and that in L^2 is of order $2^{-\alpha k}$. We are now in a position to invoke the Marcinkiewicz interpolation theorem 7.14 of Chapter IV: Let $0 < \eta < 1$ and put $1/p = (1-\eta) + \eta/2$. Then the norm $\|T_k\|$ of T_k as a mapping in L^p is of order

$$\|T_k\| \leqslant c 2^{(1-\alpha)(1-\eta)k} 2^{-\alpha\eta k} = c 2^{(1-\alpha-\eta)k} \qquad \text{and} \qquad \sum_{k=0}^{\infty} \|T_k\| < \infty$$

provided $\eta > 1 - \alpha$. Thus restricting our consideration to those η's, and consequently looking at those p's so that $1/2 \leqslant 1/p < (1+\alpha)/2$, we obtain that $Tf = \lambda_0 c_0 + \sum_{k=0}^{\infty} T_k f$ is also bounded in L^p. A similar argument applies with (4.7) replaced by

$$\| T_k f\|_\infty \leqslant c 2^{(1-\alpha)k} \|f\|_\infty \tag{4.9}$$

and by (4.8) and (4.9) we may now interpolate between L^2 and L^∞ to obtain those p's so that $(1-\alpha)/2 < 1/p \leqslant 1/2$. ■

Notice that in this case, as well as in the Hilbert transform, the set of $1/p$'s for which $\{\lambda_j\} \in M_p$ is a symmetric interval about $\frac{1}{2}$; we shall return to this observation shortly.

There is yet another way to arrive at the multiplier problem, and it is motivated by the integral, or convolution representation of H. First a definition: for a function f and $a \in T$ let $\tau_a f$ denote the translation of f given by $\tau_a f(x) = f(x + a)$, $x \in T$. More generally, if F is a distribution we define the distribution $\tau_a F$ by means of

$$\tau_a F(u) = F(\tau_{-a} u), \qquad u \in C^\infty(T). \tag{4.10}$$

We say that a linear operator T is "translation invariant" if it commutes with translations; more precisely,

$$\tau_a(Tf) = T(\tau_a f), \qquad \text{all} \quad a \in T \tag{4.11}$$

for all trigonometric polynomials f, say. We are interested in obtaining an intrinsic characterization of these operators and in this direction we have

Proposition 4.2. Assume T is a linear operator which verifies (4.11) and which, in addition, admits a continuous extension to some L^p, $1 \leqslant p \leqslant \infty$. Then there exists a sequence $\{\lambda_j\}$ so that for all trigonometric polynomials $f \sim \sum c_j e^{ijx}$,

$$Tf(x) = \sum \lambda_j c_j e^{ijx}. \tag{4.12}$$

Moreover,

$$\sup_j |\lambda_j| \leqslant \|T\|, \qquad \text{the norm of} \quad T \quad \text{in} \quad L^p. \tag{4.13}$$

Clearly, if A is the distribution $\sum \lambda_j e^{ijx}$ and f is a trigonometric polynomial, then $Tf = A * f$.

Proof. For each j let $f_j = T(e^{ijx})$, then $\{f_j\}$ is a sequence of L^p functions each with norm $\leqslant \|T\|$. By the translation invariance and linearity of T we readily see that

$$\tau_a f_j = \tau_a(T(e^{ijx})) = T(\tau_a e^{ijx}) = T(e^{ij(x+a)}) = e^{ija} f_j. \tag{4.14}$$

We compare now the Fourier coefficients of the functions in (4.14). Since by (3.1) of Chapter I we have that $c_k(\tau_a f) = e^{ika} c_k(f)$, (4.14) obtains

$$e^{ika} c_k(f_j) = e^{ija} c_k(f_j), \qquad \text{all} \quad j, k. \tag{4.15}$$

Clearly, the solution to (4.15) is $c_k(f_j) = 0, j \neq k$, and the functions $f_j = \lambda_j e^{ijx}$, all j. Thus $|\lambda_j| = \|T(e^{ijx})\|_p \leqslant \|T\|$, all j. By the linearity of T we see that for trigonometric polynomials f, $T(\sum c_j e^{ijx}) = \sum c_j T(e^{ijx}) = \sum \lambda_j c_j e^{ijx}$, and also $Tf = A * f$. ∎

Corollary 4.3. Assume T is as in Proposition 4.2, then for trigonometric polynomials f, g, $Tf * g = T(f * g) = f * Tg$.

Thus translation operators T correspond to multipliers, and vice versa; this suggests our next result.

Proposition 4.4. Suppose $\{\lambda_j\} \in l^\infty$. Then $\{\lambda_j\} \in M_p$ if and only if for all trigonometric polynomials f, g and with $1/p + 1/p' = 1$,

$$\left| \sum \lambda_j c_j(f) c_j(g) \right| \leqslant c \|f\|_p \|g\|_{p'} \tag{4.16}$$

with c independent of f and g.

Proof. Just use the convolution representation $\sum \lambda_j c_j(f) e^{ijx} = A * f$. ■

By (4.16) we see at once that for $p \geq 1$

$$\{\lambda_j\} \in M_p \quad \text{if and only if} \quad \{\lambda_j\} \in M_{p'}, \qquad 1/p + 1/p' = 1, \tag{4.17}$$

and from this is clear that

$$M_p = M_{p'}, \qquad 1/p + 1/p' = 1 \tag{4.18}$$

Moreover, we also have

Proposition 4.5. Assume $\{\lambda_j\} \in M_p$. Then $\{\lambda_j\} \in M_r$ for $\min(p, p') \leq r \leq \max(p, p')$; thus

$$M_p \subseteq M_q, \qquad 1 \leq p \leq q \leq 2. \tag{4.19}$$

Proof. By (4.17) we may assume that $1 \leq p \leq 2$ and by the Marcinkiewicz interpolation theorem 7.14 of Chapter IV we see at once that $\{\lambda_j\} \in M_r$, $p \leq r \leq p'$, and we are done. Consequently, the interval of $1/p$'s for which $\{\lambda_j\} \in M_p$ is symmetric about $\frac{1}{2}$. ■

Proposition 4.6. $M_2 = l^\infty$.

Proof. By (4.2) it suffices to show that L^2 multipliers are bounded; this follows at once from (4.13). Also combining (4.2) and (4.13) we see that if T denotes the multiplier operator associated with $\{\lambda_j\}$, then $\|T\| = \sup_j |\lambda_j|$. ■

Proposition 4.7. $M_1 = \{\{\lambda_j\}$: there is a finite measure μ so that $\lambda_j = c_j(\mu)\} = \mathcal{F}M$.

Proof. Since convolution with a finite measure μ is continuous in L, it suffices to show that if $\{\lambda_j\} \in M_1$, then $\lambda_j = c_j(\mu)$ for some finite measure μ. But this is easy since $\|\sigma_n(A)\|_1 = \|A * 2K_n\|_1 \leq c\|2K_n\|_1 = c$, all n, and by Theorem 3.5 of Chapter II the λ_j's are the Fourier coefficients of a finite measure μ so that $(1/2\pi)\int_T |d\mu| \leq c$. ■

Combining (4.19) with Propositions 4.6 and 4.7 we obtain the important relation

$$\mathcal{F}M \subseteq M_p \subseteq M_q \subseteq l^\infty, \qquad 1 \leq p \leq q \leq 2 \tag{4.20}$$

and the inclusions are proper for $1 < p < q < 2$ (this last statement may require some thought; 5.40 and 5.41 below are relevant here).

M_p carries the structure of a Banach algebra, but we will not pursue this here. We will, however, prove that it is a dual (Banach) space.

Proposition 4.8. Assume $\{\lambda_j\} \in M_p$, $1 \leq p < \infty$, $A \sim \sum \lambda_j e^{ijx}$, $\|A * f\|_p \leq c\|f\|_p$. Then, there is a sequence $\{\phi_n\} \subseteq C(T)$ so that

$$\|\phi_n * f\|_p \leq c\|f\|_p, \qquad \text{all} \quad n, f$$

and

$$\lim_{n\to\infty} \|\phi_n * f - A * f\|_p = 0, \qquad f \in L^p(T). \tag{4.21}$$

Proof. Let $\phi_n = \sigma_n(A)$. Then by Corollary 4.3 $\|\sigma_n(A) * f\|_p = \|\sigma_n(A * f)\|_p \leq c\|f\|_p$, all n, f. Furthermore, by Theorem 2.7 of Chapter II, $\|\sigma_n(A) * f - A * f\|_p = \|\sigma_n(A * f) - A * f\|_p \to 0$ as $n \to \infty$. ∎

We still need one more definition. If $f \in L^p$, $g \in L^{p'}$, then $f * g \in C(T)$ and $\|f * g\|_\infty \leq \|f\|_p \|g\|_{p'}$. The same is true for finite sums of convolutions of this kind and also for infinite sums provided that, in addition,

$$\sum_k \|f_k\|_p \|g_k\|_{p'} < \infty, \tag{4.22}$$

for then it is readily seen that $\sum_{k=1}^\infty f_k * g_k(x)$ converges absolutely and uniformly to a continuous function. For $1 \leq p \leq 2$ let $A_p(T) = A_p = \{h \in C(T)\colon h = \sum_{k=1}^\infty f_k * g_k, \text{and (4.22) holds}\}$. A_p is a Banach space normed by the infimum of the expression in (4.22) for all decompositions $\sum f_k * g_k$ of h. The space A_p^* of bounded linear functionals L on A_p is also a Banach space with norm

$$\|L\| = \sup_{h\neq 0} \frac{|L(h)|}{\|h\|_{A_p}}$$

and we have

Theorem 4.9 (Figa-Talamanca). M_p is isometric and isomorphic to A_p^*, $1 < p \leq 2$.

Proof. In order to simplify the notations we identify the multiplier $\{\lambda_j\}$ in M_p with the operator $Tf = A * f$, $A \sim \sum \lambda_j e^{ijx}$; we write in this case $T \in M_p$ and we also have $\|T\| =$ infimum over the c's in (4.16).

We construct a mapping $\Phi\colon M_p \to A_p^*$ with the following properties

(i) Φ is well defined,
(ii) Φ is an isometry, and
(iii) Φ is onto.

This is how to do it: for $T \in M_p$ and $h = \sum f_k * g_k$ in A_p put

$$\Phi(T)h = \sum_{k=1}^\infty Tf_k * g_k(0). \tag{4.23}$$

Notice that since

$$Tf_k * g_k(0) = \frac{1}{2\pi}\int_T Tf_k(x)g_k(-x)\,dx \tag{4.24}$$

the series in (4.23) converges absolutely and

$$|\Phi T(h)| \leqslant \|T\| \sum \|f_k\|_p \|g_k\|_{p'}. \tag{4.25}$$

We want to show that the expression in (4.23) is independent of the representation of h; in other words, if $h = \sum f_k * g_k = 0$, then the sum in (4.23) is also 0. Let $\phi_n = \sigma_n(A)$, then by Proposition 4.8, $\sum_k \|\phi_n * f_k\|_p \|g_k\|_{p'} < \infty$, and $\sum_k (\phi_n * f_k) * g_k(x)$ converges absolutely and uniformly as $n \to \infty$. Thus

$$\begin{aligned}\lim_{n\to\infty} \sum_k (\phi_n * f_k) * g_k(x) &= \sum_k \lim_{n\to\infty} (\phi_n * f_k) * g_k(x)\\ &= \sum_k Tf_k * g_k(x),\end{aligned} \tag{4.26}$$

where the last equality is justified by (4.21).

Moreover, since ϕ_n is a trigonometric polynomial for each fixed n, we may invoke (4.16) and the boundedness of the convolution with ϕ_n and note that

$$\begin{aligned}\sum_k (\phi_n * f_k) * g_k(x) &= \sum_k \phi_n * (f_k * g_k)(x)\\ &= \phi_n * \left(\sum_k f_k * g_k\right)(x) = \phi_n * h(x) = 0,\end{aligned}$$

which combined with (4.26) gives that $\sum_k Tf_k * g_k(0) = 0$, and $\Phi(T)h = 0$, thus proving (i).

From the proof above it readily follows that $|\Phi(T)h| \leqslant \|T\|\,\|h\|_{A_p}$ and consequently $\|\Phi(T)\| \leqslant \|T\|$. To prove the opposite inequality observe that by (4.16) given $\varepsilon > 0$ there are functions $f \in L^p$ and $g \in L^{p'}$ of norm 1 so that $\|T\| \leqslant |Tf * g(0)| + \varepsilon$. Then $h = f * g \in A_p$, $\|h\|_{A_p} \leqslant 1$, and $\|T\| \leqslant |\Phi(T)h| + \varepsilon \leqslant \|\Phi(T)\|\,\|h\|_{A_p} + \varepsilon \leqslant \|\Phi(T)\| + \varepsilon$ and (ii) also holds. To show (iii) let $h^* \in A_p^*, f \in L^p$ and for trigonometric polynomials g and $\tilde{g}(x) = g(-x)$ consider the map $g \to h^*(f * \tilde{g})$. It is clear that $|h^*(f * g)| \leqslant \|h^*\|\,\|f * \tilde{g}\|_{A_p} \leqslant \|h^*\|\,\|f\|_p \|g\|_{p'}$ so that each $f \in L^p$ determines a bounded linear functional on $L^{p'}$ of norm $\leqslant \|h^*\|\,\|f\|_p$. By Proposition 3.2 in Chapter II there is a unique L^p function Tf, say, of norm $\leqslant \|h^*\|\,\|f\|_p$ so that

$$h * (f * g) = \frac{1}{2\pi}\int_T Tf(x)g(-x)\,dx = Tf * g(0) = \Phi(T)(f * g).$$

Since Tf is clearly linear, by Proposition 4.2 it only remains to show that T is translation invariant; this is easy since for each trigonometric polynomial g, $T(\tau_a f) * g(0) = h^*(\tau_a f * \tilde{g}) = h^*(f * \tau_a \tilde{g}) = Tf * \tau_a g(0) = \tau_a(Tf) * g(0)$, and we have finished. ■

The case $p = 1$, not covered by the above result, can be done in several ways. The quickest is to observe that by 7.17 of Chapter III, $C(T) = L(T) * C(T)$, and consequently $A_1 = C(T)$; the proof that $A_1^* = \mathscr{F}M$ follows without much difficulty from this.

5. NOTES; FURTHER RESULTS AND PROBLEMS

Lusin considered the question of the a.e. existence of the p.v. integral defining the Hilbert transform (Theorem 1.1), as well as the integral in (5.16) of Chapter III, and concluded that the cancellation due to the positive and negative values in these expressions is an important fact which plays a fundamental role in the study of the convergence of Fourier series. He was not satisfied with his own proof of the a.e. convergence of Hf for $f \in L^2(T)$, since he felt that the use of the Riesz-Fischer theorem made it a complex, rather than a real, variables proof. It was Besicovitch who in 1926 first established the existence of Hf by purely real methods, and $f \in L(T)$. The interplay between $\tilde{f}$ and the a.e. convergence of Fourier series is given by the following result, also due to Lusin: In order that the Fourier series of an $L^2(T)$ function $f \sim \frac{1}{2}a_0 + \sum a_j \cos jx + b_j \sin jx$ converge a.e. in T it is necessary and sufficient that

$$\lim_{n\to\infty} \lim_{\varepsilon\to 0^+} \int_{[\varepsilon,\pi]} \frac{\tilde{f}(x+t) - \tilde{f}(x-t)}{t} \cos nt \, dt = 0. \tag{5.1}$$

As all the ingredients needed to prove this fact were discussed in this chapter, the reader is invited to furnish a proof. Inspired by (5.1) and the a.e. existence of the integral

$$\text{p.v.} \int_{[0,\pi]} \frac{g(t+x) - g(t-x)}{t} \, dt$$

for $g \in L^2(T)$, Lusin conjectured that the Fourier series of any $f \in L^2(T)$ converges a.e. He based this hypothesis on the uniform distribution of the positive and negative values of $\cos nx$ and $\sin nx$ in T (which suffices to show that the Fourier coefficients of any summable function tend to 0). As for the multiplier question, it seems to have originated with some results of Fekete in 1923. The closest to a characterization of M_p is the following result of Stein: a function $K(x)$ belongs to $V_q, 2 \leq q < \infty$ (V is for Verblunsky) if and only if $K \in L^q(T)$ and $\sup\|\sum_k K(b_k - \cdot) - K(a_k - \cdot)\|_q < \infty$. Here the sum is taken over any finite collection of nonoverlapping intervals, and the sup is taken over all such collections of intervals. Corresponding to a bounded sequence $\{\lambda_n\}$, put $K(x) = \sum_{n\neq 0}(\lambda_n/in)e^{inx}$, and let $2 < q < \infty$. Then if the multiplier corresponding to $\{\lambda_n\}$ belongs to M_p for all $q' \leq p \leq q$, $K(x) \in V_q$. Conversely, if $K \in V_q$, then the multiplier belongs to M_p, $q' < p < q$.

Further Results and Problems

5.1 Suppose that $f \in L(T)$ and that F is a primitive of f. Then

$$\tilde{F}(x) = \text{p.v.} -\frac{1}{\pi}\int_T f(x+t)|\ln|\sin(t/2)||\,dt, \qquad x \text{ a.e. in } T.$$

(*Hint*: Carry out an integration by parts in the integral

$$-\frac{1}{\pi}\int_{[\varepsilon,\pi]} \frac{F(x+t) - F(x-t)}{2\tan(t/2)}\,dt;$$

this result is Lusin's.)

5.2 For $f \in L(T)$, $|\{|\tilde{f}| > \lambda\}| = o(1/\lambda)$ as $\lambda \to \infty$. (*Hint*: Write $f = f_1 + f_2$, where $\|f_1\|_1 < \varepsilon$ and $f_2 \in L^2$; now $|\{|\tilde{f}_2| > \lambda/2\}| \leq c/(\lambda/2)^2 = o(1/\lambda)$ and for f_1 use the weak-type (1, 1) estimate.)

5.3 Assume that $f \in L(T)$ and $0 < p < 1$. Then $\int_T |\tilde{s}_n(f, x)|^p\,dx \leq c_p\|f\|_1^p$. What is the order of c_p as $p \to 1^-$? (*Hint*: Take a look at identity (3.5) in Chapter III.)

5.4 Assume that $f \in L(T)$ and $0 < p < 1$. Then $\|\tilde{s}_n(f) - \tilde{f}\|_p \to 0$ as $n \to \infty$.

5.5 Assume that $f, \tilde{f} \in L(T)$ and that $\|s_n(f)\|_1 \leq A$, all n. Then there is a constant B, independent of n, such that $\|\tilde{s}_n(f)\|_1 \leq B$. (*Hint*: Observe that $\tilde{s}_n(f, x) - \tilde{\sigma}_n(f, x) = (-s_n'(f, x))/(n+1)$, $\tilde{\sigma}_n(f, x) = \sigma_n(\tilde{f}, x)$ and by Bernstein's inequality in Chapter II, $\|s_n'(f)\|_1 \leq cn$.)

5.6 Assume that $f, \tilde{f} \in L(T)$ and that in addition $\|s_n(f) - f\|_1 \to 0$ as $n \to \infty$. Show that $\|\tilde{s}_n(f) - \tilde{f}\|_1 \to 0$ as $n \to \infty$. (*Hint*: First choose N so large that $\|s_n(f) - s_N(f)\|_1 < \varepsilon$ for $n > N$, by Bernstein's inequality

$$\|s_n'(f) - s_N'(f)\|_1 \leq c\varepsilon n;$$

next from the estimate $|\tilde{s}_n(f, x) - \tilde{\sigma}_n(f, x)| \leq |s_n'(f, x) - s_N'(f, x)|/(n+1) + |s_N'(f, x)|/(n+1)$ obtain that $\lim_{n\to\infty}\|\tilde{s}_n(f) - \tilde{\sigma}_n(f)\|_1 = 0$. Finally, recall that since $\tilde{f} \in L$, $\tilde{\sigma}_n(f, x) = \sigma_n(\tilde{f}, x)$ a.e. and $\|\sigma_n(\tilde{f}) - \tilde{f}\|_1 \to 0$ as $n \to \infty$.)

5.7 Assume that $f \in L(T)$, $0 < p < 1$. Then $\int_T |\tilde{s}_n(f, x)|^p \, dx \leq c_p \|f\|_1^p$. What is the order of c_p as $p \to 1^-$?

5.8 Assume that $f \in L(T)$ and $0 < p < 1$. Show that

$$\lim_{n\to\infty} \|\tilde{s}_n(f) - Hf\|_p = 0.$$

(*Hint*: For trigonometric polynomials g and large enough n

$$|\tilde{s}_n(f, x) - Hf(x)|^p \leq |\tilde{s}_n(f - g, x)|^p + |Hg(x) - Hf(x)|^p).$$

5.9 Suppose $f \in L(T)$. Then there is a sequence $n_k \to \infty$ so that simultaneously $s_{n_k}(f, x) \to f(x)$ a.e. and $\tilde{s}_{n_k}(f, x) \to Hf(x)$ a.e. (*Hint*: Use 5.4 and 5.8.)

5.10 (Parseval's relation) Assume $f \in L^p(T)$, $g \in L^{p'}(T)$, $1 < p, p' < \infty$, $1/p + 1/p' = 1$. Then

$$\frac{1}{2\pi} \int_T f(x) g(x) \, dx = \sum c_n(f) c_{-n}(g).$$

(*Hint*: Since $\|s_n(f) - f\|_p \to 0$ as $n \to \infty$,

$$\lim_{n\to\infty} \frac{1}{2\pi} \int_T s_n(f, x) g(x) \, dx = \frac{1}{2\pi} \int_T f(x) g(x) \, dx.$$

Compare with 5.18 in Chapter II.)

5.11 Does the conclusion in 5.10 above hold if $f \in L(T)$ and $g \in L^\infty(T)$? Does it hold if $f \in L(T)$ and g is of bounded variation?

5.12 The trigonometric series $\sum c_n e^{inx}$ is the Fourier series of an $L^p(T)$ function, $1 < p < \infty$, if and only if for every $g \in L^{p'}(T)$, $\sum c_n c_{-n}(g)$ converges. (*Hint*: We must only show that the condition is sufficient. If $\sigma_n(x)$ denote the $(C, 1)$ means of $\sum c_n e^{inx}$ and $\tau_n(g) = \tau_n$ the $(C, 1)$ means of $\sum c_n c_{-n}(g)$, then $\tau_n = (1/2\pi) \int_T \sigma_n(x) g(x) \, dx$. But τ_n is a linear functional in $L^{p'}$ of norm $\|\sigma_n\|_p$; moreover, since the series converges, it is all the more $(C, 1)$ summable and the values τ_n are bounded for each $g \in L^{p'}$, i.e., $|\tau_n| \leq c_g < \infty$. Therefore by the Banach–Steinhaus theorem 2.3 in Chapter I, the norms of the functionals are bounded, i.e., $\|\sigma_n\|_p \leq c < \infty$.)

5.13 Suppose we know that for some p, $1 < p < 2$, $f \to \tilde{f}$ is of type (p, p) when restricted to characteristic functions of measurable sets (operators with this property are said to be of restricted type (p, p)). Then show that $f \to \tilde{f}$ is of weak-type (p', p'), $1/p + 1/p' = 1$, and consequently also of type (r, r), $p < r < p'$. What conclusion follows from a restricted weak-type (p,p) assumption? (*Hint*: For a trigonometric polynomial f let $A_\lambda = \{\tilde{f} > \lambda\}$, $B_\lambda = \{\tilde{f} < -\lambda\}$. Then

$$\lambda |A_\lambda| < \int_{A_\lambda} \tilde{f}(x) \, dx = -\int_T f(x) \tilde{\chi}_{A_\lambda}(x) \, dx$$

$$\leq c \|f\|_{p'} \|\chi_{A_\lambda}\|_p = c \|f\|_{p'} |A_\lambda|^{1/p};$$

repeat the argument for B_λ.)

5.14 There is a bounded function $f(x) \sim \sum_{n=1}^{\infty} (\sin nx)/n = (\pi - x)/2$ whose conjugate $\tilde{f}(x) \sim \sum_{n=1}^{\infty} (\cos nx)/n = -\ln|\sin(x/2)|$ is unbounded.
5.15 There is a continuous function $f(x) = \sum_{n=2}^{\infty}(\sin nx)/(n \ln n)$ whose conjugate $\tilde{f}(x) \sim \sum_{n=2}^{\infty}(\cos nx)/(n \ln n)$ has a Fourier series which converges everywhere except at 0. (*Hint*: To show the continuity invoke 7.24 in Chapter III: actually, the partial sums of $\tilde{f}$ diverge to ∞ at 0, which means that the series also diverges $(C, 1)$ to ∞ there.)
5.16 Assume $f, \tilde{f} \in L^{\infty}(T)$ and that in addition $\|s_n(f)\|_{\infty} \leqslant A$, all n. Then there is a constant B such that also $\|\tilde{s}_n(f)\|_{\infty} \leqslant B$.
5.17 Assume now that $f, \tilde{f} \in C(T)$ and $\lim_{n\to\infty}\|s_n(f) - f\|_{\infty} = 0$; then $\lim_{n\to\infty}\|\tilde{s}_n(f) - \tilde{f}\|_{\infty} = 0$.
5.18 Assume $f \in C(T)$, $\|f\|_{\infty} \leqslant 1$. Then there are constants c_1, c_2 independent of n so that $|\{|s_n(f)| > \lambda\}| \leqslant c_1 e^{-c_2\lambda}$.
5.19 Suppose $f \in L \ln L$ and show that $\|s_n\|_1$,

$$\|\tilde{s}_n\|_1 \leqslant A + B \int_T |f(x)| \ln^+|f(x)|\, dx,$$

and

$$\lim_{n\to\infty} \|s_n(f) - f\|_1 = \lim_{n\to\infty} \|\tilde{s}_n(f) - \tilde{f}\|_1 = 0.$$

5.20 Parseval's relation holds for $f \in L \ln L(T)$ and $g \in L^{\infty}(T)$.
5.21 Assume $f \in C(T)$. Then $\int_T e^{\lambda|\tilde{f}(x)|}\, dx < \infty$ for each $\lambda > 0$. Show that $\|\tilde{f}\|_p = o(p)$ as $p \to \infty$. (*Hint*: Write $f = f - g + g$, where g is a trigonometric polynomial with $\|f - g\|_{\infty}$ sufficiently small.)
5.22 There is an absolute constant $\lambda_0 > 0$ such that if $\|f\|_{\infty} \leqslant 1$, then for $0 < \lambda \leqslant \lambda_0$, $\int_T e^{\lambda|s_n(f,x)|}\, dx$, $\int_T e^{\lambda|\tilde{s}_n(f,x)|}\, dx \leqslant c$, where c depends only on λ. If $f \in C(T)$, the estimate holds for each $\lambda > 0$. (*Hint*: Consider, as usual, the modified partial sums $\tilde{s}_n^*(f, x)$ of $\tilde{f}$ given by

$$\frac{1}{\pi} \int_T f(x + t) \frac{\sin nt}{2 \tan(t/2)}\, dt,$$

and observe that rewriting $\sin nt = \sin n(t + x) \cos nx - \cos n(t + x) \sin nx$ it follows that $\tilde{s}_n^*(f, x) = \tilde{g}_n(x) \sin nx - \tilde{h}_n(x) \cos nx$, where $g_n(x)$ and $h_n(x)$ are $f(x) \cos nx$ and $f(x) \sin nx$, respectively, $|g_n(x)|, |h_n(x)| \leqslant 1$. Therefore,

$$\int_T e^{\lambda|\tilde{s}_n^*(f,x)|}\, dx \leqslant \left(\int_T e^{2\lambda|\tilde{g}_n(x)|}\, dx\right)^{1/2} \left(\int_T e^{2\lambda|\tilde{h}_n(x)|}\, dx\right)^{1/2}.)$$

5.23 Assume f is as in 5.22. Then there is an absolute constant $\lambda_0 > 0$, so that if $\|f\|_\infty \leq 1$, then

$$\lim_{n\to\infty} \frac{1}{2\pi}\int_T e^{\lambda|s_n(f,x)-f(x)|}\,dx = \lim_{n\to\infty}\frac{1}{2\pi}\int_T e^{\lambda|\tilde{s}_n(f,x)-\tilde{f}(x)|}\,dx = 1.$$

If, in addition $f \in C(T)$, the conclusion obtains for each $\lambda > 0$ (*Hint*: Since $0 \leq e^u - 1 \leq ue^u$, $u > 0$, we see that

$$\begin{aligned}1 &\leq \frac{1}{2\pi}\int_T e^{\lambda|s_n(f,x)-f(x)|}\,dx \\ &\leq 1 + \frac{\lambda}{2\pi}\int_T |s_n(f,x)-f(x)|e^{\lambda|s_n(f,x)-f(x)|}\,dx \\ &\leq 1 + \lambda\|s_n(f)-f\|_p\left(\frac{1}{2\pi}\int_T e^{\lambda p'|s_n(f,x)-f(x)|}\,dx\right)^{1/p'},\end{aligned}$$

$1/p + 1/p' = 1$. If λ is sufficiently small, and $p < \infty$ sufficiently large so that $\lambda p'$ is still small, the integral above is bounded and the term $\|s_n(f) - f\|_p = o(1)$.)

5.24 If $\int_T e^{|f(x)|^\alpha}\,dx < \infty$, then $\int_T e^{\lambda|\tilde{f}(x)|^\beta}\,dx < \infty$ provided $\beta = \alpha/(\alpha+1)$ and λ is sufficiently small (*Hint*: A way to do this is by dualizing the result $\int_T |f(x)|(\ln^+|f(x)|)^\alpha\,dx < \infty$ implies $\int_T |\tilde{f}(x)|(\ln(2+|\tilde{f}(x)|))^{\alpha-1}\,dx < \infty$, $\lambda > 0$. Of course, there is always a direct proof.)

5.25 Let $A(u)$ be a continuous, nondecreasing function, $A(0) = 0$, such that $\limsup_{u\to\infty} A(2u)/A(u) = \infty$. Then there exists a function f so that $\int_T A(|f(x)|)\,dx < \infty$, yet $\int_T A(|\tilde{f}(x)|)\,dx = \infty$. (*Hint*: Construct by induction an increasing sequence of positive numbers u_k so that $A(2u_k) > 2^{2k}A(u_k) > 2$, $k \geq 1$, and $u_k \geq 2u_{k-1}$, $k \geq 2$. Next define positive integers $1 \leq n_1 < n_2 < \cdots$ so that $2^{-k-1} < 2^{-n_k}A(u_k) \leq 2^{-k}$, $k \geq 1$, and define the function

$$f(x) = \begin{cases} u_k, & x \in [-2^{-n_k}, -2^{-n_{k+1}}) \quad k = 1, 2, \ldots, \\ 0, & x \in T\backslash[2^{-n_1}, 0). \end{cases}$$

It is readily seen that $\int_T A(f(x))\,dx < \infty$. f may also be written

$$\textstyle\sum_{k=1}^{\infty} (u_k - u_{k-1})\chi_{[-2^{-n_k},0)}(x),$$

and consequently (since for $0 < \eta < \pi/2$,

$$\tilde{\chi}_{[-\eta,0)}(x) = \frac{2}{\pi}\ln\left(\frac{\sin((x+\eta)/2)}{\sin(x/2)}\right) \geq 0 \qquad \text{for} \quad 0 \leq x \leq \pi/2,$$

$c_0\ln(\eta/x)$ for $0 < x \leq \eta$) taking $c_1 = e^{-4/c_0}$ we note that for $x \in (0, c_1 2^{-n_k})$, $\tilde{f}(x) \geq (u_k/2)\tilde{\chi}_{[2^{-n_k},0)}(x) \geq (u_k/2)c_0\ln(2^{-n_k}/c_1 2^{-n_k}) = 2u_k, \qquad k = 1, 2, \ldots.$ Thus

$$\int_{[0,c_1 2^{-n_k})} A(\tilde{f}(x))\,dx \geq c_1 2^{-n_k}A(2u_k) \geq c_1 2^{2k}2^{-n_k}A(u_k) \geq c_1 2^{k-1},$$

which gives $\int_T A(|\tilde{f}(x)|)\,dx = +\infty$. A similar, but more involved, argument shows that there is a function h,

$$\int_T A(|h(x)|)\,dx < \infty \qquad \text{yet} \qquad \lim_{n\to\infty} \int_T A(|s_n(h,x) - h(x)|)\,dx = \infty;$$

these constructions are from Oswald [1982].)

5.26 Corresponding to a collection $\{I_j\}$ of disjoint subintervals of T we associate the function

$$\delta(x) = \sum_j \int_{I_j} \frac{|I_j|}{|x-t|^2 + |I_j|^2}\,dt;$$

this is closely related to the Marcinkiewicz function introduced in (1.16). Show that there is a constant c, independent of $\lambda > 0$, such that $|\{\delta > \lambda\}| \leqslant c(\sum |I_j|)e^{-\lambda}$. (*Hint*: Let $E = E_\lambda = \{\delta > \lambda\}$; if $|E| = 0$ there is nothing to prove. Otherwise set $\phi(x) = \chi_E(x)/(|E| \ln(4\pi/|E|)) \geqslant 0$, $\|\phi\|_1 = 1/\ln(4\pi/|E|)$, $\|\phi\|_\infty = 1/(|E| \ln(4\pi/|E|))$. Thus

$$\begin{aligned}
\frac{\lambda}{\ln(4\pi/|E|)} &\leqslant \int_T \delta(x)\phi(x)\,dx \\
&= \sum_j \int_{I_j} \left(\int_T \frac{|I_j|}{|x-t|^2 + |I_j|^2}\phi(x)\,dx \right) dt \\
&\leqslant c \sum_j \int_{I_j} M\phi(t)\,dt = \sum \int_{\bigcup I_j} M\phi(t)\,dt = A,
\end{aligned}$$

say (by Proposition 2.3 in Chapter IV). One way to estimate A is to observe that it does not exceed $c + c\int_T \phi(x) \ln^+ \phi(x)\,dx$ which obtains the estimate in the conclusion without the factor $\sum |I_j|$. A sharper estimate for A is

$$\begin{aligned}
&c \int_{[0,\|\phi\|_\infty)} \min\left(\sum |I_j|, \frac{\|\phi\|_1}{s} \right) ds \\
&\qquad \leqslant c\left(1/\ln\left(\frac{4\pi}{|E|}\right) + \ln \frac{((\sum |I_j|)/|E|)}{\ln(4\pi/|E|)} \right),
\end{aligned}$$

which gives $\lambda \leqslant c(1 + \ln((\sum |I_j|)/|E|))$, and we are done. The first estimate above is, for instance, in Hunt's work [1972]. The full result is in Muckenhoupt [1983], where background on the problem is given. Notice that a similar conclusion holds for

$$\delta(x) = \sum_j \int_{I_j} \frac{|I_j|^{p-1}}{|x-t|^p + |I_j|^p}\,dt, \qquad p > 1, \quad \text{as well.)}$$

5.27 Assume $f \in L^\infty(T)$, $\|f\|_\infty \leqslant \eta$, and put $E = \{|\tilde{f}| > \eta\}$; then $|E| \leqslant c(\|f\|_1/\|f\|_\infty)e^{-\eta/c\|f\|_\infty}$, where c is an absolute constant independent of f and η. (*Hint*: Let $F = \{|\tilde{f}| > \|f\|_\infty\}$. Then since for appropriate constants $c_1, c_2 > 0$, $e^u \leqslant c_1(e^u - 1 - u)$ for $u \geqslant c_2$, we see that

$$\int_F e^{c_1|\tilde{f}(x)|/\|f\|_\infty}\,dx \leqslant c_2 \sum_{k=2}^{\infty} \frac{1}{k!}\left(\frac{c_1}{\|f\|_\infty}\right)^k \int_T |\tilde{f}(x)|^k\,dx \leqslant c\|f\|_1/\|f\|_\infty.$$

Now if $\eta > \|f\|_\infty$, $E \subset F$, and we are done; this observation is Muckenhoupt's [1983].)

5.28 Assume $f \in L(T)$ and for $\eta > 0$, $\lambda > 0$, let $E = \{x \in T : |\tilde{f}(x)| > \lambda\eta,\ Mf(x) \leqslant \eta\}$. Then $|E| \leqslant c\|f\|_1 e^{-\lambda/c}/\eta$, where c is independent of f, λ and η. (*Hint*: We may assume that $\eta > \|f\|_1$, for otherwise E is empty, and that $\lambda \geqslant 2$, for otherwise the result follows from the weak-type estimate for $\tilde{f}$. We may also assume that $Mf(\pi) \leqslant \eta$ and write the open set $\{Mf > 4\eta\} = \bigcup I_j$, where the I_j's are disjoint open intervals, and put

$$g(x) = \sum_j \left(\frac{1}{|I_j|}\int_{I_j} f(t)\,dt\right)\chi_{I_j}(x) + f(x)\chi_{T\setminus\bigcup I_j}(x)$$

and

$$b = f - g.$$

Note that $Mf(x) > \eta$ for $x \in \bigcup 3I_j$, and, consequently, $E \subseteq E_1 \cup E_2$, where $E_1 = \{|\tilde{g}| > \lambda\eta/2\}$ and $E_2 = \{x \in T\setminus\bigcup 3I_j : |\tilde{b}(x)| > \lambda\eta/2\}$. To show that $|E_1| \leqslant c\|f\|_1 e^{-\lambda/c}/\eta$ we use 5.27, and to estimate $|E_2|$ we use 5.26; this proof is also Muckenhoupt's and extends a result of Hunt [1972].)

5.29 The most general multiplier is a sequence $\{\lambda_j\}$ with the property that for each $f \in C^\infty(T)$, $f \sim \sum c_j e^{ijx}$, $\sum \lambda_j c_j e^{ijx} \in D'$, we denote this class (C^∞, D'). Show that the characterization of (C^∞, D') is also as general as possible, namely, $\{\lambda_j\} \in (C^\infty, D')$ if and only if the sequence $\{\lambda_j\}$ is tempered.

5.30 Another natural question is how to characterize (C, M), i.e., to identify those sequences $\{\lambda_j\}$ which transform continuous functions into finite measures. This was done by Gaudry [1966] as follows: let $B = \{h \in C(T) : h = \sum_{j=1}^\infty f_j * g_j,\ \sum \|f_j\|_\infty \|g_j\|_\infty < \infty\}$. Then B is a Banach space and elements of its dual B^* are called quasimeasures. Then $\{\lambda_j\} \in (C, M)$ if and only if $\sum \lambda_j e^{ijx}$ is a quasimeasure. (*Hint*: The proof follows along the lines of Theorem 4.9.)

5.31 Let $\mu = \sum_{j=1}^N c_j\delta(x_j)$, where $\delta(x_j)$ denotes the Dirac measure with unit mass at x_j, and the x_j's are distinct points of T. Show that the norm of the convolution mapping $f * \mu$, $f \in L^p$, $\geqslant (\sum_{j=1}^N |c_j|^p)^{1/p}$. (*Hint*: For any $\|f\|_p = 1$ we have that the norm in question is at least $\|\sum_{j=1}^N c_j \tau_{-x_j} f\|_p$. Now choose $f = c\chi_{[0,\delta]}$, where c is a normalizing constant and $0 < \delta <$ half of the minimum distance between the x_j's. Then the functions $\tau_{-x_j} f$ have disjoint supports and the norm of the sum is the sum of the norms; this observation is Brown's [1977].)

5.32 Suppose that $1 \leq p < \infty$ and that $\{\lambda_j^k\}$ is a sequence of M_p multipliers which verify

(i) $\lim_{k \to \infty} \lambda_j^k = \lambda_j$ exists for all j, and
(ii) $|\sum_j \lambda_j^k c_j(f) c_j(g)| \leq c_k \|f\|_p \|g\|_{p'}$,

for all trigonometric polynomials g, $1/p + 1/p' = 1$, and $\lim_{k\to\infty} c_k \leq L$. Show that under these conditions $\{\lambda_j\} \in M_p$ and its multiplier norm $\leq L$.

5.33 Suppose that the sequence $\{\lambda_j\}$ is such that $(\sum_{2^k \leq |j| \leq 2^{k+1}} |\lambda_j|^2)^{1/2} \leq c2^{-\alpha k}$, c independent of k and $0 < \alpha < \frac{1}{2}$. Then $\{\lambda_j\} \in M_p$ for $2/(1+2\alpha) < p < 2/(1-2\alpha)$. This result is Hirschman's.

5.34 Suppose $F(z)$ is an analytic function in $|z| < r$ and that $\{\lambda_j\} \in M_p$, $1 \leq p \leq \infty$, with multiplier norm $A < r$. Show that $\{F(\lambda_j)\} \in M_p$ also. (*Hint*: By 4.13 $|\lambda_j| \leq A < r$; for a trigonometric polynomial $f(x) = \sum c_j e^{ijx}$, we consider

$$\sum F(\lambda_j) c_j e^{ijx} = \sum c_j \left(\sum_{n=0}^{\infty} \frac{1}{n!} \lambda_j^n F^{(n)}(0) \right) e^{ijx}$$
$$= \sum_n \frac{1}{n!} F^{(n)}(0) \sum_j \lambda_j^n c_j e^{ijx}$$

and

$$\left\| \sum F(\lambda_j) c_j e^{ijx} \right\|_p \leq \left(\sum_n \frac{1}{n!} |F^{(n)}(0)| A^n \right) \|f\|_p.)$$

5.35 Assume that $\{\lambda_j\} \in M_p$, $1 \leq p \leq \infty$, and consider the sequence $\{\lambda_{j+N}\}$, where N is a fixed integer. Show that also $\{\lambda_{j+N}\} \in M_p$, with multiplier norm equal to that of $\{\lambda_j\}$, independently of N. (*Hint*: If T denotes the multiplier corresponding to $\{\lambda_j\}$, and $f(x) = \sum c_j e^{ijx}$ is a trigonometric polynomial, then $\sum \lambda_{j+N} c_j e^{ijx} = e^{-iNx} T(e^{iNt} f)(x)$.)

5.36 Let $\{\lambda_j\}$ be the sequence of 1's for $N_1 \leq j \leq N_2$, and 0's otherwise, where N_1, N_2 are arbitrary integers. Show that $\{\lambda_j\} \in M_p$, $1 < p < \infty$, with multiplier norm independent of N_1 and N_2, and that $\{\lambda_j\} \notin M_1$. (*Hint*: Use 5.35 and appropriate results on projections.)

5.37 We say that a function ϕ defined on R is in the class V_1 if $\sum_{j=-\infty}^{\infty} |\phi(j) - \phi(j+1)| = A < \infty$. Show that if $\phi \in V_1$, then $\{\phi(j)\} \in M_p$, $1 < p < \infty$, with norm $\leq |\phi(0)| + A$. (*Hint*: For trigonometric polynomials f, g put $\psi(j) = \sum_{n=-N}^{j} c_n(f) c_n(g)$, where $N = \max(\text{degree } f, \text{degree } g)$ and $\psi(-N-1) = 0$. Then

$$I = \sum_{-N}^{N} \phi(j) c_j(f) c_j(g) = \sum_{-N}^{N} \phi(j)(\psi(j) - \psi(j-1))$$

$$I = \sum_{-N}^{N-1} (\phi(j) - \phi(j-1))\psi(j) + \phi(N)\psi(N)$$

and consequently

$$|I| \leqslant \sum_{-N}^{N-1} |\phi(j) - \phi(j-1)|\,|\psi(j)| + |\phi(N) - \phi(0)|\,|\psi(N)|.$$

The conclusion follows readily from this, since by 5.36 $|\psi(j)| \leqslant c\|f\|_p\|g\|_{p'}$, c independent of j. This result is due to Steckin.)

5.38 (Rudin–Shapiro polynomials) We define the trigonometric polynomials P_m, Q_m inductively as follows: $P_0 = Q_0 = 1$ and

$$P_{m+1}(x) = P_m(x) + e^{i2^m x}Q_m(x),$$

$$Q_{m+1}(x) = P_m(x) - e^{i2^m x}Q_m(x).$$

Thus, $P_1(x) = 1 + e^{ix}$, $P_2(x) = 1 + e^{ix} + e^{i2x} - e^{i3x}$, and so on. These polynomials have the following properties:

(i) $|P_{m+1}(x)|^2 + |Q_{m+1}(x)|^2 = 2(|P_m(x)|^2 + |Q_m(x)|^2)$, and, consequently, $|P_m(x)|^2 + |Q_m(x)|^2 = 2^{m+1}$; $\|P_m\|_\infty \leqslant 2^{(m+1)/2}$;

(ii) for $|n| < 2^m$, $c_n(P_{m+1}) = c_n(P_m)$; hence there is a sequence $\{\varepsilon_n\}_{n=0}^{\infty}$ consisting of ± 1 such that $P_m(x) = \sum_{n=0}^{2^m-1} \varepsilon_n e^{inx}$.

5.39 For $1 \leqslant p \leqslant 2$ the norm $\|P_m\|_{M_p}$ of the Rudin–Shapiro polynomial P_m as a (convolution) mapping in $L^p(T)$ is of order $2^{m(1/p-1/2)}$; more precisely, there are constants c_1, c_2 independent of m so that

$$c_1 2^{m(1/p-1/2)} \leqslant \|P_m\|_{M_p} \leqslant c_2 2^{m(1/p-1/2)}$$

(*Hint*: $\|P_m\|_{M_1} \leqslant \|P_m\|_\infty \leqslant 2^{(m+1)/2}$, $\|P_m\|_{M_2} \leqslant 1$; by the Marcinkiewicz interpolation theorem 7.14 in Chapter IV, for $1/p = 1 - \eta + \eta/2, 0 < \eta < 1$, $\|P_m\|_{M_p} \leqslant 2^{m(1-\eta)/2}$. The converse inequality follows by combining the estimates $\|P_m\|_{p'} \geqslant \|P_m\|_2 \geqslant c2^{m/2}$,

$$\|P_m\|_{p'} = \|P_m * 2D_{2^m-1}\|_{p'} \leqslant \|P_m\|_{M_{p'}}\|2D_{2^m-1}\|_{p'} \leqslant c\|P_m\|_{M_p}2^{m/p'}.)$$

5.40 For $1 < p < q < 2$, $M_p \subsetneqq M_q$. (*Hint*: The following proof uses some functional analysis, an argument which does not is given in 5.41 below. Suppose to the contrary that $M_q = M_p$. Then the injection i: $M_q \to M_p$ has a closed graph; indeed, if a sequence $\{T_n\}$ of multiplier operators converges to 0 in M_q and to T in M_p, then, for all trigonometric polynomials f we have $T_n f \to 0$ in measure and $Tf_n \to Tf$ also in measure; thus $Tf = 0$ for all such f, and $T = 0$. This proves that i has a closed graph and is continuous. On the other hand, $\|P_m\|_{M_p}/\|P_m\|_{M_q} \geqslant c2^{m(1/p-1/q)} \to \infty$ with m.)

5.41 It is often more convenient to work with Rudin–Shapiro measures, for it is then simpler to patch multipliers without making use of more sophisticated techniques such as the Littlewood–Paley theory we will discuss later on. The construction, carried out by Kahane [1970] goes as follows. We proceed again by induction; first put $\mu_0 = \nu_0 =$ Dirac delta centered at the origin, and having chosen measures μ_m, ν_m supported in a finite set S_m choose $x_m \in T$ so that $x_m + S_m$ are disjoint, and let $\mu_{m+1} = \mu_m + \nu_m(\cdot + x_m)$, $\nu_{m+1} = \mu_m - \nu_m(\cdot + x_m)$. It is easy to verify inductively that there are sequences $\{y_k\} \subseteq T$ and fixed ± 1 valued sequences $\{r_k\}$ and $\{s_k\}$ so that $\mu_m = \sum_{k=1}^{2^m} r_k \delta(y_k)$, $\nu_m = \sum_{k=1}^{2^m} s_k \delta(y_k)$. An important consideration is the norm of these measures as convolution mappings in $L^p(T)$. A first step is to estimate $\|\mu_m\|_{M_1}$ and $\|\mu_m\|_{M_2}$; clearly, $\|\mu_m\|_{M_1} \leqslant \sum_{k=1}^{2}\|r_k\delta(y_k)\|_{M_1} \leqslant 2^m$. On the other hand, by the inductive definitions and the parallelogram law, the following relation holds: $\|\mu_m * f\|_2^2 + \|\nu_m * f\|_2^2 = 2^{m+1}\|f\|_2^2$, and $\|\mu_m\|_{M_2} \leqslant 2^{(m+1)/2}$. By interpolation we see that $\|\mu_m\|_{M_p} \leqslant c2^{m/p'}$, and since the opposite inequality also holds, by 5.39 above, then $\|\mu_m\|_{M_p} \sim 2^{m/p'}$. It is now possible to construct an explicit example of a convolution operator in $M_q \backslash M_p$, $1 < p < q < 2$. Choose disjointly supported translations ρ_m of μ_m. It is readily seen that the multiplier norm properties of the ρ_m's are identical to those of the μ_m's. Finally, consider $T = \sum_{m=0}^{\infty} 2^{-m/p}\rho_m$, this series converges in M_q, and if $T_N = \sum_{m=0}^{N} 2^{-m/p}\rho_m$ denotes its sequence of partial sums, then for $2 > q > p$, $\|T_N\|_{M_q} \geqslant (\sum_{m=0}^{N} 2^{-mq/p} 2^n)^{1/q}$ and $\|T\|_{M_q} \geqslant (\sum_{m=0}^{\infty} 2^{m(1-q/p)})^{1/q}$. Now, if T were to belong to M_p, then by Theorem 7.14 of Chapter IV, the norm $\|T\|_{M_q}$ must remain bounded as $q \to p^+$, and clearly the above estimates show that this is not so. Actually, estimates in the same spirit give an interesting result of Zafran [1975]: for each p, $1 \leqslant p < 2$ there is a multiplier which is of weak-type (p, p) but not of type (p, p); for $p = 1$ the multiplier is the Hilbert transform, but for the other values of p things are a bit more complicated. This construction is due to Cowling–Fournier [1976], and the reader is encouraged to consult their work to grasp the complete scope of their methods, which extend these and other results to locally compact groups.)

5.42 Show that if the multiplier associated with $\{\lambda_j\}$ is of weak-type (p, p), $1 \leqslant p < \infty$, then $|\lambda_j| \leqslant c < \infty$. How does c depend on the weak-norm of the multiplier? (*Hint*: $\{|T(e^{inx})| > \eta\}$ is empty for η sufficiently large.)

CHAPTER

VI

Paley's Theorem and Fractional Integration

1. PALEY'S THEOREM

The time has come to consider some basic questions concerning Fourier coefficients of L^p functions; for instance, we are interested in identifying among the sequences $\{c_j\}$ with $\lim_{|j|\to\infty} c_j = 0$ those which correspond to Fourier coefficients of L^p functions and to make a more precise statement about their summability properties. The two results we have encountered so far are

$$|c_j(f)| \leq \|f\|_1 \tag{1.1}$$

and

$$\sum |c_j(f)|^2 = \|f\|_2^2. \tag{1.2}$$

(1.1) states that the operation $f \to \{c_j(f)\}$ is of type $(1, \infty)$, i.e., it maps $L(T)$ into $l^\infty(Z)$; (1.2) states that it is of type $(2, 2)$. Since the interpolation results in Chapter IV do not cover this case, we first attempt to restate the problem at hand as one to which they do apply.

Following Zygmund we introduce the weighted sequence spaces $l^p_\mu(Z) = l^p_\mu$ and wk-$l^p_\mu(Z) =$ wk-l^p_μ. l^p_μ consists of those sequences $\{c_j\}$ so that the quantity $\|c_j\|_{l^p_\mu} = (\sum_j |c_j|^p \mu(j))^{1/p} < \infty, 0 < p \leq \infty$ (with the obvious interpretation when $p = \infty$), and wk-l^p_μ consists of those sequences $\{c_j\}$ for which there is a constant $c > 0$, so that, if $\mathcal{O}_\lambda = \{j \in Z: |c_j| > \lambda\}$, then $\lambda^p \sum_{j \in \mathcal{O}_\lambda} \mu(j) \leq c$, all λ.

An important example is $\mu_p(j) = (1 + |j|)^{-p}$; note that $l^\infty \subseteq l^1_{\mu_p}$ whenever $p > 1$. With this notation (1.1) and (1.2) can be restated as follows: if to each $f \in L + L^2$, $f \sim \sum c_j e^{ijx}$, we assign the sequence $\{c_j \mu_2(j)^{-1/2}\}$, then this mapping is bounded from L^2 to $l^2_{\mu_2}$ and from L into wk-l_{μ_2}.

Only the weak-type statement requires proof; this is easy since $\mathcal{O}_\lambda = \{j \in Z\colon |c_j|\mu_2(j)^{-1/2} > \lambda\} \subseteq \{j \in Z\colon \mu_2(j)^{1/2} < \|f\|_1/\lambda\}$ and consequently

$$\sum_{j\in\mathcal{O}_\lambda} \mu_2(j) \leqslant \int_{[c\lambda/\|f\|_1,\infty)} \frac{ds}{s^2} = \frac{c\|f\|_1}{\lambda}.$$

This observation readily leads to

Theorem 1.1 (Hardy–Littlewood). Assume $f \in L^p(T)$, $f \sim \sum c_j e^{ijx}$, $1 < p < 2$. Then there is a constant c independent of f such that

$$\|c_j\|_{l^p_{\mu_{2-p}}} = \|c_j\mu_2(j)^{-1/2}\|_{l^p_{\mu_2}} \leqslant c\|f\|_p,$$

more precisely,

$$\left(\sum|c_j|^p(1+|j|)^{p-2}\right)^{1/p} \leqslant c\|f\|_p \tag{1.3}$$

with $c = O(1/(p-1))$ as $p \to 1^+$.

Proof. Just invoke the abstract version of the Marcinkiewicz interpolation Theorem 4.1 in Chapter IV. ∎

There is a statement symmetric to Theorem 1.1, namely,

Theorem 1.2. Assume $\{c_j\} \in l^q_{\mu_{2-q}}$, $2 < q < \infty$. Then there is a function $f \in L^q(T)$ such that $c_j = c_j(f)$ and $\|f\|_q \leqslant c\|c_j\|_{l^q_{\mu_{2-q}}}$ with $c = O(q)$ as $q \to \infty$, c independent of f. In other words

$$\|f\|_q \leqslant c\left(\sum|c_j|^q(1+|j|)^{q-2}\right)^{1/q}. \tag{1.4}$$

Proof. We dualize Theorem 1.1. Put

$$\sigma_n(x) = \sum_{|j|\leqslant n}\left(1 - \frac{|j|}{n+1}\right)c_j e^{ijx}$$

and note that for trigonometric polynomials g,

$$\begin{aligned}\frac{1}{2\pi}\int_T \sigma_n(x)g(x)\,dx &= \sum_{|j|\leqslant n}\left(1-\frac{|j|}{n+1}\right)c_j c_{-j}(g)\\ &= \sum_{|j|\leqslant n}(c_j\mu_2(j)^{-1/2})\left(1-\frac{|j|}{n+1}\right)(c_{-j}(g)\mu_2(j)^{-1/2})\mu_2(j)\\ &= A,\end{aligned}$$

say.

By Hölder's inequality with indices q and its conjugate p, $1 < p < 2$, we get that $|A|$ does not exceed

$$\left(\sum_{|j| \leqslant n} \left(1 - \frac{|j|}{n+1}\right)^p |c_{-j}(g)|^p \mu_{p-2}(j)\right)^{1/p} \|c_j\|_{l^q_{\mu_{2-q}}}$$

$$\leqslant c\|g\|_p \, \|c_j\|_{l^q_{\mu_{2-q}}},$$

where c is the constant of Theorem 1.1. The converse to Hölder's inequality now gives that

$$\|\sigma_n\|_q \leqslant c\|c_j\|_{l^q_{\mu_{2-q}}}$$

and the conclusion follows by Fejér's Theorem 3.3 in Chapter III. ■

To obtain the sharp version of the Hardy–Littlewood theorem we need the notion of rearranged sequence. Suppose $\{c_j\}$ is such that $\lim_{|j|\to\infty} c_j = 0$; then choose among the c_j's with $j \geqslant 0$ one with largest $|c_j|$ and rename it c_0^*. At the same time rename the corresponding character e^{ijx}, $\phi_0(x)$. If there is more than one index with largest $|c_j|$, just choose any; observe however that since $\lim_{j\to\infty} c_j = 0$ there can only be finitely many. Remove c_0^* and repeat now the process for the remaining c_j's, and so on. This procedure obtains a new sequence with $|c_0^*| \geqslant |c_1^*| \geqslant \cdots \to 0$ as well as an ONS $\{\phi_j\}_{j=0}^{\infty}$. Similarly, we obtain $\{c_j^*\}_{j=-\infty}^{-1}$, so that $|c_{-1}^*| \geqslant |c_{-2}^*| \geqslant \cdots \to 0$, and also a corresponding ONS $\{\phi_j\}_{j=-\infty}^{-1}$. Note that $\{\phi_j\}_{j=-\infty}^{\infty}$ is an ONS consisting of bounded functions, and since these were the only properties we actually used in proving Theorem 1.1 we also have

Theorem 1.3 (Paley). Assume $f \in L^p(T)$, $f \sim \sum c_j e^{ijx}$, $1 < p \leqslant 2$. Then

$$\left(\sum |c_j^*|^p (1 + |j|)^{p-2}\right)^{1/p} \leqslant c\|f\|_p \tag{1.5}$$

with c independent of f and $O(1/(p-1))$ as $p \to 1^+$.

Proof. Fix $f \in L^p(T)$, note that $\sum c_j e^{ijx} = \sum c_j^* \phi_j(x)$ and apply the version of the Hardy–Littlewood theorem corresponding to $\{\phi_j\}$. ■

Note that (1.5) is sharper than (1.3) in the sense that the expression on the left-hand side is largest among all possible rearrangements of the given c_j's. Next, to pass to unweighted sequences, we have the following.

Proposition 1.4. For $1/p + 1/q = 1$, $1 < p \leqslant 2$, $l^p_{\mu_{2-p}} \subseteq l^q$. More precisely, there is a constant $c > 0$ independent of the sequences involved, so that $\|c_j\|_{l^q} \leqslant c\|c_j\|_{l^p_{\mu_{2-p}}}$.

Proof. Since $\|c_j\|_{l_q}^q = \sum_{j=0}^{\infty}|c_j^*|^q + \sum_{j=-\infty}^{-1}|c_j^*|^q = I + J$, say, we may assume that $\{c_j\}_{j=0}^{\infty}$ is nonnegative and nonincreasing and $\{c_j\}_{j=-\infty}^{-1}$ is nonnegative and nondecreasing. To estimate I observe that

$$I = c_0^q + \sum_{k=0}^{\infty} \sum_{2^k \leq j < 2^{k+1}} c_j^q \leq c_0^q + \sum_{k=0}^{\infty} c_{2^k}^q 2^k$$

$$= c_0^q + \sum_{k=0}^{\infty} c_{2^k}^p 2^{k(p-1)} (c_{2^k}^p)^{(q-p)/p} 2^{k(2-p)}. \tag{1.6}$$

Furthermore, since a simple estimate shows that

$$(c_{2^k}^p)^{(q-p)/p} 2^{k(2-p)} \leq c\left(\sum_{2^{k-1} < j \leq 2^k} c_j^p j^{p-2}\right)^{(q-p)/p}$$
$$\leq c\|c_j\|_{l^p_{\mu_{2-p}}}^{q-p},$$

(1.6) gives

$$I \leq c\|c_j\|_{l^p_{\mu_{2-p}}}^{q-p}\left(c_0^p + \sum_{k=0}^{\infty} c_{2^k}^p 2^{k(p-1)}\right)$$
$$\leq c\|c_j\|_{l^p_{\mu_{2-p}}}^{q-p} \|c_j\|_{l^p_{\mu_{2-p}}}^p = c\|c_j\|_{l^p_{\mu_{2-p}}}^q.$$

Clearly, an identical estimate holds for J and we have finished. ■

This observation gives at once

Theorem 1.5 (Hausdorff-Young). Suppose $f \in L^p(T)$, $1 < p < 2$, $f \sim \sum c_j e^{ijx}$, and $1/p + 1/q = 1$. Then

$$\left(\sum |c_j|^q\right)^{1/q} \leq c\|f\|_p \tag{1.7}$$

with c independent of f. Moreover, if $\{c_j\} \in l^p$, then there is a function $f \in L^q(T)$ so that $c_j = c_j(f)$ and

$$\|f\|_q \leq c\|c_j\|_{l^p} \tag{1.8}$$

with c independent of $\{c_j\}$.

Proof. Suppose $f \in L^p$, then by (1.1) $\lim_{|j|\to\infty} c_j = 0$ and by Paley's theorem 1.3, $\|c_j^*\|_{l^p_{\mu_{2-p}}} \leq c\|f\|_p$. Moreover, by Proposition 1.4, $\|c_j\|_{l^q} \leq c\|c_j^*\|_{l^p_{\mu_{2-p}}}$ and (1.7) obtains. (1.8) follows now by a duality argument in the spirit of Theorem 1.2. ■

The Hausdorff-Young theorem states that in the case of Fourier sequences, estimates (1.1) and (1.2) suffice to obtain an $L^p(T)$-$l^q(Z)$ result as well, $1/p + 1/q = 1$, $1 < p < 2$. A similar conclusion holds for arbitrary sublinear operations T simultaneously of types $(1, \infty)$ and $(2, 2)$; we prove a rather general result in this direction since it serves to illustrate the role of Young's functions in interpolation. First we present some definitions.

For a function $A(s)$ defined in $s \geqslant 0$ we say that $A(s)/s$ increases provided that

(i) $0 \leqslant A(s) \leqslant \infty$, $A(0) = 0$,
(ii) $A(s)/s$ increases (in the wide sense),
(iii) A is left-continuous, and
(iv) A is nontrivial, i.e., $0 \not\equiv A(s) \not\equiv \infty$ for $s > 0$.

If $A(s)/s$ increases and we set

$$A_0(s) = \int_{[0,s)} \frac{A(u)}{u}\, du, \tag{1.9}$$

then A_0 is convex, increasing, 0 at 0, and nontrivial (in fact A_0 is positive and finite in the same set as A); A_0 was introduced by Jodeit as the regularization of A. Moreover, $A_0(s) \leqslant A(s) \leqslant A_0(2s)$. Functions which have the properties of A_0 are called Young's functions; cf. 5.2 in Chapter I. It is often convenient to use A rather than A_0 because, for example, $\min(A(s), B(s))$ satisfies (i)–(iv) but is not in general convex when A and B are convex.

Let $A(s)/s$ increase. The Young's complement $\bar{A}$ of A is given by

$$\bar{A}(u) = \sup_{s \geqslant 0}(us - A(s)) \tag{1.10}$$

and the inequality

$$su \leqslant A(s) + \bar{A}(u) \tag{1.11}$$

is called Young's inequality. The inverse of A is defined on $[0, \infty]$ by

$$A^{-1}(u) = \inf\{s\colon A(s) > u\} \qquad (\inf \varnothing = +\infty). \tag{1.12}$$

A^{-1} is positive and finite for $u > 0$, $A^{-1}(\infty) = \infty$, A^{-1} is nondecreasing and right-continuous, and $A^{-1}(u)/u$ decreases. Moreover,

$$A(A^{-1}(u)) \leqslant u \leqslant A^{-1}(A(u)),\ u \geqslant 0 \tag{1.13}$$

and $A(s) = \sup\{u\colon A^{-1}(u) < s\}$, $\sup \varnothing = 0$.

If p and q are conjugate exponents, then the relation $s^{1/p}s^{1/q} = s$ is the prototype of

$$s \leqslant A^{-1}(s)\bar{A}^{-1}(s) \leqslant 2s. \tag{1.14}$$

For our purpose there is a special operation to mention, namely,

$$RAR(s) = \begin{cases} 0, & s = 0, \\ \sup\limits_{u<s}(1/A(1/u)), & s > 0, \end{cases} \tag{1.15}$$

where $1/0 = \infty$ and $1/\infty = 0$. Then RAR is left-continuous and nontrivial, and $RAR(s)/s$ increases. The effect of this operation is to reverse the behavior of A at 0 with that at ∞, and vice versa. Since

$$(RAR)^{-1}(u) = 1/A^{-1}(1/u), \tag{1.16}$$

it is readily seen that

$$\overline{RAR}(s) \sim R\bar{A}R(s) \tag{1.17}$$

in the sense that there are constants c_1, c_2 independent of s so that $\overline{RAR}(c_1 s) \leq R\bar{A}R(s) \leq \overline{RAR}(c_2 s)$. Let now

$$A(s) = \int_{[0,s)} a(u)\, du, \tag{1.18}$$

where (i) a is nonnegative, continuous, and strictly increasing from 0 to ∞ with u and (ii) $a(u)/u$ is nonincreasing. If we denote by $\bar{a}(u)$ the inverse function to $\bar{a}(u)$ and put

$$B(s) = \int_{[0,s)} \frac{1}{\bar{a}(1/u)}\, du, \tag{1.19}$$

then the equivalence

$$R\bar{A}R(s/4) \leq B(s) \leq R\bar{A}R(s) \tag{1.20}$$

holds.

We are now ready to interpolate. Let (X, μ) and (Y, υ) be arbitrary measure spaces and assume that T is a sublinear mapping defined in $L(X) + L^2(X)$ with values in the ν-measurable functions on Y. The question we address is the following: if T is a sublinear operation simultaneously of types $(1, \infty)$ and $(2, 2)$ with norm ≤ 1, i.e., if

$$\|Tf\|_{L^\infty_\nu} \leq \|f\|_{L_\mu} \tag{1.21}$$

and

$$\|Tf\|_{L^2_\nu} \leq \|f\|_{L^2_\mu} \tag{1.22}$$

and the function $A(s)$ given by (1.18) above is such that $A(s)/s$ increases and $A(s)/s^2$ decreases, what integrability condition obtains for Tf provided that $\int_X A(|f(x)|)\, d\mu(x) \leq 1$, say? In this direction we have

Theorem 1.6. Under the circumstances described above and if $A(s)/s \to 0$ with s and $A(s)/s \to \infty$ also with s, then

$$\int_Y B\left(\frac{|Tf(y)|}{2}\right) d\nu(y) \leq \int_X A(|f(x)|)\, d\mu(x) \tag{1.23}$$

provided that B is defined by (1.19) and the right-hand side of (1.23) does not exceed 1.

Proof. The proof is clearest when we treat the function B as an unknown and discover its exact form by making use of (1.18), (1.21) and (1.22) to insure that (1.23) actually holds. Let then $B'(u) = b(u) \geqslant 0$, suppose that $\eta(u) = b(u)/u$ is nonincreasing, and put $\eta = \eta(0^+)$. By Fubini's theorem and the sublinearity of T we get

$$\int_{[0,\infty)} \nu(\{|Tf| > 2s\})b(s)\,ds = \int_{[0,\infty)} s\nu(\{|Tf| > 2s\}) \int_{[0,s)} d\eta(u)$$
$$+\,\eta \int_{[0,\infty)} s\nu(\{|Tf| > 2s\})\,ds$$
$$= \int_{[0,\infty)} \int_{[u,\infty)} s\nu(\{|Tf| > 2s\})\,ds\,d\eta(u) + K$$
$$= J + K, \tag{1.24}$$

say. We bound J first. Observe that with the notation $f_z(x) = \min(z, |f(x)|)\operatorname{sgn} f(x)$, $f^z(x) = f(x) - f_z(x)$, it readily follows that J does not exceed

$$\int_{[0,\infty)} \int_{[u,\infty)} s\nu(\{|Tf^z| > s\})\,ds\,d\eta(u)$$
$$+ \int_{[0,\infty)} \int_{[u,\infty)} s\nu(\{|Tf_z| > s\})\,ds\,d\eta(u) = J_1 + J_2,$$

say, where z is to be regarded as a monotone function of u, yet to be chosen. To treat J_1, note that by (1.21), $\|Tf^z\|_\infty \leqslant \|f^z\|_1$, so that $\nu(\{|Tf^z| > s\}) = 0$ if $s \geqslant z$ and $\|f^z\|_1 = \int_{[z,\infty)} \mu(\{|f| > s\})\,ds < z$. Since

$$\int_X A(|f(x)|)\,d\mu(x) = \int_{[0,\infty)} a(s)\mu(\{|f| > s\})\,ds \leqslant 1,$$

it readily follows that $\|f^z\|_1 \leqslant 1/a(z)$. Thus if we define now $z(u)$ by the relation

$$1/a(z(u)) = u \tag{1.25}$$

it follows that $J_1 = 0$.

Moreover, since

$$\int_{[u,\infty)} s\nu(\{|Tf_z| > s\})\,ds \leqslant \int_{[0,\infty)} s\mu(\{|f_z| > s\})\,ds$$
$$= \int_{[0,z)} s\mu(\{|f_z| > s\})\,ds,$$

we also have

$$J_2 \leqslant \int_{[0,\infty)} s \int_{[0,z^{-1}(s))} d\eta(u)\, \mu(\{|f| > s\})\, ds.$$

In case $\eta = 0$, there is no term K, and

$$s \int_{[0,z^{-1}(s))} d\eta(u) = s\eta(z^{-1}(s)) \leqslant a(s)$$

if $\eta(s) \leqslant 1/s\bar{a}(1/s)$, that is, if

$$b(s) \leqslant 1/\bar{a}(1/s). \tag{1.26}$$

Hence in this case (1.23) holds since

$$\int_Y B(|h(y)|)\, d\nu(y) \leqslant \int_{[0,\infty)} \nu(\{|h| > s\})/\bar{a}(1/s)\, ds.$$

Now suppose we define $b(s)$ as in (1.19), i.e., with equality in place of inequality in (1.26). Then $\eta = \lim_{s\to 0} b(s)/s = \lim_{s\to\infty} \eta(s)/s = \inf_s \eta(s)/s$. In case $\eta > 0$, we have $A(s) \geqslant \frac{1}{2}\eta s^2$ so that $f \in L^2(X)$ and by (1.22)

$$K \leqslant \frac{1}{4}\eta \int_{[0,\infty)} s\mu(\{|f| > s\})\, ds < \infty.$$

On the other hand, returning to the estimate for the J's, when $\eta > 0$ we note that $s \int_{[0,u^{-1}(s))} d\eta(v) = s(\eta(u^{-1}(s)) - \eta)$; also the negative term and the finiteness of K allow some cancellation. That is,

$$J + K \leqslant J + 4K = \int_{[0,\infty)} a(s)\mu(\{|f| > s\})\, ds,$$

(1.23) holds, and the proof is complete. ■

In addition to the Hausdorff-Young theorem, which follows with the choice $A(s) = s^p$, $1 < p < 2$, Theorem 1.6 gives the following two results of Hardy and Littlewood.

Example 1.7. Let $A(s) = s \ln(1 + s)$ so that $\int_T A(|f(x)|)\, dx < \infty$ means that $f \in L \ln L(T)$. Then $\bar{A}(s) \sim s(e^s - 1)$ and

$$R\bar{A}R(s) \sim \frac{s}{(e^{1/s} - 1)} \sim \begin{cases} e^{-1/s} & \text{for small values,} \\ s^2 & \text{for large values.} \end{cases}$$

Theorem 1.6 then gives that, if $f \sim \sum c_j e^{ijx}$ and $\eta > 0$, then

$$\sum_j e^{-\eta/|c_j|} < \infty \qquad \text{whenever} \quad \int_T |f(x)| \ln^+ |f(x)|\, dx < \infty.$$

Let now $A(s) = s/\ln(1+s)$; since RAR is A in the preceeding example,

$$R\bar{A}R \sim s(e^s - 1) \sim \begin{cases} s^2 & \text{for small values,} \\ e^s & \text{for large values,} \end{cases}$$

and in this case we get that, if $\{c_j\}$ is a sequence such that

$$\sum \frac{|c_j|}{\ln(1 + 1/|c_j|)} < \infty,$$

then there is a function $f \in L$ so that $c_j = c_j(f)$ and $\int_T e^{k|f(x)|}\,dx < \infty$ for each $k > 0$.

2. FRACTIONAL INTEGRATION

We have seen that if $f \in L$, $f \sim \sum c_j e^{ijx}$, $c_0 = 0$, then the Fourier series of its indefinite integral is $C_0 + \sum_{j\neq 0}(1/(ij))c_j e^{ijx}$. It is therefore natural to consider also the "fractional" analogue of the above formula, namely,

$$\sum_{j\neq 0} \frac{1}{|j|^\alpha} c_j e^{ijx}, \qquad 0 < \alpha < 1. \tag{2.1}$$

We see at once that (2.1) is the Fourier series of a function with the same integrability properties as f, since by Proposition 4.4 in Chapter III there is a summable function g with $c_j(g) = |j|^{-\alpha}$, $j \neq 0$. But we expect a better behavior from the series in (2.1), and, to get an idea of what this might be, we investigate whether we actually know something else about g.

Consider the integrable function $h(x) = |x|^{-\eta}$, $0 < \eta < 1$. We note that its Fourier coefficients are given by the even sequence

$$\begin{aligned} c_j(h) &= \frac{1}{\pi}\int_{[0,\pi)} \frac{1}{x^\eta}\cos jx\,dx \\ &= \frac{1}{|j|^{1-\eta}}\frac{1}{\pi}\int_{[0,|j|\pi)} \frac{1}{s^\eta}\cos s\,ds, \qquad j \neq 0, \end{aligned} \tag{2.2}$$

and since the integral in (2.2) has a limit $L \neq 0$ as $|j| \to \infty$, we see that essentially $c_j(h) \sim L/|j|^{1-\eta}$ for $|j|$ large. Thus $g(x) \sim |x|^{\alpha-1}$ near the origin, and in this case $g \in L^r(T)$ for $r < 1/(1-\alpha)$. Therefore by Young's convolution theorem (cf. 4.30 below), (2.1) is the Fourier series of an L^q function, with $1/q > 1/p - \alpha$, whenever $f \in L^p(T)$ and $1 \leq p < 1/\alpha$; this still is not sharp. The precise statement is the following: for $0 < \alpha < 1$-consider the Riesz fractional integral I_α given by

$$I_\alpha f(x) = \int_T \frac{f(x-t)}{|t|^{1-\alpha}}\,dt, \qquad x \in T. \tag{2.3}$$

We then have

Theorem 2.1 (Hardy–Littlewood, Sobolev). Assume that $f \in L^p(T)$, $1 < p < 1/\alpha$. Then $I_\alpha f \in L^q(T)$, $1/q = 1/p - \alpha$ and there is a constant $c = c(\alpha, p, q)$ independent of f such that

$$\|I_\alpha f\|_q \leq c\|f\|_p. \tag{2.4}$$

Proof. Since $|I_\alpha f(x)| \leq I_\alpha(|f|)(x)$, we may suppose that $f \geq 0$. For $0 < \eta < \pi$ write

$$I_\alpha f(x) = \left(\int_{|t|<\eta} + \int_{\eta \leq |t| < \pi}\right) \frac{f(x-t)}{|t|^{1-\alpha}}\, dt = J + K, \tag{2.5}$$

say. To estimate J note that $\phi(t) = |t|^{\alpha-1}\chi_{(-1,1)}(t)$ is integrable, and, if $\phi_\eta(t) = (1/\eta)\phi(t/\eta)$, then $J = \eta^\alpha \int_T f(x-t)\phi_\eta(t)\, dt$. Thus, by (a simple variant of) Proposition 2.3 in Chapter IV,

$$J \leq c_1 \eta^\alpha Mf(x), \tag{2.6}$$

where $c_1 = 2/\alpha$ is independent of f. To bound K, note that since $(1-\alpha)p' > 1$, Hölder's inequality implies that

$$\begin{aligned} |K| &\leq \left(\int_{\eta \leq |t| < \pi} |t|^{(\alpha-1)p'}\, dt\right)^{1/p'} \|f\|_p \\ &\leq 2^{1/p'} \frac{(\eta^{1-(1-\alpha)p'})}{(1-\alpha)p' - 1}^{1/p'} \|f\|_p = c_2 \eta^{\alpha - 1/q}\|f\|_p, \end{aligned} \tag{2.7}$$

where $c_2 = (2q/p')^{1/p'}$.

Therefore, adding (2.6) and (2.7), we see that

$$I_\alpha f(x) \leq c(\eta^\alpha Mf(x) + \eta^{\alpha - 1/p}\|f\|_p). \tag{2.8}$$

Now we seek to choose η to minimize the right-hand side of (2.8), a good choice being that η which makes both summands equal, i.e.,

$$\eta = (\|f\|_p / Mf(x))^p. \tag{2.9}$$

Note that this choice of η is all right as long as it does not exceed π; however, if it does, it is because $Mf(x) \leq \pi^{-1/p}\|f\|_p$ and for these x's, by setting $\eta = \pi$ in (2.8), we see that

$$I_\alpha f(x) \leq c\|f\|_p. \tag{2.10}$$

On the other hand, (2.9) gives that

$$I_\alpha f(x) \leq c\|f\|_p^{\alpha p} Mf(x)^{1-\alpha p}. \tag{2.11}$$

Since we prefer not to have to determine which estimate applies, we just add (2.10) and (2.11) and obtain that for all x in T

$$I_\alpha f(x) \leqslant c(\|f\|_p^{\alpha p} Mf(x)^{1-\alpha p} + \|f\|_p). \tag{2.12}$$

This is all we need to complete the proof. Indeed, since $(1-\alpha p)q = p$, by (2.2) it readily follows that

$$\|I_\alpha f\|_q \leqslant c(\|f\|_p^{\alpha p}\|Mf\|_p^{1-\alpha p} + \|f\|_p), \tag{2.13}$$

and since by the maximal theorem $\|Mf\|_p \leqslant c\|f\|_p$ for $1 < p$, the proof is complete. ■

Theorem 2.1 is best possible in the sense that an estimate of the form $\|I_\alpha f\|_r \leqslant c\|f\|_p$ implies $r \leqslant q$. This is readily seen as follows: for $f(t) = \chi_{(-\eta,\eta)}(t)$ and $|x| \leqslant \eta/2$ we note that

$$I_\alpha f(x) = \int_{|t|\leqslant\eta} \frac{1}{|x-t|^{1-\alpha}}\,dt \geqslant \int_{|x-t|\leqslant\eta/2} \frac{1}{|x-t|^{1-\alpha}}\,dt = c\eta^\alpha,$$

whence

$$c\eta^{1/r+\alpha} \leqslant \left(\int_{|x|\leqslant\eta/2} I_\alpha f(x)^r\,dx\right)^{1/r} \leqslant c\|f\|_p^p = c\eta^{1/p}$$

and this can only hold as $\eta \to 0^+$ provided that $(1/r) + \alpha \geqslant 1/p$.

We must still consider $p = 1$ and $p = 1/\alpha$. The estimate (2.11) suffices to handle the case $p = 1$ quite easily, namely,

Theorem 2.2. Suppose $f \in L(T)$ and $1/q = 1 - \alpha$; then $I_\alpha f \in$ wk-$L^q(T)$ and there is a constant c, independent of f, so that for $\lambda > 0$

$$|\{I_\alpha f > \lambda\}| \leqslant c(\|f\|_1/\lambda)^q. \tag{2.14}$$

Furthermore, if $f \in L \ln L(T)$, then $I_\alpha f \in L^q(T)$ and there are constants A, B so that

$$\|I_\alpha f\|_q \leqslant A + B\int_T |f(t)| \ln^+|f(t)|\,dt. \tag{2.15}$$

Proof. To prove (2.14) we may assume that $\lambda > 2c\|f\|_1$, where c is the constant in (2.12) above, for otherwise the conclusion is obvious. Then $\{I_\alpha f > \lambda\} \subseteq \{c\|f\|_1^\alpha Mf^{1-\alpha} > \lambda\}$ and by the maximal theorem

$$\begin{aligned} |\{I_\alpha f > \lambda\}| &\leqslant c|\{M > (\lambda/c\|f\|_1^\alpha)^{1/(1-\alpha)}\}| \\ &\leqslant c\|f\|_1\|f\|_1^{\alpha/(1-\alpha)}\lambda^{1/(1-\alpha)} = c(\|f\|_1/\lambda)^q. \end{aligned}$$

Also by (2.12), Theorem 5.3 in Chapter IV, and Young's inequality $u^\alpha v^{1-\alpha} \leqslant \alpha u + (1-\alpha)v$ (cf. 5.2 in Chapter I), we get $\|I_\alpha f\|_q \leqslant c\|f\|_1^\alpha \|Mf\|_1^{1-\alpha} + c\|f\|_1 \leqslant c\|f\|_1 + c\|Mf\|_1$ and we are done. ■

Whereas the result concerning the weak-type is sharp, namely, there is an integrable function f so that $I_\alpha f \notin L^q$, $1/q = 1 - \alpha (f(t) = 1/t \ln(1/t)^{1+1/q}$ for $0 < t < \frac{1}{2} = 0$ otherwise will do) the $L \ln L \to L^q$ result is not, cf. 4.28.

We consider next whether there is a maximal operator which in some sense controls the Riesz potentials as the Hardy–Littlewood maximal operator controls the Hilbert transform. First note that for $f \geqslant 0$ and $\pi > \delta > 0$

$$I_\alpha f(x) \geqslant \int_{|t| \leqslant \delta} \frac{f(x-t)}{|t|^{1-\alpha}}\, dt \geqslant \frac{1}{\delta^{1-\alpha}} \int_{|t| \leqslant \delta} f(x-t)\, dt;$$

therefore a natural candidate for the task at hand is

$$M_\eta f(x) = \sup_{x \in I} \frac{1}{|I|^{1-\eta}} \int_I |f(t)|\, dt, \qquad 0 < \eta < 1 \tag{2.16}$$

with $\eta = \alpha$. The continuity properties of M_η are readily established. Indeed, assume that $1 \leqslant p \leqslant 1/\eta$ and $\int_T |f(t)|^p\, dt = 1$. By Hölder's inequality we see that

$$\begin{aligned} \frac{1}{|I|^{1-\eta}} \int_I |f(t)|\, dt &\leqslant |I|^{\eta - 1/p} \left(\int_I |f(t)|^p\, dt \right)^{1/p} \\ &\leqslant |I|^{\eta - 1/p} \left(\int_I |f(t)|^p\, dt \right)^{\varepsilon/p}, \end{aligned} \tag{2.17}$$

for any $0 < \varepsilon < 1$ since the integral $\leqslant 1$. By choosing $\varepsilon = 1 - \eta p$, or $\varepsilon/p = 1/p - \eta$, we get at once from (2.17) that

$$M_\eta f(x) \leqslant M(|f|^p)(x)^{1/p - \eta} \tag{2.18}$$

and consequently

$$|\{M_\eta f > \lambda\}| \leqslant c/\lambda^{1/(1/p - \eta)}, \tag{2.19}$$

i.e., M_η is a sublinear operation of weak-type (p, q) with $1/q = 1/p - \eta$, $1 \leqslant p \leqslant 1/\eta$. This result can be improved to give the type (p, q) estimate for $1 < p \leqslant 1/\eta$ ($p = 1/\eta$ we already have) and this requires, of course, an interpolation result.

Theorem 2.3. Let $0 < p_0 \leqslant q_0 \leqslant \infty$, $0 < p_1 \leqslant q_1 \leqslant \infty$, $p_0 < p_1$, $q_0 \neq q_1$. If the sublinear operator T defined in $L^{p_0}(X) + L^{p_1}(X)$ is simultaneously of weak-type (p_0, q_0) with norm $\leqslant c_0$ and of type (p_1, q_1) with norm $\leqslant c_1$, then T is of type (p, q), where $1/p = (1-\eta)/p_0 + \eta/p_1$, $1/q = (1-\eta)/q_0 + \eta/q_1$ with norm $\leqslant c c_0^{1-\eta} c_1^{\eta}$, $0 < \eta < 1$; here c is a constant independent of f which remains bounded as $\eta \to 1^-$ but tends to infinity as $\eta \to 0^+$.

Proof. The proof follows along what are well-known lines by now. With the notation of Theorem 1.6 we put $f = f_z + f^z$ and choose z as the monotone function given by

$$z^{-1}(s) = s^{p(1/p_0 - 1/p_1)/q(1/q_0 - 1/q_1)}.$$

The same ideas as in Theorem 1.6 work here. ■

As for the relation between I_α and M_α it is given by

Theorem 2.4 (Welland). Suppose that $f \in L(T)$, $0 < \varepsilon < \alpha < \alpha + \varepsilon < 1$. Then there is a constant c independent of f so that

$$|I_\alpha f(x)| \leqslant c(M_{\alpha-\varepsilon} f(x) M_{\alpha+\varepsilon} f(x))^{1/2}. \tag{2.20}$$

Proof. Suppose a function ϕ defined in R verifies $|\phi(u)| \leqslant \psi(u)$, where ψ is an even, nonincreasing function in $[0, \infty)$, $\int_R (\psi(u)/|u|^\gamma)\, du < \infty$, $0 < \gamma < 1$. Then an argument identical to that of Proposition 2.3 in Chapter IV gives that

$$\sup_{s>0} \left| \int_T f(x-t) \frac{1}{s^{1-\gamma}} \phi\left(\frac{t}{s}\right) dt \right| \leqslant c_\gamma M_\gamma f(x).$$

This observation implies, in the notation of Theorem 2.1, that $J \leqslant c\eta^\varepsilon M_{\alpha-\varepsilon} f(x)$. Also, a simple computation shows that $K \leqslant c\eta^{-\varepsilon} M_{\alpha+\varepsilon} f(x)$ and, consequently,

$$|I_\alpha f(x)| \leqslant c(\eta^\varepsilon M_{\alpha-\varepsilon} f(x) + \eta^{-\varepsilon} M_{\alpha+\varepsilon} f(x)). \tag{2.21}$$

Minimizing (2.21) gives that the optimal choice of $\eta = (M_{\alpha+\varepsilon} f(x)/M_{\alpha-\varepsilon} f(x))^{\varepsilon/2}$; furthermore, note that unlike Theorem 2.1 this choice always works since $M_\delta f(x) \leqslant (2\pi)^\gamma M_{\delta-\gamma} f(x)$. It is now clear that (2.20) holds and the proof is complete. ■

The estimate (2.20) gives Theorem 2.1, but the proof is conceptuallly different since it relies on a "nondiagonal" interpolation result.

We conclude this section by discussing briefly a result that arises naturally in connection with the restriction of Riesz and other potentials defined in Euclidean n-space to lower-dimensional manifolds. More precisely, we consider integral operators of the form

$$F(y) = Tf(y) = \int_X k(x, y) f(x)\, d\mu(x), \tag{2.22}$$

where k is a positive kernel, such as $|x-y|^{\alpha-1}$ when $n=1$, and are interested in the $(L^p(x), L^q(Y))$ continuity properties of T. We impose the following restrictions on k, namely,

$$|k|_1 = \sup_{y\in Y}\sup_{\lambda>0} \lambda^r \mu(\{x \in X: k(x,y) > \lambda\}) < \infty \tag{2.23}$$

and

$$|k|_2 = \sup_{x\in X}\sup_{\lambda>0} \lambda^s \nu(\{y \in Y: k(x,y) > \lambda\}) < \infty. \tag{2.24}$$

We then have

Theorem 2.5 (Adams). Assume ν is a regular Borel measure on Y so that $\nu(K) < \infty$ for each compact K. If k satisfies (2.23) and (2.24) above, $1 \leq s$, $r < \infty$ (not both s and r simultaneously 1), $p < q$ and $s/p' + r/q = 1$, then there is a constant c independent of f such that

$$\lambda^q \nu(\{F > \lambda\}) \leq c|k|_1^{q/p'}|k|_2 \left(\int_X |f(x)|^p \, d\mu(x)\right)^{q/p}. \tag{2.25}$$

Consequently, interpolating (2.25) gives that the strong inequality also holds in the same range of p's and q's.

Proof. We may assume $f \geq 0$. Let $E_\lambda = \{y \in Y: F(y) > \lambda\}$ and let K be an arbitrary compact subset of E_λ. Then by familiar arguments it follows that

$$\begin{aligned} \lambda\nu(K) &\leq \int_K F(y)\, d\nu(y) = \int_X f(x) \int_K k(x,y)\, d\nu(y)\, d\mu(x) \\ &= \int_X f(x) \int_{[0,\infty)} \nu(\{y \in K: k(x,y) > \lambda\})\, d\lambda\, d\mu(x) \\ &= \int_{[0,\infty)} \int_X f(x)\nu(\{y \in K: k(x,y) > \lambda\})\, d\mu(x)\, d\lambda \\ &= I, \end{aligned} \tag{2.26}$$

say. Denote $\nu(\{y \in K: k(x,y) > \lambda\}) = \phi(K, x, \lambda)$ and observe that by (2.24) this quantity $\leq \min(\nu(K), \lambda^{-s}|k|_2)$. Similarly,

$$\begin{aligned} \int_X \phi(K, x, \lambda)\, d\mu(x) &= \int_K \int_X \chi_{\{y\in K: k(x,y)>\lambda\}}\, d\mu(x)\, d\nu(y) \\ &= \int_K \mu(\{x \in X: k(x,y) > \lambda\})\, d\nu(y) \\ &\leq \lambda^{-r}|k|_1 \nu(K). \end{aligned}$$

This is all we need. Indeed, the above estimates give that the innermost integral in (2.26) does not exceed

$$\left(\int_X f(x)^p \phi(K, x, \lambda)\, d\mu(x)\right)^{1/p}\left(\int_X \phi(K, x, \lambda)\, d\mu(x)\right)^{1/p'}$$
$$\leqslant \min(\nu(K), \lambda^{-s}|k|_2)^{1/p}(\lambda^{-r}|k|_1\nu(K))^{1/p'}\|f\|_{L^p_\mu},$$

and, consequently,

$$\lambda\nu(K) \leqslant c\nu(K)^{1-(1/s)(1-r/p')}|k|_1^{1/p'}|k|_2^{(1/s)(1-r/p')}\|f\|_{L^p_\mu},$$

which on account of the relation between p and q gives (2.25) but with K in place of $\{F > \lambda\}$ there. The desired statement follows now by the regularity of ν. ■

The restriction $p < q$ is essential in the above proof, but not for the validity of the result; the reader should consult the work of Kerman and Sawyer [1985] in this direction. Nevertheless, returning to the trace results, we introduce the following definitions. A positive Borel measure ν on T is said to be β-dimensional, $0 \leqslant \beta \leqslant 1$, if for each subinterval $I \subseteq T$, $\nu(I) \leqslant c|I|^\beta$, with c independent of I. Also, denote by L^p_α the class of functions f so that there exists ϕ, $\|\phi\|_p < \infty$, and $f(x) = \int_T \phi(t)/|x-t|^{1-\alpha}\, dt$. We then have

Proposition 2.6. Assume ν is a β-dimensional Borel measure. Then $L^p_\alpha \subseteq L^q(\nu)$, provided that $\beta/q = 1/p - \alpha$.

Proof. Note that $|x|^{1-\alpha} \in \text{wk-}L^{1/(1-\alpha)}$ and apply Theorem 2.5. ■

3. MULTIPLIERS

Now we turn our attention briefly to the spaces M^p_q consisting of bounded L^p-L^q multipliers; more precisely, we say that a sequence $\{\lambda_j\} \in M^p_q$ if and only if for every trigonometric polynomial f,

$$\left\|\sum \lambda_j c_j(f) e^{ijx}\right\|_q \leqslant c\|f\|_p \tag{3.1}$$

with c independent of f; the infimum over the c's in (3.1) is called the norm of the multiplier in M^p_q. As indicated in the case $p = q$, M^p_q can be identified with the (Banach) space of bounded linear transformations T from L^p into L^q which commute with translations. There are some simple identifications and inclusions, for instance

Proposition 3.1. Suppose $1 \leqslant p < \infty$, $1 < q \leqslant \infty$, $1/p + 1/p' = 1/q + 1/q' = 1$. Then there is an isometric linear isomorphism of M^p_q onto $M^{q'}_{p'}$.

Proof. Assume $1 < p, q < \infty$ first, and corresponding to $T \in M_q^p$, we define the adjoint $T' : L^{q'} \to L^{p'}$ as the operator verifying

$$\int_T f(x) T'g(x)\, dx = \int_T Tf(x) g(x)\, dx$$

for all trigonometric polynomials f, g. It is clear that T' is bounded from $L^{q'}$ to $L^{p'}$ with the same norm as T, and it is also simple to verify that it commutes with translations. By the reflexivity of the Lebesgue spaces involved we see that $T \to T'$ is onto and we are finished in this case. The remaining cases require an *ad hoc* argument which is left for the reader to carry out (cf. 4.34). ■

As in (4.13) in Chapter V it is readily seen that if $\{\lambda_j\} \in M_q^p$ and $Tf \sim \sum \lambda_j c_j(f) e^{ijx}$, then,

$$\sup_j |\lambda_j| \leq \|T\|. \tag{3.2}$$

In fact since the underlying space is of finite measure the following also holds: if $2 \leq p < \infty$ and $1 < q \leq 2$, then the following is true:

(i) T is a translation invariant operator from L^p into L^q;
(ii) There exists a unique sequence $\{\lambda_j\}$ verifying (3.2) so that $Tf \sim \sum \lambda_j c_j(f) e^{ijx}$ for each trigonometric polynomial f.

This result may be restated as follows: if $2 \leq p_1, p_2 < \infty$ and $1 < q_1, q_2 \leq 2$, then there is an isometric linear isomorphism of $M_{q_1}^{p_1}$ onto $M_{q_2}^{p_2}$.

As for the symmetry we expect to hold in this case as well, we have the following result. Let $1 \leq p < \infty$, $1 < q \leq \infty$. If $1 \leq r, s \leq \infty$ is such that the point $(1/r, 1/s)$ in the square $0 \leq a, b \leq 1$ is symmetric with respect to the line $a + b = 1$ to the point $(1/p, 1/q)$, then there exists an isometric linear isomoprhism of M_q^p onto M_s^r.

Some inclusions, such as $M_q^p \subseteq M_s^r$, $p \leq r$, $q > s$ are trivial, and also some noninclusion results hold, as follows. Let $1 \leq p, q, r, s \leq \infty$, and suppose that $p \leq q$ and that $\min(r, s') < \min(p, q')$. Then M_q^p is not included in M_s^r. For further results in this direction, as well as extensions to more general settings the reader is referred to Larsen's monograph [1971] and to Cowling and Fournier's article [1976].

Examples of multipliers in the (most interesting) case $p \leq q$ are easy to come by; indeed by Young's inequality $L^r \subset M_q^p$, $1/q + 1 = 1/p + 1/r$ (the inclusion is proper), and a simple argument combining Theorems 1.3 and 1.5 shows that, for $p \leq 2 \leq q$ and $1/q = 1/p - \alpha$, any sequence $\{\lambda_j\}$ such that $|\lambda_j| = O(|j|^{-\alpha})$ as $|j| \to \infty$ is in M_q^p. In some particular instances, such as M_p^1, it is possible to characterize these spaces (cf. 4.35).

Finally, we point out that M_q^p is also a dual space; in fact, let $A_q^p = \{h \in L^r(T): h = \sum_{k=1}^{\infty} f_k * g_k, \sum \|f_k\|_p \|g_k\|_{q'} < \infty\}$; A_q^p is endowed with the natural norm. We then have

Theorem 3.2 (Figa-Talamanca-Gaudry). For $1 \leq p, q < \infty$, the space M_q^p is isometrically isomorphic to $(A_q^p)^*$, the dual of A_q^p.

4. NOTES; FURTHER RESULTS AND PROBLEMS

In 1912, W. H. Young extended Parseval's theorem to the relation

$$\left(\sum |c_j|^q\right)^{1/q} \leq \left(\frac{1}{2\pi}\int_T |f(t)|^p \, dt\right)^{1/p}$$

between a function f and its sequence of Fourier coefficients $\{c_j\}$; the exponents p and q had to verify the relation $1/p + 1/q = 1$ with q an even integer. His proof went roughly as follows: Given a trigonometric polynomial $f(x) = \sum c_j e^{ijx}$, consider $f * \bar{f}(x) = \sum |c_j|^2 e^{ijx}$ and observe that $(\sum |c_j|^4) = (1/2\pi)\int_T |f * \bar{f}(x)|^2 \, dx$. By Young's convolution theorem we have $\|f * \bar{f}\|_2^2 \leq c\|f\|_{4/3}^4$, and, consequently, we obtain the above estimate for $q = 4$, except for the constant $c > 1$ appearing on our right-hand side. To obtain the whole range of Young's original proof we keep convolving with f. Note in turn that, for appropriate indices p, q, r, Young's convolution theorem can be obtained from the Fourier transform inequality. Young's original statement is sharp in T, and in fact Hardy and Littlewood characterized the extremal functions: they are multiples of the characters. However, on the real line a much sharper inequality was obtained by Babenko in 1961 for the special values of q = even integer. Recently, Beckner obtained sharp L^p inequalities for both the n-dimensional Fourier transform (the constant is $(A_p)^n$, where $A_p = (p^{1/p}/p'^{1/p'})^{1/2}$) and for Young's convolution theorem (the constant for $L^p * L^q \subseteq L^r$ is $(A_p A_q A_{r'})^n$).

In 1913, Young published the dual to his original result in which sum and integral switch sides; it was ten years later when F. Hausdorff showed that the only restriction was $q \geq 2$. At about the same time F. Riesz observed that the trigonometric systems could be replaced by an arbitrary, uniformly bounded ONS. The "complex" interpolation theorem of M. Riesz had all these results as a first application. The generalization of the result to the case when the powers p and q are replaced by an appropriately related pair of positive, increasing functions was first considered by Mullholland; the complex method of interpolation did not apply in this context. The Marcinkiewicz interpolation theorem with powers replaced by convex functions was first considered by Zygmund.

For $f \in L(T)$ we set $F_1(x) = \int_{(-\pi,x]} f(t)\, dt$, and in general $F_\alpha(x) = \int_{(-\pi,x]} F_{\alpha-1}(t)\, dt$, $\alpha = 2, \ldots$. It is readily seen that actually

$$F_\alpha(x) = \frac{1}{\Gamma(\alpha)} \int_{(-\pi,x]} \frac{f(t)}{(x-t)^{1-\alpha}}\, dt,$$

and consequently we may extend this notion, with this definition, to all values of $\alpha > 0$; this concept is due to Riemann and Liouville. The proof given here of the Hardy–Littlewood, Sobolev inequality is due to Hedberg and is interesting because it reduces the result to a "main diagonal" interpolation result, which is the easiest case.

Suppose that J_α denotes the Bessel potential of order α in n-dimensional Euclidean space. Stein showed that if $f \geq 0$, then

$$\left(\int_{R^n} |J_\alpha f(x)|^p\, d\mu(x) \right)^{1/p} \leq c\|f\|_p,$$

where $\alpha > (n-k)/p$ and $\mu = \mu_k$ is the restriction of Lebesgue measure to R^k, considered as a subspace of R^n. In this case, the inequality can be restated as

$$\int_{R^k} |J_\alpha f(x_1, \ldots, x_k, 0, \ldots, 0)|^p\, dx_1 \cdots dx_k \leq c \int_{R^n} f(x)^p\, dx_1 \cdots dx_n$$

and the result therefore refers to the restriction, or trace of $J_\alpha f$ to a lower dimensional space.

Further Results and Problems

4.1 Assume $1 < \eta < 2$ and $\{\lambda_n\}_{n=1}^\infty$ is a sequence of positive numbers. Show that, if $\sum n^{\eta-1}\lambda_n < \infty$, there is a continuous function g so that the λ_n's are the Fourier sine coefficients of g and $x^{-\eta}g(x) \in L[0, \pi]$. (*Hint*: Note that since $\sum \lambda_n < \infty$, $\sum \lambda_n \sin nx$ converges uniformly to a continuous function g. Use Fatou's lemma to bound

$$\int_{[0,\pi]} x^{-\eta}|g(x)|\, dx \leq \sum \lambda_n \int_{[0,\pi]} x^{-\eta}|\sin nx|\, dx$$
$$= \sum n^{1-\eta}\lambda_n \int_{[0,n\pi]} x^{-\eta}|\sin x|\, dx.$$

The proof gives more, namely, $|x-a|^{-\eta}(g(x) - g(x)) \in L[0, \pi]$ for all $a \in [0, \pi]$. A similar result obtains for cosine series for $1 < \eta < 3$ now, and the corresponding functions f verify $x^{-\eta}(f(x) - f(0)) \in L[0, \pi]$ and also

$|x-a|^{-\eta}(f(x)-f(a)) \in L[0, \pi]$ when $1 < \eta < 2$. This result, as well as the next eight, is from Boas's book [1967].)

4.2 Assume $0 \leqslant \eta \leqslant 1$ and λ_n decreases to 0. Then $\sum \lambda_n \sin nx$ converges to $g(x)$ and $x^{-\eta}g \in L[0, \pi]$ if and only if $\sum n^{\eta-1}\lambda_n$ converges. (*Hint*: To show the necessity, note that the partial sums of $\sum n^{\eta-1} \sin nx$ are uniformly bounded and of order $O(x^{-\eta})$ for each η; separate the cases $0 < \eta < 1$ and $\eta = 0, 1$. Thus if $x^{-\eta}g(x)$ is integrable, by the dominated convergence theorem we get that

$$\frac{\pi}{2}\left|\sum_{n=N}^{M} n^{\eta-1}\lambda_n\right| = \left|\int_{[0,\pi]}\left\{\sum_{n=N}^{M} n^{\eta-1}\sin nx\right\}g(x)\,dx\right| \to 0 \qquad \text{as} \quad N, M \to \infty.$$

To prove the necessity, we sum by parts and invoke 4.1 above. The corresponding result for cosines is also true.)

4.3 Is the converse to 4.1 true?

4.4 Hardy's inequality 4.6 in Chapter I also holds in this general form: if $p, r > 1$ and f is integrable and positive, then

$$\int_{[0,A]} x^{-r}\left(\int_{[0,x]} f(t)\,dt\right)^p dx \leqslant c\int_{[0,A]} x^{-r}(xf(x))^p\,dx;$$

the best constant c is $(p/(r-1))^p$. State and prove a similar result for sequences.

4.5 If $\eta < 2$, then $t^{\eta-2}\int_{[0,t]} s^{-\eta}\sin s\,ds$ decreases in $(0, \pi)$ and the same is true for $t^{\eta-3}\int_{[0,t]} s^{-\eta}(1-\cos s)\,ds$ if $\eta < 3$. (*Hint*: Elementary calculus.)

4.6 If $\lambda_n \geqslant 0$, $p > 1$, $s > 0$, and $c < sp - 1$, then

$$\sum_{n=1}^{\infty} n^c\left(\sum_{k=n}^{\infty}\lambda_k\right)^p < \infty \tag{4.1}$$

implies

$$\sum_{n=1}^{\infty} n^{c-sp}\left(\sum_{k=1}^{n} k^s\lambda_k\right)^p < \infty. \tag{4.2}$$

Moreover, if $c > -1$, then (4.2) implies (4.1). (*Hint*: Sum by parts and invoke Hardy's inequality for sequences at the appropriate places.)

4.7 If the λ_n's are the Fourier sine coefficients of the continuous function g, $1 < p < \infty$, $1/p < \eta < (1/p)+1$ and $\lambda_n \geqslant 0$, then $|x-a|^{-\eta} \times (g(x)-g(a)) \in L^p[0, \pi)$ for every a, $0 \leqslant a < \pi$, if and only if $\sum_1^{\infty} n^{p\eta-p-2}(\sum_{k=1}^{n} k\lambda_k)^p < \infty$ (or equivalently $\sum n^{p\eta-2}(\sum_{k=n}^{\infty}\lambda_k)^p < \infty$). (*Hint*: We do the necessity first; assume $x^{-\eta}g(x) \in L^p[0, \pi]$. By 4.1 and 4.3 above, $\sum n^{\eta-1}\lambda_n$ converges. Now if $\varepsilon > 0$,

$$\left|\int_{(\varepsilon,x)} s^{-\eta}\sin ns\,ds\right| = n^{\eta-1}\left|\int_{(\varepsilon n,xn)} s^{-\eta}\sin s\,ds\right| = O(n^{\eta-1}),$$

uniformly in ε and x. Hence $\int_{(\varepsilon,x)} s^{-\eta} g(s)\,ds = \sum \lambda_n \int_{(\varepsilon,x)} s^{-\eta} \sin ns\,ds$ converges uniformly in ε, and by letting $\varepsilon \to 0^+$ we obtain

$$\int_{[0,x]} s^{-\eta} g(s)\,ds = \sum \lambda_n \int_{[0,x)} s^{-\eta} \sin ns\,ds \tag{4.3}$$

and each term on the right-hand side of (4.3) is positive. Applying 4.4 to the left-hand side of (4.3) we get that

$$\int_{[0,\pi]} (\sum \lambda_n x^{-1} \int_{[0,x)} s^{-\eta} \sin ns\,ds)^p\,dx < \infty,$$

which implies that also

$$\int_{[0,\pi)} \left(\sum n\lambda_n x^{-1}(nx)^{\lambda-2} \int_{[0,nx)} s^{-\eta} \sin u\,du\right)^p < \infty. \tag{4.4}$$

Now, we only decrease the integral in (4.4) by replacing the sum by $\sum_{n=1}^{1/x}$ and then replacing $(nx)^{\eta-2} \int_{[0,nx)} s^{-\eta} \sin u\,du$ by its minimum (here use 4.5); thus, $\int_{[0,\pi)}(\sum_{n=1}^{1/x} n\lambda_n x^{1-\eta})^p\,dx < \infty$. Conversely, for any $N > 0$,

$$\left(\int_{[0,\pi)} \left|t^{-\eta} \sum_{k=1}^{N} \lambda_k \sin kt\right|^p dt\right)^{1/p} \leqslant \left(\int_{[0,\pi)} \left|t^{-\eta} \sum_{k=1}^{1/t} \lambda_k \sin kt\right|^p dt\right)^{1/p} + \left(\int_{[0,\pi)} \left|t^{-\eta} \sum_{k=1/t}^{N} \lambda_k \sin kt\right|^p dt\right)^{1/p},$$

and both terms are easily estimated. There is a corresponding statement for cosines.)

4.8 Assume that λ_n decreases to 0 and denote by f either the Fourier cosine or sine series with coefficients λ_n (cosine and sine results involving L^p integrability are usually equivalent because the series are conjugate to each other). Then for $1 < p < \infty$ and $-1/p' < \eta < 1/p$, $x^{-\eta} g(x) \in L^p[0, \pi]$ if and only if $\sum n^{p\eta+p-2} \lambda_n^p$ converges. (*Hint*: We deduce the cosine version from the sine version of 4.7 above; the sine version is obtained similarly. To show the necessity, assume for simplicity that $\lambda_0 = 0$, then

$$F(x) = \int_{[0,x)} f(t)\,dt = \sum n^{-1} \lambda_n \sin nx$$

and by 4.4, $x^{-\eta-1} F(x) \in L^p[0, \pi)$. Thus by 4.7 $\sum n^{p\eta-2}(\sum_{k=1}^{n} \lambda_k)^p < \infty$, which gives the desired conclusion since the λ_k's decrease. To prove the sufficiency note that $2f(x) \sin x = \sum(\lambda_n - \lambda_{n+2}) \sin((n+1)x)$; therefore, we may invoke 4.7 again (with $\eta + 1$ in place of η there) and obtain $x^{-\eta-1}(xf(x)) \in L^p[0, \pi)$, i.e., $x^{-\eta} f(x) \in L^p[0, \pi)$, provided that

$$\sum n^{p\eta+p-2} \left(\sum_{k=n}^{\infty} |\lambda_k - \lambda_{k+2}|\right)^p$$

converges. But since $\sum_{k=n}^{\infty}|\lambda_k - \lambda_{k+2}| = \lambda_n + \lambda_{n+1}$ we are done.)

4.9 Assume $f(x)$ is a positive, decreasing function in $[0, \pi)$, $1 < p < \infty$ and $-1/p' < \eta < 1/p$. Furthermore, assume that the λ_n's are either the Fourier cosine or sine coefficients of f. Show that $\sum n^{-\eta p}|\lambda_n|^p$ converges if and only if $x^{p\eta+p-2}f(x)^p$ is integrable over $[0, \pi)$. (*Hint*: Dualize 4.8.)

4.10 Assume $w(x) > 0$ for $x \in T$ is such that for all $\lambda > 0$, $\int_{\{w(x)>\lambda\}} w(x)^{-2}\,dx \leq A/\lambda^2$, with A independent of λ. If $1 < p \leq 2$ and $\{c_j\} \in l^p$, then there is $f \in L(T)$ such that $c_j(f) = c_j$ and

$$\left(\int_T |f(x)|^p w(x)^{p-2}\,dx\right)^{1/2} \leq c\left(\sum |c_j|^p\right)^{1/p}.$$

Also if $2 \leq q < \infty$ and $f \sim \sum c_j e^{ijx}$, then

$$\left(\sum |c_j|^q\right)^{1/q} \leq \left(\int_T |f(x)|^q w(x)^{q-2}\,dx\right)^{1/q}.$$

Can you think of "rearranged" statements such as Paley's theorem?

4.11 The Hausdorff–Young theorem and Paley's theorem are not equivalent; indeed, let $c_0 = c_1 = c_{-1} = 0$, $c_j = (|j| \ln|j|)^{-3/4}$, $|j| \geq 2$. Then $\sum c_j^{4/3} = \infty$, but $\sum c_j^{4/3}|j|^{4/3-2} < \infty$.

4.12 There is a continuous function f so that $\sum |c_j(f)|^{2-\varepsilon} = \infty$ for each $\varepsilon > 0$. (*Hint*: Put $f(x) = \sum_{j=1}^{\infty} j^{-2}2^{-j/2}(P_j(x) - P_{j-1}(x))$, where the P_j's are the Rudin–Shapiro polynomials introduced in 5.38 of Chapter V. The result, which, in particular, shows that there is no Hausdorff–Young theorem for $p > 2$, is Carleman's, the example is from Katznelson's book.)

4.13 The Hausdorff–Young inequality is best possible in the sense that if for $f \in L^p(T)$, $1 < p < 2$, $\sum |c_n(f)|^r < \infty$, then $r \geq p'$. (*Hint*: Take a look at the function $|x|^{-\eta}$, $0 < \eta < 1$, which is in $L^p(T)$ provided that $\eta p < 1$, and its Fourier coefficients $c_j \sim |j|^{-(1-\eta)}$, which are in l^r only when $r(1-\eta) > 1$.)

4.14 Assume that $f \in L \ln L(T)$, $f \sim \sum c_j e^{ijx}$, and show that there are constants A, B such that

$$\sum \frac{|c_j|}{1+|j|} \leq A + B\int_T |f(x)| \ln^+|f(x)|\,dx.$$

Is the conclusion still true if we replace $|c_j|$ by $|c_j^*|$ in the above sum?

4.15 If $f \sim \sum c_j e^{ijx}$ and $|c_j| \leq 1/|j| + 1$, then $e^{\lambda|f(x)|}$ is integrable, provided $0 < \lambda$ is sufficiently small. (*Hint*: Use Hausdorff–Young to estimate $\|f\|_p$, $p = 2, 3, \ldots$.)

4.16 Suppose $\{\lambda_j\} \in M_q^p$, $1 \leq p \leq 2$, $1 < q \leq 2$. Show that $\sum(1+|j|)^{-q'/p'-\varepsilon}|\lambda_j|^{q'} < \infty$ for each $\varepsilon > 0$. (*Hint*: There is an L^p function with $c_j(f) \sim |j|^{-1/p'-\delta}$, $n \neq 0$. Since we assume that $\sum \lambda_j c_j(f)e^{ijx} \in L^q(T)$ we may invoke the Hausdorff–Young theorem.)

4.17 Assume that T is a sublinear operator simultaneously of weak-type $(2,2)$ and of type $(1,\infty)$, in both cases with norm $\leqslant 1$. Let $B(t)=\int_{[0,t)} b(s)\,ds$, where b is positive and nondecreasing and $A(t)=\int_{[0,t)} a(s)\,ds$, where a is continuous and strictly increasing from 0 to ∞ with s. Show that if $t\int_{[0,t)} b(s)(ds/s^2)\leqslant 1/2\bar{a}(2/t)$, for $t>0$, then $\int_Y B(|Tf(y)|/2)\,d\nu(y)\leqslant 1$ whenever $\int_X A(|f(x)|)\,d\mu(x)\leqslant 1$. (*Hint*: $\bar{a}$ is defined in Theorem 1.6; the result is from Jodeit–Torchinsky [1971].)

4.18 Assume that $A(s)$ is as in Theorem 1.6, that $\{n_k\}$ is a fixed Hadamard sequence, i.e., $n_{k+1}/n_k\geqslant\lambda>1$, and that $\int_T A(|f(x)|)\,dx\leqslant 1$. Show that $\lim_{k\to\infty} s_{n_k}(f,x)=f(x)$ a.e. (*Hint*: $\sum_j R\bar{A}R(|c_j(s_{n_k}(f)-\sigma_{n_k}(f))|/2)<\infty$ (cf. 7.25 in Chapter IV).)

4.19 The real Laplace transform F of f given for $x>0$ by $F(x)=\int_{[0,\infty)} e^{-xu}f(u)\,du$ is a linear mapping simultaneously of types $(1,\infty)$ and (2.2).

4.20 The "local" control that $M_\alpha f$ exhibits over $I_\alpha f$ in (2.20) may also be expressed in terms of the so-called "good-λ inequalities," about which we will have more to say later on; these results are quite important in the understanding of the weighted norm inequality problem. In this setting, the good-λ inequality assumes the following form: Let $I\subseteq T$, $\lambda>0$, and, for positive constants ε, δ, let $E=\{x\in I: I_\alpha f(x)>\varepsilon\lambda,\ M_\alpha f(x)\leqslant\delta\lambda\}$. Then there are constants ε_0 and k, depending only on α, so that if f is a nonnegative function and I contains a point where $I_\alpha f(x)\leqslant\lambda$, then $|E|\leqslant k(\delta/\varepsilon)^{1/(1-\alpha)}|I|$ for all $\varepsilon>\varepsilon_0$. (*Hint*: Put $g=f\chi_{2I}$, $h=f-g$. We may assume there is a point $t\in I$ so that $M_\alpha f(t)\leqslant\delta\lambda$, for otherwise E is empty and there is nothing to prove. Also observe that by Theorem 2.2 there is a constant c (depending only on α) so that

$$|\{I_\alpha g>\varepsilon\lambda/2\}|\leqslant 2^{1/(1-\alpha)}c\left(\int_{2I}\frac{f(x)\,dx}{\varepsilon\lambda}\right)^{1/(1-\alpha)}$$
$$\leqslant c\left(\frac{M_\alpha f(t)}{\varepsilon\lambda}\right)^{1/(1-\alpha)}|I|\leqslant k\left(\frac{\delta}{\varepsilon}\right)^{1/(1-\alpha)}|I|.$$

Next let $s\in I$ be the point where $I_\alpha f(x)<\lambda$. Since for some constant $L>1$ and for $x\in I$ and $y\notin 2I$ we have that $|s-y|\leqslant L|x-y|$, it is immediate to see that $I_\alpha h(x)\leqslant L^{1-\alpha}I_\alpha f(s)\leqslant L\lambda$. If we pick now $\varepsilon_0=2L$, then for $\varepsilon\geqslant\varepsilon_0$ the set $\{x\in I: I_\alpha h(x)>\varepsilon\lambda/2\}$ is empty. This result is from Muckenhoupt–Wheeden [1974].)

4.21 Assume that $f\geqslant 0$. Then for $0<\alpha,\ \delta<1$, $\|I_{\alpha\delta}f\|_r\leqslant c\|f\|_p^{1-\delta}\|I_\alpha f\|_q^{\delta}$, where $1/r=(1-\delta)/p+\delta/q$, $1<p<q\leqslant\infty$. (*Hint*: As in (2.5) bound the integral defining $I_{\alpha\delta}f(x)$ by $J+K$; estimate K by

$$\eta^{\alpha(\delta-1)}\int_{|t-x|\geqslant\eta} f(t)|x-t|^{\alpha-1}\,dt\leqslant\eta^{\alpha(\delta-1)}I_\alpha f(x)$$

and minimize. This result and the next two are from Hedberg's paper [1972].)

4.22 Suppose that $f \geqslant 0$, then $\|I_{\alpha\delta}(f^t)\|_r \leqslant c\|f\|_p^{t-\delta}\|I_\alpha f\|_q^\delta$, where $0 < \alpha$, $\delta < 1$, $0 < p < \infty$, $0 < q \leqslant \infty$, $\delta < t < \delta + (1-\delta)p$ and $1/r = (t-\delta)/p + \delta/q$. (*Hint*: We consider only the case $t > 1$ as $t = 1$ is Problem 4.21 and $0 < t < 1$ uses similar ideas. As before, bound the integral defining $I_{\alpha\delta}(f^t)$ by $J + K$; $J \leqslant c\eta^{\alpha\delta}M(f^t)(x)$. To estimate K choose $s < p$ so that $t < \delta + (1-\delta)s$ and use Hölder's inequality to show that $K \leqslant c\eta^{-\alpha\delta/(t-1)}M(f^s)(x)$; now minimize.)

4.23 Suppose that $f \in L^p$ and let $p = 1/\alpha > 1$. Then there is a constant c (which may be computed explicitly; this is left for the reader) so that for each interval $I \subseteq T$ and $\varepsilon > 0$

$$\int_I \exp\left(c\left|\frac{|I_\alpha f(x)|}{\|f\|_p} - \varepsilon\right|^{p'}\right) dt \leqslant c_\varepsilon |I|.$$

(*Hint*: Assume $\|f\|_p = 1$, estimate $|I_\alpha f(x)| \leqslant J + c(\ln(|I|/2\eta))^{1-1/p}$, and minimize.)

4.24 The condition $q_0 \neq q_1$ is essential in the interpolation theorem 2.3. (*Hint*: For $f:[0,1] \to C$ consider the mapping

$$f = Tf(x) - F(x) = x^{-1/2}\int_{[0,x]} f(t)\, dt.$$

Then T is of weak-type $(p, 2)$ when $1 \leqslant p \leqslant 2$, but is not of strong-type $(p, 2)$ for any p in that range.)

4.25 For f's as above consider the continuity properties of the mappings $f(x) \to Tf(x) = x^\eta \int_{[0,x)} t^\alpha f(t)\, dt$, where α, η are real numbers and not necessarily positive. Construct examples of operators of weak-type (p, q) but not of type (p, q) for any $1 \leqslant p \leqslant q < \infty$, and consider whether the interpolation theorem holds for $q < p$.

4.26 Suppose T is an operator with the following property: for each $\lambda > 0$, T may be written as $T_1 + T_2$, where T_1 maps $L^p(x)$ into $L^\infty(Y)$ with norm $\leqslant \lambda/2$ and T_2 is of weak-type (p, p) with norm $\leqslant c\lambda^{1-q/p}$, $q > 0$. Show that T is of weak-type (p, q). (*Hint*: For $\|f\|_p = 1$ we have $\nu(\{|Tf| > \lambda\}) \leqslant \nu(\{|T_1 f| > \lambda/2\}) + \nu(\{|T_2 f| > \lambda/2\}) \leqslant 0 + c\lambda^{p-q}\lambda^{-p} = c\lambda^{-q}$. This gives a different proof of Theorem 2.1, 2.2.)

4.27 Assume T is a sublinear operator of weak-type (p_0, q_0), $0 < p_0 \leqslant q_0 < \infty$. A sufficient condition that T also be of weak-type (p, q), where $1/p = 1/q = 1/p_0 - 1/q_0$, $p \geqslant p > 1$, is that for every sequence $\{I_j\}$ of pairwise disjoint intervals and every function h in $L^p(T)$ supported in $\bigcup I_j$ and such that $\int_{I_j} h(x)\, dx = 0$, all j, the following estimate holds: $|\{x \in T \setminus \bigcup 2I_j : |Th(x)| > \lambda\}| \leqslant c(\|h\|_p/\lambda)^q$. (*Hint*: We may restrict ourselves to nonnegative functions f with $\|f\|_p = 1$ and λ large, >1, say. By the Calderón–Zygmund decomposition of f^p at level λ we obtain a sequence $\{I_j\}$

of pairwise disjoint intervals such that

$$\lambda^q \sim \frac{1}{|I_j|}\int_{I_j} f(x)^p\, dx,$$

$$\frac{1}{|I_j|}\int_{I_j} f(x)\, dx \leqslant (2\lambda^q)^{1/p},$$

$$f(x)^p \leqslant \lambda^q \qquad \text{a.e. in} \quad T\setminus\bigcup I_j$$

and

$$\sum|I_j| \leqslant 1/\lambda^q.$$

Let now $g(x) = \sum_j f_{I_j}\chi_{I_j}(x) + f(x)\chi_{T\setminus\bigcup I_j}(x)$, $b(x) = f(x) - g(x)$. The estimate for the term involving b is easy; also $|\{|Tg| > \lambda\}| \leqslant c(\|g\|_{p_0}/\lambda)^{q_0}$, which together with $g(x) \leqslant 2\lambda^{q/p}$ a.e., $\|g\|_p \leqslant 1$, and the relations between p and q, gives the desired conclusion. This result is from Baishanski and Coifman's paper [1978].)

4.28 Suppose T is a sublinear operator simultaneously of weak-type $(1, q)$ and (p, q_1), with $1 < q$, $p < q_1$, $q < q_1$. Then there are constants A, B independent of f such that

$$\int_T |Tf(x)|^q\, dx \leqslant A + B\left(\int_T |f(x)|(\ln^+|f(x)|)^{1/q}\, dx\right)^q. \tag{4.5}$$

(*Hint*: By (4.16) in Chapter IV, it readily follows that there are constants c_1, c_2, c_3 so that for $\lambda > 0$,

$$|\{|Tf| > \lambda\}| \leqslant c_1\left(\frac{1}{\lambda^{q_1}}\left(\int_{\{|f|\leqslant c_2\lambda\}} |f(x)|^p\, dx\right)^{q_1/p} + \frac{1}{\lambda^q}\left(\int_{\{|f|> c_3\lambda\}} |f(x)|\, dx\right)^q\right).$$

Once we show that

$$J = \int_{[c_3,\infty)} \lambda^{q-1}\frac{1}{\lambda^q}\left(\int_{\{|f|> c_3\lambda\}} |f(x)|\, dx\right)^q d\lambda$$

is dominated by the right-hand side of (4.5) we are done, and this is not hard to do. The result is O'Neil's [1966]; however, in the particular case of the fractional integral, it had already been noted by Zygmund.)

4.29 Let $0 \leqslant 1/q_i = \beta_i \leqslant 1/p_i = \alpha_i < \infty$, $i = 0, 1$, $\alpha_0 \neq \alpha_1$, $\beta_0 \neq \beta_1$, and let $y = \varepsilon x + \gamma$ be the equation of the line passing through the points (α_i, β_i). Assume that the generalized Young's functions A, B are given by $A(u) = \int_{[0,u)} a(s)\, ds$, $B(u) = \int_{[0,u)} b(s)\, ds$, with a and b monotone and further assume that, if $M = \max(q_0, q_1)$ and $m = \min(q_0, q_1)$, then $B(u)/u^M$ decreases and $B(u)/u^m$ increases and

$$\int_{[0,u)} (B(s)/s^{m+1})\, ds = O(B(u)/u^m)$$

and

$$\int_{[u,\infty)} (B(s)/s^{M+1})\, ds = O(B(u)/u^M).$$

Then if $B^{-1}(u) = A^{-1}(u^\varepsilon)u^\gamma$, there is a constant c, independent of f, so that $\int_Y B(|Tf(y)|/c)\, d\nu(y) \leq 1$ whenever $\int_X A(|f(x)|)\, d\mu(x) \leq 1$. (*Hint*: Set the monotone function $z^{-1}(s) = B^{-1}(A(s)^{1/\varepsilon})$.)

4.30 (Young's Convolution Theorem. $L^p * L^r \subseteq L^q$, $1/q + 1 = 1/p + 1/r$. (*Hint*: Fix $f \in L^p(T)$, $1 < p < \infty$, then convolution with f gives a bounded mapping from L into L^p and from $L^{p'}$ into L^∞; now interpolate. Other proofs give a better constant.)

4.31 More generally, wk-$L^p * L^r \subseteq L^q$, same range of p, q, r's as in 4.30.

4.32 Assume that $1/p + 1/q > 1$, $1 < p$, $q \leq 2$; then $L^p * L^q$ is not included in $\bigcup_{s>r} L^s$, $1/r + 1 = 1/p + 1/q$. (*Hint*: For $k > 0$ put $a_n = 1/(k+1)^{2/p}2^{(k+1)/p'}$ for $2^k \leq n < 2^{k+1}$, and extend a_n as an even sequence for all n's and similarly for $\{b_n\}$ with q in place of p. Since $\sum a_n^p |n|^{p-2}$, $\sum b_n^q |n|^{q-2} < \infty$ there are functions $f \sim \sum a_n e^{inx} \in L^p$ and $g \sim \sum b_n e^{inx} \in L^q$. However, since $\sum (a_n b_n)^s |n|^{s-2} = \infty$ for each $s > r$, $f * g$ cannot be in L^s. This example is from Quek and Yap's paper [1983].)

4.33 If $\alpha > 0$, $0 < \varepsilon < 1$, $1 < p < \varepsilon/\alpha$, $1 \leq q \leq \infty$ and $f \in L^p$ and $M_{\varepsilon/p} f \in L^q$, then $I_\alpha f \in L^r$ and $\|I_\alpha f\|_r \leq c\|M_{\varepsilon/p} f\|_q^{\alpha p/\varepsilon} \|f\|_p^{1-\alpha p/\varepsilon}$. (This result is from Adams' work [1975].)

4.34 Show that $M_p^1 = \mathscr{F}L^p$, $1 < p \leq \infty$, i.e. $\{\lambda_j\} \in M_p^1$ if and only if there is a function $\phi \in L^p(T)$ so that $c_j(\phi) = \lambda_j$, all j. (*Hint*: The sufficiency is obvious; as for the necessity, if $A \sim \sum \lambda_j e^{ijx}$, then $\|A * K_n\|_p \leq c\|K_n\|_1$ for all n. What does the statement imply concerning A_p^1?)

4.35 The space of multiplier sequences from $L^p(T)$ into $C(T)$ is $\mathscr{F}L^{p'}$, $1/p + 1/p' = 1$, $1 \leq p < \infty$, and so is M_∞^p. What does the statement say about A_∞^p?

CHAPTER

VII

Harmonic and Subharmonic Functions

1. ABEL SUMMABILITY, NONTANGENTIAL CONVERGENCE

We refer here to yet another classic, and very important, summability method. This method requires the identification of T with ∂D, the boundary of the unit disk $D = \{z \in C : |z| < 1\}$ in the complex plane.

Given $0 \leqslant \alpha < \pi/2$ we define the set $\Omega_\alpha(0)$ with vertex at 1, that is, e^{i0}, and opening α as the convex hull of the disk of radius $\sin \alpha$ and $\{1\}\backslash\{1\}$,

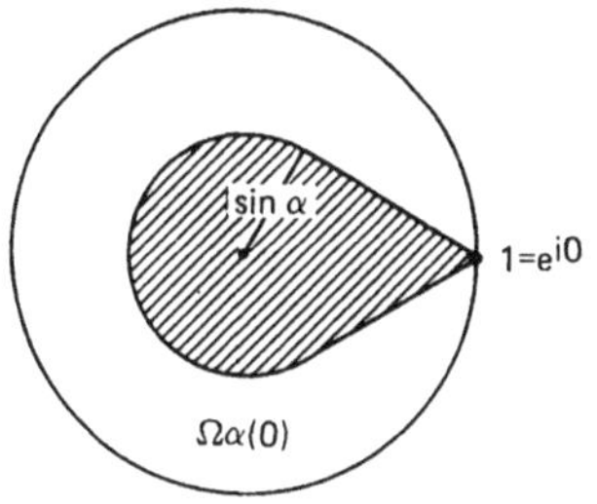

This set corresponds to the notion of "nontangential" approach to 1. It is readily seen that points z in $\Omega_\alpha(0)$ satisfy the condition $1 \leqslant |1 - z|/(1 - |z|) \leqslant 2 \max(1/(1 - \sin \alpha), 1/\cos \alpha)$. Thus it is equivalent, and often simpler, to consider instead the "cone"

$$\Gamma_\alpha(0) = \{z \in D: |1 - z|/(1 - |z|) \leqslant \alpha\}, \qquad \alpha \geqslant 1. \tag{1.1}$$

We are now ready for

Definition 1.1. Given $\alpha \geqslant 1$ and a function $f(z)$ defined in D, we say that

f converges (Abel) nontangentially of order α to L as $z \to 1$ provided that

$$\lim_{z\to 1, z\in\Gamma_\alpha(0)} f(z) = L,$$

and we denote this by $\lim_{z\to 1} f(z) = L(A_\alpha)$.

When $\alpha = 1$ we call the approach "radial," for then z takes values on the radius joning 0 to 1. Similarly, in the case of series we have

Definition 1.2. Let $\{c_j\}$ be a sequence of complex numbers. We say that the series $\sum_{j=0}^{\infty} c_j$ is (Abel) nontangentially convergent of order $\alpha(\geq 1)$ to s, and denote this by $\sum_{j=0}^{\infty} c_j = s(A_\alpha)$, provided that for $f(z) = \sum_{j=0}^{\infty} c_j z^j$, $\lim_{z\to 1} f(z) = s(A_\alpha)$. The usual algebraic properties hold in this case as well. For instance, if s_0 and s_1 are finite numbers and $\sum c_j = s_0(A_\alpha)$, $\sum d_j = s_1(A_\alpha)$, then also $\sum(c_j + d_j) = s_0 + s_1(A_\alpha)$.

It would be reassuring to know that convergent series also converge nontangentially to the same limit; in fact more is true.

Proposition 1.3. Suppose $\sum_{j=0}^{\infty} c_j = s(C, 1)$. Then also $\sum_{j=0}^{\infty} c_j = s(A_\alpha)$, $1 \leq \alpha < \infty$.

Proof. Let s_n and σ_n denote the partial sums and Cesàro means of order n of $\{c_j\}$; our assumption implies that $s_n = o(n)$. In the first place, observe that

$$\sum_{j=0}^{n} c_j z^j = (1-z)\sum_{j=0}^{n-1} s_j z^j + s_n z^n. \tag{1.2}$$

Therefore, if the limit of either side in (1.2) exists as $n \to \infty$, then the limit of the other side also exists and they are equal. Moreover, since $s_n z^n = o(1)$, then also

$$\sum_{j=0}^{\infty} c_j z^j = (1-z)\sum_{j=0}^{\infty} s_j z^j \tag{1.3}$$

provided either series converges. Summing by parts once again, and since $n\sigma_{n-1} z^{n-1} = o(1)$ for $z \in D$ as $n \to \infty$, a similar argument gives

$$\sum_{j=0}^{\infty} c_j z^j = (1-z)^2 \sum_{j=0}^{\infty} (j+1)\sigma_j z^j = f(z), \tag{1.4}$$

say, again provided either sum converges. Now, it is readily seen that the right-hand side of (1.4) converges absolutely for $|z| < 1$ with sum $\leq c/(1-|z|)^2$, and so the left-hand side is also finite. Next we show that $f(z)$ tends to s nontangentially. In the process we make use of the identity

$$1 = (1-z)^2 \sum_{j=0}^{\infty} (j+1) z^j, \qquad |z| < 1. \tag{1.5}$$

More precisely, given $\varepsilon > 0$ we show there exists $\delta > 0$ so that $|f(z) - s| \leqslant \varepsilon$ provided $z \in \Gamma_\alpha(0)$ and $|1 - z| \leqslant \delta$. Indeed, combining (1.4) and (1.5), it readily follows that

$$f(z) - s = (1 - z)^2\left(\sum_{j=0}^{N} + \sum_{j=N+1}^{\infty}\right)(j + 1)(\sigma_j - s)z^j = I + J,$$

say. Now, since

$$\sum_{j=N+1}^{\infty} (j + 1)r^j \leqslant \sum_{j=0}^{\infty}(j + 1)r^j = \frac{1}{(1 - r)^2},$$

if we choose N so that $|\sigma_j - s| \leqslant \varepsilon/2\alpha^2$ for $j \geqslant N + 1$, then it readily follows that for z in $\Gamma_\alpha(0)$,

$$|J| \leqslant \left(\frac{2\alpha^2}{|1 - z|^2\varepsilon}\right) \sum_{j=N+1}^{\infty} (j + 1)|z| \leqslant \frac{\varepsilon}{2}.$$

Now that N has been fixed, we note that

$$|I| \leqslant c|1 - z|^2 \sum_{j=0}^{N}(j + 1) \leqslant c_N\delta^2 \leqslant \frac{\varepsilon}{2}$$

provided δ is small enough. ■

We are interested in applying Proposition 1.3 to Fourier series, more specifically to Lebesgue's theorems 4.2 in Chapter II and 6.3 in Chapter III. Since the expression appearing in the Definition 1.2 is one sided, given $f \in L$, $f \sim \sum c_j e^{ijt}$, we introduce the notations

$$C_0(t) = c_0, \qquad C_j(t) = c_{-j}e^{-ijt} + c_j e^{ijt}, \qquad j \geqslant 1 \tag{1.6}$$

and

$$\tilde{C}_0(t) = 0, \qquad \tilde{C}_j(t) = (-i)(-c_{-j}e^{-ijt} + c_j e^{ijt}), \qquad j \geqslant 1. \tag{1.7}$$

Proposition 1.3 asserts that actually

$$\sum_{j=0}^{\infty} C_j(t) = f(t)(A_\alpha) \qquad \text{almost every} \quad t \quad \text{in} \quad T \tag{1.8}$$

and

$$\sum_{j=0}^{\infty} \tilde{C}_j(t) = Hf(t)(A_\alpha) \qquad \text{almost every} \quad t \quad \text{in} \quad T. \tag{1.9}$$

We may, and do, assume that (1.8) and (1.9) hold simultaneously. More precisely, if $z = re^{ix} \in \Gamma_\alpha(0)$, then for almost every t in T

$$\lim_{z\to 1, z\in\Gamma_\alpha(0)} \sum_{j=0}^{\infty} C_j(t)r^j e^{ijx} = f(t) \tag{1.10}$$

and

$$\lim_{z\to 1, z\in\Gamma_\alpha(0)} \sum_{j=0}^{\infty} \tilde{C}_j(t) r^j e^{ijx} = Hf(t). \tag{1.11}$$

We would like to unravel (1.10) and (1.11) and express them in terms of the original c_j's. First, since $\Gamma_\alpha(0)$ is symmeric, that is, $z \in \Gamma_\alpha(0)$ if and only if $\bar{z} \in \Gamma_\alpha(0)$, and as is readily seen

$$\sum_{j=0}^{\infty} C_j(t) r^j e^{ijx} = \sum_{j=-\infty}^{\infty} c_j r^{|j|} e^{i|j|x} e^{ijt},$$

then for almost every t in T the following is true:

$$\lim_{re^{\pm ix}\to 1, re^{\pm ix}\in\Gamma_\alpha(0)} \sum_{j=-\infty}^{\infty} c_j r^{|j|} e^{\pm i|j|x} e^{ijt} = f(t). \tag{1.12}$$

Similarly, by (1.11) now, and for the same t's,

$$\lim_{re^{\pm ix}\to 1, re^{\pm ix}\in\Gamma_\alpha(0)} \sum_{j=-\infty}^{\infty} (\operatorname{sgn} j) c_j r^{|j|} e^{\pm i|j|x} e^{ijt} = iHf(t). \tag{1.13}$$

Whence, by subtracting the "−" statement in (1.13) from the "+" statement in (1.12) and rearranging the expressions involved, we obtain

$$\lim_{re^{ix}\to 1, re^{ix}\in\Gamma_\alpha(0)} \left(\sum c_j r^{|j|} e^{ijx} e^{ijt} - \sum (\operatorname{sgn} j) c_j r^{|j|} e^{-ijx} e^{ijt}\right)$$

$$= f(t) - iHf(t), \qquad \text{a.e. in } \quad T. \tag{1.14}$$

Similarly, by adding the "−" statement in (1.12) and the "+" statement in (1.13) and rearranging the expressions involved, we obtain

$$\lim_{re^{ix}\to 1, re^{ix}\in\Gamma_\alpha(0)} \left(\sum c_j r^{|j|} e^{ijx} e^{ijt} - \sum (\operatorname{sgn} j) c_j r^{|j|} e^{-ijx} e^{ijt}\right)$$

$$= f(t) + iHf(t), \qquad \text{a.e. in } \quad T. \tag{1.15}$$

We have thus arrived at one of the most interesting and important results in this chapter, namely,

Theorem 1.4. Suppose $f \in L(T)$, $f \sim \sum c_j e^{ijt}$. Then for almost every t in T

$$\lim_{re^{ix}\to 1, re^{ix}\in\Gamma_\alpha(0)} \sum c_j r^{|j|} e^{ijx} e^{ijt} = f(t) \tag{1.16}$$

and

$$\lim_{re^{ix}\to 1, re^{ix}\in\Gamma_\alpha(0)} \sum (-i)(\operatorname{sgn} j) c_j r^{|j|} e^{ijx} e^{ijt} = Hf(t). \tag{1.17}$$

Proof. (1.16) follows by adding (1.14), (1.15), and (1.17), by subtracting (1.14) from (1.15), and invoking the symmetry of $\Gamma_\alpha(0)$ to change $-x$ into x. ■

2. THE POISSON AND CONJUGATE POISSON KERNELS

Expressions (1.16) and (1.17) in Theorem 1.4 correspond to the convolution of f with the functions $P(z)$ and $Q(z)$ with (absolutely convergent) Fourier series given by

$$P(z) = \sum_{j=-\infty}^{\infty} r^{|j|} e^{ijt}, \qquad z = re^{it}, \qquad 0 < r < 1 \tag{2.1}$$

and

$$\tilde{P}(z) = Q(z) = \sum_{j=-\infty}^{\infty} (-i)(\operatorname{sgn} j) c_j r^{|j|} e^{ijt}, \qquad z = re^{it}, \qquad 0 < r < 1 \tag{2.2}$$

respectively; these functions are called the Poisson and conjugate Poisson kernel. When convenient, we view these functions as defined in T and denote them by $P_r(t)$ and $Q_r(t)$, i.e., as a family of kernels indexed by $r, 0 \leq r < 1$. The first task at hand is to obtain an explicit expression for these kernels. By summing the geometric series in (2.1) we get

$$P_r(t) = \sum_{j=0}^{\infty} r^j e^{ijt} + \sum_{j=1}^{\infty} r^j e^{-ijt} = \frac{1}{1 - re^{it}} + \frac{re^{-it}}{1 - re^{-it}} \tag{2.3}$$

which easily reduces to

$$P_r(t) = \frac{1 - r^2}{1 - 2r\cos t + r^2} = \frac{1 - r^2}{(1 - r)^2 + 2r(1 - \cos t)} \,. \tag{2.4}$$

Similarly, by summing (2.2) we see that

$$\begin{aligned} Q_r(t) &= (-i)\sum_{j=1}^{\infty} r^j e^{ijt} - (-i)\sum_{j=1}^{\infty} r^j e^{-ijt} \\ &= (-i)\left(\frac{re^{it}}{1 - re^{it}} - \frac{re^{-it}}{1 - re^{-it}}\right), \end{aligned} \tag{2.5}$$

which also readily reduces to

$$Q_r(t) = \frac{2r\sin t}{1 - 2r\cos t + r^2} = \frac{2r\sin t}{(1 - r)^2 + 2r(1 - \cos t)} \,. \tag{2.6}$$

The interaction between P and Q is given by the expression $P + iQ$, which on account of (2.3) and (2.5) equals

$$\frac{1}{1 - re^{it}} + \frac{re^{it}}{1 - re^{it}} = \frac{1+z}{1-z}. \tag{2.7}$$

This is an analytic function of z in D; we will have more to say about this later on.

By (2.4) we note that $P_r(t)$ is a periodic, positive, even function of $t \in T$, monotone in $[0, \pi)$, which is dominated either by

$$\frac{1-r^2}{(1-r)^2} = \frac{1+r}{1-r} \leqslant \frac{2}{1-r} \tag{2.8}$$

or, on account of estimate (1.16) in Chapter I, by

$$\frac{1-r^2}{2r(1-\cos t)} \leqslant \frac{1-r}{2r(\sin(t/2))^2} \leqslant \frac{\pi(1-r)}{2rt^2}. \tag{2.9}$$

Also, by (2.1),

$$\frac{1}{2\pi}\int_T P_r(t)\,dt = 1. \tag{2.10}$$

The reader may have noted the similarity of Fejér's kernel $K_n(t)$ with the Poisson kernel $P_r(t)$ for $r = 1 - 1/(n+1)$; this hints at a similarity of results as well. For instance by (the continuous version of) Proposition 2.3 in Chapter IV there is a constant c, independent of f, so that

$$\sup_{r<1} |f * P_r(x)| \leqslant cMf(x). \tag{2.11}$$

But on account of statement (1.16) of Theorem 1.4 we expect a stronger result, of a nontangential nature, to hold. Before considering this point we list some properties of the conjugate Poisson kernel. By (2.6), $Q_r(t)$ is an odd function and $|Q_r(t)|$ is dominated by either

$$cr|t|/(1-r)^2 \qquad \text{or} \qquad c/|t|. \tag{2.12}$$

Also by (2.5)

$$\int_T Q_r(t)\,dt = 0. \tag{2.13}$$

As for $\|Q_r\|_1$, it does not remain bounded as $r \to 1^-$ (cf. 6.4); this is all we need to know about the kernels.

Now some definitions. Let $\Gamma_\alpha(x) = \{z \in D: e^{-ix}z \in \Gamma_\alpha(0)\}$, $1 \leqslant \alpha < \infty$. In other words $\Gamma_\alpha(x)$ is the cone $\Gamma_\alpha(0)$ rotated so that its vertex lies at e^{ix} rather than at 1; a similar definition is given for $\Omega_\alpha(x)$.

Definition 2.1. Assume $f \in L$ and $1 \leq \alpha < \infty$. We introduce the nontangential maximal function of $f * P_r$ of order α at x by

$$N_\alpha(f * P_r, x) = \sup_{z=re^{it}, z \in \Gamma_\alpha(x)} |f * P_r(t)|. \tag{2.14}$$

N_1 is called the radial maximal function.

We then have

Theorem 2.2. Assume that $f \in L$ and $1 \leq \alpha < \infty$. Then there is a constant c_α, independent of f, such that

$$N_\alpha(f * P_r, x) \leq c_\alpha Mf(x). \tag{2.15}$$

Proof. The radial maximal function was already considered in (2.11), and the nontangential case is a straightforward computation. Indeed, since each $z = re^{it}$ in $\Gamma_\alpha(x)$ is of the form $re^{i(x+s)}$, $re^{is} \in \Gamma_\alpha(0)$, it readily follows that

$$f * P_r(t) = I = \frac{1}{2\pi}\int_T f(u)P_r(x+s-u)\,du = \frac{1}{2\pi}\int_T f(x+u)P_r(s-u)\,du,$$

and consequently

$$|I| \leq \frac{1}{2\pi}\left(\int_{[-\pi,0)} + \int_{[0,\pi)}\right)|f(x+u)|P_r(s-u)\,du = J + K, \tag{2.16}$$

say. As the estimate for both summands in (2.16) is obtained in a similar way, we only do K. Let $F_x(u) = F(u) = \int_{[0,u)}|f(x+v)|\,dv$; then F is absolutely continuous and $F'(u) = |f(x+u)|$ a.e. Thus, by integrating by parts the integral in K, we note that it equals

$$F(u)P_r(s-u)]_0^\pi - \int_{[0,\pi)} f(u)P_r'(s-u)\,du = K_1 + K_2,$$

say. K_1 offers no difficulty: since $F(0) = 0$, $F(\pi) \leq cMf(x)$ and $P_r(s-\pi) \leq c_\alpha$, we see at once that $K_1 \leq c_\alpha Mf(x)$, which is of the right order. Moreover, since for $0 < u < \pi$

$$F(u)/\sin(u/2) \leq cMf(x),$$

also

$$K_2 \leq c\left(\int_{[0,\pi)} \sin\left(\frac{u}{2}\right)P_r'(s-u)\,du\right)Mf(x). \tag{2.17}$$

To bound the integral in (2.17) we distinguish two cases, namely, $0 < r \leq \frac{1}{2}$, and $\frac{1}{2} < r < 1$. The latter is pretty easy since then the integral does not exceed

$$\int_T |P_r'(u)|\,du = 2\int_{[-\pi,0)} P_r'(u)\,du \leq 2P_r(0) \leq \frac{4}{1-r} \leq 8.$$

As for the former we observe that the integral there is dominated by

$$\int_T \left|\sin\left(\frac{(s-u)}{2}\right)\right| |P_r'(u)|\, du$$
$$\leq c|s| \int_T |P_r'(u)|\, du + c\int_T |u|\,|P_r'(u)|\, du = K_3 + K_4,$$

say. To estimate K_4 we note that the integrand $uP_r'(u)$ is even and, consequently,

$$K_4 = -2\int_{[0,\pi)} uP_r'(u)\, du = -2uP_r(u)]_0^\pi + \int_{[0,\pi)} P_r(u)\, du$$
$$= -2\pi P_r(\pi) + 2\pi \leq 2\pi. \tag{2.18}$$

Finally, to bound K_3, observe that $\frac{1}{2}|s| \leq r|s| \leq c|r - re^{is}| \leq c|r-1| + c|1 - re^{is}|$; and consequently

$$K_3 \leq c(1-r)\frac{1}{1-r} + c\frac{|1-re^{is}|}{1-r} \leq c + c_\alpha. \tag{2.19}$$

Whence combining (2.18) and (2.19) gives $K_2 \leq c_\alpha Mf(x)$, which is also of the right order. ■

Corollary 2.3. Assume that $f \in L^p(T)$, $1 \leq p \leq \infty$. Then there is a constant $c = c_{\alpha,p}$ independent of f so that

$$\|N_\alpha(f * P_r)\|_p \leq c\|f\|_p, \qquad 1 < p \leq \infty \tag{2.20}$$

and

$$\lambda|\{N_\alpha(f * P_r) > \lambda\}| \leq c\|f\|_1. \tag{2.21}$$

Corollary 2.4. Suppose $f \in L^p(T)$, $1 < p \leq \infty$. Then there is a constant $c = c_\alpha$ independent of f so that

$$N_\alpha(f * Q_r, x) \leq cM(\tilde{f})(x) \tag{2.22}$$

and consequently for $1 < p < \infty$,

$$\|N_\alpha(f * Q_r)\|_p \leq c\|f\|_p. \tag{2.23}$$

Proof. Since $\sigma_n(f) * Q_r(x) = \tilde{\sigma}_n(f) * P_r(x)$ and $1 < p$, it follows that also $f * Q_r(x) = \tilde{f} * P_r(x)$ and the conclusion follows from Theorem 2.2 and Corollary 2.3. ■

As for convergence in L^p we have

Proposition 2.5. Assume $f \in L^p(T)$, $1 \leq p < \infty$. Then

(i) $\|f * P_r\|_p \leq \|f\|_p$, $r < 1$,
(ii) $\lim_{r\to 1}\|f * P_r - f\|_p = 0$, and
(iii) $\lim_{z = re^{ix}\to 1, z\in\Gamma_\alpha(0)}\|f * P_r(x + \cdot) - f\|_p = 0$.

A similar result holds for $p = \infty$ provided that $f \in C(T)$.

Proof. (i) and (ii) have a counterpart, with identical proof, in the case of Fejér's kernel. As for (iii), in case $p < \infty$, the proof is a simple application of Fatou's lemma since, by Theorem 2.2, $|f * P_r(x+t) - f(t)| \le cMf(t)$ and, by Theorem 1.4, $\lim_{z=re^{ix}\to 1, z\in\Gamma_\alpha(0)} f * P_r(x+t) = f(t)$ a.e. The case $p = \infty$ also follows easily. ∎

As for the conjugate Poisson Kernel, we have

Theorem 2.6. Assume $f \in L(T)$, then

$$\lim_{z=re^{ix}\to 1, z\in\Gamma_\alpha(0)} (f * Q_r(x+t) - H_{1-r}f(t)) = 0 \qquad \text{a.e.} \tag{2.24}$$

and there is a constant $c = c_\alpha$ independent of f such that

$$\sup_{z=re^{ix}, z\in\Gamma_\alpha(0)} |f * Q_r(x+t) - H_{1-r}f(t)| \le cMf(t) \qquad \text{a.e.} \tag{2.25}$$

Consquently, if $f \in L^p(T)$, $1 < p < \infty$, then also

$$\lim_{z=re^{ix}\to 1, z\in\Gamma_\alpha(0)} \|f * Q_r(x+\cdot) - H_{1-r}f\|_p = 0. \tag{2.26}$$

Proof. (2.24) and (2.25) obtain by arguments similar to those invoked in the case of Fejér's kernel, combined now with the ideas in Theorem 2.2 to incorporate the non-tangential convergence; we sketch the proof of the radial case. First, observe that, since $Q_1(x) = (2\sin x)/[2(1-\cos x)] = 1/\tan(x/2)$, $x \ne 0$, we have

$$\begin{aligned} f * Q_r(t) - H_{1-r}f(t) &= \frac{1}{2\pi}\int_{|x|\le 1-r} f(t-x)Q_r(x)\,dx \\ &\quad + \frac{1}{2\pi}\int_{1-r<|x|<\pi} f(t-x)(Q_r(x) - Q_1(x))\,dx \\ &= I_1(t) + I_2(t), \end{aligned} \tag{2.27}$$

say.

By (2.12) we see that

$$|I_1(t)| \le \frac{c}{1-r}\int_{|x|\le 1-r} |f(t-x)|\,dx$$

and consequently $|I_1(t)| \le cMf(t)$ and $|I_1(t)| = o(1)$, as $r \to 1^-$, at each Lebesgue point t of f. Moreover, since $Q_r(x) - Q_1(x) = ((1-r)/(1+r))Q_1(x)P_r(x)$ is an odd and monotone decreasing function in $(0, \pi)$, and by (2.12),

$$\frac{1-r}{1+r}Q_1(x) \le \frac{1-r}{1+r}\frac{1}{\sin(x/2)} < \pi, \qquad \text{for} \quad 0 < x < \pi,$$

then we may estimate

$$|I_2(t)| \leq c\int_T |f(t-x)|P_r(x)\,dx \leq cMf(t).$$

Also an argument similar to Theorem 6.3 in Chapter III obtains that $I_2(t) = o(1)$, as $r \to 1^-$. ■

Corollary 2.7. Suppose $f \in L(T)$. Then there is a constant $c = c_\alpha$ independent of f so that

$$N_\alpha(f * Q_r, t) \leq c(H^*f(t) + Mf(t)). \tag{2.28}$$

3. HARMONIC FUNCTIONS

The Laplacian Δ is the second order partial differential operator given by

$$\Delta = \frac{1}{r^2}\frac{\partial^2}{\partial t^2} + \frac{\partial^2}{\partial r^2} + \frac{1}{r}\frac{\partial}{\partial r}. \tag{3.1}$$

It is readily seen that $\Delta(r^{|j|}e^{ijt}) = 0$, all j, and consequently by superposition $\Delta(P_r(t)) = \Delta(f * P_r(t)) = 0$, $f \in L$, $re^{it} \in D$. This operator is also familiar in cartesian coordinates, i.e., in terms of $x = r\cos t$, $y = r\sin t$, and it is given by

$$\Delta = \frac{\partial^2}{\partial x^2} + \frac{\partial^2}{\partial y^2}. \tag{3.2}$$

In what follows we use (3.1) and (3.2) interchangeably. Functions u which verify $\Delta u = 0$ (in D) are called harmonic (in D). We begin by showing that in some sense the only harmonic functions are those which arise as Poisson integrals. More precisely, we have

Proposition 3.1. Assume $u \in C^2(D)$ is harmonic. Then there is a sequence of complex numbers $\{c_j\}$ such that

$$u(re^{it}) = \sum_{j=-\infty}^{\infty} c_j r^{|j|} e^{ijt}, \qquad re^{it} \in D. \tag{3.3}$$

Proof. By assumption the partial derivatives $\partial u/\partial x$, $\partial u/\partial y$ exist and are continuous in D. We construct a function $v(x, y)$ such that for $(x, y) \in D$ (or more precisely $x + iy \in D$),

$$\frac{\partial}{\partial x}u(x,y) = \frac{\partial}{\partial y}v(x,y), \qquad \frac{\partial}{\partial y}u(x,y) = -\frac{\partial}{\partial x}v(x,y). \tag{3.4}$$

A pair of functions (u, v) verifyinng (3.4) is called a Cauchy–Riemann pair and the function v is called a harmonic conjugate, or conjugate, to u; our claim is that to each harmonic function u in D we may assign a conjugate v. The construction in D, or any simply connected domain for that matter, is quite easy. Indeed, fix (x_0, y_0) in D and note that v is well defined by means of the path integral

$$v(x, y) - v(x_0, y_0) = \int_{(x_0, y_0)}^{(x,y)} dv, \tag{3.5}$$

where the path in (3.5) is totally contained in D, provided that dv is an exact differential for then the integral in (3.5) is path independent. But if we put

$$dv = -\frac{\partial}{\partial y} u(x, y)\, dx + \frac{\partial}{\partial x} u(x, y)\, dy$$

then this occurs since

$$\frac{\partial}{\partial y}\left(-\frac{\partial}{\partial y} u\right) = \frac{\partial}{\partial x}\left(\frac{\partial}{\partial x} u\right),$$

as u is harmonic. Thus (3.5) determines a conjugate v to u and by (3.4) the function $f(z) = u(z) + iv(z)$ is analytic in D. Therefore, there is a sequence of complex constants $\{d_j\}$ such that

$$f(z) = \sum_{j=0}^{\infty} d_j z^j, \qquad |z| < 1.$$

Moreover, since

$$u(z) = \operatorname{Re} f(z) = \frac{1}{2}\left(\sum_{j=0}^{\infty} d_j z^j + \sum_{j=0}^{\infty} \bar{d}_j \overline{z^j}\right),$$

it readily follows that

$$u(re^{it}) = \sum c_j r^{|j|} e^{ijt}, \qquad re^{it} \in D,$$

with

$$c_j = \tfrac{1}{2}\bar{d}_{-j}, \quad j < 0, \qquad c_0 = \tfrac{1}{2}(d_0 + \bar{d}_0), \qquad \text{and} \qquad c_j = \tfrac{1}{2} d_j, \quad j > 0.$$

By the way, the proof also shows that $v(re^{it}) = \sum c'_r r^{|j|} e^{ijt}$, with $c'_j = (-i)(\operatorname{sgn} j) c_j$, $j \neq 0$, $c'_0 = (-i/2)(d_0 - \bar{d}_0)$. ■

Remark 3.2. If $f(z) = u(z) + iv(z)$ is analytic in a region containing D, by the results discussed in this chapter it follows that

$$f(z) = \frac{1}{2\pi} \int_T \frac{1 + re^{i(t-x)}}{1 - re^{i(t-x)}} u(e^{ix})\, dx + iv(0),$$

where $z = re^{it}, 0 \leq r < 1$. This formula exhibits the close connection between an analytic function f and its real part u.

Next we discuss the analog to Fejér's theorem

Theorem 3.3. Suppose u is harmonic in D and that, in addition, for some $1 < p \leq \infty$,

$$\sup_{r<1}\left(\frac{1}{2\pi}\int_T |u(re^{it})|^p\,dt\right)^{1/p} = A < \infty.$$

Then there is a function $f \in L^p(T)$, $\|f\|_p \leq A$, such that $u(re^{it}) = f * P_r(t)$,

$$\lim_{re^{ix}\to 1, re^{ix}\in\Gamma_\alpha(0)} u(re^{i(x+t)}) = f(t) \qquad \text{a.e.}$$

and

$$\lim_{re^{ix}\to 1, re^{ix}\in\Gamma_\alpha(0)} \|u(re^{i(x+\cdot)}) - f\|_p = 0, \qquad 1 < p < \infty. \tag{3.6}$$

Proof. By Theorem 1.4 and Proposition 2.5 it suffices to show that $u(re^{it}) = f * P_r(t)$, $f \in L^p(T)$, $\|f\|_p \leq A$. By Proposition 3.1 it follows that $u(re^{it}) = \sum c_j r^{|j|} e^{ijt}$, and we are reduced to showing that the c_j's are the Fourier coefficients of an L^p function f. Let $\sigma_n(t)$ denote the Cesàro means of order n of the c_j's. By Fejér's theorem 3.3 in Chapter II it is enough to prove that $\|\sigma_n\|_p \leq A$, all n. We write

$$\sigma_n(t) = (\sigma_n(t) - \sigma_n * P_r(t)) + \sigma_n * P_r(t) = I(t) + J(t),$$

say, and take a closer look at each summand. Observe that

$$\begin{aligned} I(t) &= \sum_{|j|\leq n}\left(1 - \frac{|j|}{n+1}\right)(1 - r^{|j|})c_j e^{ijt} \\ &= (1-r)\sum_{1\leq|j|\leq n}\left(1 - \frac{|j|}{n+1}\right)\left(\frac{1-r^{|j|}}{1-r}\right)c_j e^{ijt} \\ &= (1-r)I_1(t), \end{aligned}$$

say. Since $I_1(t)$ is a trigonometric polynomial of degree n, we have that

$$\|I\|_p \leq (1-r)\|I_1\|_p = o(1), \qquad \text{as} \quad r \to 1^-. \tag{3.7}$$

As for $J(t)$, note that it actually equals $u(r\cdot) * 2K_n(t)$ and, consequently,

$$\|J\|_p \leq \|u(r\cdot)\|_p \|2K_n\|_1 \leq A, \qquad \text{all} \quad r, n. \tag{3.8}$$

Thus, by combining (3.7) and (3.8), we get that $\|\sigma_n\|_p \leqslant A$ and we have finished. As is usually the case, (3.6) holds for $p = \infty$ provided $f \in C(T)$. ■

As for $p = 1$ we have

Theorem 3.4. Suppose u is harmonic in D and that, in addition,

$$\sup_{r<1} \frac{1}{2\pi} \int_T |u(re^{it})|\, dt \leqslant A < \infty.$$

Then there is a finite Borel measure μ with total variation $\|\mu\| \leqslant A$, such that $u(re^{it}) = \mu * P_r(t)$. Furthermore, if

$$\phi(t) = \int_{[0,t)} d\mu(x), \qquad t \in T,$$

then also

$$\lim_{re^{ix}\to 1, re^{ix}\in\Gamma_\alpha(0)} u(re^{i(x+t)}) = \phi'(t) \qquad \text{a.e.}$$

If $u(re^{it}) \geqslant 0$, then there is a positive measure μ such that $u(re^{it}) = \mu * P_r(t)$. Finally if $\{u(re^{ix})\}_{r<1}$ is Cauchy in $L(T)$ as $r \to 1$, then there is an integrable function f such that $u(re^{it}) = f * P_r(t)$; f satisfies (3.6) with $p = 1$.

Proof. The existence of μ and the statement concerning the nontangential convergence follow as in Theorem 3.3. Note that we cannot assert in general that $u(re^{it}) = \phi' * P_r(t)$; the simplest example of this is the Poisson kernel $P_r(t)$ itself which has nontangential limits 0 at every point save 0 and yet corresponds to μ = Dirac delta at the origin.

Now, if u is a nonnegative harmonic function, $u(re^{it}) = \sum c_j r^{|j|} e^{ijt}$, then

$$\int_T |u(re^{it})|\, dt = \int_T u(re^{it})\, dt = 2\pi c_0 < \infty$$

and $u(re^{it}) = \mu * P_r(t)$. Therefore

$$u(r\cdot) * 2K_n(t) = \sigma_n(\mu) * P_r(t) \geqslant 0, \qquad \text{all} \quad n, r$$

and

$$\sigma_n(\mu, t) = \lim_{r\to 1^-} \sigma_n(\mu) * P_r(t) \geqslant 0 \qquad \text{a.e.} \tag{3.9}$$

By Theorem 3.5 in Chapter II, (3.9) implies that μ is a nonnegative measure.

Finally, in case $\{u(re^{it})\}$ is Cauchy in $L(T)$ as $r \to 1^-$, we infer two facts, to wit: since $L(T)$ is complete there is $f \in L$ such that $\lim_{r\to 1^-}\|u(r\cdot) - f\|_1 = 0$ and by the first part of the theorem there is a measure μ so that $u(re^{it}) = \mu * P_r(t)$. Since for each fixed s, $0 < s < 1$, $P_s(x - t) \in L^\infty(T)$, it follows at once that $\lim_{r\to 1} u(r\cdot) * P_s(x) = f * P_s(x)$. On the other hand, since (as is readily seen by direct examination of the Fourier series) $P_r * P_s = P_{rs}$, $0 < r$, $s < 1$, we also have that $u(r\cdot) * P_s(x) = \mu * P_{rs}(x)$, which tends to $\mu * P_s(x) = u(se^{ix})$ as $r \to 1^-$. Thus $u(se^{ix}) = f * P_s(x)$ and we are done. ∎

It is a natural question to consider what integrability properties, if any, are satisfied by the conjugates to harmonic functions verifying the assumptions of Theorem 3.3 or 3.4. The answer is straightforward.

Theorem 3.5. Suppose that u is harmonic in D and that, in addition, for some $1 \leq p < \infty$,

$$\sup_{r<1}\left(\frac{1}{2\pi}\int_T |u(re^{it})|^p\,dt\right)^{1/p} = A < \infty.$$

Let v denote the conjugate to u so that $v(0) = 0$. Then there is a constant $c = c_{\alpha,p}$ independent of u such that

$$\int_T N_\alpha(v,x)^p\,dx \leq cA^p, \qquad 1 < p < \infty, \tag{3.10}$$

and, for all $\lambda > 0$,

$$\lambda|\{N_\alpha(v) > \lambda\}| \leq cA, \qquad p = 1. \tag{3.11}$$

Proof. Suppose first $p > 1$; then $u(re^{it}) = f * P_r(t)$, $f \in L^p(T)$, and by inspection of the Fourier series, $v(re^{it}) = f * Q_r(t) = Hf * P_r(t)$. Whence $N_\alpha(v,x) \leq N_\alpha(Hf * P_r, x) \leq cM(Hf)(x)$ and

$$\|N_\alpha(v)\|_p \leq c\|M(Hf)\|_p \leq c\|f\|_p, \qquad 1 < p < \infty.$$

In case $p = 1$, we note that $f * Q_r(t) = (f * Q_r(t) - H_{1-r}f(t)) + H_{1-r}f(t)$, and consequently by Theorem 2.6 $N_\alpha(v,x) \leq c(Mf(x) + H^*f(x))$. Thus (3.11) also holds and the proof is complete. ∎

Two remarks about the above result: the estimate (3.10) points to a satisfactory state of affairs when $1 < p < \infty$ (a proof of a similar statement which does not rely on the Hilbert transform is sketched in below) and the estimate (3.11) does not. More precisely, if we denote by $H^p(D)$ the Hardy H^p space consisting of those harmonic functions u in D such that

$$\sup_{r<1}\left(\frac{1}{2\pi}\int_T |u(re^{it})|^p\,dt\right)^{1/p} \leq A < \infty, \qquad 0 < p \leq \infty, \tag{3.12}$$

and call the infimum over the constants A in (3.12) the H^p-norm of u, then for $1 < p < \infty$ the conjugate v, $v(0) = 0$, of a harmonic function u in $H^p(D)$ is also in $H^p(D)$ and its norm does not exceed a multiple of the norm of u; for $p = 1$ we only have a weak-type result. In particular, if we now denote by $H^p_a(D)$ the analytic Hardy H^p space consisting of those analytic functions f such that

$$\sup_{r<1}\left(\frac{1}{2\pi}\int_T |f(re^{it})|^p\,dt\right)^{1/p} \leq A < \infty, \qquad 0 < p < \infty \tag{3.13}$$

and call the infimum over the constants A in (3.13) the H^p_a norm of f, then for $1 < p < \infty$ each $u \in H^p(D)$ is the real part of an analytic function $f \in H^p_a(D)$ and the H^p_a norm of f does not exceed $c(H^p$ norm of $u)$; we cannot make a similar assertion for $p = 1$. Since it is quite an interesting problem to determine under what circumstances a harmonic function u is the real part of an analytic function f in $H^1_a(D)$, or $H^p_a(D)$, $0 < p < 1$, for that matter, we develop now some further properties of harmonic functions to address this question. Clearly, this is connected to the question of when $Hf \in L^p(T)$, $0 < p \leq 1$, the case $p = 1$ being particularly interesting.

4. FURTHER PROPERTIES OF HARMONIC FUNCTIONS AND SUBHARMONIC FUNCTIONS

We seek now to use to advantage the fact that harmonic functions satisfy Laplace's equation; we begin by showing the mean value property they verify.

Proposition 4.1. (Mean Value Property). Let u be harmonic in D and suppose that $D(z_0, r) = \{z \in D : |z_0 - z| < r\}$ is totally contained in D. Then

$$u(z_0) = \frac{1}{2\pi}\int_T u(z_0 + re^{it})\,dt. \tag{4.1}$$

Proof. Assume first $z_0 = 0$; then by Proposition 3.1

$$u(re^{it}) = \sum c_j r^{|j|} e^{ijt} \qquad \text{and} \qquad \frac{1}{2\pi}\int_T u(re^{it})\,dt = c_0 = u(0).$$

If $z_0 \neq 0$, let $U(z) = u(z + z_0)$; then U is also harmonic in $D(0, r)$, now, and $u(z_0) = U(0) = (1/2\pi)\int_T U(re^{it})\,dt = (1/2\pi)\int_T u(z_0 + re^{it})\,dt$. ∎

Corollary 4.2. Assume u and $D(z_0, r)$ are as above, then

$$u(z_0) = \frac{1}{\pi r^2}\int_{D(z_0,r)} u(z)\,dx\,dy. \tag{4.2}$$

Proof. By Proposition 4.1 we get

$$u(z_0)\frac{r^2}{2} = \int_{[0,r)}\left(\frac{1}{2\pi}\int_T u(z_0 + se^{it})\,dt\right)s\,ds$$
$$= \frac{1}{2\pi}\int_T\int_{[0,r)} u(z_0 + se^{it})s\,ds\,dt$$

whence the conclusion follows passing from polar to Cartesian coordinates. ■

It is interesting to note that the converse to Proposition 4.1 is also true (cf. 6.19). An important consequence of the mean value property is the distribution of the maximum and minimum values of a harmonic function in a connected domain.

Proposition 4.3. Assume that a continuous, real-valued function u verifies the mean value property for each disk $D(z_0, r) \subseteq D$, and that $A = \sup_{z\in D} u(z) < \infty$. Then either u is identically equal to A in D or else $u(z) < A$ for every z in D.

Proof. Let $\mathcal{U} = \{z \in D\colon u(z) = A\}$; by the continuity of u, $\mathcal{U}$ is closed in D; we show now it is also open. By (4.2) it follows that for each z_0 interior to D, and sufficiently small r,

$$u(z_0) \leq \frac{1}{\pi r^2}\int_{D(z_0,r)} u(z)\,dx\,dy \leq A. \tag{4.3}$$

Therefore, if for some such z_0 we have that $u(z_0) = A$, then, by (4.3) and the continuity of u, $u(z)$ is identically equal to A for z in $D(z_0, r)$ and $\mathcal{U}$ is open relative to D. Since D is connected, either $\mathcal{U}$ is D or empty. In the former case u is constant and the latter $u(z) < A$ for each z in D. ■

A closer look at the above proof shows that the full strength of the mean value property was not invoked to obtain the maximum principle, but rather the fact that the value of a function at a point is dominated by the average of its values near that point. This is an important property which is satisfied, for instance, by $u = \max(u_1, u_2)$, where u_1, u_2 are harmonic in D (note that u is not necessarily harmonic); it is therefore natural to investigate such functions further.

Definition 4.4. A real valued function s is said to be subharmonic in D if

(i) $-\infty \leq s(z) < \infty$, $z \in D$,
(ii) s is upper semicontinuous, i.e.,

$$s(z_0) \geq \limsup_{z\to z_0} s(z), \qquad z_0 \in D,$$

and

(iii) if $z_0 \in D$ and $D(z_0, r) \subset D$, then

$$s(z_0) \leqslant \frac{1}{\pi r^2} \int_{D(z_0,r)} s(z)\, dx\, dy. \tag{4.4}$$

First, we try to establish sufficient conditions which will assure that a function is subharmonic; an important tool in this endeavour is Green's theorem. There are some restrictions concerning the domain D in the statement of this theorem, but we shall not be concerned with them since in our applications D is either a disk or a difference of disks. The statement of Green's theorem is the following: Assume D is a sufficiently smooth domain and u, v are $C^2(D)$ functions; then

$$\int_D (u\,\Delta v - v\,\Delta u)\, dx\, dy = -\int_{\partial D} \left(u \frac{\partial}{\partial n} v - v \frac{\partial}{\partial n} u \right) ds, \tag{4.5}$$

where ds denotes the element of arc length along the boundary ∂D of D and $\partial/\partial n$ denotes the directional derivative along the inward normal into D.

We are then in a condition to prove

Proposition 4.5. Assume $u \in C^2(D)$. Then u is subharmonic in D if and only if $\Delta u(z) \geqslant 0$, $z \in D$.

Proof. By setting $v = 1$ in Green's theorem (4.5), we note that for $D(z_0, r) \subset D$

$$\int_{D(z_0,r)} \Delta u(z)\, dx\, dy = -\int_T \frac{\partial}{\partial n} u(z_0 + re^{it})\, dt. \tag{4.6}$$

Let now $I(r) = (1/2\pi r) \int_T u(z_0 + re^{it})\, dt$; a simple computation shows that $(rI(r))' = (1/2\pi) \int_T u'(z_0 + re^{it}) e^{it}\, dt$, and, since the inward normal to T at t is $(-e^{it})$ by (4.6), we see that

$$\int_{D(z_0,r)} \Delta u(z)\, dx\, dy = (rI(r))'. \tag{4.7}$$

If the Laplacian of u is $\geqslant 0$, then by (4.7), $rI(r)$ is nondecreasing, and, since its value at 0 is $u(z_0)$, we obtain that

$$u(z_0) \leqslant \frac{1}{2\pi} \int_T u(z_0 + re^{it})\, dt$$

and (4.4) follows by integration as in Corollary 4.2. Conversely, suppose there is a point z_0 in D where $\Delta u(z_0) < 0$. Then by continuity the same holds in a small neighborhood $D(z_0, r)$ of z_0, and, by (4.7), $rI(r)$ decreases for small enough r; this implies that for those r's

$$\frac{1}{2\pi}\int_T u(z_0 + re^{it})\,dt < u(z_0)$$

which, by integration, contradicts (4.4) at z_0. ■

As a first application of this result we have

Proposition 4.6. Let $f(z)$ be analytic in D. Then $\ln|f(z)|$ is subharmonic there.

Proof. It is a tedious but straightforward computation. Let $f(z) = u(z) + iv(z)$ and put

$$H(z) = \ln(u(z)^2 + v(z)^2 + \eta)^{1/2}, \qquad \eta > 0.$$

We show that $\Delta H(z) \geqslant 0$ for z in D. Since u, v are harmonic and (u, v) is a Cauchy–Riemann pair, a direct computation shows that

$$\Delta H(z) = \frac{2((\partial/\partial x)u(z))^2 + (\partial/\partial x)v(z))^2)\eta}{((u(z))^2 + (v(z))^2 + \eta)^2} \geqslant 0.$$

Whence for each $\eta > 0$ and $D(z_0, r) \subset D$

$$\ln(|f(z_0)|^2 + \eta)^{1/2} \leqslant \frac{1}{\pi r^2}\int_{D(z_0,r)} \ln(|f(z)|^2 + \eta)^{1/2}\,dx\,dy$$

and the desired conclusion follows by Fatou's lemma. ■

Corollary 4.7. Let $f(z)$ be analytic in D. Then $|f(z)|^\varepsilon$ is subharmonic there for each $\varepsilon > 0$.

Proof. Note that Jensen's inequality, 5.1 in Chapter I, gives at once that if s is subharmonic and ϕ convex and increasing and continuous at $t = -\infty$, then, also, $\phi(s)$ is subharmonic; the desired conclusion follows from Proposition 4.6 with $\phi(\lambda) = e^{\varepsilon\lambda}$, $\varepsilon > 0$. ■

Before we go on we need one more property of subharmonic functions, namely,

Proposition 4.8. Suppose s is continuous and suharmonic in D and $0 < r$, $\eta < 1$. Then

$$s(\eta re^{it}) \leqslant s(\eta\,\cdot) * P_r(t), \qquad re^{it} \in D. \tag{4.8}$$

Proof. For a fixed $\eta < 1$ and $z = re^{it}$, let $u(z) = s(\eta\,\cdot) * P_r(t)$; then, by Proposition 2.5, u is harmonic in D and continuous up to the boundary.

In addition $s(\eta z) - u(z)$ is subharmonic in D, $\leqslant 0$ at the boundary, and consequently by the maximum principle also $\leqslant 0$ throughout D; this gives (4.8) and we have finished. ■

The first application we give of these results is to the identification, among all harmonic functions in D, of those which are the real part of analytic functions in $H^p_a(D)$, $0 < p < \infty$. This result is due to Hardy and Littlewood. We also identify it as half of the Burkholder–Gundy–Silverstein theorem, and it provides the key to the development of a purely real variable theory of Hardy spaces.

Theorem 4.9. Assume $f = u + iv \in H^p_a(D)$, $0 < p < \infty$, H^p_a norm of $f \leqslant A$. Then there is a constant c_α independent of f such that

$$\|N_\alpha(u)\|_p, \|N_\alpha(v)\|_p \leqslant c_\alpha A. \tag{4.9}$$

Proof. Let $s(re^{it}) = |f(re^{it})|^{\eta p}$, $0 < \eta < 1$; by Corollary 4.7, s is continuous and subharmonic in D. Moreover,

$$\sup_r \frac{1}{2\pi} \int_T s(re^{it})^{1/\eta}\, dt \leqslant A^p < \infty \tag{4.10}$$

and $\{s(re^{it})\}$ is bounded in $L^{1/\eta}(T)$; therefore by Proposition 3.2 in Chapter II there are a sequence $r_k \to 1^-$ and a function $h \in L^{1/\eta}(T)$, $\|h\|_{1/\eta} \leqslant A^{\eta p}$, such that

$$\lim_{k\to\infty} \frac{1}{2\pi} \int_T s(r_k e^{it})\phi(t)\, dt = \frac{1}{2\pi} \int_T h(t)\phi(t)\, dt \tag{4.11}$$

for $\phi \in L^{(1/\eta)'}(T)$. Since for each $r < 1$, $P_r(x - \cdot) \in L^{(1/\eta)'}$, we may substitute it for ϕ in (4.11) and thus obtain

$$\lim_{k\to\infty} s(r_k \cdot) * P_r(x) = h * P_r(x), \qquad 0 < r < 1, \quad x \in T. \tag{4.12}$$

Moreover, since, by Proposition 4.8, $s(rr_k e^{ix}) \leqslant s(r_k \cdot) * P_r(x)$ and, by continuity, $\lim_{k\to\infty} s(rr_k e^{ix}) = s(re^{ix})$, from (4.12) it follows that $s(re^{ix}) \leqslant h * P_r(x)$, or

$$|f(re^{ix})|^p \leqslant (h * P_r(x))^{1/\eta}, \qquad 0 < r < 1, \quad x \in T. \tag{4.13}$$

Thus, taking sup in (4.13) over $re^{ix} \in \Gamma_\alpha(t)$ and by Theorem 2.2, we get that $N_\alpha(|f|, t)^p \leqslant N_\alpha((h * P_r)^{1/\eta}, t) \leqslant c_\alpha (Mh(t))^{1/\eta}$. Whence $\|N_\alpha(u)\|_p$, $\|N_\alpha(v)\|_p \leqslant \|N_\alpha(|f|)\|_p \leqslant c\alpha \|Mh\|^{\eta/p}_{1/\eta} \leqslant cA$, (4.9) holds and we are done. ■

This extremely interesting result has many important consequences. We identify a couple of them as they relate to what we can say about those f's for which $Hf \in L(T)$.

Proposition 4.10. Assume $f \in L(T)$, $Hf \in L(T)$. Then $N_\alpha(f * P_r)$ and $N_\alpha(Hf * P_r)$ are integrable and there is a constant $c = c_\alpha$ independent of f such that

$$\|N_\alpha(f * P_r)\|_1, \|N_\alpha(Hf * P_r)\|_1 \leqslant c(\|f\|_1 + \|Hf\|_1). \tag{4.14}$$

Proof. Let $F(z) = f * P_r(t) + iHf * P_r(t)$, $z = re^{it}$. Then $F \in H^1_a(D)$ with norm $\leq (\|f\|_1 + \|Hf\|_1)$. Indeed, since, as is readily seen, $Hf * P_r(t) = f * Q_r(t)$, then $(f * P_r, f * Q_r)$ is a Cauchy-Riemann pair, $F(z) = f * P_r(t) + iHf * P_r(t)$ is analytic, and for $r < 1$

$$\frac{1}{2\pi}\int_T |F(re^{it})|\, dt \leqslant \frac{1}{2\pi}\int_T |f * P_r(t)|\, dt + \frac{1}{2\pi}\int_T |Hf * P_r(t)|\, dt$$
$$\leqslant \|f\|_1 + \|Hf\|_1,$$

The conclusion follows now by Theorem 4.9. ■

Corollary 4.11. Suppose $f \geqslant 0$ is integrable and $Hf \in L(T)$. Then $f \in L \ln L(T)$.

Proof. By Proposition (4.1) $N_\alpha(f * P_r)$ is also integrable. Moreover, since by (2.4) $P_r(t) \geqslant c/(1-r)$ for $|t| \leqslant 1 - r$, then

$$\frac{1}{2(1-r)}\int_{|t|\leqslant 1-r} f(x-t)\, dt \leqslant cf * P_r(x), \qquad 0 < r < 1,$$

and $Mf(x) \leqslant cN_1(f * P_r, x) \leqslant cN_\alpha(f * P_r, x)$. Thus Mf is also integrable and the desired conclusion follows by Theorem 5.4 in Chapter IV. ■

It is natural to consider whether Theorem 4.9 and Proposition 4.10 admit a converse. More precisely, if a harmonic function u verifies $N_\alpha(u) \in L^p(T)$, $0 < p \leqslant 1$, is $u = \operatorname{Re} f$, $f \in H^p_a(T)$? Or if $N_\alpha(f * P_r) \in L(T)$, does it follow that $Hf \in L(T)$? We prefer to postpone the discussion of these questions until Chapter XIV where we do the real variable theory of the Hardy spaces. In the meantime, and to provide a glimpse of things to come, we show

Theorem 4.12. Assume $u \geqslant 0$ is harmonic in D and $N_\alpha(u) \in L^p(T)$, $0 < p < 1$. Then there is an analytic function $f \in H^p_a(D)$ so that $u = \operatorname{Re} f$ and H^p_a norm of $f \leqslant cu(0) \leqslant c\|N_\alpha(u)\|_p$, c independent of u.

Proof. Let v be the conjugate to u so that $v(0) = 0$; we show that the analytic function $f = u + iv$ verifies all the required conditions. For $\eta > 0$ put $F_\eta(z) = (u(z)^2 + v(z)^2 + \eta^2)^{p/2}$ and $G_\eta(z) = -(u(z) + \eta)^p$; a simple computation shows that

$$\Delta F_\eta = p\left(\left(\frac{\partial}{\partial x}u\right)^2 + \left(\frac{\partial}{\partial x}v\right)^2\right)(u^2 + v^2 + \eta^2)^{p/2-2}(p(u^2 + v^2) + 2\eta^2)$$

and

$$\Delta G_\eta = (1-p)p\left(\left(\frac{\partial}{\partial x}u\right)^2 + \left(\frac{\partial}{\partial x}v\right)^2\right)(u+\eta)^{p-2}.$$

Thus there is a constant $c = c_p = O(1/1-p)$ so that

$$0 \leqslant \Delta F_\eta \leqslant c\,\Delta G_\eta$$

and

$$0 \leqslant \int_D \Delta F_\eta(z)\,dx\,dy \leqslant c\int_D \Delta G_\eta(z)\,dx\,dy. \tag{4.15}$$

We may now invoke Green's theorem on both sides of (4.15) and obtain that

$$\int_T \frac{\partial}{\partial r}F_\eta(re^{it})\,dt \leqslant c\int_T \frac{\partial}{\partial r}G_\eta(re^{it})\,dt. \tag{4.16}$$

Whence, by integrating (4.16) with respect to r, it now follows that

$$\begin{aligned}
&\int_T\int_{[0,R)} \frac{\partial}{\partial r}F_\eta(re^{it})\,dr\,dt\\
&\quad= \int_T (u(Re^{it})^2 + v(Re^{it})^2 + \eta^2)^{p/2}\,dt - 2\pi(u(0)^2+\eta^2)^{p/2}\\
&\quad\leqslant c2\pi\left((u(0)+\eta)^p - \int_T (u(Re^{it})+\eta)^p\,dt\right).
\end{aligned} \tag{4.17}$$

Thus letting $\eta \to 0$ in (4.17) we get

$$\int_T |f(Re^{it})|^p\,dt \leqslant cu(0)^p$$

with c independent of R, and we are done. ■

5. HARNACK'S AND MEAN VALUE INEQUALITIES

Nonnegative harmonic functions satisfy an important inequality which restates the maximum principle in strong terms, namely,

Theorem 5.1 (Harnack's Inequality). Let u be a nonnegative harmonic function in D and let $D(z_0, r) \subset D$. Then

$$\sup_{D(z_0,r)} u(z) \leqslant \left(\frac{1+r}{1-r}\right)^2 \inf_{D(z_0,r)} u(z).$$

Proof. By the monotonicity of the Poisson kernel in $[0, \pi)$ it follows that for $0 \leqslant t \leqslant \pi$

$$\frac{1-r}{1+r} = \frac{1-r^2}{(1+r)^2} \leqslant P_r(t) \leqslant \frac{1-r^2}{(1-r)^2} = \frac{1+r}{1-r}. \tag{5.1}$$

Furthermore, since by Theorem 3.4 there is a nonnegative measure μ such that $u(re^{it}) = \mu * P_r(t)$, $u(0) = (1/2\pi)\mu(T)$, integrating (5.1) yields

$$\frac{1-r}{1+r}\frac{1}{2\pi}\mu(T) \leqslant u(re^{it}) \leqslant \frac{1+r}{1-r}\frac{1}{2\pi}\mu(T).$$

Thus

$$\sup_{D(z_0,r)} u(re^{it}) \leqslant \frac{1+r}{1-r}u(0) \leqslant \left(\frac{1+r}{1-r}\right)^2 \inf_{D(z_0,r)} u(re^{it}). \quad \blacksquare$$

As for harmonic functions of arbitrary sign, in addition to the mean value property, there is a mean value inequality due to Hardy and Littlewood; we begin by proving a version which follows at once from the maximum principle

Proposition 5.2. Assume u is harmonic in D and let $D(z_0, r) \subset D(z_0, R) \subset D$. Then there is a constant c independent of r, R and u such that

$$\sup_{D(z_0,r)} |u| \leqslant \frac{c}{(R-r)^2} \int_{D(z_0,R)\setminus D(z_0,r/2)} |u(z)|\, dx\, dy. \tag{5.2}$$

Proof. By the maximum principle the sup on the left-hand side of (5.2) is attained at a point z_1 with $|z_1 - z_0| = r$, and by Corollary 4.2

$$u(z_1) = \frac{1}{\pi\rho^2} \int_{D(z_1,\rho)} u(z)\, dx\, dy, \qquad \rho < R - r. \tag{5.3}$$

We distinguish two cases according to the relative sizes of r and R; if $R \leqslant \frac{3}{2}r$ we put $\rho = R - r$ in (5.3), and if $R > \frac{3}{2}r$ we put $\rho = r/2$ there and note that in either case $D(z_1, \rho) \subseteq D(z_0, R)\setminus D(z_0, r/2)$. $\blacksquare$

Observe that by Hölder's inequality the right-hand side of (5.2) may be replaced by $(c/(R-r)^{p-2})\|u\|_{L^p(D(z_0,R)\setminus D(z_0,r/2))}$ whenever $1 < p < \infty$; the remarkable fact is that it may also be replaced by a similar expression for $0 < p < 1$ as well.

Theorem 5.3. Suppose u is a measurable function defined in $D(z_0, R_1)$ which verifies

$$\sup_{D(z_0,r)} |u| \leqslant \frac{c}{(R-r)^{2/p_0}} \left(\int_{D(z_0,R)\setminus D(z_0,r/2)} |u(z)|^{p_0}\, dx\, dy \right)^{1/p_0} \tag{5.4}$$

for all $0 < r < R < R_1$ and some $0 < p_0 < \infty$ with c independent of r and R. Then there is a constant $c = c_p$ so that (5.4) holds for $0 < p < p_0$ as well. More precisely,

$$\sup_{D(z_0,r)} |u| \leqslant \frac{c_p}{(R-r)^{2/p}} \left(\int_{D(z_0,R)\setminus D(z_0,r/2)} |u(z)|^p \, dx \, dy \right)^{1/p} \tag{5.5}$$

for all $0 < r < R_1$, with c_p independent of r and R.

Proof. Note that, if $0 < \rho < R_1$, it follows by (5.4) that $\sup_{D(x_0,\rho)}|u| < \infty$. Moreover, since $|u|^{p_0} = (|u|^{1-p/p_0}|u|^{p/p_0})^{p_0}$ we also have

$$\begin{aligned} &\left(\int_{D(z_0,R)\setminus D(z_0,r/2)} |u(z)|^{p_0} \, dx \, dy \right)^{1/p_0} \\ &\qquad \leqslant \left(\sup_{D(z_0,R)} |u| \right)^{1-p/p_0} \left(\int_{D(z_0,R)\setminus D(z_0,r/2)} |u(z)|^p \, dx \, dy \right)^{1/p_0}. \end{aligned} \tag{5.6}$$

Whence combining (5.4) and (5.6) we get

$$\begin{aligned} \sup_{D(z_0,r)} |u| \leqslant \frac{c}{(R-r)^{2/p_0}} &\left(\sup_{D(z_0,R)} |u| \right)^{1/p/p_0} \\ &\times \left(\int_{D(z_0,R)\setminus D(z_0,r/2)} |u(z)|^p \, dx \, dy \right)^{1/p_0} \end{aligned} \tag{5.7}$$

for all $0 < r < R < R_1$. Fix now $r_0 < R_0 < R_1$; we will show that (5.5) holds with $r = r_0$, $R = R_0$ there.

Let $I = (\int_{D(z_0,R_0)\setminus D(z_0,r_0/2)} |u(z)|^p \, dx \, dy)^{1/p}$; then dividing both sides of (5.7) by I we see that for $r_0 \leqslant r < R \leqslant R_0$,

$$\begin{aligned} \sup_{D(z_0,r)} (|u|/I) &\leqslant \frac{c}{(R-r)^{2/p_0}} \left(\sup_{D(z_0,R)} (|u|/I) \right)^{1-p/p_0} \\ &\qquad \times \left(\frac{\int_{D(z_0,R)\setminus D(z_0,r/2)} |u(z)|^p \, dx \, dy}{I^p} \right)^{1/p_0} \\ &\leqslant \frac{c}{(R-r)^{2/p_0}} \left(\sup_{D(z_0,R)} (|u|/I) \right)^{1-p/p_0}, \end{aligned} \tag{5.8}$$

since

$$\int_{D(z_0,R)\setminus D(z_0,r)\text{-}} u(z)^p \, dx \, dy \Big/ I^p \leqslant 1$$

(as $R_0 \geqslant R$ and $r \geqslant r_0$). With the notation $v(z) = |u(z)|/I$, (5.8) may be rewritten as

$$\sup_{D(z_0,r)} v \leqslant \frac{c}{(R-r)^{2/p_0}} \left(\sup_{D(z_0,R)} v \right)^{1-p/p_0} \tag{5.9}$$

for $r_0 \leqslant r < R \leqslant R_0$. Let $\tau = r_0/R_0 < 1$, and consider the sequence

$$r_0 = R_0\tau, \qquad r_k = R_0\left(\tau + (1-\tau)\sum_{j=1}^{k} 2^{-j}\right), \qquad k = 1, 2, \ldots. \tag{5.10}$$

Note that $\{r_k\}$ is increasing, $r_k - r_{k-1} = R_0(1-\tau)/2^k$, and $\lim_{k\to\infty} r_k = R_0$. We apply now (5.9) successively with $R = r_k$, $r = r_{k-1}$, $k \geqslant 1$ and obtain

$$\sup_{D(z_0,r_0)} v \leqslant \frac{c}{(R_0(1-\tau)/2)^{2/p_0}}\left(\sup_{D(z_0,r_1)} v\right)^{1-p/p_0}$$

$$\leqslant \frac{c}{(R_0(1-\tau)/2)^{2/p_0}}\left(\frac{c}{(R_0(1-\tau)/2^2)^{2/p_0}}\right)^{1-p/p_0}\left(\sup_{D(z_0,r_2)} v\right)^{(1-p/p_0)^2},$$

and in general

$$\leqslant \frac{c^{\phi(k)}}{(R_0(1-\tau))^{2/p_0\phi(k)}} 2^{2/p_0\psi(k)}\left(\sup_{D(z_0,r_k)} v\right)^{(1-p/p_0)^k}, \tag{5.11}$$

where for $k = 1, 2, \ldots$

$$\phi(k) = \sum_{j=0}^{k}(1-p/p_0)^j \qquad \text{and} \qquad \psi(k) = \sum_{j=0}^{k}(j+1)(1-p/p_0)^j.$$

This is all we need to complete the proof; indeed, since as is readily seen

$$\lim_{k\to\infty}\phi(k) = p_0/p, \qquad \lim_{k\to\infty}\psi(k) = (p_0/p)^2$$

and

$$\lim_{k\to\infty}\left(\sup_{D(z_0,r_k)} v\right)^{(1-p/p_0)^k} = 1,$$

by letting $k \to \infty$, (5.11) gives at once

$$\sup_{D(z_0,r_0)} v \leqslant \frac{c^{p_0/p}}{(R(1-\tau))^{2/p_0\cdot p_0/p}} 2^{2/p_0(p_0/p)^2}\cdot 1$$

$$= \frac{c}{(R_0 - r_0)^{2/p}},$$

which, on account of the definition of v, is precisely what we wanted to show. ■

Because of its importance, we emphasize Theorem 5.3 in the particular instance of harmonic functions. The reader should also consider the corresponding statement for nonnegative subharmonic functions as well. This is

Theorem 5.4 (Hardy–Littlewood). Assume u is harmonic in D and let $D(z_0, r) \subset D(z_0, R) \subset D$. If $0 < p < \infty$, then there is a constant $c = c_p$

independent of r, R, and u such that

$$\sup_{D(z_0,r)} |u| \leqslant \frac{c}{(R-r)^{2/p}} \left(\int_{D(z_0,R)\setminus D(z_0,r/2)} |u(z)|^p \, dx\, dy \right)^{1/p}.$$

We will have many an occasion to apply Theorem 5.4. In this section we use it to show that the radial maximal function of a harmonic function controls the nontangential maximal function.

Proposition 5.5. Assume u is harmonic in D. Then for each $\alpha \geqslant 1$ and $0 < \eta < 1$, we have

$$N_\alpha(u, x) \leqslant cM(N_1(u, \cdot)^\eta)(x)^{1/\eta}, \qquad x \in T, \tag{5.12}$$

where $c = c_{\eta,\alpha}$ is independent of u and x. Consequently,

$$\|N_\alpha(u)\|_p \leqslant c\|N_1(u)\|_p, \qquad 0 < p < \infty. \tag{5.13}$$

Proof. Since (5.13) follows at once from (5.12) (just pick $p/\eta < 1$) we only show (5.12). Also by rotation it suffices to do the case $x = 0$, i.e., $e^{i0} = 1$. To estimate $|u(w)|^\eta$, $w \in \Gamma_\alpha(0)$, we consider two cases, namely, w far from or near to the boundary; we only discuss the latter here because it is the one which offers some difficulty. Thus assume $|1 - w| \leqslant 1/2\alpha$ and note that $D(w, (1-|w|)/2) \subseteq D(1, (\alpha + \frac{1}{2})(1-|w|)) \cap D = D_1$; to see this let $z \in D(w, (1-|w|)/2)$ and estimate $|1 - z| \leqslant |1 - w| + |w + z| \leqslant (\alpha + \frac{1}{2})(1 - |w|)$. Therefore, by Theorem 5.4 with $D(w, (1-|w|)/2)$, we see that

$$\begin{aligned}
|u(w)|^\eta &\leqslant \frac{c}{(1-|w|)^2} \int_D |u(z)|^\eta \, dx\, dy = \frac{c}{(1-|w|)^2} \int_{D_1} |u(re^{it})|^\eta r\, dr\, dt \\
&\leqslant \frac{c}{(1-|w|)^2} \int_{[-(1-|w|),1-|w|]} \int_{[1-(1-|w|)/2,1)} N_1(u,t)^\eta r\, dr\, dt \\
&\leqslant \frac{c}{(1-|w|)} \int_{[-(1-|w|),1-|w|]} N_1(u,t)^\eta \, dt \\
&\leqslant cM(N_1(u,\cdot)^\eta)(0),
\end{aligned}$$

with c independent of w, and we have finished. ■

6. NOTES; FURTHER RESULTS AND PROBLEMS

The first sep in solving a steady state heat conduction problem is to find a differential equation which governs the situation; we will do this in the disk D. The natural coordinates are polar and the temperature at the point re^{it} is denoted by $u(r, t)$. Consider, then, any section of D given by $0 < r_0 < r < r_1 \leq 1$, $0 \leq t_0 < t < t_1 < 2\pi$. Since we are considering a steady state, the rate at which heat flows into this section must be zero for otherwise the average temperature would change with time. Now it is a basic postulate of heat conduction that the rate at which heat crosses a curve C is proportional to the integral along C of the normal derivative $\partial u/\partial n$, or the derivative of u with respect to arc length along any curve perpendicular to C. When C is the side $t = t_1$ we take the perpendicular curves to be given by $r =$ const. Then, since the length of a circular arc is the angle times the radius, in this case we have $\partial u/\partial n = r^{-1}(\partial u/\partial t)(r, t_1)$ and the rate at which heat flows into the section along the boundary $t = t_1$ is

$$k \int_{(r_0, r_1)} r^{-1} \frac{\partial}{\partial t} u(r, t_1)\, dr,$$

where k is the conductivity. Adding the expressions corresponding to all the boundaries and setting the net flow equal to zero we get

$$\int_{(r_0, r_1)} r^{-1}\left(\frac{\partial u}{\partial t}(r, t_1) - \frac{\partial u}{\partial t}(r, t_2)\right) dr$$
$$+ \int_{(t_0, t_1)} \left(t_1 \frac{\partial u}{\partial r}(r_1, t) - t_0 \frac{\partial u}{\partial r}(r_0, t)\right) dt = 0.$$

Upon dividing by $(t_1 - t_0)$ and letting $t_1 \to t_0$ and then dividing by $(r_1 - r_0)$ and letting $r_1 \to r_0$ we get that

$$\frac{1}{r_0} \frac{\partial^2}{\partial t^2} u(r_0, t_0) + \frac{\partial}{\partial r}\left(r \frac{\partial}{\partial r} u\right)(r_0, t_0) = 0$$

for any point $r_0 e^{it_0}$ in D; this is the polar-coordinate form of Laplace's equation, the main subject of this section. Laplace's influence on mathematics, as well as in our daily affairs (as a member of the Bureau de Consultation des Arts et Métiers he was quite influential in the design and adoption of the metric system), is a lasting one. F. Riesz made significant contributions to many fields of mathematics, including the theory of Hardy spaces, but the subject he created and developed is that of subharmonic functions. The proof of Theorem 5.3 given here is based on some idea of Chipot [1984].

Further Results and Problems

6.1 The converse to Proposition 1.3 is not true, namely, that there is a series which converges radially but is not $(C, 1)$ summable. (*Hint*: For t

fixed let $c_j = j \sin jt$. Then

$$f(r) = \sum_{j=1}^{\infty} jr^j \sin jt = r(1-r^2)\sin t/(1-2r\cos t + r^2)$$

and therefore $\lim_{r\to 1^-} f(r)$ exists and equals zero for every real t. If the series in question is $(C, 1)$ summable at t, then $\lim_{j\to\infty} c_j/j = 0$, and, consequently, $\lim_{j\to\infty} \sin jt = 0$; this in turn implies that $t = k\pi$ for some integer k. For instance by taking $t = \pi/2$ we get that $1 + 0 - 3 + 0 + 5 \cdots$ is Abel summable to zero but is not $(C, 1)$ convergent.)

6.2 Suppose that $\sum c_j = s(A_\alpha)$ and $1 < \beta$; show that $\lim_{j\to\infty} c_j/j^\beta = \lim_{j\to\infty} s_j/j^\beta = 0$.

6.3 (Fatou) If, for some $t \in T, \phi'(t)$ exists (in the notation of Theorem 3.4) and is ∞, then also $\lim_{r\to 1} \mu * P_r(t) = \infty$; the result still holds for nontangential convergence provided the measure μ is nonnegative.

6.4 Show that $\|Q_r\|_1 \sim \ln(1/1-r)$, as $r \to 1$.

6.5 Suppose that $f \in L \ln L(T)$, and let $u(z) = f * P_r(t)$, $z = re^{it}$; show that $N_\alpha(u)$ is integrable. Is the same true for $N_\alpha(v)$? Here v is the conjugate to u with $v(0) = 0$.

6.6 Let f be as in 6.5. Show that

$$\lim_{z=re^{ix}\to 1, z\in\Gamma_\alpha(0)} \|f * P_r(x+\cdot) - f\|_1 = 0.$$

Is it also true that

$$\lim_{z=re^{ix}\to 1, z\in\Gamma_\alpha(0)} \|f * Q_r(x+\cdot) - Hf\|_1 = 0?$$

6.7 (Fejér–F. Riesz) Suppose $f \in L^p(T)$, $1 < p < \infty$, and put $F(re^{ix}) = \sup_{\rho<r}|f * P_\rho(x)|$, $0 < r < 1$. Show that there exists a constant $c = c_p$ independent of f and $x \in T$ such that $\int_{[0,1)} F(re^{ix})^p \, dr \leqslant c \int_T |f(t)|^p \, dt$. Moreover, if $f \in L \ln L(T)$, then $F \in L([0, 1))$. (*Hint*: Note that $F(re^{ix}) \leqslant (\sup_{t\in T, \rho<r} P_\rho(t))\|f\|_1 \leqslant c\|f\|_1/(1-r) \in wk - L([0, 1))$ and $\sup_r F(re^{ix}) \leqslant (\sup_{\rho<r}\|P_\rho\|_1)\|f\|_\infty$; now interpolate. For a similar proof and n-dimensional extensions, see Sagher's work [1977]).

6.8 Show that u is harmonic in D if and only if there is a sequence $\{c_j\}$ such that $u(re^{it}) = \sum c_j r^{|j|} e^{ijt}$ and $\limsup_{|j|\leftarrow\infty} |c_j|^{1/j} = 1$.

6.9 Show that u is harmonic in D if and only if there is a distributon F such that $u(re^{it}) = F * P_r(t)$.

6.10 (The Dirichlet problem) Given an $L^p(T)$ function f, $1 \leqslant p \leqslant \infty$, find a function u in D so that $\Delta u = 0$ in D and u converges nontangentially to f a.e. What is the corresponding statement for f in $C(T)$? (*Hint*: Put $u(re^{it}) = f * P_r(t)$.)

6.11 If u is harmonic in D, then $\int_T (\partial/\partial n)(u(e^{it}))\, dt = 0$. (*Hint*: Use Green's theorem.)

6.12 (The Neumann Problem) Given an L^p function g, $1 \leqslant p \leqslant \infty$, $\int_T g(t)\,dt = 0$, find a function u in D so that $\Delta u = 0$ and $-(\partial u/\partial r)(re^{it})$ converges nontangentially to g a.e. Is there a corresponding statement for g in $C(T)$? (*Hint*: $u(z) = \sum c_j r^{|j|} e^{ijt}$, where the c_j's are chosen so that the (formally) differentiated series converges nontangentially to $-g$; note that in the limit we get $(\partial/\partial n)u$, the directional derivative of u along the inward normal into D.)

6.13 State and solve the Dirichlet and Neumann problems in arbitrary disks $D(z_0, r)$.

6.14 (M. Riesz) Suppose u is harmonic in D and $N_\alpha(u) \in L^p(T)$, $1 < p < \infty$. Then there is a function $f \in H^p_a(D)$ so that $u = \operatorname{Re} f$ and H^p_a norm of $f \leqslant c\|N_\alpha(u)\|_p$, c independent of u. (*Hint*: The reader may assume $u \geqslant 0$; the proof of Theorem 4.2 works here with $G_\eta(z) = (u(z) + \eta)^p$.)

6.15 Let f be analytic in D; if $\operatorname{Re} f \geqslant 0$, then $f \in H^p_a(D)$ for $0 < p < 1$ and its H^p_a norm $\leqslant c|f(0)|^p$. (*Hint*: Since $\operatorname{Re} f = \mu * P_r$ we may invoke Proposition 2.10 in Chapter IV.)

6.16 For each $0 < p < 1$ there is a function $u \in H^p(D)$ so that its conjugate v, $v(0) = 0$, is not in $H^q(D)$ for any q. (*Hint*: In fact there is an analytic function $f(z) = u(z) + iv(t)$ such that $u \in H^p(D)$ for all $p < 1$, yet $v \notin H^q(D)$ any q; the function f is $\sum_{n=1}^\infty \varepsilon_n z^{2^n}/(1 - z^{2^{n+1}})$, for some choice $\varepsilon_n = \pm 1$. This example is from Duren's [1970] book.)

6.17 Assume $u \in C^2(D)$ and $D(z_0, r) \subset D$; then

$$u(z_0) = \frac{1}{2\pi}\int_T u(z_0 + re^{it})\,dt - \frac{1}{2\pi}\int_{D(z_0,r)} \ln\left(\frac{r}{|z - z_0|}\right) \Delta u(z)\,dx\,dy.$$

(*Hint*: Apply Green's theorem to u and $v(z) = \ln(r/|z - z_0|)$ in $D(z_0, R) \backslash D(z_0, \varepsilon)$; then let $\varepsilon \to 0$.)

6.18 Suppose $f \in L$, $f \sim \sum c_j e^{ijt}$ is such that $\|(\partial/\partial r)(f * P_r)\|_p = O((1 - r)^{1/p}\phi(r))$, $1 < p \leqslant 2$, $\phi \in L([0,1))$; then $\sum |c_j| < \infty$. (*Hint*: For $j \neq 0$, $2|j| \int_{[0,1)} r^{2|j|-1}\,dr = 1$; thus

$$\begin{aligned}
\sum_{j\neq 0} |c_j| &= 2\int_{[0,1)} \left(\sum_{j\neq 0} |c_j|\,|j| r^{|j|-1} r^{|j|}\right) dr \\
&\leqslant 2\int_{[0,1)} \left(\sum (|c_j|\,|j| r^{|j|-1})^{p'}\right)^{1/p'} \left(\sum r^{|j|p}\right)^{1/p} dr \\
&\leqslant c\int_{[0,1)} \left\|\frac{\partial}{\partial r}(f * P_r)\right\|_p (1/(1 - r))^{1/p}\,dr,
\end{aligned}$$

the last step being due to Hausdorff–Young. When $p = 2$ this is a variant of a well-known result due to Bernstein.)

6.19 Suppose u satisfies the mean value property (4.1). Then u is a C^∞, harmonic function in D. (*Hint*: For each z_0 in D we show that u is $C^\infty(D(z_0, R))$, $D(z_0, R) \subset D$; first let ϕ be a C^∞, radial function, i.e., $\phi(re^{it}) = \phi(r)$, supported in $D(z_0, (1-|z_0|)/2) \subset D$ with $\int_D \phi(z)\,dx\,dy = 1$ (cf. 4.12 in Chapter I) and note that for $R < (1-|z_0|)/2$ and w in $D(z_0, R)$,

$$\frac{1}{2\pi}\int_D u(w+z)\phi(x)\,dx\,dy = \int_{[0,1)} \frac{1}{2\pi}\int_T u(w+re^{it})\,dt\,\phi(r)r\,dr$$
$$= u(w)\int_{[0,1)} \phi(r)\,dr = u(w);$$

changing variables $z \to z - w$ shows that the differentiation can be done on the ϕ and consequently $u \in C^\infty(D)$. Moreover, since

$$\int_T \frac{\partial}{\partial r} u(z_0 + re^{it})\,dt = \frac{\partial}{\partial r}\left(\int_T u(z_0 + re^{it})\,dt\right)$$
$$= \frac{\partial}{\partial r}(2\pi u(z_0)) = 0,$$

by Green's theorem and the Lebesgue differentiation theorem we see that

$$0 = \frac{1}{\pi r^2}\int_{D(z_0,r)} \Delta u(z)\,dx\,dy \to \Delta u(z_0)$$
$$\text{as}\quad r \to 0, \qquad \text{everywhere in}\quad D.$$

The idea in the first part of the proof can be used to estimate the growth of the derivatives of u as well.)

6.20 (Weyl's Lemma) Suppose u is a locally integrable function in D whose distributional Laplacian vanishes there. Show that u is a $C^\infty(D)$ function. (*Hint*: If ϕ is a $C^\infty(D)$ function with small support and $\int_D \phi(z)\,dx\,dy = 1$, it is readily seen that $U(w) = \int_D u(w+z)\phi(z)\,dx\,dy$ is harmonic in a disk $D' \subset D$ (D' depends on the support of ϕ), and by Proposition 4.1 $U(w) = (1/2\pi r)\int_T U(w + re^{it})\,dt$ for w in D' and all r sufficiently small; but U converges to u uniformly on compact subsets of D: thus u is continuous in D and satisfies the mean value property.)

6.21 If $f \in C^\infty(D)$ and u is a locally integrable function D whose distributional Laplacian equals f in D, then u is a $C^\infty(D)$ function.

6.22 Suppose u is harmonic in $\{z = x + iy \in D\colon y > 0\}$ and $u(x) = 0$; show that u can be extended to be harmonic in D by the formula $u(x - iy) = -u(x + iy)$, $y > 0$. (*Hint*: It is clear that the extension is continuous in D and harmonic, except, perhaps, at $y = 0$. For $x \in D$ then let $D(x, r) \subset D$ and solve the Dirichlet problem $\Delta v = 0$ in $D(x, r)$, $v(z) = u(z)$ for $|z - x| = r$; by the explicit expression of v given by 6.13 it follows at once that $v = u$ in $D(x_0, r)$.)

6.23 Suppose that u is harmonic in $D\backslash\{z_0\}$ and $|u(z)| = O(\ln(1/|z - z_0|))$ as $z \to z_0$. Show that u can be redefined at z_0 so as to be harmonic in D. (*Hint*: Let $D(z_0, r) \subset D$ and solve the Dirichlet problem $\Delta v = 0$ in $D(z_0, r)$, $v(z) = u(z)$ for $|z - z_0| = r$; if we can show that $u = v$ in $D(z_0, r)\backslash\{z_0\}$ we can remove the singularity by putting $u(z_0) = v(z_0)$. To do this, consider the function $u(z) - v(z) + \varepsilon \ln|z - z_0|$ in $D(z_0, r)\backslash D(z_0,\delta)$, sufficiently small δ, and invoke the maximum principle.)

6.24 (Harnack) Suppose $\{u_n\}$ is a monotonic increasing sequence of harmonic functions in D; then either $u_n(z)$ diverges to $+\infty$ everywhere in D or else $u_n(z) \to u(z)$ uniformly in every compact subset of D, and $u(z)$ is harmonic in D. (*Hint*: Clearly, $u(z)$ exists everywhere in D as either a finite or infinite limit. Suppose that there is a z_0 in D so that $u(z_0) < \infty$. Then for $n > m$ sufficiently large we have $u_n(z_0) - u_m(z_0) < \varepsilon$ and by (a simple variant of) Harnack's inequality it follows that $0 < u_n(z) - u_m(z) < c_r\varepsilon$ for z in $D(z_0, r) \subset D$. Thus $u_n(z)$ converges uniformly there to a finite and continuous limit $u(z)$; similarly, we get that, if $u(z_0) = \infty$, then $u(z) = \infty$ for z in $D(z_0, r) \subset D$. Thus the sets where $u(z) = \infty$ and $u(z) < \infty$ are both open in D and one of them must be empty. That $u(z)$ is harmonic follows easily from the Poisson integral representation of each u_n which holds in the interior of D).

6.25 (Green's function) The function $G(z, w)$ is said to be a (classical) Green's function of z with respect to the domain D and the point w of D if (i) G is harmonic in $D\backslash\{w\}$, (ii) G is continuous up to the boundary of $D\backslash\{w\}$ and assumes the value 0 at the boundary; (iii) $G + \ln|z - w|$ is harmonic at $z = w$ and hence everywhere. Show that if a Green's function exists it is unique. Also show that for disk D the Green's function is given by $G(z, w) = \ln(|z - w'|\,|w|/|z - w|)$, $w \neq 0$, $G(z, 0) = \ln(1/|z|)$, where $w' = w/|w|^2$.

6.26 If u is harmonic in D, then $|u|^k$ is subharmonic there for $k \geq 1$. If, on the other hand f is an analytic in D, then $(\ln^+|f|)^k$ is subharmonic in D for $k \geq 1$.

6.27 Assume that u is subharmonic in D and that μ is a nonnegative Borel measure on a compact subset of K of D; show that $\int_D u(z + w)\, d\mu(w)$ is subharmonic in D and harmonic in $D\backslash K$.

6.28 Let u be an upper semicontinuous function in D, then u is subharmonic in D if and only if for every harmonic function v defined in an open subset $\mathcal{O}$ of D and every w in the boundary of $\mathcal{O}$, $\limsup_{z\to w, z\in\mathcal{O}}(u(z) - v(z)) \leq 0$ implies $u(z) - v(z) \leq 0$ for all z in 0. (*Hint*: If u is subharmonic, then the condition follows by a maximum principle argument. Conversely let $D(z_0, r) \subset D$ and note that since u is upper semicontinuous there is a sequence $\{u_n\}$ of continuous functions decreasing to u on $\partial D(z_0, r)$ as $n \to \infty$. Let $U_n(z)$ denote the harmonic function in $D(z_0, r)$ with boundary values $u_n(z)$; U_n is continuous up to the boundary, $u(z_0) \leqslant U_n(z_0)$ and

$$u(z_0) \leqslant \lim_{n\to\infty} \frac{1}{2\pi} \int_T u_n(z_0 + re^{it})\, dt = \frac{1}{2\pi} \int_T u(z_0 + re^{it})\, dt).$$

6.29 A subharmonic function s in D is said to have a harmonic majorant if there is a harmonic function $u(z)$ such that $s(z) \leqslant u(z)$ throughout D. Show that this occurs in D if and only if $\sup_{r<1}(1/2\pi)\int_T s(re^{it})\, dt < \infty$. (*Hint*: The necessity of the condition follows by Harnack's theorem 6.24; the converse follows by 6.28. What is the least harmonic majorant of s?)

6.30 Suppose $u \geqslant 0$ is continuous in D, has continuous second order partial derivatives in the subset $\mathcal{U} = \{z \in D: u(z) > 0\}$, and satisfies $\Delta u(z) \geqslant 0$ in $\mathcal{U}$. Show that u is subharmonic in D. (*Hint*: The proof is a combination of the idea in 6.22 and 6.28; once all is said and done the reader will note that the following more general result has actually been proven: If $u \geqslant 0$ is continuous in D and subharmonic in $\mathcal{U}$, then u is subharmonic in D.)

6.31 (Littlewood Subordination Theorem) Suppose that $s(z)$ is subharmonic in D and that f is analytic in D and verifies $|f(z)| \leqslant |z|$ there. Then

$$\int_T s(f(re^{it}))\, dt \leqslant \int_T s(re^{it})\, dt, \qquad 0 < r < 1.$$

(*Hint*: The reader may assume that $f(z) \not\equiv e^{i\lambda}z$, some real λ, for otherwise there is noting to prove; thus $|f(z)| < r$ for $|z| \leqslant r$, and, if $u(z)$ denotes the harmonic extension of s to $D(0, r)$, it follows that $u(f(z))$ is harmonic in $|z| \leqslant r$ and by the mean value property

$$\int_T u(f(re^{it}))\, dt = 2\pi u(f(0)) = 2\pi u(0) = \int_T s(re^{it})\, dt$$

by 6.28. Also by 6.28 $s(z) \leqslant u(z)$ for z in $D(0, r)$ and so

$$\int_T s(f(re^{it}))\, dt \leqslant \int_T u(f(re^{it})\, dt = \int_T s(re^{it})\, dt.)$$

6.32 Suppose that $u \in H^p(D)$ and show that $|u(re^{it})| \leqslant c/(1-r)^{2/p}$, where c depends only on the H^p norm of u. (*Hint*: Use Theorem 5.3.)

6.33 Suppose $u \geqslant 0$ is a measurable function defined in $D(z_0, R_1)$ which verifies

$$(R-r)^{2/p_0} \Big/ \left(\int_{D(z_0,R)\setminus D(z_0,r/2)} u(z)^{p_0}\, dx\, dy \right)^{1/p_0} \leqslant c \inf_{D(z_0,r)} u$$

for al $0 < r < R < R_1$ and some $0 < p_0 < \infty$, with c independent of r and R. Then the same conclusion holds for $0 < p < p_0$ in place of p_0 on the left-hand side of the above inequality; the constants c on the right-hand side will depend only on p. Could we replace the inf in the right-hand side by $1/(\int_{D(z_0,r)} u(z)^q\,dx\,dy)^{1/q}$, for $q > p_0$, and an appropriate power of $(R-r)$ on the left-hand side?

6.34 Suppose $u \geqslant 0$ is a measurable function defined in $D(z_0, R_1)$ which verifies

$$\left(\int_{D(z_0,r)} u(z)^q\,dx\,dy\right)^{1/q} \leqslant \frac{c}{(R-r)^{2(1/p_0-1/q)}}\left(\int_{D(z_0,R)} u(z)^{p_0}\,dx\,dy\right)^{1/p_0}$$

for all $0 < r < R < R_1$ and some $0 < p_0 < q < \infty$, with c independent of r and R. Then the same conclusion holds for $0 < p < p_0$ in place of p_0 on the right-hand side of the above inequality; the constant c depends solely on p.

6.35 Assume $u \geqslant 0$ is a measurable function defined in $D(z_0, R_1)$ which verifies

$$\sup_{D(z_0,r)} u(z) \leqslant c\frac{R^{2\varepsilon}}{(R-r)^{2\varepsilon}}\left(\frac{1}{R^2}\int_{D(z_0,R)} u(z)\,dx\,dy + \chi R^{1/\alpha}\right)$$

for some $\varepsilon > 0$ and all $0 < r < R < R_1$, and constants c, α, χ independent of r, R. Then the same conclusion holds with the expression on the right-hand side replaced by

$$c\frac{R^{2\varepsilon/p}}{(R-r)^{2\varepsilon/p}}\left(\left(\frac{1}{R^2}\int_{D(z_0,R)} u(z)^p\,dx\,dy\right)^{1/p} + \chi R^{1/\alpha}\right), \qquad 0 < p < 1$$

and a constant c which depends only on p.

CHAPTER

VIII

Oscillation of Functions

1. MEAN OSCILLATION OF FUNCTIONS

We introduce in this section a maximal function which has become extremely important in various areas of analysis including harmonic analysis, PDEs, and function theory. The spaces generated by this maximal function are also of interest since, in the scale of Lebesgue spaces, they may be considered an appropriate substitute for $L^\infty(T)$ and beyond. Of course the notion of "appropriate" is a matter of personal choice, but from our point of view an appropriate substitute for $L^\infty(T)$ is a space which is preserved by a wide class of important operators such as the Hardy-Littlewood maximal function and the Hilbert transform and which can be used as an end point in interpolating L^p spaces. In this sense the John-Nirenberg class $BMO(T)$ we consider below fits the bill.

We introduce this space as follows: for $f \in L(T)$ let

$$\tilde{M}^{\#}f(x) = \sup_I \inf_c \frac{1}{|I|}\int_I |f(t) - c|\, dt, \tag{1.1}$$

where c above varies over the complex constants and I is an interval containing x, $|I| \leq 2\pi$. This definition can actually be simplified. Suppose that f is real valued. Then so will be the constant c which minimizes the integral in (1.1). By elementary considerations we expect c to be among those values for which

$$\frac{d}{dc}\left(\int_I |f(t) - c|\, dt\right) = -\int_I \frac{f(t) - c}{|f(t) - c|}\, dt = 0.$$

Since the integrand above equals 1 for $f(t) > c$ and -1 for $f(t) < c$ we have, in particular, that $|\{f > c\}| = |\{f < c\}|$. This actually means that any such constant c verifies simultaneously $|\{f > c\}| \leq |I|/2$ and $|\{f < c\}| \leq |I|/2$. In other words $c = m_f(I)$ is a median value of f over I. These considerations can be formalized, and extended to the case when f is complex valued by introducing the median values $m_f(I) = m_{\mathrm{Re} f}(I) + i m_{\mathrm{Im} f}(I)$. Is there, however, a simpler way to choose c? The answer is contained in the next result.

Proposition 1.1. If f_I denotes the average of f over I, then for any constant c

$$\int_I |f(t) - f_I|\, dt \leq 2 \int_I |f(t) - c|\, dt. \tag{1.2}$$

Proof. Since

$$|f(t) - f_I| \leq |f(t) - c| + |c - f_I| \leq |f(t) - c| + (1/|I|) \textstyle\int_I |f(y) - c|\, dy,$$

the conclusion follows at once upon integrating over I. ∎

Corollary 1.2. For $f \in L(T)$ and $I \subseteq T$ we have

$$\inf_c \int_I |f(t) - c|\, dt \leq \int_I |f(t) - f_I|\, dt \leq 2 \inf_c \int_I |f(t) - c|\, dt. \tag{1.3}$$

In view of this corollary we may redefine the "sharp" maximal function in (1.1) by the equivalent expression

$$M^{\#} f(x) = \sup_I \frac{1}{|I|} \int_I |f(t) - f_I|\, dt, \tag{1.4}$$

where I is an open interval containing x, $|I| \leq 2\pi$. Clearly, $M^{\#} f$ is a measurable, subadditive function. Let now

$$\|f\|_* = \|M^{\#} f\|_\infty \tag{1.5}$$

and put $BMO(T) = BMO = \{f \in L(T) \colon \|f\|_* < \infty\}$.

This is the John–Nirenberg space of functions with bounded mean oscillation. Endowed with the norm given in (1.5), BMO becomes a Banach space provided we identify functions which differ a.e. by a constant; clearly, $\|f\|_* = 0$ for $f(t) = c$ a.e. in T. Bounded functions f are in BMO and $\|f\|_* \leq 2\|f\|_\infty$; however, observe that $\|\chi_I\|_* = \frac{1}{2}$. On the other hand, does BMO contain unbounded functions? The standard example that this is the case is $f(t) = \ln|t|, |t| < \pi$; we sketch the proof of this fact. Let $I = (a, b) \subset T$. We show that for an appropriate choice of c_I,

$$\frac{1}{|I|} \int_I |\ln|t| - c_I|\, dt \leq 1, \tag{1.6}$$

which in turn implies that $\|\ln|\cdot|\|_* \le 2$. To prove (1.6) we consider three cases, namely, (i) $0 < a < b$, (ii) $-b < a < b$, and (iii) the rest. In case (i), we pick $c_I = \ln b$ and note that

$$\int_I |\ln|t| - \ln b|\, dt = \int_{(a,b)} (\ln b - \ln t)\, dt$$
$$= (b - a) - a(\ln b - \ln a).$$

Therefore,

$$\frac{1}{|I|}\int_I |\ln|t| - \ln b|\, dt = 1 - a\frac{(\ln b - \ln a)}{b - a},$$

and (1.6) follows at once since $0 < a < b$. In case (ii) we may restrict ourselves to $-b < a < 0 < b$. Again pick $c_I = \ln b$ and note that

$$\int_I |\ln|t| - \ln b|\, dt = \int_{(a,-a)} |\ln|t| - \ln b|\, dt$$
$$+ \int_{(-a,b)} (\ln b - \ln t)\, dt = J + K,$$

say. The above computation shows that $K = (b + a) + a(\ln b - \ln(-a))$. As for J, since the integrand is an even function, it equals

$$2 \lim_{\varepsilon \to 0^+} \int_{(\varepsilon,-a)} (\ln b - \ln t)\, dt = 2(-a \ln b + a \ln(-a) - a).$$

Thus $J + K = (b - a) + a(\ln b - \ln(-a))$ and

$$\frac{1}{|I|}\int_I |\ln|t| - \ln b|\, dt = 1 - (-a)\frac{(\ln b - \ln(-a))}{b + a}\frac{(b + a)}{(b - a)}. \tag{1.7}$$

Since $a < 0$, the right-hand side in (1.7) is ≤ 1, and (1.6) holds in this case as well. The remaining cases can be reduced to either (i) or (ii) since we are dealing with an even function.

Now that we know that BMO functions are not necessarily bounded, the question is how large thay can be; we take another look at $\ln|t|$. Fix $(0, b) = I \subset T$ and consider those t's in I where $\ln|t|$ is large, i.e., consider

$$\mathcal{O}_\lambda = \{t \in I : |\ln|t| - c_I| > \lambda\}, \qquad \lambda > 0,$$

where $c_I = (\ln|\cdot|)_I$. We are interested in $\mathcal{O}_\lambda$ for large values of λ. Clearly, $\mathcal{O}_\lambda = \{t \in I : t > e^{\lambda + c_I}\} \cup \{t \in I : t < e^{-\lambda + c_I}\}$. Obviously, the first set in $\mathcal{O}_\lambda$ is empty for λ large, and for those λ's we get

$$|\mathcal{O}_\lambda| \le |\{t \in I : t < e^{-\lambda + c_I}\}| = e^{-\lambda} e^{c_I}.$$

By Jensen's inequality, $e^{c_I} \leqslant (1/|I|) \int_I e^{\ln t}\, dt = |I|/2$ and consequently $|\mathcal{O}_\lambda| \leqslant \frac{1}{2}|I|e^{-\lambda}$.

The remarkable fact is that a similar estimate holds for arbitrary f's in BMO and $I \subseteq T$. More precisely, we have

Theorem 1.3 (John–Nirenberg Inequality). Assume that $f \in BMO$ and $I \subseteq T$. Then there are constants c_1, $c_2 > 0$, independent of f and I, so that

$$|\{t \in I: |f(t) - f_I| > \lambda\}| \leqslant c_1 e^{-c_2\lambda/\|f\|_*}|I| \tag{1.8}$$

for all $\lambda > 0$.

Proof. By replacing f by $(f - f_I)/\|f\|_*$ if necessary, we may assume that $f_I = 0$ and $\|f\|_* = 1$; we must then prove that

$$|\mathcal{O}_\lambda| = |\{t \in I: |f(t)| > \lambda\}| \leqslant c_1 e^{-c_2\lambda}|I|. \tag{1.9}$$

This is achieved by the use of the Calderón–Zygmund decomposition. First, since $(1/|I|) \int_I |f(t)|\, dt \leqslant \|f\|_* = 1$, we may invoke the Calderón–Zygmund decomposition for $f(t)\chi_I(t)$ at level 2. We thus obtain (a first generation of) open, disjoint subintervals $\{I_j^1\}$ of I such that

(i) $|f(t)| \leqslant 2$ a.e. in $I \setminus \bigcup I_j^1$,
(ii) $2 < (1/|I_j^1|) \int_{I_j^1} |f(t)|\, dt \leqslant 4$, and
(iii) $\sum |I_j^1| \leqslant \frac{1}{2}\int_I |f(t)|\, dt = \frac{1}{2}|I|((1/|I|)\int_I |f(t)|\, dt) \leqslant \frac{1}{2}|I|$.

Next we consider each I_j^1 individually. To simplify notations, fix such an interval, call it I^1, and consider the function $(f(t) - f_{I^1})\chi_{I^1}(t)$. Since

$$\frac{1}{|I^1|}\int_{I^1} |f(t) - f_{I^1}|\, dt \leqslant \|f\|_* = 1,$$

we may invoke the Calderón–Zygmund decomposition of $(f(t) - f_{I^1}(t)\chi_{I^1}(t)$ at level 2 and obtain (a second generation of) open, disjoint subintervals $\{I_j^2\}$ of I^1 such that

(i) $|f(t) - f_{I^1}| \leqslant 2$ a.e. in $I^1 \setminus \bigcup I_j^2$,
(ii) $2 < (1/|I_j^2|) \int_{I_j^2} |f(t) - f_{I^1}|\, dt \leqslant 4$, and
(iii) $\sum |I_j^2| \leqslant \frac{1}{2}\sum \int_{I_j^2} |f(t) - f_{I^1}|\, dt \leqslant \frac{1}{2}|I^1|$.

Moreover, considering all I^1's now we also have

(i′) for each I^1, $|f(t)| \leqslant |f(t) - f_{I^1}| + |f_{I^1}| \leqslant 2.4$ a.e. in $I^1 \setminus \bigcup I_j^2$, and
(ii′) $\sum_{\text{all } j\text{'s}} |I_j^2| \leqslant \frac{1}{2}\sum_{\text{all } I^1\text{'s}} |I^1| \leqslant (\frac{1}{2})^2|I|$.

We continue with this process and obtain at the nth step a family of open, disjoint subintervals $\{I_j^n\}$ of I^{n-1} such that

$$|f(t)| \leqslant 4n \qquad \text{a.e. in} \quad I^{n-1} \setminus \bigcup I_j^n \tag{1.10}$$

and

$$\sum |I_j^n| \leqslant (\tfrac{1}{2})^n |I|. \tag{1.11}$$

These estimates are all we need. Suppose first that $\lambda > 4$ and let $n \geqslant 1$ be the integer so that $4n < \lambda \leqslant 4(n+1)$. By (1.10) it readily follows that $\mathcal{O}_\lambda \subseteq \{t \in I : |f(t)| > 4n\} \subseteq \bigcup I_j^n$ and by (1.11) that

$$|\mathcal{O}_\lambda| \leqslant (\tfrac{1}{2})^n |I|. \tag{1.12}$$

Moreover, since $2^{-n} = e^{-n \ln 2}$ and $\lambda \leqslant 8n$, we note that $2^{-n} \leqslant e^{-c_2\lambda}$ with $c_2 = (\ln 2)/8$. Thus by (1.12)

$$|\mathcal{O}_\lambda| \leqslant e^{-c_2\lambda} |I| \tag{1.13}$$

in this case. If, on the other hand, $0 < \lambda \leqslant 4$, then

$$|\mathcal{O}_\lambda| \leqslant |I| \leqslant c_1 e^{-c_2\lambda} |I| \tag{1.14}$$

provided that $c_1 e^{-c_2\lambda} \geqslant 1$ when $\lambda \leqslant 4$; to ensure that this occurs we set $c_1 = e^{4(\ln 2)/8} = \sqrt{2} > 1$. By combining (1.13) and (1.14), we have that in all cases $|\{t \in I : |f(t)| > \lambda\}| \leqslant \sqrt{2}\, e^{-(\ln 2/8)\lambda} |I|$, as we wanted to show. ∎

The converse to the John–Nirenberg inequality also holds, namely,

Proposition 1.4. Assume that $f \in L(T)$ and that there are constants $c_1, c_2 > 0$ so that $|\{t \in I : |f(t) - f_I| > \lambda\}| \leqslant c_1 e^{-c_2\lambda} |I|$ for $I \subseteq T$ and $\lambda > 0$. Then for $0 < c < c_2$, $e^{c|f(t) - f_I|} \in L(T)$ and

$$\int_I e^{c|f(t) - f_I|}\, dt \leqslant \frac{c_1 c}{c_2 - c} |I|. \tag{1.15}$$

Proof. We note that the left-hand side in (1.15) equals

$$c \int_{[0,\infty)} |\{t \in I : |f(t) - f_I| > \lambda\}| e^{c\lambda}\, d\lambda \leqslant c c_1 |I| \int_{[0,\infty)} e^{-(c_2 - c)\lambda}\, d\lambda,$$

and we are done. ∎

Corollary 1.5. Suppose that $f \in L(T)$. Then $f \in BMO$ with norm $\|f\|_*$ if and only if

$$\sup_{I \subseteq T} \left(\frac{1}{|I|} \int_I |f(t) - f_I|^p\, dt \right)^{1/p} \leqslant k_p \|f\|_*, \qquad k_p \sim p, \quad 1 \leqslant p < \infty. \tag{1.16}$$

Proof. To obtain the necessity of the condition observe that by the John–Nirenberg inequality and Proposition 1.4 we have

$$\int_I e^{c|f(t) - f_I|/\|f\|_*}\, dt \leqslant \frac{c_1 c}{c_2 - c} |I|, \tag{1.17}$$

where c_1, c_2 are the constants in Theorem 1.3 and $0 < c < c_2$. From (1.17) it follows at once that for $p \geq 1$

$$c^p \int_I |f(t) - f_I|^p \, dt \leq (p!)\|f\|_*^p |I|,$$

and we have finished. Conversely, if (1.16) holds for any $p \geq 1$, it also holds for $p = 1$ and the desired conclusion follows from the John-Nirenberg inequality. ■

2. THE MAXIMAL OPERATOR AND *BMO*

In this section we prove that the Hardy–Littlewood maximal operator is bounded in *BMO*.

Theorem 2.1 (Bennett-DeVore-Sharpley). Assume $f \in BMO$. Then $Mf \in BMO$ and there is a constant c independent of f such that

$$\|Mf\|_* \leq c\|f\|_*. \tag{2.1}$$

Proof. Since $Mf(x) = M(|f|)(x)$ and $\||f|\|_* \leq 2\|f\|_*$ (cf. 6.12), we may assume that $f \geq 0$. We must then show there is a constant c, independent of the subinterval $J \subseteq T$ and f, so that

$$\frac{1}{|J|} \int_J |Mf(x) - (Mf)_J| \, dx \leq c\|f\|_*. \tag{2.2}$$

Fix an interval J and for each x in J divide those intervals $I \subseteq T$ containing x into two families according to their relative size with J; more precisely, let $\mathscr{I}_1(x) = \{I \subseteq T\colon x \in I \text{ and } I \subseteq 3J \cap T\}$ and $\mathscr{I}_2(x) = \{I \subseteq T\colon x \in I \text{ and } I \cap (T \backslash 3J) \neq \varnothing\}$. If we set now

$$F_1(x) = \sup_{I \in \mathscr{I}_1(x)} \frac{1}{|I|} \int_I f(t) \, dt \tag{2.3}$$

and

$$F_2(x) = \sup_{I \in \mathscr{I}_2(x)} \frac{1}{|I|} \int_I f(t) \, dt \tag{2.4}$$

it clearly follows that $Mf(x) = \max(F_1(x), F_2(x))$. Furthermore, since $\int_J (Mf(x) - (Mf)_J) \, dx = 0$, if $\mathcal{O} = \{x \in J\colon Mf(x) > (Mf)_J\}$, we readily see that

$$\frac{1}{|J|} \int_J |Mf(x) - (Mf)_J| \, dx = \frac{2}{|J|} \int_{\mathcal{O}} (Mf(x) - (Mf)_J) \, dx. \tag{2.5}$$

Thus, if we set $\mathscr{U}_1 = \{x \in \mathcal{O}: F_1(x) \geq F_2(x)\}$ and $\mathscr{U}_2 = \mathcal{O}\backslash\mathscr{U}_1$, we may rewrite the right-hand side of (2.5) as

$$\frac{2}{|J|}\int_{\mathscr{U}_1}(F_1(x) - (Mf)_J)\,dx + \frac{2}{|J|}\int_{\mathscr{U}_2}(F_2(x) - (Mf)_J)\,dx = A + B,$$

say, and (2.2) will hold for J provided we show that that for some absolute constant c,

$$A, B \leq c\|f\|_*. \tag{2.7}$$

We consider A first. Let now I denote the interval $3J \cap T$. Since $f_I \leq \inf_{x\in I} Mf(x) \leq \inf_{x\in J} Mf(x)$, then clearly $f_I \leq (Mf)_J$, and consequently we may invoke the Calderón–Zygmund decomposition of $f\chi_I$ at level $(Mf)_J$. We thus obtain a sequence of open, disjoint subintervals $\{I_j\}$ of I verifying

(i) $f(t) \leq (Mf)_J$ a.e. in $I\backslash\bigcup I_j$,
(ii) $(Mf)_J < (1/|I_j|)\int_{I_j} f(t)\,dt \leq 2(Mf)_J$, all j,
(iii) $\bigcup I_j \subseteq I$,

and there is an additional property we emphasize: If I_j' is the "ancestor" interval corresponding to I_j, i.e., I_j' is the dyadic subinterval of I which when subdivided gave rise to I_j, then

(iv) $|I_j'| = 2|I_j|$ and $(1/|I_j'|)\int_{I_j'} f(t)\,dt \leq (Mf)_J$.

We may then consider (a variant of) the Calderón–Zygmund decomposition $g + b$ of $f\chi_I$ obtained by putting

$$g(t) = \sum_j f_{I_j'}\chi_{I_j}(t) + f(t)\chi_{I\backslash\bigcup I_j}(t) \qquad \text{and} \qquad b(t) = \sum_j (f(t) - f_{I_j'})\chi_{I_j}(t).$$

By (i) and (iv) above it follows that

$$g(t) \leq (Mf)_J \tag{2.8}$$

in contrast to the usual decomposition where $g(t) \leq 2(Mf)_J$. Also by Corollary 1.15 and (iii) and (iv)

$$\begin{aligned}\int_I |b(t)|^2\,dt &= \sum_j \int_{I_j} |f(t) - f_{I_j'}|^2\,dt \leq c\Big(\sum |I_j'|\Big)\|f\|_*^2 \\ &\leq c|J|\,\|f\|_*^2.\end{aligned} \tag{2.9}$$

This is all we need. Indeed, by (2.3) we readily see that $F_1(x) \leq M(f\chi_I)(x) \leq Mg(x) + Mb(x)$ and by (2.8) and (2.9) that

$$\begin{aligned}\int_{\mathscr{U}_1} F_1(x)\,dx &\leq \int_{\mathscr{U}_1} Mg(x)\,dx + \int_{\mathscr{U}_1} Mb(x)\,dx \\ &\leq |\mathscr{U}_1|\,\|Mg\|_\infty + |\mathscr{U}_1|^{1/2}\left(\int_{\mathscr{U}_1} Mb(x)^2\,dx\right)^{1/2} \\ &\leq |\mathscr{U}_1|(Mf)_J + c|J|^{1/2}\|b\|_2 \\ &\leq |\mathscr{U}_1|(Mf)_J + c|J|\,\|f\|_*.\end{aligned} \tag{2.10}$$

Thus passing the first summand from the right to the left-hand side in (2.10), we see at once that $A \leqslant c\|f\|_*$, which is the A statement in (2.7).

To bound B we actually prove the stronger estimate

$$F_2(x) - (Mf)_J \leqslant c\|f\|_*, \qquad x \in \mathscr{U}_2. \tag{2.11}$$

Fix x in $\mathscr{U}_2$ and let Q be any interval in $\mathscr{I}_2(x)$; clearly $|Q| \geqslant |J|$. Consider now the subinterval $Q \cup J$ of T, $|Q \cup J| \leqslant 2|Q|$. As above we note that $f_{Q\cup J} \leqslant (Mf)_J$ and consequently

$$f_Q - (Mf)_J \leqslant f_Q - f_{Q\cup J} \leqslant \frac{1}{|Q|}\int_Q |f(t) - f_{Q\cup J}|\,dt$$

$$\leqslant \frac{1}{|Q|}|Q \cup J|\frac{1}{|Q\cup J|}\int_{Q\cup J}|f(t) - f_{Q\cup J}|\,dt \leqslant 2\|f\|_*.$$

Taking sup over $Q \in \mathscr{I}_2(x)$ we obtain (2.11), which in turn implies the B statement in (2.7), and we have finished. ∎

3. THE CONJUGATE OF BOUNDED AND *BMO* FUNCTIONS

As we remarked in 2.7 of Chapter V, if $f(x) = \chi_{[0,1]}(x)$, then $\tilde{f} \in BMO$. This statement holds in general. In other words,

Theorem 3.1. Suppose $f \in L^\infty(T)$. Then $\tilde{f} \in BMO(T)$ and there is a constant c independent of f so that $\|\tilde{f}\|_* \leqslant c\|f\|_\infty$.

The proof of this theorem is essentially contained in that of Theorem 3.3, which we do shortly. However, this is an excellent opportunity for the reader to prove this result directly. At any rate, before we consider our next result we need an observation concerning *BMO* functions.

Proposition 3.2. Suppose $f \in BMO$ and $I \subseteq T$, then

$$|f_{2I} - f_I| \leqslant 2\|f\|_*, \qquad 2I \subseteq T \tag{3.1}$$

and

$$|f_{2^kI} - f_I| \leqslant 2k\|f\|_*, \qquad 2^kI \subseteq T. \tag{3.2}$$

Proof. Since (3.2) follows from (3.1) on account of the observation

$$|f_{2^kI} - f_I| \leqslant \sum_{j=1}^{k}|f_{2^jI} - f_{2^{j-1}I}|,$$

it suffices to prove (3.1). But this is easy since

$$|f_{2I} - f_I| = \frac{1}{|I|}\left|\int_I (f(t) - f_{2I})\,dt\right|$$
$$\leqslant \frac{2}{2|I|}\int_{2I} |f(t) - f_{2I}|\,dt \leqslant 2\|f\|_*. \quad \blacksquare$$

We are now ready to prove

Theorem 3.3 (Spanne, Stein). Assume that $f \in BMO$. Then $\tilde{f} \in BMO$ and there is a constant c, independent of f, such that

$$\|\tilde{f}\|_* \leqslant c\|f\|_*. \tag{3.3}$$

Proof. Given an interval $I \subseteq T$, we show that there is a constant $c_I = c(I, \tilde{f})$ so that

$$\frac{1}{|I|}\int_I |\tilde{f}(t) - c_I|\,dt \leqslant c\|f\|_*, \tag{3.4}$$

where c is independent of I and f; this clearly implies $\|\tilde{f}\|_* \leqslant 2c\|f\|_*$ and we have finished. First, since $f \in BMO$, then $f \in L^2$ and $\tilde{f}$ is a well defined, L^2 function. Fix an interval I and put

$$f(t) = (f(t) - f_I)\chi_{2I}(t) + (f(t) - f_I)\chi_{T\setminus 2I}(t) + f_I$$
$$= f_1(t) + f_2(t) + f_I,$$

say. Since $(f_I)^{\sim} = 0$, we have $\tilde{f}(t) = \tilde{f}_1(t) + \tilde{f}_2(t)$, and it suffices to show that (3.4) holds with f replaced by f_1 and f_2, with suitable constants $c_{I,1}$ and $c_{I,2}$, respectively. The estimate for $\tilde{f}_1$, with $c_{I,1} = 0$, is immediate since

$$\frac{1}{|I|}\int_I |\tilde{f}_1(t)|\,dt \leqslant \left(\frac{1}{|I|}\int_I |\tilde{f}_1(t)|^2\,dt\right)^{1/2}$$
$$\leqslant \left(\frac{1}{|I|}\|\tilde{f}_1\|_2^2\right)^{1/2} \leqslant \left(\frac{1}{|I|}\|f_1\|_2^2\right)^{1/2}$$
$$= \left(\frac{1}{|I|}\int_{2I} |f(t) - f_I|^2\,dt\right)^{1/2}$$
$$\leqslant \left(\frac{1}{|I|}\int_{2I} |f(t) - f_{2I}|^2\,dt\right)^{1/2} + |f_{2I} - f_I| = A + B,$$

say. By Corollary 1.5, $A \leqslant c\|f\|_*$, and, by Proposition 2.2, $B \leqslant 2\|f\|_*$. Thus

$$\frac{1}{|I|}\int_I |\tilde{f}_1(t)|\,dt \leqslant c\|f\|_* \tag{3.5}$$

with c independent of I and f, as we wanted to show.

To estimate $\tilde{f}_2(t)$ let t_I denote the center of I and put

$$c_{I,2} = -\frac{1}{\pi}\int_T \frac{f_2(t)}{2\tan((t_1 - t)/2)}\,dt,$$

where the integral is absolutely convergent, since f_2 vanishes in a neighborhood of I. With this choice for $c_{I,2}$ and the notation $k(t) = -1/(\pi 2\tan(t/2))$, we have

$$\begin{aligned}\int_I |\tilde{f}_2(t) - c_{I,2}|\,dt &= \int_I \left|\int_T (f(x) - f_I)\chi_{T\setminus 2I}(x)(k(t-x) - k(t_I - x))\,dx\right| dt\\ &\leqslant \int_I\int_{T\setminus 2I} |f(x) - f_I||k(t-x) - k(t_I - x)|\,dx\,dt\\ &= \int_{T\setminus 2I} |f(x) - f_I|\left(\int_I |k(t-x) - k(t_I - x)|\,dt\right)dx. \qquad (3.6)\end{aligned}$$

The integrand of the innermost integral in (3.6) was denoted $k(t, x, t_I)$ in (1.10) of Chapter V and estimated in (1.13) there by $c(|I|/|x - t_I|^2)$. Thus the innermost integral in (3.6) is of order $c|I|^2/|x - t_I|^2$ and the right-hand side of (3.6) does not exceed a multiple of

$$|I|^2 \int_{T\setminus 2I} (|f(x) - f_I|/|x - t_I|^2)\,dx. \qquad (3.7)$$

If k_0 now denotes the largest integer k so that $2^k I \subseteq T$, the integral in (3.7) is bounded by

$$\begin{aligned}&\sum_{k=1}^{k_0} \int_{2^{k+1}I\setminus 2^k I} (|f(x) - f_I|/|x - t_I|^2)\,dx\\ &\qquad \leqslant c\sum_{k=1}^{k_0} \frac{1}{(2^k|I|)^2}\int_{2^{k+1}I} |f(x) - f_I|\,dx\\ &\qquad = c\sum_{k=1}^{k_0} A_k, \qquad (3.8)\end{aligned}$$

say. We examine each A_k more closely; since

$$\begin{aligned}\int_{2^{k+1}I} |f(x) - f_I|\,dx &\leqslant \int_{2^{k+1}I} |f(x) - f_{2^{k+1}I}|\,dx + c(2^k|I|)|f_{2^{k+1}I} - f_I|\\ &\leqslant c2^k|I|(\|f\|_* + k\|f\|_*) \leqslant ck2^k|I|\,\|f\|_*,\end{aligned}$$

we see at once that

$$A_k \leqslant c\left(\frac{k}{2^k}\|f\|_*\right)\Big/|I|, \qquad \text{all} \quad k. \qquad (3.9)$$

Whence replacing the estimate (3.9) in (3.8), by (3.7) we see at once that (3.6) is dominated by

$$\left(c\sum_{k=1}^{k_0}\frac{k}{2^k}\right)|I|\,\|f\|_* \leq c|I|\,\|f\|_*$$

In other words

$$\frac{1}{|I|}\int_I|\tilde{f}_2(t)-c_{I,2}|\,dt \leq c\|f\|_*, \tag{3.10}$$

(3.4) follows by adding (3.5) and (3.10), and we have finished. ■

Theorem 3.4. Assume $f \in BMO$. Then there is a constant c, independent of f, such that

$$\|H^*f\|_* \leq c\|f\|_*.$$

Proof. By Theorem 2.13 in Chapter V, $H^*f(x) \leq c(Mf(x) + M(Hf)(x))$, and the conclusion follows at once from this on account of Theorems 3.3 and 2.1. ■

4. Wk-$L^p(T)$ AND K_f. INTERPOLATION

Let $I \subseteq T$ and $(f - f_I) \in L^p(I)$, $1 \leq p < \infty$. If $\{I_j\}$ is a finite collection of open, disjoint subintervals of I, then by Corollary 1.2 and Jensen's inequality it readily follows that

$$\sum|I_j|\left(\frac{1}{|I_j|}\int_{I_j}|f(t)-f_{I_j}|\,dt\right)^p \leq 2^p\int_I|f(t)-f_I|^p\,dt. \tag{4.1}$$

The consideration of a converse to (4.1), which also corresponds to taking l^p rather than l^∞ norms on the sequences $\{(1/|I_j|)\int_{I_j}|f(t)-f_{I_j}|\,dt\}$, as was done in the definition of BMO, leads to

Theorem 4.1 (John–Nirenberg). Suppose that $f \in L(I)$ and that for some $1 < p < \infty$

$$K_f = \sup\left(\sum|I_j|\left(\frac{1}{|I_j|}\int_{I_j}|f(t)-f_{I_j}|\,dt\right)^p\right)^{1/p} < \infty, \tag{4.2}$$

where the sup is taken over all finite partitions of subintervals $\{I_j\}$ of I. Then $(f - f_I) \in$ wk-$L^p(I)$ with norm $\leq cK_f$, c independent of f.

Proof. By replacing f by $(f-f_I)/K_f$ if necessary, we may assume that $f_I = 0$ and $K_f = 1$; consequently, the conclusion will follow once we show

$$\lambda^p|\mathcal{O}_\lambda| = \lambda^p|\{t \in I: |f(t)| > \lambda\}| \leq c^p \tag{4.3}$$

with c independent of f, λ, and I. The idea of the proof is to set things so that the Calderón-Zygmund decomposition can get the job done. To start out we must have a number $\mu > |f|_I$; in our case, since

$$\frac{1}{|I|}\int_I |f(t)|\,dt \leq \frac{K_f}{|I|^{1/p}} = \frac{1}{|I|^{1/p}}, \tag{4.4}$$

$\mu \geq 1/|I|^{1/p}$ will do. For such μ, invoke the Calderón-Zygmund decomposition at level μ and obtain (a first generation of) open, disjoint subintervals I_j^1 of I such that

(i) $|f(t)| \leq \mu$ a.e. in $I \setminus \bigcup I_j^1$,
(ii) $\mu < (1/|I_j^1|)\int_{I_j^1}|f(t)|\,dt \leq 2\mu$, all j, and
(iii) $\sum|I_j^1| \leq (1/\mu)\int_I|f(t)|\,dt$.

Fix an I_j^1, call it I^1, and consider $(f(t)-f_{I^1})\chi_{I^1}(t)$. Since

$$\frac{1}{|I^1|}\int_{I^1}|f(t)-f_{I^1}|\,dt \leq \frac{1}{|I^1|}\int_{I^1}|f(t)|\,dt + |f_{I^1}|$$
$$\leq 2\mu + 2\mu = 2^2\mu,$$

we may invoke the Calderón-Zygmund decomposition of this function at level $2^2\mu$. We then get (a second generation of) open, disjoint subintervals $\{I_j^2\}$ of I^1 such that

(i) $|f(t)-f_{I^1}| \leq 2^2\mu$ a.e. in $I^1 \setminus \bigcup I_j^2$
(ii) $2^2\mu < (1/|I_j^2|)\int_{I_j^2}|f(t)-f_{I^1}|\,dt \leq 2^3\mu$, all j
(iii) $\sum|I_j^2| \leq (1/2^2\mu)\sum\int_{I_j^2}|f(t)-f_{I^1}|\,dt \leq (1/2^2\mu)\int_{I^1}|f(t)-f_{I^1}|\,dt$.

We would like to rewrite (i) and (iii) in terms of f taking into consideration all the I^1's. It is readily seen that we have

(i′) $|f(t)| \leq |f(t)-f_{I^1}| + |f_{I^1}| \leq 2^2\mu + 2\mu \leq 2^3\mu$, a.e. in $I^1 \setminus \bigcup I_j^2$,

and summing over all j's corresponding to all I^1's also

(iii′) $\sum_{\text{all } j\text{'s}}|I_j^2| \leq (1/2^2\mu)\sum_{\text{all } I^1\text{'s}}\int_{I^1}|f(t)-f_{I^1}|\,dt$.

We take a closer look at the sum on the right-hand side of (iii′). With $1/p + 1/p' = 1$, we note that it does not exceed

$$|I^1|^{1/p}\left(\frac{1}{|I^1|}\int_{I_1}|f(t)-f_{I1}|\,dt\right)|I^1|^{1/p'} \leq K_f\left(\sum|I^1|\right)^{1/p'} \leq \left(\frac{1}{\mu}\int_I|f(t)|\,dt\right)^{1/p'}.$$

Thus (iii′) actually states

$$\sum_{\text{all } j\text{'s}} |I_j^2| \leqslant (1/2^2\mu)\left((1/\mu)\int_I |f(t)|\,dt\right)^{1/p'}.$$

Clearly, this process may be iterated as we did inthe proof of the John-Nirenberg inequality: in general, and in the kth step we invoke the Calderón-Zygmund decomposition for the function $(f - f_{I^{k-1}})\chi_{I^{k-1}}$ at level $2^{2(k-1)}\mu$ and we get a collection of open, disjoint subintervals $\{I_j^k\}$ of I^{k-1} so that

$$|f(t)| \leqslant 2^{2k-1}\mu \qquad \text{a.e. in} \quad I^{k-1}\backslash \bigcup I_j^k \tag{4.5}$$

and

$$\sum_{\text{all } j\text{'s}} |I_j^k| \leqslant \left(\int_I |f(t)|\,dt\right)^{(k-1)/p'} \Big/ \phi(\mu, k), \tag{4.6}$$

where $1/p + 1/p' = 1$ and

$$\phi(\mu, k) = (2^{2(k-1)}\mu)(2^{2(k-2)}\mu)^{1/p'}(2^{2(k-3)}\mu)^{(1/p')^2}\cdots\mu^{(1/p')^{k-1}}.$$

To complete the proof all we need is a good estimate for the right-hand side of (4.6); note that

$$\frac{1}{\phi(\mu, k)} = \frac{2^{2/p'(\sum_{j=1}^{\infty} j/(p')^{j-1})}}{(2^{2(k-1)}\mu)^{1+1/p'+\cdots+(1/p')^{k-1}}} \tag{4.7}$$

Moreover, since $\sum_{j=1}^{\infty} j/(p')^{j-1} = 1/(1-1/p')^2 = p^2$,

$$\sum_{j=0}^{\infty} 1/(p')^j = p \qquad \text{and} \qquad \sum_{j=k}^{\infty} 1/(p')^j = p/(p')^k,$$

the right-hand side of (4.7) can also be written as

$$2^{2p^2/p'}(2^{2(k-1)}\mu)^{p/(p')^k}/(2^{2(k-1)}\mu)^p. \tag{4.8}$$

Furthermore, since $(k-1)/(p')^k \leqslant c_p = c$ for all $k \geqslant 1$, the numerator in (4.9) does not exceed $c\mu^{p/(p')^k}$ and consequently the right-hand side of (4.6) is bounded by

$$c\left(\mu^{p/p'}\int_I |f(t)|\,dt\right)^{1/(p')^{k-1}} \Big/ (2^{2(k-1)}\mu)^p. \tag{4.9}$$

We pick now $\mu = 2/|I|^{1/p}$ and observe that with this choice,

$$\mu^{p/p'}\int_I |f(t)|\,dt \leqslant 2^{p/p'}K_f \leqslant c,$$

which finally obtains that for this choice of μ the right-hand side of (4.6) is bounded by

$$c/(2^{2(k-1)}\mu)^p. \tag{4.10}$$

We are now ready to prove (4.3). First assume $\lambda > 2\mu$ and let k be the largest integer $\geqslant 1$ so that $2^{2k-1}\mu < \lambda$. In this case, and on account of (4.5) we have that $\mathcal{O}_\lambda \subseteq \bigcup I_j^k$, where the union is taken over all j's and k's. By (4.10) then

$$|\mathcal{O}_\lambda| \leqslant c/(2^{(2k-1)}\mu)^p \leqslant c/\lambda^p, \tag{4.11}$$

which is what we wanted to show. It only remains to consider the case $\lambda \leqslant 2\mu = 4/|I|^{1/p}$; but now $\lambda^p|\mathcal{O}_\lambda| \leqslant \lambda^p|I| \leqslant 4$ and we are done. ■

Not only is Theorem 4.1 of interest in itself but it also has important applications. The one we consider here is to interpolate with *BMO* as an endpoint space.

Theorem 4.2 (Stampacchia). Suppose T is a linear operator which is bounded in $L^{p_0}(I)$, $1 \leqslant p_0 < \infty$ and which is of type $(\infty, *)$ there, i.e., $\|Tf\|_{BMO(I)} \leqslant c\|f\|_{L^\infty(I)}$; then T maps $L^p(I)$ into itself, $p_0 < p < \infty$.

Proof. Let $\{I_j\}$ be a finite partition of disjoint subintervals of I, and put

$$\tau f(x) = \sum\left(\frac{1}{|I_j|}\int_{I_j}|Tf(t) - (Tf)_{I_j}|\,dt\right)\chi_{I_j}(x).$$

We claim that τ is a sublinear mapping simultaneously of types (p_0, p_0) and (∞, ∞) with norm, in each case, bounded independently of the particular partition of I. This is easy to check in the ∞ case since $\|\tau f\|_\infty \leqslant \|Tf\|_* \leqslant c\|f\|_\infty$ (all norms are taken over I) and for p_0 note that since the I_j's are disjoint, by (4.1) we have

$$\left(\int_I |\tau f(x)|^{p_0}\,dx\right)^{1/p_0} = \|\tau f\|_{p_0} = \left(\sum|I_j|\left(\frac{1}{|I_j|}\int_{I_j}|Tf(t) - (Tf)_{I_j}|\,dt\right)^{p_0}\right)^{1/p_0}$$

$$\leqslant 2\left(\int_I |Tf(t) - (Tf)_I|^{p_0}\,dt\right)^{1/p_0} \leqslant c\|Tf\|_{p_0} + |I|^{1/p_0}|Tf|_I.$$

Moreover, since $|I|^{1/p_0}|Tf|_I \leqslant \|Tf\|_{p_0}$, also $\|\tau f\|_{p_0} \leqslant c\|f\|_{p_0}$, and our claim is verified. By the Marcinkiewicz interpolation theorem in Chapter IV we obtain that τ is bounded on $L^p(I)$, $p_0 < p < \infty$, with norm bounded independently of the partition $\{I_j\}$ as this is the case for $p = p_0, \infty$. Moreover, since the I_j's are disjoint we readily see that

$$\|\tau f\|_p = \left(\sum|I_j|\left(\frac{1}{|I_j|}\int_{I_j}|Tf(t) - (Tf)_{I_j}|\,dt\right)^p\right)^{1/p}$$

and consequently for each f in $L^p(I)$, $K_{Tf} \leqslant c\|f\|_p$, with c independent of f. By Theorem 4.1, $Tf - (Tf)_I$ is in wk-$L^p(I)$ with norm $\leqslant c\|f\|_p$ for $p_0 < p < \infty$. In other words the linear mapping $f \to Sf = Tf - (Tf)_I$ is of weak-type

(p, p) for $p_0 < p < \infty$. By the Marcinkiewicz interpolation theorem, S is actually of type (p, p) for the same range of p's. Next we show that T is bounded in $L^p(I)$, recall that by assumption T is bounded in $L^{p_0}(I)$. Observe that

$$\begin{aligned}\|Tf\|_p &\leqslant \|Sf\|_p + |I|^{1/p}|Tf|_I \\ &\leqslant c\|f\|_p + |I|^{1/p}\left(\frac{1}{|I|}\int_I |Tf(t)|^{p_0}\,dt\right)^{1/p_0} \\ &\leqslant c\|f\|_p + c|I|^{1/p}\left(\frac{1}{|I|}\int_I |f(t)|^{p_0}\,dt\right)^{1/p_0} \\ &\leqslant c\|f\|_p,\end{aligned}$$

and we have finished. ■

Remark 4.3. A particularly interesting application of Theorem 4.2 is to the conjugate operator as this is a linear operation simultaneously of types (2, 2) and $(\infty, *)$. This proof is not simpler than the one we gave in Chapter V since it relies on Theorem 4.2, which in turn makes use of the Calderón-Zygmund decomposition; in other words, both proofs depend on the same basic principle.

5. LIPSCHITZ AND MORREY SPACES

We say that f is Lipschitz of order α in I, $0 < \alpha < 1$, and denote this by $f \in \Lambda_\alpha(I)$ or $\mathrm{Lip}_\alpha(I)$ if there is a constant c so that $|f(x) - f(y)| \leqslant c|x - y|^\alpha$ for every x, y in I; the smallest such constant c is called the Lip α norm of f and is denoted by $\|f\|_{\Lambda_\alpha}$. Endowed with this norm $\mathrm{Lip}_\alpha(I)$ becomes a Banach space provided we identify functions which differ by a constant (a.e.). Since this norm in some sense measures the oscillation of f over I, we are interested in expressing it in an equivalent form that will enable us to apply the techniques developed in this chapter to $\mathrm{Lip}_\alpha(I)$ as well. We begin by proving

Theorem 5.1 (Campanato, Meyers). Suppose that $f \in L(I)$ and $0 < \alpha < 1$. Then the following four statements are equivalent in the sense that each implies the other three and the constants appearing on the right-hand side of each are also equivalent.

(i) $|f(x) - f(y)| \leqslant c_1|x - y|^\alpha$, all x, y in I,
(ii) $(1/|J|^{1+\alpha}) \int_J |f(t) - f_J|\,dt \leqslant c_2$, all $J \subseteq I$,
(iii) $|f(x) - f_J| \leqslant c_3|J|^\alpha$, all $x \in J$, $J \subseteq I$, and
(iv) $((1/|J|^{1+\alpha p}) \int_J |f(t) - f_J|^p\,dt)^{1/p} \leqslant c_4$, all $J \subseteq I$, $1 \leqslant p < \infty$.

Proof. The equivalence is to be understood to mean that each f which satisfies (ii), (iii) or (iv) can be modified in a set of measure 0 so as to coincide with a continuous function which verifies (i) as well; with this in mind we proceed with the proof. The only implication which is not trivial is (ii) implies (i). So assume $x < y$ are points in I and let $J = [x, y]$, $|J| = y - x$. We then have

$$|f(x) - f(y)| \leqslant |f(x) - f_J| + |f_J - f(y)| = A + B, \tag{5.1}$$

say. We only consider A in (5.1), since the estimate for B is identical. We construct a sequence of subinterval $\{J_k\}$ of J which tends to x as follows: put $J_1 = J$, J_2 the left half of J_1, i.e., the half which contains x, $|J_2| = \frac{1}{2}|J_1|$, and so on. Clearly, for $k \geqslant 2$,

$$A \leqslant |f(x) - f_{J_k}| + \sum_{n=1}^{k-1} |f_{J_{n+1}} - f_{J_n}| = A_1 + A_2, \tag{5.2}$$

say. Next note that

$$\begin{aligned} A_2 &\leqslant \sum_{n=1}^{k-1} \frac{1}{|J_{n+1}|} \int_{J_{n+1}} |f(t) - f_{J_n}|\, dt \\ &\leqslant 2 \sum_{n=1}^{k-1} \frac{1}{|J_n|} \int_{J_n} |f(t) - f_{J_n}|\, dt \\ &\leqslant 2c_2 \sum_{n=1}^{\infty} \frac{|J|^\alpha}{2^{(n-1)\alpha}} = c_\alpha c_2 |J|^\alpha. \end{aligned}$$

Thus, substituting this estimate in (5.2) we see that

$$A \leqslant |f(x) - f_{J_k}| + c_\alpha c_2 |J|^\alpha, \qquad \text{all} \quad k > 2. \tag{5.3}$$

Now by Theorem 2.2 in Chaper IV, for almost every x in I we have $\lim_{n\to\infty} |f(x) - f_{J_k}| = 0$ and consequently by (5.3) we readily see that $A \leqslant cc_2|x - y|^\alpha$ for almost every x in I. A similar estimate holds for B, provided, of course, that y is also a point where Theorem 2.2 of Chapter IV applies. Therefore by (5.1) we get that

$$|f(x) - f(y)| \leqslant cc_2|x - y|^\alpha, \quad \text{for almost every } x, y \quad \text{in} \quad I. \tag{5.4}$$

Since it is evident that f may be redefined so that (5.4) holds everywhere in I, the proof is complete. ■

An interesting application of the above result is

Theorem 5.2 (Privalov). Assume $0 < \alpha < 1$ and $f \in \mathrm{Lip}_\alpha(T) = \mathrm{Lip}_\alpha$. Then $\tilde{f} \in \mathrm{Lip}_\alpha$ and there is a constant c, independent of f, so that

$$\|\tilde{f}\|_{\Lambda_\alpha} \leqslant c\|f\|_{\Lambda_\alpha}. \tag{5.5}$$

Proof. The proof is similar to that of Theorem 3.3. In the notation of that theorem, and by (iv) with $p = 2$ above, we note that

$$\frac{1}{|I|}\int_I |\tilde{f}_1(t)|\,dt \leq c_4|I|^\alpha + |f_{2I} - f_I|$$

and since $|f_{2I} - f_I| \leq 2c_4 2^\alpha |I|^\alpha$ we get that

$$\frac{1}{|I|}\int_I |\tilde{f}_1(t)|\,dt \leq cc_4|I|^\alpha.$$

Similarly, $\sum A_k$ in (3.8) of that theorem is readily estimated as follows: first, since

$$\sum \int_{2^{k+1}I} |f(x) - f_{2^{k+1}I}|\,dx \leq c_4(2^{k+1}|I|)^{1+\alpha}$$

and

$$2^k|I|\,|f_{2^{k+1}I} - f_I| \leq cc_4 k 2^k |I|^{1+\alpha},$$

we see at once that $A_k \leq cc_4(2^{k(\alpha-1)} + k2^{-k})|I|^{\alpha-1}$. Therefore, also,

$$\int_I |\tilde{f}_2(t) - c_{I,2}|\,dt \leq cc_4|I|^{1+\alpha},$$

$c_4 \sim \|f\|_{\Lambda_\alpha}$, and we have finished. ■

There is yet another important family of spaces defined in terms of oscillations and we will consider it here briefly since, in some sense, it provides a glimpse of what, in addition to Lip_α, lies beyond *BMO*. Given $1 \leq p < \infty$ and $0 \leq \eta < 1 + p$ let $\mathcal{L}_{p,\eta}(T) = \mathcal{L}_{p,\eta}$ denote the class of $f \in L(T)$ such that

$$\int_I |f(t) - f_I|^p\,dt \leq c^p|I|^\eta, \qquad \text{all} \quad I \subseteq T. \tag{5.6}$$

The smallest constant c in (5.6) above is denoted by $|f|_{p,\eta}$ and is called the norm of f in $\mathcal{L}_{p,\eta}$. Endowed with this norm, $\mathcal{L}_{p,\eta}$ becomes a Banach space provided we identify functions which differ by a constant. For each fixed p these spaces coincide with familiar ones for some values of η; for instance, if $\eta = 0$, $\mathcal{L}_{p,\eta} \sim L^p$; if $\eta = 1$, $\mathcal{L}_{p,\eta} \sim BMO$, and if $1 < \eta < 1 + p$, then $\mathcal{L}_{p,\eta} \sim \mathrm{Lip}_{(\eta-1)/p}$. In the range $0 < \eta < 1$ they are called Morrey spaces and it is possible to show that also $\int_I |f(t)|^p\,dt \leq c|I|^\eta$ for each $I \subseteq T$. Some relations among these spaces, such as $\mathcal{L}_{p,\eta} \subseteq \mathcal{L}_{p_1,\eta_1}$ if $p \geq p_1$ and $(\eta - 1)/p = (\eta_1 - 1)p_1$ are quite easy to prove. How about interpolation properties and the action of the conjugate operator on these spaces? The reader may think about these questions, or consult Peetre's work [1969].

6. NOTES; FURTHER RESULTS AND PROBLEMS

In the early 1960s John and Nirenberg introduced a space, which they called *BMO*, in order to be able to use the fact that a function which can be approximated in every subcube I of a cube I_0 in the L mean by a constant a_I with an error independent of I differs also in the L^p norm from a_I in I by an error of the same order of magnitude, $1 < p < \infty$; this is of course one of the meanings of the John–Nirenberg inequality. The proof we give here of this result follows ideas of Calderón (see Neri [1971] and Garnett [1981]). An interesting description of the role of *BMO* in the theory of elasticity, as well as the introduction of the local sharp maximal function $M^\#_{0,s}$ considered in problem 6.1, is in John's work [1964].

Further Results and Problems

6.1 Given a measurable function f defined in T and x in T we put $M^\#_{0,s}f(x) = \sup_I \inf_c \inf\{A \geq 0: |\{y \in I: |f(y) - c| > A\}| < s|I|\}$, where I is an interval containing x and $0 < s \leq \frac{1}{2}$ (this restriction on s is necessary since for $s > \frac{1}{2}$, $M^\#_{0,s}f \equiv 0$ for any function which assumes only two values); some properties of this function, such as $M^\#_{0,s}(f+g)(x) \leq 2(M^\#_{0,s/2}f(x) + M^\#_{0,s/2}g(x))$ are easy to show. An important relation is the following: if $x \in I$, then $|\{y \in I: |f(y) - m_f(I)| > 4M^\#_{0,s}f(x)\}| < s|I|$; prove it. (*Hint*: To show the last relation, given $\varepsilon > 0$ let $c_I = a + ib$ be such that $|\{y \in I: |f(y) - c_I| \leq M^\#_{0,s}f(x) + \varepsilon\}| \geq (1-s)|I|$; then

$$|\{y \in I: |\mathrm{Re}\, f(y) - a| > M^\#_{0,s}f(x) + \varepsilon\}| < s|I|$$

and a similar inequality holds for $|I_m f(y) - b|$ as well. Thus $a - (M^\#_{0,s}f(x) + \varepsilon) \leq \mathrm{Re}(m_f(I)) \leq a + (M^\#_{0,s}f(x) + \varepsilon)$ and similarly for b and $\mathrm{Im}(m_f(I))$, and $|c_I - m_I(f)| \leq M^\#_{0,s}f(x) + \varepsilon$. Let now $I_1 = \{y \in I: |f(y) - c_I| \geq |m_f(I) - c_I|\}$, $I_2 = I \setminus I_1$; since it is readily seen that $|\{y \in I_2: |f(y) - m_f(I)| > 2(M^\#_{0,s}f(x) + \varepsilon)\}| = 0$, the choice $\varepsilon = M^\#_{0,s}f(x)$ completes the proof. Clearly an equivalent statement is $|\{y \in I: |f(y) - m_f(I)| > 4 \inf_I M^\#_{0,s}f\}| < s|I|$; this result and the next 5 are from Strömberg's paper [1979a].)

6.2 Let f be measurable, $0 < s \leq \frac{1}{2}$, $I \subseteq J$, $|J| \leq 2^k|I|$. Then

$$|m_f(I) - m_f(J)| \leq ck \inf_I M^\#_{0,s}f.$$

(*Hint*: Cf. (3.2).)

6.3 Assume f is measurable $I \subseteq T$, $0 < s \leq \frac{1}{2}$, and $\lambda, \beta > 0$; then if $|m_f(I)| \leq \lambda$ there is a collection $\{I_j\}$ of disjoint, dyadic subintervals of I such that no interval I_j is totally contained in $\{y \in I: M^\#_{0,s}f(y) > \beta\}$, $\lambda <$

$|m_f(I_j)| \leq \lambda + c\beta$ (c is an absolute constant $\leq 10\sqrt{2}$) and $|f(y)| \leq \lambda$ a.e. in $I \setminus (\bigcup I_j \cup \{y \in I: M^{\#}_{0,s}f(y) > \beta\})$. This statement corresponds to the Calderón-Zygmund decomposition of f and is proved similarly.

6.4 Suppose f is measurable, $0 < s \leq \frac{1}{2}$ and $I \subseteq T$. Then for β, $\lambda > 0$, $|\{y \in I: |f(y) - m_f(I)| > \beta\}| \leq c_1 e^{-c\beta/\lambda}|I| + |\{y \in I: M^{\#}_{0,s}f(y) > \lambda\}|$, where c, c_1 are absolute constants. (*Hint*: This statement corresponds to the John-Nirenberg inequality and is proved in a similar fashion. An interesting variant of this result is the following: for f, β, λ as above, s sufficiently small, and $0 < p < \infty$, $|\{y \in I: |f(y) - m_f(I)| > \beta,$

$$M^{\#}_{0,s}f(y) \leq \lambda\beta\}| \leq c_1 e^{-c/\lambda} \|f - m_f(I)\|^p_{L^p(I)}/\beta^p,$$

6.4 corresponds to $p = 0$. This result is due to Jawerth and Torchinsky [1985].)

6.5 Suppose f is measurable, $0 < s \leq \frac{1}{2}$, and $M^{\#}_{0,s}f \in L^\infty(T)$. Then $f \in BMO$ and $\|f\|_* \sim \|M^{\#}_{0,s}f\|_\infty$. One of the interesting features of this result is that our original assumption only involved the measurability, and not the integrability, of f.

6.6 Suppose $f \in L$, $0 < s < 1$. Then $M^{\#}_{0,s}Hf(x) \leq (c/s)M^{\#}f(x)$, $x \in T$. (*Hint*: Fix I and let $x \in I$; then put $f_1(y) = (f(y) - m_f(I))\chi_{2I}(y)$ and $f_2(y) = (f(y) - m_f(I))\chi_{T\setminus 2I}(y)$. Since $Hf(x) = Hf_1(x) + Hf_2(x)$, we treat each term separately: it is easy to show that for y in I $|Hf_2(y) - c| \leq c_1 M^{\#}f(x)$ and by the weak-type (1, 1) property of H that $\lambda|\{y \in I: |Hf_1(y)| > \lambda\}| \leq cM^{\#}f(x)|I|$.)

6.7 A subadditive operator T is of weak-type (1, 1) and maps L^∞ into BMO if and only if for every interval I, $f \in L + L^\infty$ and $\lambda > 0$, $|\{y \in I: |Tf(y) - m_{Tf}(I)| > \lambda\}| \leq c_1 e^{-c\lambda/\|f\|_\infty} \min(|I|, \|f\|_1/\lambda)$. (*Hint*: The proof of the necessity makes use of the variant of the John-Nirenberg inequality given in problem 6.4. This result is from Jawerth-Torchinsky [1985], where also another necessary and sufficient condition is given, namely, $|\{M^{\#}_{0,s}Tf > \lambda\}| \leq c_1|\{Mf > c\lambda\}|$, c_1, c constants which depend on T but not on f and $\lambda > 0$.)

6.8 Prove that BMO is complete. (*Hint*: It suffices to show that if $\int_T f_n(t)\, dt = 0$, $n \geq 1$, and $\sum \|f_n\|_* < \infty$, then $\lim_{N\to\infty} \sum_{n=1}^N f_n$ exists in BMO; it is not hard to show that the limit in L actually coincides with that in BMO.)

6.9 Show that $\|f\|_1 + \|f\|_*$ gives an equialent norm in BMO, where now functions are identified when they coincide a.e. in T.

6.10 Does there exist a BMO function f so that $\tilde{f} \in L^\infty(T)$? (*Hint*: If $f \in L^2(T)$, then $\tilde{\tilde{f}} = -f + c_0(f)$.)

6.11 Show that $|\ln|x||^p \notin BMO$ for $p > 1$; also $\chi_{(0,1)}(x) \ln|x| \notin BMO$.

6.12 Show that $M^{\#}(|f|)(x) \leq 2M^{\#}f(x)$. (*Hint*: By Proposition 1.1, $\int_I ||f(t)| - |f|_I|\, dt \leq 2\int_I ||f(t)| - |f_I||\, dt \leq 2\int_I |f(t) - f_I|\, dt$).

6.13 Suppose $f \in BMO$ and show that $|f(t)|^\alpha$ also belongs to BMO for $0 < \alpha \leq 1$; this is in contrast to Problem 6.4.

6.14 Suppose f is real valued and show that for each real constant c, $M^{\#}(\max(f, c))(x) \leq \frac{3}{2}M^{\#}f(x)$ and $M^{\#}(\min(f, c))(x) \leq \frac{3}{2}M^{\#}f(x)$. (*Hint*: $\max(f(t), c) = \frac{1}{2}(f(t) + c) + \frac{1}{2}|f(t) - c|$ and $\min(f(t), c) = \frac{1}{2}|f(t) + c| - \frac{1}{2}|f(t) - c|$.)

6.15 Suppose $f \in BMO(I)$, $\|f\|_* = 1$, $\int_I f(t)\,dt = 0$; then f can be written as $g + h$ where $h(t) = \sum_{j=1}^{\infty} a_j\chi_{I_j}(t)$, the I_j's are dyadic subintervals of I, $|a_j| \leq c$, all j, and $\|g\|_\infty \leq c_1$; c, c_1 are absolute constants. (*Hint*: We take a closer look at the John–Nirenberg inequality; with the notation of Theorem 1.3, let $f_1 = f\chi_{I\setminus\bigcup I_j}$, $f_2 = \sum a_j\chi_{I_j}$, $a_j = f_{I_j}$, $|a_j| \leq 2$. Next apply the Calderón–Zygmund decomposition to each $f^1 = (f - a_j)\chi_{I_j}$ and obtain f_1^1, f_2^1 as before; repeat this process and set g equal to the sum of all the f_1^j's, h equal to the sum of all the f_2^j's. This result is related to some of Varoupoulos [1977].)

6.16 Here is another way to build up BMO functions: given a sequence $\{I_j\}$ of dyadic (not necessarily disjoint) open subintervals of T, we say that a sequence of functions $\{a_j\}$ is "adapted" to the I_j's provided (i) a_j vanishes off $3I_j$, (ii) $|a_j(x) - a_j(t)| \leq |x - t|$, all j, (iii) $|a_j(x)| \leq c_1$, $|a_j'(x)| \leq c_2/|I|$ a.e., all j. If, in addition, $\sum_{j;I_j\subseteq I}|I_j| \leq c_3|I|$, for all $I \subseteq T$, then for any adapted sequence of functions we have $\|\sum a_j\|_* \leq c(c_1 + c_2)c_3$. (*Hint*: Given I, separate those I_j's so that $3I_j \cap I \neq \phi$ into two families, $\partial_1 = \{I_j: |I_j| \leq |I|\}$ and ∂_2 the rest, and put

$$f(t) = \sum_{j;I_j\in\partial_1} a_j(t), \qquad g(t) = \sum_{j;I_j\in\partial_2} a_j(t).$$

Clearly,

$$\frac{1}{|I|}\int_I |f(t)|\,dt \leq \frac{1}{|I|}\sum_{j;I_j\in\partial_1} c_1|I_j| \leq 6c_1c_3.$$

To estimate $A = (1/|I|)\int_I |g(t) - c_I|\,dt$, let

$$E_k = \{I_j \in \partial_2 : 2^k|I| < |I_j| \leq 2^{k+1}|I|\}$$

and note that each E_k contains at most c_1c_3 cubes and bound

$$A \leq \sum_{k=0}^{\infty}\sum_{j:I_j\in E_k}\frac{1}{|I|}\int_I |a_j(t) - a_j(x_j)|\,dt, \qquad \text{where} \quad x_j \in I_j.$$

What this means is that we chose $c_I = \sum_j a_j(x_j)$. The result is due to Garnett and Jones [1982] and it admits a converse.)

6.17 Suppose $f \in BMO$. Then for $1 \leq p < \infty$

$$\sup_{z=re^{it}\in D}\left(\frac{1}{2\pi}\int_T |f(x) - f * P_r(t)|^p P_r(x - t)\,dx\right)^{1/p} \leq c\|f\|_* \qquad (6.1)$$

with c independent of f. Conversely, if $f \in L$ and (6.1) holds for any $1 \leq p < \infty$, then $f \in BMO$. (*Hint*: The proof of the first statement is similar in spirit to Theorem 3.3. On the other hand, suppose (6.1) holds for $p = 1$, say; if $z = re^{it}$, I = interval centered at t with $|I| = 2\pi(1-r)$ and $x \in I$, then $P_r(x-t) \geq c/|I|$, and, consequently, $(1/|I|)\int_I |f(x) - f * P_r(t)|\,dx \leq c$, all I. Note that if (6.1) holds for p's < 1, we may also infer that $f \in BMO$ by invoking 6.5 above.)

6.18 With the same notation and assumptions as in 7.14 of Chapter III and 7.29 of Chapter IV, show that $s^* \in BMO$ and $\|s^*\|_* \leq c(\sum |r_k|^2)^{1/2}$. (*Hint*: Given $I \subseteq T$, let N be an integer such that $2\pi/\lambda_{N+1} < |I| \leq 2\pi/\lambda_N$. There are two cases. First, if such an N does not exist, it is because $|I| > 2\pi/\lambda_0$, and the bound for $(1/|I|)\int_I s^*(t)\,dt$ obtains at once from 7.29 in Chapter IV. If, on the other hand, such an N exists let t be the center of I and put $c_I = \sup\{|s_0(t)|, \ldots, |s_N(t)|\}$; we claim that in this case

$$\left(\frac{1}{|I|}\int_I |s^*(x) - c_I|^2\,dx\right)^{1/2} \leq c_\theta\left(\sum |r_k|^2\right)^{1/2}. \tag{6.2}$$

To show (6.2) we make repeated use of the following fact: if $u_n, v_n \geq 0$ and $|u_n - v_n| \leq 1$, then $|\sup_{n\geq 0} u_n - \sup_{n\geq 0} v_n| \leq 1$. Let then $\alpha(x) = \sup\{|s_0(t)|, \ldots, |s_N(t)|, \quad |s_N(t) + R_{N+1}(x)|, \ldots, |s_N(t) + R_{N+k}(x)|, \ldots\}$ where $R_j(x) = \sum_{k=j}^{\infty} r_k f_k(\lambda_k x)$; by 7.29 in Chapter IV $|s_k(x) - s_k(t)| \leq c(\sum |r_k|^2)^{1/2}$ if $k \leq N$, $x \in I$, and, by the above remark, $|\alpha(x) - s^*(x)| \leq c(\sum |r_k|^2)^{1/2}$ as well. If now $\beta(x) = \sup\{|R_{N+1}(x)|, \ldots, |R_{N+k}(x)|, \ldots\}$, then $|\alpha(x) - c_I| \leq \beta(x)$ and finally $|s^*(x) - c_I| \leq \beta(x) + c(\sum |r_k|^2)^{1/2}$ where $((1/|I|)\int_I \beta^2(x)\,dx)^{1/2} \leq c(\sum |r_k|^2)^{1/2}$; thus (6.2) holds and we are done. The result is Meyer's.)

6.19 If $f \in BMO$, then there exists a constant $\varepsilon > 0$ such that

$$\sup_{I \subseteq T} (|\{x \in I : |f(x) - f_I| > \lambda\}|/|I|) \leq e^{-\lambda/\varepsilon}, \tag{6.3}$$

whenever $\lambda > \lambda(\varepsilon, f)$; indeed, by the John–Nirenberg inequality, we have $\varepsilon = c\|f\|_*$, $\lambda(\varepsilon, f) = c_1\|f\|_*$. Let now $\varepsilon(f) = \inf\{\varepsilon > 0 : (6.3) \text{ holds}\}$ and show that

$$\lim_{p\to\infty} \frac{1}{p}\left(\sup_{I \subseteq T} \frac{1}{|I|}\int_I |f(x) - f_I|^p\,dx\right)^{1/p} = \frac{\varepsilon(f)}{\pi}.$$

Moreover, also by the John–Nirenberg inequality there is $\eta > 0$ such that

$$\sup_{I \subseteq T}\left(\frac{1}{|I|}\int_I e^{\eta|f(x) - f_I|}\,dx\right) < \infty, \tag{6.4}$$

and let $\eta(f) = \sup\{\eta > 0 : (6.4) \text{ holds}\}$. Show that $\eta(f) = 1/\varepsilon(f)$. These results are from Garnett and Jones' work.

6.20 Let ϕ be a positive, nondecreasing function defined on $[0, 2\pi]$ and put

$$\phi(f, I) = \frac{1}{|I|\phi(|I|)} \int_I |f(x) - f_I|\, dx, \quad \|f\|_{*,\phi} = \sup_{I \subseteq T} \phi(f, I)$$

and finally $BMO_\phi = \{f \in L: \|f\|_{*,\phi} < \infty\}$. The following results are essentially due to Spanne [1965].

(i) if $I \subseteq J$ and

$$\rho(f, s) = \sup_{|I| \leqslant s} \frac{1}{|I|} \int_I |f(x) - f_I|\, dx, \qquad \text{then } |f_I - f_J| \leqslant c \int_{[|I|/2, |J|]} \rho(f, s) \frac{ds}{s};$$

(ii) If $I(x, r)$ = interval centered at x, length $2r$, then $\|f - f_{I(\cdot, r)}\|_* \leqslant c\rho(f, r)$;

(iii) If $f \in BMO_\phi$ and $\tau_s f(x) = f(x + s)$, $s \geqslant 0$, then $\|f - \tau_s f\|_* = O(\rho(s))$;

(iv) If $\phi(t)/t$ is nonincreasing, then $f(x) = \int_{(|x|, 1]} \phi(t)/t\, dt \in BMO_\phi$ (this is in a way the largest function in the space).

6.21 Assume $\phi(t) = t^\eta$, $0 < \eta < 1$, and suppose that $M_\phi^\# f(x) = \sup_{x \in I} \phi(f, I) \in L^p(T)$, $1 < p < 1/\eta$; then $M^\# f \in L^r(T)$, $1/r = 1/p - \eta$ and there is a constant c independent of f such that $\|M^\# f\|_r \leqslant c\|M_\phi^\# f\|_p$.

6.22 Let $I = [0, 1]$. We say that a function g is a "pointwise multiplier" of $BMO_\phi(I)$ provided that $gf \in BMO_\phi(I)$ whenever $f \in BMO_\phi(I)$ and there is a constant c independent of f such that $\|gf\|_{*,\phi} \leqslant c\|f\|_{*,\phi}$. Assume that $\phi(r)/r$ is nonincreasing and put $\psi(r) = \phi(r)/\int_{(r,1]} \phi(s)/s\, ds$; then g is a pointwise multiplier of $BMO_\phi(I)$ if and only if $g \in BMO_\psi(I) \cap L^\infty(I)$. (*Hint*: Since $\rho(f, r) \leqslant \phi(r)\|f\|_{*,\phi}$, by 6.20 it follows that for any I with $|I| = r < \frac{1}{2}$ we have

$$\frac{1}{r} \int_I |f(x)g(x) - f_I g_I|\, dx \leqslant \phi(r)\|g\|_\infty \|f\|_{*,\phi} + |f_I| \frac{1}{|I|} \int |g(x) - g_I|\, dx$$

$$\leqslant c\phi(r)\|f\|_{*,\phi}(\|g\|_* + \|g\|_{BMO,\psi}).$$

Conversely, let $I = I(t, r)$ be an interval as above, let f be the function introduced in 6.20 (iv), and put $h(x) = \max(f(x - t), \int_{(r,1]} \phi(s)/s\, ds)$; then h is continuous and $\|h\|_{*,\phi} \leqslant c\|f\|_{*,\phi}$. Since $gh \in BMO_\phi$, a simple computation shows that g is as it should be. For $\phi(t) = 1$, the result is due to Stegenga [1976] and for more general ϕ's to Janson [1976].)

6.23 Given ϕ as in 6.20 let $\tilde{\phi}(r) = r \int_{[r,\infty)} \phi(s)/s^2\, ds$; one easily checks that $\tilde{\phi}$ is nondecreasing, $\tilde{\phi}(r) \to 0$ as $r \to 0$, $\tilde{\phi}(2r) \leqslant 2\tilde{\phi}(r)$ and $\tilde{\phi}(r) \geqslant \phi(r)$. Show that, if $f \in BMO_\phi$, then $Hf \in BMO_{\tilde{\phi}}$ and there is a constant c independent of f such that $\|Hf\|_{*,\tilde{\phi}} \leqslant c\|f\|_{*,\phi}$. (*Hint*: The result is Peetre's [1966] and the proof is similar to that of Theorem 5.2.)

6.24 Sarason introduced the class $VMO(T)$ of functions of vanishing mean oscillation in T, consisting of those f's in $BMO(T)$ for which $\lim_{|I|\to 0}(1/|I|)\int_I |f(x)-f_I|\,dx = 0$. Obviously, VMO contains every continuous function in T, and since, as is easily verified, VMO is a closed subspace of BMO, it also contains the BMO-closure of $C(T)$; in fact, VMO is precisely that closure and in many ways VMO bears the same relation to BMO as C does to L^∞. There are many reasons to consider VMO. Here is one: we are already aware that the conjugate $\tilde{f}$ of $f \in C(T)$ is not necessarily continuous (cf. 5.14 in Chapter V); however, it is in VMO. A quick proof of this result goes as follows. If $f \in C(T)$, then we can write it as the sum of a trigonometric polynomial and a function of small L^∞ norm; by Theorem 3.1, $\tilde{f}$ is then the sum of a trigonometric polynomial and a function with small BMO-norm. Thus $\tilde{f}$ is in the BMO closure of $C(T)$ as we wished to show. For further applications the reader should consult Sarason [1979].

6.25 Prove that $VMO(T) = \bigcup BMO_\phi(T)$.

6.26 $f \in VMO(T)$ if and only if $\lim_{r\to 1}\|f - f * P_r(\cdot)\|_* = 0$.

6.27 With the notation of 6.18, show that $s^* \in VMO$.

6.28 Let $f \in VMO$. Then $M^\# f \in C(T)$. (*Hint*: It suffices to show that for $\lambda > 0$, $\{M^\# f < \lambda\}$ is open; to do this, suppose $M^\# f(t) < \lambda$ and we will show that the same holds in an interval about t. Since f is in VMO, there exists $\delta > 0$ such that $\sup_{|I|\le\delta}((1/|I|)\int_I |f(y)-f_I|\,dy) < M^\# f(t)$; now let J be an interval centered at t whose length is very small compared to δ, let $x \in J$, and, finally, let I be an interval containing x. If $|I| < \delta$, then $(1/|I|)\int_I |f(y)-f_I|\,dy \le M^\# f(x) < \lambda$ by the choice of δ; suppose then that $|I| \ge \delta$ and let $K = I \cup J$. Then K contains t and

$$\frac{1}{|I|}\int_I |f(y)-f_I|\,dy \le \frac{|K|}{|I|}\left(\frac{1}{|K|}\int_K |f(y)-f_K|\,dy + |f_K - f_I|\right)$$

$$\le \frac{|K|}{|I|}(M^\# f(t) + |f_K - f_I|).$$

Since $|J|$ is very small compared to $|I|$, the intervals K and I are almost the same and $M^\# f(x) < \lambda$ as well. The same proof works for

$$M_p^\# f(x) = \sup_{x\in I}\left(\frac{1}{|I|}\int_I |f(y)-f_I|^p\,dy\right)^{1/p}, \qquad 1 < p < \infty,$$

and this result has an important application, namely, f is an extreme point of the unit ball of VMO normed by $\|M_p^\# f\|_\infty$, $1 < p < \infty$, if and only if $M_p^\# f(t) = 1$. These results are from Axler and Shields [1982].)

6.29 BMO is the natural substitute for L^∞ also in the case of fractional integrals. More precisely, if $0 < \alpha < 1$ and $p = 1/\alpha$, then I_α maps $L^p(T)$

into $BMO(T)$ and there is a constant c independent of f such that $\|I_\alpha f\|_* \leq c\|f\|_p$. (*Hint*: With the notation of (2.16), we prove that $M(I_\alpha f)^\#(x) \leq cM_\alpha f(x)$. Fix $I = I(x, r)$ and put $f = f\chi_{2I} + f\chi_{T\setminus 2I} = g + h$, say. Note that

$$\left|\int_I I_\alpha g(y)\,dy\right| \leq \int_{2I} |g(t)| \int_I \frac{1}{|y-t|^{1-\alpha}}\,dy\,dt$$
$$\leq cr\frac{1}{r^{1-\alpha}}\int_{2I} |g(t)|\,dt \leq crM_\alpha f(x);$$

thus $|(I_\alpha f)_I| \leq cM_\alpha f(x)$ and consequently $\int_I |I_\alpha g(y) - (I_\alpha g)_I|\,dy \leq crM_\alpha f(x)$. By the mean value theorem, we get that for t in I, $|I_\alpha h(x) - I_\alpha h(t)| \leq cr\int_{T\setminus 2I} |y-x|^{\alpha-2}|f(y)|\,dy \leq cM_\alpha f(x)$, a.e. in t, and the right estimate for this term follows by integrating over I. In fact the ideas in this proof give a more complete picture of the general situation when $f \geq 0$, for in that case we have that (i) $I_\alpha f \in BMO$ if and only if $M_\alpha f \in L^\infty$ and (ii) $I_\alpha f$, $M_\alpha f$ and $M(I_\alpha f)^\#$ have comparable norms in L^p, $1 < p < \infty$. These results are discussed in Adams [1975].)

6.30 Let $f \sim \sum c_j e^{ijx}$, $f \in \mathrm{Lip}_\alpha(T)$, $0 < \alpha < 1$, show that $c_j = O(|j|^{-\alpha}\,\|f\|_{\Lambda_\alpha})$. This result is sharp, indeed, if $f(t) = \sum_{n=1}^\infty 2^{-n\alpha}\cos 2^n t$, then $f \in \mathrm{Lip}_\alpha(T)$, and $c_j \sim j^{-\alpha}$, $j = 2^n$. (*Hint*: As in 5.16 in Chapter I we consider $g(t) = \sum_{n=1}^{|k|} d_n\chi_{(a_{n-1}, a_n)}(t)$, where $a_n = 2\pi n/|k|$, $|k| \neq 0$, $n = 1, \ldots, |k|$. Use that $c_k(g) = 0$ and the choice $d_n = f_{(a_{n-1}, a_n)}$ to obtain the first assertion. Next, observe that if $I = (a, b) \subseteq T$, then

$$f(t) - f_I = \sum_{n=1}^\infty 2^{-n\alpha}\left(\cos 2^{nt} - \frac{(\sin 2^n b - \sin 2^n a)}{2^n(b-a)}\right)$$
$$= \sum_{n=1}^\infty 2^{-n\alpha}A(n, t, I),$$

say. Now $|A(n, t, I)| \leq 2$, but we need a better estimate when $|I|$ is small compared with 2^n. In that case, $|A(n, t, I)| \leq c(2^n|I|)$. This gives $|f(t) - f_I| \leq c|I|^\alpha$, $t \in I$, and we have finished. Note that, if $\alpha = 1$, the above estimate becomes $|f(t) - f_I| \leq c|I|\ln(2\pi/|I|)$ instead.)

6.31 Let $f \in \mathrm{Lip}_\alpha$, $0 < \alpha < 1$ and show that $s_n(f, x) - f(x) = O(n^{-\alpha}\ln n)$ as $n \to \infty$, uniformly in x; this result cannot be improved.

6.32 Let $f \in L$. Then $\sigma_n(f, x) - f(x) = O(n^{-\alpha})$ if and only if $f \in \mathrm{Lip}_\alpha$, $0 < \alpha < 1$.

6.33 Let $0 < \alpha$, $\beta < \alpha + \beta < 1$, and consider I_β as an operator on Lip_α. For what value of γ is the result I_β: $\mathrm{Lip}_\alpha \to \mathrm{Lip}_\gamma$ true?

6.34 Let T be a linear operator such that $T: L^{p_0} \to \mathscr{L}_{p_0, \eta_0}$ and $T: L^{p_1} \to \mathscr{L}_{p_1, \eta_1}$. Then $T: L^p \to \mathscr{L}_{p, \eta}$ whenever $1/p = (1-\theta)/p_0 + \theta/p_1$ and $\eta = (1-\theta)\eta_0 + \theta\eta_1$, $0 \leq \theta \leq 1$. (*Hint*: The proof is straightforward once we consider the operator $\tau_I f = Tf - (Tf)_I$: the result is Stampacchia's.)

CHAPTER

IX

A_p Weights

1. THE HARDY–LITTLEWOOD MAXIMAL THEOREM FOR REGULAR MEASURES

Let μ be a nonnegative Borel measure in R^n, finite on bounded sets. For this measure we pose the question of whether it differentiates μ-locally integrable functions f; in other words, if $x \in R^n$ and I is an open cube containing x, does the statement $\lim_{\mu(I)\to 0}(1/\mu(I)) \int_I f(y)\, d\mu(y) = f(x)$ hold μ a.e.? Here $|I|$ denotes the measure of I. As in the case of the Lebesgue measure in Chapter IV we opt to approach this question by first considering a weak-type result for the corresponding maximal function. More precisely, if for $x \in R^n$ and f locally in $L_\mu(R^n)$ we put

$$M_\mu f(x) = \sup \frac{1}{\mu(I)} \int_I |f(y)|\, d\mu(y), \tag{1.1}$$

where the I's are open cubes containing x, is the mapping $f \to M_\mu f$ of weak-type $(1, 1)$?

We can go about answering this as follows: let $\mathcal{O}_\lambda = \{M_\mu f > \lambda\}$; $\mathcal{O}_\lambda$ is open since to each x in $\mathcal{O}_\lambda$ there corresponds an open cube I_x containing x such that

$$\frac{1}{\mu(I_x)} \int_{I_x} |f(y)|\, d\mu(y) > \lambda \tag{1.2}$$

and consequently $I_x \subseteq \mathcal{O}_\lambda$. In fact

$$\mathcal{O}_\lambda = \bigcup_{x \in \mathcal{O}_\lambda} I_x. \tag{1.3}$$

We want to estimate $\mu(\mathcal{O}_\lambda)$ in terms of the μ-measure of the (complicated) set on the right-hand side of (1.3); it is apparent that we need some control over this set. So, suppose, in addition, that μ is regular; in other words, if $\mathcal{U}$ is μ-measurable, then

$$\mu(\mathcal{U}) = \sup_{K \subseteq \mathcal{U},\, K \text{ compact}} \mu(K). \tag{1.4}$$

If K is a compact subset of $\mathcal{O}_\lambda$ now, there are finitely many I_x's, $I_{x_1}, \ldots, I_{x_m}$, say, so that $K \subseteq \bigcup_{j=1}^m I_{x_j}$; in fact, we may avoid unnecessary overlaps by discarding any cube I_{x_k} such that $I_{x_k} \subseteq \bigcup_{j\neq k} I_{x_j}$. However, even once this is done we may still be left with quite a bit of overlap. To handle this we proceed as follows. Since we are dealing with a finite number of cubes, there is one with largest sidelength (if there is more than one just pick any); separate it and rename it I_1. Now, if any of the remaining cubes, say I, intersects I_1, since sidelength $I \leq$ sidelength of I_1, it follows that $I \subseteq 3I_1$, the cube concentric with I_1 with sidelength three times that of I_1; all these cubes I can be discarded as well. We are thus left with a finite collection of open cubes, each one disjoint with I_1. Repeat for this family the procedure used to select I_1, that is select a cube with largest sidelength, call it I_2, and discard all other cubes which intersect it. After a finite number of steps we are left with a collection $\{I_1, \ldots, I_k\}$ of disjoint open cubes so that $K \subseteq \bigcup_{j=1}^k 3I_j$. Thus

$$\mu(K) \leq \sum_{j=1}^k \mu(3I_j). \tag{1.5}$$

Under what circumstances can we replace $\mu(3I_j)$ by $\mu(I_j)$ in (1.5)? This can be done for the so called doubling measures, namely those measures μ for which $\mu(2I) \leq c\mu(I)$, all open cubes I, c independent of I. In case μ is doubling then from (1.5) it follows that

$$\mu(K) \leq c^2 \sum_{j=1}^k \mu(I_j) \tag{1.6}$$

where c is the doubling constant of μ. But the I_j's are special cubes; in particular they all satisfy (1.2). Whence combining (1.6) and (1.2), and since the I_j's are pairwise disjoint, we get

$$\begin{aligned} \lambda\mu(K) &\leq c^2 \sum_{j=1}^k \int_{I_j} |f(y)|\, d\mu(y) = c^2 \int_{\bigcup I_j} |f(y)|\, d\mu(y) \\ &\leq c^2 \|f\|_{L^1_\mu}. \end{aligned} \tag{1.7}$$

Finally, on account of (1.4), (1.7) gives

$$\lambda\mu(\mathcal{O}_\lambda) \leq c^2 \|f\|_{L^1_\mu} \tag{1.8}$$

and the mapping is of weak-type $(1, 1)$ with norm $\leqslant c^2$. Summing up, we have proved

Theorem 1.1. Let μ be a nonnegative Borel measure in R^n which in addition is finite on bounded sets, doubling, and regular. Then the mapping $f \to M_\mu f$ is of weak-type $(1, 1)$, with norm $\leqslant c^2$, c = doubling constant of μ.

Corollary 1.2. Let μ be as in Theorem 1.1. then there is a constant $c = c_p$ independent of f such that

$$\|M_\mu f\|_{L^p_\mu} \leqslant c\|f\|_{L^p_\mu}, 1 < p < \infty. \tag{1.9}$$

Proof. M_μ is of weak-type $(1, 1)$, and is bounded in L^∞_μ. Thus the Marcinkiewicz interpolation theorem applies. ∎

Corollary 1.3. Let μ be as in theorem 1.1, and let f be locally in $L^1_\mu(R^n)$. Then

$$\lim_{\substack{\mu(I)\to 0\\ I\supset\{x\}}} \frac{1}{\mu(I)} \int_I f(y)\, d\mu(y) = f(x) \qquad \mu\text{-a.e.}$$

Proof. Since $\mu(K) < \infty$ for compact sets K, continuous functions are dense in $L^1_\mu(R^n)$; the proof is therefore entirely analogous to Theorem 2.2 in Chapter IV. ∎

2. A_p WEIGHTS AND THE HARDY–LITTLEWOOD MAXIMAL FUNCTION

The boundedness of the Hardy–Littlewood maximal function is an essential ingredient in the consideration of the $L^p(T)$ behavior of the various operators we have considered thus far. It is natural, then, to expect that the $L^p_\mu(R^n)$ behavior of the Hardy–Littlewood maximal operator will have important applications in the study of weighted norm inequalities for similar, n-dimensional, operators. We open our discussion with the study of the necessary conditions; more precisely, suppose there is a constant $k = k_{p,\mu}$ independent of f such that for some p, $1 \leqslant p < \infty$, and all $f \in L^p_\mu(R^n)$ and $\lambda > 0$

$$\lambda^p \mu(\{Mf > \lambda\}) \leqslant k^p \|f\|^p_{L^p_\mu}. \tag{2.1}$$

What can we, then, say about μ? Simple examples show, for instance, that μ cannot have atoms, but in fact much more is true.

Theorem 2.1. Assume μ is a nonnegative Borel measure, finite on bounded sets and assume that for some $1 \le p < \infty$ (2.1) holds. Then

(i) μ is absolutely continuous with respect to Lebesgue measure, i.e., there is a nonnegative, locally integrable function w such that $d\mu(x) = w(x)\,dx$.

(ii) for all open cubes I and $f \in L^p_\mu(R^n)$

$$\frac{1}{|I|}\int_I |f(y)|\,dy \le c\left(\frac{1}{\mu(I)}\int_I |f(y)|^p\,d\mu(y)\right)^{1/p}, \tag{2.2}$$

where $c = c_{\mu,k}$ is independent of f.

(iii) (A_p condition) w satisfies Muckenhoupt's $A_p(R^n) = A_p$ condition, or $w \in A_p$; i.e., there is a constant $c = c_{\mu,k}$ independent of the open cube I so that

$$\left(\frac{1}{|I|}\int_I w(y)\,dy\right)\left(\frac{1}{|I|}\int_I w(y)^{-1/(p-1)}\,dy\right)^{p-1} \le c, \qquad 1 < p < \infty \tag{2.3}$$

or

$$\frac{1}{|I|}\int_I w(y)\,dy \le c \operatorname{ess\,inf}_I w, \qquad p = 1. \tag{2.4}$$

The infimum over the constants on the right-hand side of (2.3) and (2.4) is called the A_p, or A_1, constant of w; moreover, the A_1 constant of w is less than or equal to ck and the A_p constant of w is less than or equal to ck^{p-1}, $p > 1$. A statement which depends on the A_p constant of a weight w rather than on the weight itself is called "independent in A_p." Theorem 3.1 below is an example of this.

(iv) (Strong Doubling) For each open cube I and measurable subset E of I,

$$\mu(I) \le c(|I|/|E|)^p \mu(E), \tag{2.5}$$

where $c = c_{\mu,k}$ is independent of E and I.

Proof. (i) Suppose the Lebesgue measure $|E|$ of E is 0. We show that also $\mu(E) = 0$. By the regularity of the measures involved we may assume that E is compact and that, given $\varepsilon > 0$, there is an open set $\mathcal{O} \supset E$ such that $\mu(\mathcal{O}\setminus E) < \varepsilon$. Let $f(y) = \chi_{\mathcal{O}\setminus E}(y)$; then $f \in L^p_\mu(R^n)$ and $\|f\|^p_{L^p_\mu} = \mu(\mathcal{O}\setminus E) < \varepsilon$. Next observe that $Mf(x) = 1$ for $x \in E$: indeed, to each $x \in E$ there corresponds an open cube $I \subset \mathcal{O}$, $x \in I$, and consequently $1/|I| \int_I f(y)\,dy = |(\mathcal{O}\setminus E) \cap I|/|I| = 1$ since $|E| = 0$. By (2.1) it then follows that $\mu(E) \le \mu(\{Mf \ge \frac{1}{2}\}) \le 2^p k^p \|f\|^p_{L^p_\mu} \le c\varepsilon$, which gives the desired result as ε is arbitrary.

(ii) Fix an open cube I and consider for $f \in L^p_\mu$ the quantity $(1/|I|)\int_I |f(y)|\,dy = |f|_I$, which we may assume >0. Since

$$\inf_{x \in I} M(f\chi_I)(x) \geq |f|_I, \tag{2.6}$$

$|f|_I$ must be finite for each I, for otherwise (2.1) cannot hold unless μ is the 0 measure. Thus, if we put $\mathcal{O} = \{M(f\chi_I) > |f|_I/2\}$, by (2.1) and (2.6) it follows that $\mu(I) \leq \mu(\mathcal{O}) \leq k^p(1/|f|_I)^p \|f\chi_I\|^p_{L^p_\mu}$, which is equivalent to (2.2).

(iii) and (iv) Since there is no *a priori* reason why w cannot vanish on a set of positive measure, we introduce the measure $d\nu(y) = d\mu(y) + \varepsilon\,dx$, $\varepsilon > 0$, to avoid unnecessary technical difficulties. Clearly, ν is also absolutely continuous with respect to Lebesgue measure, $d\nu(y) = v(y)\,dy$, $v > 0$, and, more importantly, (2.1) also holds with μ replaced by ν with constant independent of ε. Assume $p > 1$ first. In order to estimate $\int_I v(y)^{-1/(p-1)}\,dy$ we note that it equals $\|1/v\|^{p'}_{L^{p'}_\nu}$, $1/p + 1/p' = 1$, which by the converse to Hölder's inequality may be estimated by

$$\sup_{\|f\|_{L^p_\nu} \leq 1} \left| \int_I \frac{f(y)}{v(y)} v(y)\,dy \right|^{p'} = \sup_{\|f\|_{L^p_\nu} \leq 1} \left| \int_I f(y)\,dy \right|.$$

Now by (ii), which also holds for ν, it follows that for such f's

$$\left| \int_I f(y)\,dy \right| \leq ck|I| \left(\frac{1}{\nu(I)} \int_I |f(y)|^p\,d\nu \right)^{1/p} \leq \frac{ck|I|}{\nu(I)^{1/p}}$$

and consequently

$$\frac{1}{|I|} \int_I v(y)^{-1/(p-1)}\,dy \leq ck \frac{1}{|I|} \left(\frac{|I|}{\nu(I)^{1/p}} \right)^{p'}. \tag{2.7}$$

Unraveling (2.7) gives (2.3), that is, the A_p condition for v. Next we show that (2.5) holds for v; indeed, note that

$$\begin{aligned} |E| &= \int_E \left(\frac{v(y)^{1/p}}{v(y)^{1/p}} \right) dy \\ &\leq \left(\int_E v(y)\,dy \right)^{1/p} \left(\int_E v(y)^{-1/(p-1)}\,dy \right)^{1/p'} \\ &\leq \nu(E)^{1/p} \left(\int_I v(y)^{-1/(p-1)}\,dy \right)^{1/p'}, \end{aligned}$$

which by (2.7) does not exceed $ck\nu(E)^{1/p}(|I|/\nu(I)^{1/p})$. In other words

$$\nu(I) \leq ck^p(|I|/|E|)^p \nu(E). \tag{2.8}$$

Since the constant in (2.8) is independent of ε, we may let $\varepsilon \to 0$ there and obtain that $\mu(I) \leq ck^p(|I|/|E|)^p \mu(E)$ as well; i.e., (iv) holds. Moreover, if there is a set E with $|E| > 0$ so that $\mu(E) = 0$, then by (2.5) μ is the 0 measure since it vanishes on any open cube I containing E. Thus $d\mu(x) = w(x)\,dx$, $w > 0$ a.e., and the above argument may be repeated with w in place of v; this completes the proof when $p > 1$.

The case $p = 1$ is rather simple since by putting $f = \chi_E$, $|E| > 0$, in (ii), we see that $(|E|/|I|) \leq ck(\mu(E)/\mu(I))$ or equivalently

$$\mu(I)/|I| \leq ck(\mu(E)/|E|), \tag{2.9}$$

which is (iv). Furthermore, by choosing E to be a sequence of open cubes converging to a Lebesgue point x of w so that $w(x) \sim \operatorname{ess\,inf}_I w$, (2.9) gives (2.4), and the proof is complete. ■

3. A_1 WEIGHTS

As we have seen, A_1 is a necessary condition for the Hardy–Littlewood maximal operator M to map $L_\mu(R^n)$ into wk-$L_\mu(R^n)$, but is it also sufficient?

Theorem 3.1 (Muckenhoupt). Suppose $w \in A_1$. Then M maps $L_\mu(R^n)$ into wk-$L_\mu(R^n)$, with norm independent in A_1.

Proof. First note that if $w \in A_1$, then μ is doubling with doubling constant $\leq c(A_1$ constant of $w)$; indeed, since $(1/|2I|)\int_{2I} d\mu(y) \leq c \operatorname{ess\,inf}_I w \leq c(1/|I|)\int_I d\mu(y)$, it readily follows that $\mu(2I) \leq c\mu(I)$, $c \leq 2^n(A_1$ constant of $w)$. Moreover, since

$$\begin{aligned}\frac{1}{|I|}\int_I |f(y)|\,dy &= \frac{\mu(I)}{|I|}\frac{1}{\mu(I)}\int_I |f(y)|\,dy \\ &\leq c\frac{1}{\mu(I)}(\operatorname{ess\,inf}_I w)\int_I |f(y)|\,dy \\ &\leq c\frac{1}{\mu(I)}\int_I |f(y)|\,d\mu(y),\end{aligned}$$

we also have that $Mf(x) \leq cM_\mu f(x)$, $c = A_1$ constant of w. Thus $\{Mf > \lambda\} \subseteq \{M_\mu f > \lambda/c\}$, and by Theorem 1.2 $\lambda\mu(\{Mf > \lambda\}) \leq (\lambda/c)\mu(\{M_\mu f > \lambda/c\}) \leq c^3\|f\|_{L_\mu}$. ■

Some observations concerning A_1 are obvious: for instance, A_1 is the limiting A_p condition as $p \to 1^+$ and an equivalent way of stating A_1 is

$$Mw(x) \leq cw(x) \qquad \text{a.e.} \tag{3.1}$$

But what are the A_1 weights? Can we give some examples or even characterize them? As a first step we consider powers of $|x|$, i.e., $|x|^\eta$. When $n = 1$ and $\eta > 0$, by letting $I = (0, b)$ we note that $(1/b)\int_{(0,b)} x^\eta\, dx = 1b^\eta/(\eta+1) \to \infty$ as $b \to \infty$, whereas $\inf_I x^\eta = 0$. Thus positive powers of $|x|$ are ruled out, but how about negative powers? We must have $-n < \eta \leq 0$ for otherwise $|x|^\eta$ is not locally integrable, but this is essentially the only restriction. Indeed, we have

Proposition 3.2. Suppose $-n < \eta \leq 0$. Then $|x|^\eta \in A_1$; more precisely, there is a constant c independent of I such that

$$\frac{1}{|I|}\int_I |x|^\eta\, dx \leq c \inf_I(|x|^\eta). \tag{3.2}$$

Proof. Fix a cube I and let I_0 denote the translate of I centered at 0; we consider two mutually exclusive cases, to wit (i) $2I_0 \cap I \neq \varnothing$ and (ii) $2I_0 \cap I = \varnothing$. In case (i) we have that $6I_0 \supseteq I$ and $(1/|I|)\int_I |x|^\eta\, dx \leq (1/|I|)\int_{6I_0} |x|^\eta\, dx \leq c|I|^{\eta/n}$, where c is a (dimensional) constant, independent of I; clearly, (3.2) holds in this case. Case (ii) is easier, for then $|x| \sim |y|$ for x, y in I; indeed, we have $|x| \leq |x-y| + |y| \leq c|I|^{1/n} + |y| \leq c|y|$, and the opposite inequality follows by exchanging x and y above. Thus $|y|^\eta \leq c\inf_I |x|^\eta$, all $y \in I$, and averaging over y in I, (3.2) holds for case (ii) as well. ∎

Next we consider functions which behave like $|x|^\eta$, $-n < \eta \leq 0$, and show that they also are A_1 weights.

Proposition 3.3 (Coifman–Rochberg). Let μ be a nonnegative Borel measure so that $M\mu(x)$ is not identically ∞. Then for each $0 \leq \varepsilon < 1$, $M\mu(x)^\varepsilon \in A_1$, with A_1 constant which depends only on ε.

Proof. Recall that $M\mu(x) = \sup_{x\in I}(1/|I|)\mu(I)$. For a fixed open cube I we estimate $(1/|I|)\int_I M\mu(x)^\varepsilon\, dx$ by $A^\varepsilon = (\inf_I M\mu)^\varepsilon$ as follows: for each x in I we divide those open cubes Q containing x into two families by setting $\mathcal{J}_1 = \{Q\colon |Q| \leq |2I|\}$ and $\mathcal{J}_2 = \{Q\colon |Q| > |2I|\}$. Thus

$$\begin{aligned} M\mu(x) &\leq \sup_{Q\in\mathcal{J}_1} \frac{1}{|Q|}\int_Q d\mu(y) + \sup_{Q\in\mathcal{J}_2} \frac{1}{|Q|}\int_Q d\mu(y) \\ &= A(x) + B(x), \end{aligned} \tag{3.3}$$

say. The estimate for $B(x)$ is readily obtained; since for $Q \in \mathcal{J}_2$ we have $3Q \supseteq I$, it follows that

$$\frac{1}{|Q|}\int_Q d\mu(y) \leq \frac{c}{|3Q|}\int_{3Q} d\mu(y) \leq c \inf_{3Q} M\mu \leq cA,$$

and

$$B(x) \leq cA \tag{3.4}$$

with c independent of μ. As for $A(x)$, let μ_1 denote the restriction of μ to $6I$, i.e., $d\mu_1(y) = \chi_{6I}(y)\,d\mu(y)$, and note that

$$A(x) \leqslant M\mu_1(x). \tag{3.5}$$

Thus on account of (3.3), (3.4), and (3.5) we get that

$$\frac{1}{|I|}\int_I M\mu(x)^\varepsilon\,dx \leqslant \frac{1}{|I|}\int_I M\mu_1(x)^\varepsilon\,dx + cA^\varepsilon$$

and it suffices to prove the desired estimate with $M\mu$ replaced by $M\mu_1$. But by (a simple variant of the Lebesgue measure version of) Theorem 1.1 and 7.5 in Chapter IV, we readily see that

$$\frac{1}{|I|}\int_I M\mu_1(x)^\varepsilon\,dx \leqslant \frac{1}{|I|}c(\text{wk-}L\text{ norm of }M\mu_1)^\varepsilon|I|^{1-\varepsilon}$$

$$\leqslant c\left(\frac{1}{|I|}\int_{6I} d\mu\right)^\varepsilon \leqslant cA^\varepsilon,$$

with c depending only on ε, and we have finished. ■

The interesting fact is that the converse to Proposition 3.3 also holds, namely,

Theorem 3.4 (Coifman–Rochberg). Assume $w \in A_1$. Then there are functions b and f and $0 \leqslant \varepsilon < 1$ so that

(i) $0 < A \leqslant b(x) \leqslant B < \infty$ a.e.
(ii) $f \in L_{\text{loc}}(R^n)$, $Mf(x)^\varepsilon$ is finite a.e.

and

$$w(x) = b(x)Mf(x)^\varepsilon. \tag{3.6}$$

The proof of this theorem relies on the so-called "reverse Hölder" property of w; this property is of independent interest and plays an important role in the theory of weights.

Theorem 3.5 (Reverse Hölder). Suppose $w \in A_1$. Then there is a positive number η so that

$$\left(\frac{1}{|I|}\int_I w(x)^{1+\eta}\,dx\right)^{1/(1+\eta)} \leqslant c\frac{1}{|I|}\int_I w(x)\,dx, \qquad \text{all } I, \tag{3.7}$$

where $c = c_\eta$ is independent in A_1 and independent of I, but not, of course, of η; we abbreviate (3.7) by $w \in RH_{1+\eta}$.

Now, suppose that Theorem 3.5 has been proved. Then by (3.7) also $w^{1+\eta} \in A_1$, $M(w^{1+\eta})(x) \leqslant cw(x)^{1+\eta}$ a.e. and Theorem 3.4 holds with $b(x) = w(x)/M(w^{1+\eta})(x)^{1/(1+\eta)}$, $f(x) = w(x)^{1+\eta}$, and $\varepsilon = 1/(1+\eta)$. It thus only remains to prove Theorem 3.5, which we do forthwith. In order to assure that the various integral expressions we consider are finite, we introduce the function $v(x) = \min(w(x), N)$; clearly, $v \in A_1$, (A_1 constant of v) $\leqslant$ (A_1 constant of w), independently of N: indeed, for a given I let $A = \inf_I w$ and consider two cases, namely, (i) $A \geqslant N$ and (ii) $A < N$. We then have

$$\frac{1}{|I|}\int_I v(y)\,dy \leqslant \begin{cases} N \leqslant \inf_I v, & \text{in case (i)}, \\ \dfrac{1}{|I|}\displaystyle\int_I w(y)\,dy \leqslant c\inf_I w \leqslant c \inf_I v, & \text{in case (ii)}, \end{cases}$$

thus proving the claim. We show (3.7) with w replaced by v first; let $\eta > 0$ and observe that

$$\begin{aligned}\int_I v(y)^{1+\eta}\,dy &= \left(\int_{\{y\in I: v(y)>v_I\}} + \int_{\{y\in I: v(y)\leqslant v_I\}}\right) v(y)^{1+\eta}\,dy \\ &= A + B, \end{aligned} \tag{3.8}$$

say. Clearly,

$$B \leqslant v^{1+\eta}|I|, \tag{3.9}$$

which is the right estimate. The bound for A is not so readily obtained, and in the course of the proof we must keep track of the various constants appearing to be sure that they only depend on the A_1 constant of v, and η of course.

First observe that with the notation $\mathcal{O}_t = \{y \in I: v(y) > t\}$ we have

$$\begin{aligned} A &= (1+\eta)\int_{[v_I,\infty)} t^\eta |\mathcal{O}_t|\,dt \\ &= -(1+\eta)\int_{[v_I,\infty)} t^\eta \left(\int_{[t,\infty)} |\mathcal{O}_s|\,ds\right)'\,dt \\ &= -(1+\eta)\left(t^\eta \int_{[t,\infty)} |\mathcal{O}_s|\,ds\right)\Bigg]_{v_I}^{\infty} \\ &\quad + (1+\eta)\eta \int_{[v_I,\infty)} t^{\eta-1}\int_{[t,\infty)} |\mathcal{O}_s|\,ds\,dt = C + D, \end{aligned} \tag{3.10}$$

say. Note that, since $v \leqslant N$, $|\mathcal{O}_s| = 0$ for $s \geqslant N$, and consequently C equals

$$\begin{aligned}(1+\eta)\eta v_I^\eta \int_{\{y\in I: v(y)>v_I\}} v(y)\,dy &\leqslant (1+\eta)\eta v_I^\eta v_I |I| \\ &= (1+\eta)\eta v_I^{1+\eta}|I|, \end{aligned} \tag{3.11}$$

which is also of the right order. Next we show that for an appropriate choice of η, D is dominated by the (finite) quantity

$$\frac{1}{2}\int_I v(y)^{1+\eta}\,dy, \tag{3.12}$$

which may then be passed to the left-hand side of (3.8) to obtain the desired conclusion. We consider the innermost integral in D first. It equals

$$\int_{\{v>t\}} v(y)\,dy, \qquad t > v_I. \tag{3.13}$$

Now since $t > v_I$, we may invoke the (n-dimensional version of the) Calderón–Zygmund decomposition of v at level t, thus obtaining a collection of open, disjoint subcubes $\{I_j\}$ of I with the following properties

(i) $v(y) \leq t$ a.e. in $I \setminus \bigcup I_j$ and
(ii) $t \leq (1/|I_j|)\int_{I_j} v(y)\,dy \leq 2^n t$, all j.

From (i) it is clear that $\{v > t\} \subseteq \bigcup I_j$ and therefore the integral in (3.13) does not exceed

$$\sum \int_{I_j} v(y)\,dy = \sum |I_j| v_{I_j}. \tag{3.14}$$

Furthermore,

$$v_{I_j} \leq \begin{cases} c\inf_{I_j} v, & \text{since } v \in A_1, \\ 2^n t, & \text{since } I_j \text{ is a Calderón–Zygmund cube.} \end{cases}$$

Therefore by combining these bounds we get that $v_{I_j} \leq (c\inf_{I_j} v)^{1-\varepsilon}(2^n t)^{\varepsilon}$, all j, $0 < \varepsilon < 1$. Thus each summand in (3.14) is dominated by $ct^{\varepsilon}|I_j|(\inf_{I_j} v)^{1-\varepsilon} \leq ct^{\varepsilon}\int_{I_j} v(y)^{1-\varepsilon}\,dy$ and (3.14) does not exceed

$$ct^{\varepsilon}\int_{\bigcup I_j} v(y)^{1-\varepsilon}\,dy. \tag{3.15}$$

We need one last observation: from the left-hand side inequality in (ii) it follows that $\bigcup I_j \subseteq \{Mv > t\}$, and, since $v \in A_1$ and $Mv(x) \leq cv(x)$ a.e., we also have $\bigcup I_j \subseteq \{v > t/c\}$. Consequently, (3.14) is bounded by

$$ct^{\varepsilon}\int_{\{v>t/c\}} v(y)^{1-\varepsilon}\,dy, \qquad 0 < \varepsilon < 1$$

and the same is true of (3.13). Whence

$$\begin{aligned} D_2 &\leqslant c\eta(1+\eta)\int_{[v_I,\infty)} t^{\eta+\varepsilon-1}\int_{\{v>t/c\}} v(y)^{1-\varepsilon}\,dy\,dt \\ &= c\eta(1+\eta)\int_{\{v>v_I/c\}} v(y)^{1-\varepsilon}\int_{[v_I,\,cv(y))} t^{\eta+\varepsilon-1}\,dt\,dy \\ &\leqslant c\eta\frac{(1+\eta)}{(\varepsilon+\eta)}\int_I v(y)^{1-\varepsilon}(cv(y))^{\eta+\varepsilon}\,dy \\ &\leqslant c\eta\frac{(1+\eta)}{(\varepsilon+\eta)}\int_I v(y)^{1+\eta}\,dy. \end{aligned} \tag{3.16}$$

First fix $0<\varepsilon<1$ and then choose $\eta>0$ sufficiently small so that $c\eta(1+\eta)/(\varepsilon+\eta)<\frac{1}{2}$; this is clearly possible. Thus D is dominated by (3.12) and (3.7) holds with v in place of w there. This is a minor inconvenience since by Fatou's lemma

$$\begin{aligned} \left(\frac{1}{|I|}\int_I w(y)^{1+\eta}\,dy\right)^{1/1+\eta} &\leqslant \liminf_{N\to\infty}\left(\frac{1}{|I|}\int_I v(y)^{1+\eta}\,dy\right)^{1/1+\eta} \\ &\leqslant c\liminf_{N\to\infty}\left(\frac{1}{|I|}\int_I v(y)\,dy\right) \leqslant c\frac{1}{|I|}\int_I w(y)\,dy, \end{aligned}$$

and the proof is complete. ■

4. A_p WEIGHTS, $p > 1$

As we have seen, A_p is necessary for the Hardy-Littlewood maximal operator M to map $L^p_\mu(R^n)$ into wk-$L^p_\mu(R^n)$ and a simple argument similar to Theorem 3.1 shows it is also sufficient. However, a stronger result holds.

Theorem 4.1 (Muckenhoupt). Suppose $w\in A_p$, $1<p<\infty$. Then M maps $L^p_\mu(R^n)$ continuously into itself, with norm independent in A_p.

Proof. For a nonnegative function f in $L^p_\mu(R^n)$ and an integer $-\infty<k<\infty$, put $A_k=\{y\in R^n\colon 2^k<Mf(y)\leqslant 2^{k+1}\}$ and let $\mathcal{U}_k$ be a compact subset of A_k; we estimate $\int_{\bigcup\mathcal{U}_k} Mf(y)^p\,d\mu(y)$ by $c\|f\|^p_{L^p_\mu}$, where c is independent of f and the $\mathcal{U}_k$'s, and depends only on the A_p constant of w. A simple limiting argument then gives the desired result. To each $y\in A_k$ we assign an open cube I_y containing y so that

$$2^k<\frac{1}{|I_y|}\int_{I_y} f(x)\,dx \qquad (\leqslant 2^{k+1}). \tag{4.1}$$

Since $\mathcal{U}_k \subseteq A_k$, there are finitely many I_y's, none of which is contained in the union of the others, $\{I_{j,k}\}_{j=1}^{n(k)}$ say, so that each cube verifies (4.1) and $\mathcal{U}_k \subseteq \bigcup_{j=1}^{n(k)} I_{j,k}$. Moreover, since

$$J = \int_{\bigcup \mathcal{U}_k} Mf(y)^p \, d\mu(y) = \sum \int_{\mathcal{U}_k} Mf(y)^p \, d\mu(y) \leq \sum 2^{kp} \mu(\mathcal{U}_k) \quad (4.2)$$

we must estimate the right-hand side of (4.2); a good estimate depends on our ability to avoid unnecessary overlaps of the $I_{j,k}$'s. This is achieved as follows: put $E_{1,k} = I_{1,k} \cap \mathcal{U}_k$, $E_{2,k} = (I_{2,k} \backslash I_{1,k}) \cap \mathcal{U}_k$, and, in general,

$$E_{j,k} = (I_{j,k} \backslash \bigcup_{i<j} I_{i,k}) \cap \mathcal{U}_k, \qquad j = 1, 2, \ldots, n(k) \quad (4.3)$$

For each k the $E_{j,k}$'s are clearly disjoint, and by (4.3) it follows that

$$\bigcup_j E_{j,k} = \left(\bigcup_j I_{j,k} \right) \cap \mathcal{U}_k = \mathcal{U}_k, \qquad \text{all} \quad k. \quad (4.4)$$

Therefore, by (4.4) and (4.1) we also have

$$\begin{aligned} J &\leq \sum_{j,k} 2^{kp} \mu(E_{j,k}) \\ &\leq \sum_{j,k} \mu(E_{j,k}) \left(\frac{1}{|I_{j,k}|} \int_{I_{j,k}} f(x) \, dx \right)^p. \end{aligned} \quad (4.5)$$

Let $v(x) = w(x)^{-1/(p-1)}$, $d\nu(x) = v(x)\, dx$; then the right-hand side in (4.5) can be rewritten as

$$\sum_{j,k} \mu(E_{j,k}) \left(\frac{\nu(I_{j,k})}{|I_{j,k}|} \right)^p \left(\frac{1}{\nu(I_{j,k})} \int_{I_{j,k}} \frac{f(x)}{v(x)} \, d\nu(x) \right)^p. \quad (4.6)$$

What we are attempting to do here is to bring a combination of the A_p condition and Theorem 1.1 into play. Let m be the measure on $Z^+ \times Z$ given by

$$m((j, k)) = \mu(E_{j,k}) \left(\frac{\nu(I_{j,k})}{|I_{j,k}|} \right)^p. \quad (4.7)$$

With this notation the expression (4.6) becomes $\|\{a_{j,k}\}\|_{l_m^p}^p$, where the sequence $a_{j,k} = (1/\nu(I_{j,k})) \int_{I_{j,k}} (f(x)/v(x))\, d\nu(x)$. Note that this expression also equals

$$\int_{[0,\infty)} m(\mathcal{O}_\lambda) \, d\lambda^p, \quad (4.8)$$

where $\mathcal{O}_\lambda = \{(j, k) \in Z^+ \times Z: a_{j,k} > \lambda\}$. What we need then is a good estimate for $m(\mathcal{O}_\lambda)$; we use the notation $I(\lambda) = \bigcup_{(j,k)\in\mathcal{O}_\lambda} I_{j,k}$. Observe that for each (j, k) in $\mathcal{O}_\lambda$ by the A_p condition we have

$$\begin{aligned}
\left(\frac{\nu(I_{j,k})}{|I_{j,k}|}\right)^p &\leq c\left(\frac{|I_{j,k}|}{\mu(I_{j,k})}\right)^{p'} = c\left(\frac{1}{\mu(I_{j,k})}\int_{I_{j,k}} dx\right)^{p'} \\
&\leq c\left(\frac{1}{\mu(I_{j,k})}\int_{I_{j,k}} \chi_{I(\lambda)}(x)\frac{1}{w(x)}\, d\mu(x)\right)^{p'} \\
&\leq c \inf_{I_{j,k}} M\mu(\chi_{I(\lambda)}/w)^{p'} \leq c \inf_{E_{j,k}} M_\mu(\chi_{I(\lambda)}/w)^{p'}. \qquad (4.9)
\end{aligned}$$

Thus, substituting (4.9) in (4.7) we obtain

$$\begin{aligned}
m((j, k)) &\leq c\mu(E_{j,k}) \inf_{E_{j,k}} M_\mu(\chi_{I(\lambda)}/w)^{p'} \\
&\leq c \int_{E_{j,k}} M_\mu(\chi_{I(\lambda)}/w)(x)^{p'}\, d\mu(x)
\end{aligned}$$

and summing over (j, k) in $\mathcal{O}_\lambda$ we get that

$$m(\mathcal{O}_\lambda) \leq c\sum_{j,k}\int_{E_{j,k}} M_\mu\left(\frac{\chi_{I(\lambda)}}{w}\right)(x)^{p'}\, d\mu(x). \qquad (4.10)$$

Since the $E_{j,k}$'s are pairwise disjoint we may replace the right-hand side of (4.10) by $c\int_{R^n} M_\mu(\chi_{I(\lambda)}/w)(x)^{p'}\, d\mu(x)$ and invoke Theorem 1.1 to estimate this quantity by

$$c\int_{R^n}\left(\frac{\chi_{I(\lambda)}(x)}{w(x)}\right)^{p'} d\mu(x). \qquad (4.11)$$

Moreover, since $w(x)^{-p'}\, d\mu(x) = w(x)^{1-p'}\, dx = d\nu(x)$ the expression in (4.11) is $c\nu(I(\lambda))$ and

$$m(\mathcal{O}_\lambda) \leq c\nu(I(\lambda)). \qquad (4.12)$$

Now, the $I_{j,k}$'s whose union is $I(\lambda)$ are special cubes; in particular, by the definition of $\mathcal{O}_\lambda$, if $(j, k) \in \mathcal{O}_\lambda$, then $a_{j,k} > \lambda$. In other words, $\lambda < (1/\nu(I_{j,k}))\int_{I_{j,k}}(f(x)/v(x))\, d\nu(x)$. Thus, each such $I_{j,k}$ and also, consequently, $I(\lambda)$ is contained in $\{M_\nu(f/v) > \lambda\}$ and by (4.12) we see that

$$m(\mathcal{O}_\lambda) \leq c\nu(\{M_\nu(f/v) > \lambda\}). \qquad (4.13)$$

This is all we need to complete the proof. Indeed, on account of (4.8) $J \leq c\int_{[0,\infty)} \nu(\{M_\nu(f/v) > \lambda\})\, d\lambda^p$, which by Theorem 1.1 is dominated by $c\int_{R^n}(f(x)/v(x))^p\, d\nu(x) = c\int_{R^n} f(x)^p\, d\mu(x)$, since $v(x)^{-p}\, d\nu(x) = d\mu(x)$. This completes the proof. ∎

Corollary 4.2. M maps $L^p_\mu(R^n)$ into itself if and only if M maps $L^p_\mu(R^n)$ into wk-$L^p_\mu(R^n)$, $1 < p < \infty$.

The question once again is, what are the A_p weights? Some properties, such as $w \in A_p$ if and only if $w^{-1/(p-1)} \in A_{p'}$, $1/p + 1/p' = 1$, $1 < p < \infty$, are readily verified, but we need some examples and if possible a characterization of these weights.

Proposition 4.3. Assume $w_1, w_2 \in A_1$ and let $w(x) = w_1(x)w_2(x)^{1-p}$, $1 < p < \infty$; then $w \in A_p$.

Proof. Hölder's inequality. ∎

Corollary 4.4. $|x|^\eta \in A_p$, $1 < p < \infty$, if and only if $-n < \eta < n(p-1)$. In addition of being A_p weights, the w's in Proposition 4.3 actually verify some additional properties. More precisely,

Proposition 4.5. Assume $w(x) = w_1(x)w_2(x)^{1-p}$, w_1, $w_2 \in A_1$, $1 < p < \infty$. Then

(i) (Open ended property) $w \in A_{p-\varepsilon}$, some $\varepsilon > 0$.
(ii) (Reverse Hölder) $w \in RH_{1+\eta}$, some $\eta > 0$.
(iii) (Reverse doubling) There is $\delta > 0$ such that for all open cubes I and measurable subsets E of I, $\mu(E)/\mu(I) \leq c(|E|/|I|)^\delta$, where c is independent in A_p, also independent of E, I.

(iv) $\int_{R^n}(1+|x|)^{-np}\, d\mu(x) \leq c\mu(I_0)$,

where I_0 denotes the unit cube in R^n and c is independent in A_p.

Proof. (i) Since $w_2 \in A_1$, by Theorem 3.5, $w_2 \in RH_{1+\eta}$; put now $\varepsilon = (\eta/(1+\eta))(p-1) > 0$ and note that $p - \varepsilon - 1 = (p-1)/(1+\eta)$. It is then readily seen that

$$\left(\frac{1}{|I|}\int_I (w_1(y)w_2(y)^{1-p})^{-1/(p-\varepsilon-1)}\, dy\right)^{p-\varepsilon-1}$$
$$\leq \frac{1}{(\inf_I w_1)}\left(\frac{1}{|I|}\int_I w_2(y)^{1+\eta}\, dy\right)^{(p-1)/(1+\eta)}$$
$$\leq c\frac{1}{\inf_I w_1}\left(\frac{1}{|I|}\int_I w_2(y)\, dy\right)^{p-1}, \tag{4.14}$$

and also

$$\frac{1}{|I|}\int_I w_1(y)w_2(y)^{1-p}\, dy \leq \frac{1}{(\inf_I w_2)^{p-1}}\frac{1}{|I|}\int_I w_1(y)\, dy. \tag{4.15}$$

Whence, by multiplying (4.14) and (4.15), it follows that $w_1 w_2^{1-p} \in A_{p-\varepsilon}$, as we wanted to show.

(ii) Since $w_2 \in A_1$ from Hölder's inequality (applied to $(1/|I|)\int_I w_2(y)^{1/p'}/w_2(y)^{1/p'}\,dy$) it follows that for some constant $c>0$ independent of I

$$c \leqslant (\inf_I w_2)^{p-1}\left(\frac{1}{|I|}\int_I w_2(y)^{1-p}\,dy\right). \tag{4.16}$$

Let $w_1 \in RH_{1+\eta}$. Then by (4.16)

$$\begin{aligned}\left(\frac{1}{|I|}\int_I (w_1(y)w_2(y)^{1-p})^{1+\eta}\,dy\right)^{1/1+\eta} &\leqslant (\inf_I w_2)^{1-p}\left(\frac{1}{|I|}\int_I w_1(y)^{1+\eta}\,dy\right)^{1/1+\eta}\\ &\leqslant c\left(\frac{1}{|I|}\int_I w_2(y)^{1-p}\,dy\right)\left(\frac{1}{|I|}\int_I w_1(y)\,dy\right)\\ &\leqslant c\left(\frac{1}{|I|}\int_I w_1(y)w_2(y)^{1-p}\,dy\right),\end{aligned}$$

as we wished to show.

(iii) Let $w \in RH_{1+\eta}$. Then by Hölder's inequality

$$\begin{aligned}\mu(E) &\leqslant \left(\int_E w(y)^{1+\eta}\,dy\right)^{1/1+\eta}|E|^{\eta/1+\eta} \leqslant c|I|^{1/1+\eta}\left(\frac{1}{|I|}\int_I w(y)\,dy\right)|E|^{\eta/1+\eta}\\ &= c(|E|/|I|)^{\eta/1+\eta}\mu(I).\end{aligned}$$

(iv) By (i) and the doubling property (iv) of Theorem 2.1,

$$\mu(2^k I_0) \leqslant c2^{nk(p-\varepsilon)}\mu(I_0), \qquad k \geqslant 1. \tag{4.17}$$

Thus

$$\begin{aligned}\int_{R^n}\frac{d\mu(x)}{(1+|x|)^{np}} &\leqslant c\left(\mu(I_0) + \sum_{k=1}^{\infty} 2^{-nkp}\int_{2^{k+1}I_0\setminus 2^k I_0} d\mu(x)\right)\\ &\leqslant c\left(1+\sum_{k=1}^{\infty} 2^{-nkp}2^{nk(p-\varepsilon)}\right)\mu(I_0),\end{aligned}$$

and we have finished. ■

5. FACTORIZATION OF A_p WEIGHTS

This section is devoted to proving a remarkable fact, namely, the converse to Proposition 4.3. Before we proceed with the proof of this factorization result we need some preliminary observations; we start with a definition.

Definition 5.1. We say that an operator T is admissible provided it verifies the following four properties.

(i) There exists r, $1 < r < \infty$, so that T is bounded in $L^r(R^n)$.
(ii) T is positive, i.e., $Tf(x) \geqslant 0$ for every f in $L^r(R^n)$.
(iii) T is positively homogeneous, i.e., $T(\lambda f)(x) = \lambda Tf(x)$ a.e. for each $\lambda > 0$.
(iv) T is subadditive, i.e., $T(f+g)(x) \leqslant Tf(x) + Tg(x)$ a.e.

Some examples of admissible operators include $|f(x)|$, $Mf(x)$, and, more important for our purposes, $(M(|f|^p/v^\eta)(x)v(x)^\eta)^{1/p}$ for appropriate $1 < p < \infty$, $0 < \eta \leqslant 1$.

We verify this last example and in the process we find the necessary conditions on v for this to hold. First, we must find r, the choice $r = p/\eta$ being a natural one; in this case we have $\|Tf\|_r^r = \int_{R^n} M(|f|^p/v^\eta)(x)^{1/\eta} v(x)\,dx$ and, if we assume that $v \in A_{1/\eta}$, then this expression may be estimated by $c\int_{R^n} (|f(x)|^{p/\eta}/v(x))v(x)\,dx = c\|f\|_r^r$ and (i) holds; the verification of (ii) and (iii) are immediate. Thus it only remains to check (iv). Fix x and let I denote an open cube containing x; then by Minkowski's inequality we readily see that

$$\left(\frac{1}{|I|}\int_I \left(\frac{|f(y)+g(y)|^p}{v(y)^\eta}\right) dy\right)^{1/p}$$
$$\leqslant \left(\frac{1}{|I|}\int_I \left(\frac{|f(y)|^p}{v(y)^\eta}\right) dy\right)^{1/p} + \left(\frac{1}{|I|}\int_I \left(\frac{|g(y)|^p}{v(y)^\eta}\right) dy\right)^{1/p}$$
$$\leqslant M(|f|^p/v^\eta)(x)^{1/p} + M(|g|^p/v^\eta)(x)^{1/p},$$

and since I is arbitrary it follows that $M(|f+g|^p/v^\eta)(x)^{1/p} \leqslant M(|f|^p/v^\eta)(x)^{1/p} + M(|g|^p/v^\eta)(x)^{1/p}$, and (iv) follows at once.

An important property of admissible mappings T is that they also are σ-subadditive; more precisely, we have

Proposition 5.2. Suppose T is admissible and r is the index in property (i) in the definition of T. If $\{f_j\}, f$ are $L^r(R^n)$ functions with $\lim_{N\to\infty}\sum_{j=1}^N f_j = f$ in L^r, then $Tf(x) \leqslant \sum_{j=1}^\infty Tf_j(x)$ a.e.

Proof. Since $f = (f - \sum_{j=1}^N f_j) + \sum_{j=1}^N f_j$, from properties (ii) and (iv) it follows that

$$Tf(x) \leqslant T\left(f - \sum_{j=1}^N f_j\right)(x) + \sum_{j=1}^N Tf_j(x)$$
$$\leqslant T\left(f - \sum_{j=1}^N f_j\right)(x) + \sum_{j=1}^\infty Tf_j(x), \qquad (5.1)$$

Moreover, since $\|T(f-\sum_{j=1}^N f_j)\|_r \leqslant c\|f-\sum_{j=1}^N f_j\|_r \to 0$ as $N\to\infty$, there is a sequence $N_k\to\infty$ such that $\lim_{N_k\to\infty} T(f-\sum_{j=1}^{N_k} f_j)(x)=0$ a.e. This is the sequence of N's we choose in (5.1), and by letting $N_k\to\infty$ there we get that $Tf(x)\leqslant \sum_{j=1}^\infty Tf_j(x)$ a.e. ■

We need one more definition.

Definition 5.3. Given an admissible mapping T, we say that a nonnegative function w is in $A_1(T)$ if

$$Tw(x)\leqslant cw(x) \qquad \text{a.e.} \tag{5.2}$$

We then have

Proposition 5.4. Suppose that T_1 and T_2 are admissible mappings with the same r in (i). Then there exists a function ϕ in $L^r(R^n)$ such that ϕ is simultaneously in $A_1(T_1)$ and $A_1(T_2)$.

Proof. Put $T=T_1+T_2$ (T is also an admissible mapping) and let $A\geqslant\|T\|$, the norm of T in $L^r(R^n)$. For an arbitrary, nonnegative function g in $L^r(R^n)$ put $\phi(x)=\sum_{j=0}^\infty T^jg(x)/A^j$. We show that this ϕ will do. In the first place, $\phi\in L^r(R^n)$ since $\sum_{j=0}^\infty\|T^jg\|_r/A^j\leqslant(\sum_{j=0}^\infty(\|T\|/A)^j)\|g\|_r<\infty$. Moreover, by the σ-additivity of T_1 we see that

$$\begin{aligned}
T_1\phi(x) &= T_1\left(\sum_{j=0}^\infty \frac{T^jg}{A^j}\right)(x)\\
&\leqslant \sum_{j=0}^\infty \frac{T_1(T^jg)(x)}{A^j} \leqslant \sum_{j=0}^\infty \frac{T^{j+1}g(x)}{A^j}\\
&= A\sum_{j=1}^\infty \frac{T^jg(x)}{A^j} \leqslant A\phi(x) \qquad \text{a.e.}
\end{aligned}$$

This proves that ϕ is in $A_1(T_1)$ and a similar argument gives that ϕ is in $A_1(T_2)$ as well. ■

It is now a simple matter to prove the decomposition theorem for the A_p weights.

Theorem 5.5 (Jones). Suppose $w\in A_p$, $1<p<\infty$. Then there are weights w_1, w_2 in A_1 so that $w(x)=w_1(x)w_2(x)^{1-p}$.

Proof. Since $w\in A_p$ we also have that $w^{-1/(p-1)}\in A_{p'}$, $1/p+1/p'=1$. Let $r=pp'$ and set $T_1f(x)=(M(|f|^{p'}/w^{1/p})(x)w(x)^{1/p})^{1/p'}$ and $T_2f(x)=(M(|f|^pw^{1/p})(x)w(x)^{-1/p})^{1/p}$. By the remarks at the beginning of this section it follows that T_1 and T_2 are admissible and by Proposition 5.4 there is a nonnegative, locally integrable function ϕ simultaneously in $A_1(T_1)$ and $A_1(T_2)$. This means that $T_1\phi(x) \leq c\phi(x)$, or

$$M(\phi^{p'}w^{-1/p})(x) \leq c\phi^{p'}(x)w(x)^{-1/p} \tag{5.3}$$

and $T_2\phi(x) \leq c\phi(x)$, or

$$M(\phi^p w^{1/p})(x) \leq c\phi^p(x)w(x)^{1/p}. \tag{5.4}$$

In other words, $\phi^p w^{1/p}$ and $\phi^{p'}w^{-1/p}$ are A_1 weights. Put now $w_1 = \phi^p w^{1/p}$, $w_2 = \phi^{p'}w^{-1/p}$ and note that since $p + p'(1-p) = 0$ we have $w_1 w_2^{1-p} = \phi^{p+p'(1-p)}w^{1/p}w^{-(1-p)/p} = w$. ■

Remark 5.6. It goes without saying that by Theorem 5.5, A_p weights satisfy properties (i)-(iv) in Proposition 4.5.

6. A_p AND *BMO*

As both the A_p condition and the definition of *BMO* deal with the averaging of functions it is natural to consider whether there is any connection between these concepts.

Proposition 6.1. Assume w is a nonnegative, locally integrable function. Then $\ln w \in BMO$ if and only if there is $\eta > 0$ such that $w^\eta \in A_2$.

Proof. We show the necessity first; by the John–Nirenberg inequality there are constants $\eta \leq k/\|\ln w\|_*$ and c, independent of I, such that

$$\frac{1}{|I|}\int_I e^{\eta|\ln w(y) - (\ln w)_I|}\,dy \leq c. \tag{6.1}$$

By removing the absolute values in (6.1) we also have

$$\frac{1}{|I|}\int_I e^{\pm\eta(\ln w(y) - (\ln w)_I)}\,dy \leq c. \tag{6.2}$$

Whence multiplying the + and − estimates in (6.2) it follows that

$$\left(\frac{1}{|I|}\int_I e^{\eta\ln w(y)}\,dy\right)e^{-\eta(\ln w)_I}\left(\frac{1}{|I|}\int_I e^{-\eta\ln w(y)}\,dy\right)e^{\eta(\ln w)_I} \leq c \tag{6.3}$$

that is, w^η is in A_2. Conversely, assume such an η exists and note that

$$\frac{1}{|I|}\int_I e^{\eta|\ln w(y) - (\ln w)_I|}\,dy \leq \left(\frac{1}{|I|}\int_I e^{\eta\ln w(y)}\,dy\right)e^{-\eta(\ln w)_I}$$

$$+\left(\frac{1}{|I|}\int_I e^{-\eta\ln w(y)}\,dy\right)e^{\eta(\ln w)_I} = A + B,$$

say. Since both summands are handled in a similar fashion we only do A. By Jensen's inequality

$$A = \left(\frac{1}{|I|}\int_I e^{\eta \ln w(y)}\,dy\right) e^{(1/|I|)\int_I \eta \ln(1/w(y))\,dy}$$
$$\leqslant \left(\frac{1}{|I|}\int_I w(y)^{\eta}\,dy\right)\left(\frac{1}{|I|}\int_I w(y)^{-\eta}\,dy\right)$$
$$\leqslant (A_2 \quad \text{constant of} \quad w^{\eta}),$$

and we have finished. ■

A similar statement applies to A_p, namely,

Corollary 6.2. Assume w is a nonnegative, locally integrable function, and for some $\eta > 0$, $w^{\eta} \in A_p$, $1 \leqslant p < \infty$. Then $\ln w \in BMO$.

Proof. If $p \leqslant 2$, then also $w^{\eta} \in A_2$, and the conclusion follows by Proposition 6.1. If, on the other hand, $p > 2$, then $w^{-\eta/(p-1)} \in A_{p'}$, $p' < 2$, and again by Proposition 6.1 $\ln(w^{-\eta/(p-1)}) \in BMO$. ■

Proposition 6.3. Assume w is a nonnegative, locally integrable function. Then $w \in A_p$ if and only if

$$\left(\frac{1}{|I|}\int_I e^{(\ln w(y) - (\ln w)_I)}\,dy\right)$$
$$\times\left(\frac{1}{|I|}\int_I e^{-(\ln w(y) - (\ln w)_I)/p-1}\,dy\right)^{p-1} \leqslant c \tag{6.4}$$

with c independent of I.

As the proof should be obvious by now we omit it. Note however that by Jensen's inequality each factor in (6.4) is at least 1 and consequently the membership of w in A_p is equivalent to two separate conditions, to wit

$$\sup_I\left(\frac{1}{|I|}\int_I e^{(\ln w(y) - (\ln w)_I)}\,dy\right) \leqslant c$$

and

$$\sup_I\left(\frac{1}{|I|}\int_I e^{-(\ln w(y) - (\ln w)_I)/p-1}\,dy\right)^{p-1} \leqslant c.$$

An interesting application of these results is to evaluating the distance from BMO to L^{∞}, more precisely, an estimate of the expression $\inf_{g \in L^{\infty}}\|\phi - g\|_{*}$, $\phi \in BMO$. What is relevant here is the quantity $\eta(\phi)$ defined as follows. Let $\eta > 0$, verify

$$\sup_I \frac{1}{|I|}\int_I e^{\eta|\phi(y) - \phi_I|}\,dy < \infty, \tag{6.5}$$

and put $\eta(\phi) = \sup\{\eta\colon (6.5) \text{ holds}\}$. Two properties of this quantity are readily verified, namely, by the John–Nirenberg inequality $\eta(\phi) \geq c/\|\phi\|_*$, $c > 0$, and $\eta(\phi - g) = \eta(\phi)$ for each bounded function g. We then have

Theorem 6.4 (Garnett–Jones). There are absolute (dimensional) constants c_1, c_2 such that

$$c_1/\eta(\phi) \leq \inf_{g\in L^\infty}\|\phi - g\|_* \leq c_2/\eta(\phi). \tag{6.6}$$

Proof. The left inequality in (6.6) follows at once from the comments preceding the statement of the theorem and we say no more. Next let $\phi \in BMO$ and pick η so that $\eta(\phi)/2 < \eta < \eta(\phi)$; by Proposition 6.1 $e^{\eta\phi} \in A_2$ and consequently by Theorem 5.4 there are A_1 weights w_1, w_2 such that $e^{\eta\phi(y)} = w_1(y)/w_2(y)$, or

$$\eta\phi(y) = \ln w_1(y) - \ln w_2(y). \tag{6.7}$$

Now since on account of Proposition 3.3 $Mw_1(x)^\varepsilon \in A_1$, $0 < \varepsilon < 1$, with A_1 constant independent of w_1, by Proposition 6.1 it follows that $\|\ln w_1\|_* \leq$ absolute constant, and similarly for w_2. Furthermore, since $w_1(y) \leq Mw_1(y) \leq cw_1(y)$ a.e., the function $g_1(y) = \ln(w_1(y)/Mw_1(y)) \in L^\infty$, and similarly for $g_2(y) = \ln(w_2(y)/Mw_2(y))$. Thus rewriting (6.7) as $(\ln Mw_1(y) - \ln Mw_2(y)) + (\ln(w_1(y)/Mw_1(y)) - \ln(w_2(y)/Mw_2(y)) = b(y) + g(y)$, say we see at once that $\phi(y) = b(y)/\eta + g(y)/\eta$, $\|g/\eta\|_\infty \leq \infty$, $\|\phi - g/\eta\|_* \leq \text{absolute const}/\eta \leq c/\eta(\phi)$. Therefore the right inequality in (6.6) also holds and the proof is complete. ■

7. AN EXTRAPOLATION RESULT

This section is devoted to an important extrapolation property the A_p weights verify; first we need some definitions and preliminary results.

Definition 7.1. We say that the pair (w, v) of nonnegative, locally integrable functions w, v satisfies the A_p condition, $1 < p < \infty$, and we write $(w, v) \in A_p$, if for all open cubes I and a constant c independent of I,

$$\left(\frac{1}{|I|}\int_I w(y)\,dy\right)\left(\frac{1}{|I|}\int_I v(y)^{-1/(p-1)}\,dy\right)^{p-1} \leq c. \tag{7.1}$$

The infimum over the c's on the right-hand side of (7.1) is called the A_p constant of (w, v), and a statement involving the pair (w, v) is said to be independent in A_p if it only depends on the A_p constant of the pair, rather than on the particular functions involved. Similarly, we say that $(w, v) \in A_1$ provided that

$$\frac{1}{|I|}\int_I w(y)\,dy \leqslant c \operatorname{ess\,inf}_I v, \qquad \text{all} \quad I, \tag{7.2}$$

where c is independent of I; the statement independent in A_1 has the obvious meaning. An example of a result independent in A_p is the following: let $d\mu(y) = w(y)\,dy$, $d\nu(y) = v(y)^{-1/(p-1)}\,dy$, ν regular and doubling, then the weak type estimate

$$\lambda^p \mu(\{Mf > \lambda\}) \leqslant c \int_{R^n} |f(y)|^p\,d\nu(y) \tag{7.3}$$

holds provided $(w, v) \in A_p$, with the constant c in (7.3) independent in A_p. We do the case $p > 1$; first observe that, by Hölder's inequality and (7.1), for $f \geqslant 0$ we have

$$\begin{aligned}\frac{1}{|I|}\int_I f(y)\,dy &\leqslant \left(\frac{1}{|I|}\int_I f(y)^p v(y)\,dy\right)^{1/p}\left(\frac{1}{|I|}\int_I v(y)^{-1/(p-1)}\,dy\right)^{1/p'}\\ &\leqslant c^{1/p}\left(\frac{1}{|I|}\int_I f(y)^p\,d\nu(y)\right)^{1/p}(|I|/\mu(I))^{1/p}, \qquad \text{all} \quad I,\end{aligned} \tag{7.4}$$

where c is the A_p constant of (w, v). Whence by (7.4) it follows that, if $f_I > \lambda$, then also

$$\mu(I) \leqslant c\lambda^{-p}\int_I f(y)^p\,d\nu(y), \qquad \text{all} \quad I. \tag{7.5}$$

Let now K be an arbitrary compact subset of $\{Mf > \lambda\}$, by the estimate (7.5), and as in (1.6) we see that $\mu(K) \leqslant c\lambda^{-p}\int_{R^n} f(y)^p\,d\nu(y)$ with c independent in A_p, and we have finished.

Another result of interest to us is

Proposition 7.2. Suppose $0 < \eta \leqslant 1$, $1 < p < \infty$ and $w \in A_p$. Let $g \in L_\mu^{p'/\eta}(R^n)$ and consider $G(y) = (M(g^{1/\eta}w)(y)/w(y))^\eta$. Then

(i) $\|G\|L_\mu^{p'/\eta} \leqslant c\|g\|L_\mu^{p'/\eta}$ and,
(ii) $(gw, Gw) \in A_{\eta+p(1-\eta)}$.

Furthermore, both the constant c in (i) and the $A_{\eta+p(1-\eta)}$ constant of the pair (gw, Gw) are independent in A_p.

Proof. Statement (i) has essentially been proved in Section 5. As for (ii), let $q = \eta + p(1-\eta)$ and note that $q - 1 = (p-1)(1-\eta) \geqslant 0$, or $q \geqslant 1$. If $\eta = 1$, then also $q = 1$ and since $G = M(gw)/w$ we have that $(gw, Gw) \in A_1$ with A_1 constant 1. Let then $0 < \eta < 1$, on account of (7.1) we must show that

$$\left(\frac{1}{|I|}\int_I g(y)w(y)\,dy\right)$$
$$\times\left(\frac{1}{|I|}\int_I (M(g^{1/\eta}w)(y)/w(y))^{-\eta/q-1}w(y)^{-1/q-1}\,dy\right)^{q-1} \leqslant c \quad (7.6)$$

for a constant c independent of I and independent in A_p. Now, by Hölder's inequality with indices $1/\eta$ and its conjugate $1/1-\eta$ we see at once that

$$\frac{1}{|I|}\int_I g(y)w(y)\,dy \leqslant \left(\frac{1}{|I|}\int_I g(y)^{1/\eta}w(y)\,dy\right)^{\eta}\left(\frac{1}{|I|}\int_I w(y)\,dy\right)^{1-\eta}. \quad (7.7)$$

Also since for each y in I $M(g^{1/\eta}w)(y) \geqslant (1/|I|)\int_I g(x)^{1/\eta}w(x)\,dx$ and $q - 1 = (p-1)(1-\eta)$, the other integral in (7.6) is dominated by

$$\left(\frac{1}{|I|}\int_I g(x)^{1/\eta}w(x)\,dx\right)^{-\eta}\left(\frac{1}{|I|}\int_I w(y)^{(\eta-1)/(q-1)}\,dy\right)^{q-1}$$
$$=\left(\frac{1}{|I|}\int_I g(x)^{1/\eta}w(x)\,dx\right)^{-\eta}\left(\frac{1}{|I|}\int_I w(y)^{-1(p-1)}\,dy\right)^{(p-1)(1-\eta)}. \quad (7.8)$$

Whence, by multiplying (7.7) and (7.8), we get that the left-hand side of (7.6) is bounded by

$$\left(\frac{1}{|I|}\int_I w(y)\,dy\right)^{1-\eta}\left(\frac{1}{|I|}\int_I w(y)^{-1/(p-1)}\,dy\right)^{(p-1)(1-\eta)}$$
$$\leqslant (A_p \text{ constant of } w)^{1-\eta},$$

thus (ii) holds and we are done. ■

Remark 7.3. Proposition 7.2 may be restated as follows: assume $1 \leqslant p_0 < p$ and $w \in A_p$; then to each nonnegative function g in $L_\mu^{(p/p_0)'}(R^n)$ there corresponds a function $G \geqslant g$ such that $\|G\|L_\mu^{(p/p_0)'} \leqslant c\|g\|L_\mu^{(p/p_0)'}$ and

$(gw, Gw) \in A_{p_0}$, with both c and the A_{p_0} constant of the pair (gw, Gw) independent in A_p. Actually a stronger result holds, namely,

Proposition 7.4. Assume $1 \leqslant p_0 < p$, $w \in A_p$; then to each nonnegative function g in $L_\mu^{(p/p_0)'}(R^n)$ we may assign $G \geqslant g$ such that $\|G\| L_\mu^{(p/p_0)'} \leqslant c\|g\| L_\mu^{(p/p_0)'}$ and $Gw \in A_{p_0}$, with both c and the A_{p_0} constant of Gw independent in A_p.

Proof. We proceed by induction. Let $g_0 = g$ and put g_1 in place of G in Remark 7.3. Here, g_1 verifies the estimate $\|g_1\| L_\mu^{(p/p_0)'} \leqslant c\|g\| L_\mu^{(p/p_0)'}$, and by (7.3) the inequality

$$\lambda^{p_0} \int_{\{Mf>\lambda\}} g_0(y) w(y)\, dy \leqslant k \int_{R^n} |f(y)|^{p_0} g_1(y) w(y)\, dy$$

holds for each f in $L_\mu^{(p/p_0)'}$, $\lambda > 0$ with constants c, k independent in A_p. We can use g_1 in place of g_0 and so on; in general, given g_j, we obtain $g_{j+1} \geqslant g_j$ such that

$$\|g_{j+1}\| L_\mu^{(p/p_0)'} \leqslant c\|g_j\| L_\mu^{(p/p_0)'} \leqslant c^{j+1}\|g_0\| L_\mu^{(p/p_0)'} \tag{7.9}$$

and the estimate

$$\lambda^{p_0} \int_{\{Mf>\lambda\}} g_j(y) w(y)\, dy \leqslant k \int_{R^n} |f(y)|^{p_0} g_{j+1}(y) w(y)\, dy \tag{7.10}$$

holds for every f in $L_\mu^{(p/p_0)'}$, $\lambda > 0$, with constants c and k independent in A_p. Now put $G(y) = \sum_{j=0}^{\infty} (c+1)^{-j} g_j(y)$, where c is the constant in (7.9); since $(c+1)^{-j}\|g_j\| L_\mu^{(p/p_0)'} \leqslant (c/c+1)^j \|g\| L_\mu^{(p/p_0)'}$ the series defining G converges in $L_\mu^{(p/p_0)'}$ and we readily see that $G \geqslant g$ and $\|G\| L_\mu^{(p/p_0)'} \leqslant (c+1)\|g\| L_\mu^{(p/p_0)'}$. Multiplying each inequality (7.10) by $(c+1)^{-j}$ and summing over j we also get

$$\lambda^{p_0} \int_{\{Mf>\lambda\}} G(y) w(y)\, dy \leqslant k(c+1) \int_{R^n} |f(y)|^{p_0} G(y) w(y)\, dy,$$

i.e., the Hardy–Littlewood maximal function maps $L_{Gw}^{p_0}(R^n)$ into wk-$L_{Gw}^{p_0}(R^n)$ with norm independent in A_p. We may then invoke Theorem 2.1 part (iii) to infer that actually $Gw \in A_{p_0}$ with A_{p_0} constant independent in A_p. ∎

Proposition 7.4 has a counterpart for the case $p < p_0$ as well, namely,

Proposition 7.5. Assume $1 < p < p_0$ and $w \in A_p$; then to each nonnegative function g in $L_\mu^{p/(p_0-p)}(R^n)$ we may assign $G \geqslant g$ such that $\|G\| L_\mu^{p/(p_0-p)} \leqslant c\|g\| L_\mu^{p/(p_0-p)}$ and $G^{-1}w \in A_{p_0}$, with both c and the A_{p_0} constant of $G^{-1}w$ independent in A_p.

Proof. We dualize Proposition 7.4; our assumptions are equivalent to $1 < p_0' < p'$, $v = w^{-1/(p-1)} \in A_{p'}$, and $A_{p'}$ constant of $v = A_p$ constant of w. Let $d\nu(x) = v(x)\,dx$. Then by Proposition 7.4 we conclude that to each nonnegative h in $L_\nu^{(p'/p_0')'}(R^n)$ there corresponds $H \geq h$ such that $\|H\|L_\nu^{(p'/p_0')'} \leq c\|h\|L_\nu^{(p'/p_0')'}$ and $Hv \in A'_{p_0}$ with c and the $A_{p_0'}$ constant of Hv independent in A_p. But $(p'/p_0')' = (p_0 - 1)p/(p_0 - p)$ so that $h \in L_\nu^{(p'/p_0')'}$ is an equivalent way of saying $h^{p_0-1}w^{-(p_0-p)/(p-1)} \in L_\nu^{p/(p_0-p)}$. Also $Hv \in A'_{p_0}$ if and only if $(Hv)^{-1/(p_0'-1)} = (H^{p_0-1}w^{-(p_0-p)/(p-1)})^{-1}w \in A_{p_0}$. Thus given a nonnegative function g in $L_\mu^{p/(p_0-p)}$, we put $g = h^{p_0-1}w^{-(p_0-p)/(p-1)}$ with h in $L_\nu^{(p'/p_0')'}$, obtain the corresponding function H from Proposition 7.4 and then define $G = H^{p_0-1}w^{-(p_0-p)/(p-1)}$. ■

We are now ready to prove the extrapolation theorem alluded to at the beginning of this section.

Theorem 7.6 (Rubio de Francia). Assume T is a sublinear operator which verifies the following property: there is a p_0, $1 \leq p_0 < \infty$, such that for every $w \in A_{p_0}$, $d\mu(x) = w(x)\,dx$,

$$\|Tf\|L_\mu^{p_0} \leq c\|f\|L_\mu^{p_0}, \tag{7.11}$$

where c is independent of f and independent in A_{p_0}. Then for every p with $1 < p < \infty$, and every w in A_p, T also satisfies the inequality

$$\|Tf\|_{L_\mu^p} \leq c\|f\|_{L_\mu^p}, \tag{7.12}$$

where c is independent of f and independent in A_p.

Proof. We consider two cases; first suppose $1 \leq p_0 < p$, and let $w \in A_p$ and $f \in L_\mu^p$. As is readily seen

$$\|Tf\|_{L_\mu^p}^{p_0} = \||Tf|^{p_0}\|_{L_\mu^{p/p_0}} = \sup \int_{R^n} |Tf(y)|^{p_0} g(y)\,d\mu(y), \tag{7.13}$$

where the sup is taken over nonnegative g in $L_\mu^{(p/p_0)'}$ with $\|g\|L_\mu^{(p/p_0)'} \leq 1$. Fix such a function g and assign to it the function $G \geq g$ of Proposition 7.4. Then by (7.11) and the properties of G given in that proposition we see that the integrals in (7.13) involving g are dominated by

$$\int_{R^n} |Tf(y)|^{p_0} G(y)\,d\mu(y) \leq c\int_{R^n} |f(y)|^{p_0} G(y)\,d\mu(y)$$

$$\leq c\||f|^{p_0}\|_{L_\mu^{p/p_0}}\|G\|_{L_\mu^{(p/p_0)'}} \leq c\|f\|_{L_\mu^p}^{p_0},$$

where the constant c is independent of g and independent in A_p. Thus (7.12) holds and the discussion of this case is complete. Next suppose that

$1 < p < p_0$ and again let $w \in A_p$ and $f \in L^p_\mu$. Put $g(x) = \|f\|_{L^p_\mu}^{p-p_0} |f(x)|^{p_0-p}$ where $f(x) \neq 0$, and $g(x) = 0$ otherwise. Now note that

$$\int_{\{f \neq 0\}} |f(x)|^{p_0} g(x)^{-1} \, d\mu(x) = \|f\|_{L^p_\mu}^{p_0} \tag{7.14}$$

and $\|g\|_{L^{p/(p_0-p)}_\mu} = 1$. We are then in a position to invoke Proposition 7.5 and obtain a function $G \geqslant g$ with the properties given there. Observe that

$$\begin{aligned} \|Tf\|_{L^p_\mu}^{p_0} &= \left(\int_{R^n} \left(\frac{|Tf(x)|^p}{G(x)^{p/p_0}} \right)^{p/p_0} G(x)^{p/p_0} \, d\mu(x) \right)^{p_0/p} \\ &\leqslant \|G\|_{L^{p/(p_0-p)}_\mu} \int_{R^n} |Tf(x)|^{p_0} G(x)^{-1} \, d\mu(x) \\ &\leqslant c \int_{\{f \neq 0\}} |f(x)|^{p_0} g(x)^{-1} \, d\mu(x) = c \|f\|_{L^p_\mu}^{p_0} \end{aligned}$$

where the constant c is independent in A_p. ∎

8. NOTES; FURTHER RESULTS AND PROBLEMS

As expected, weighted inequalities are important in considering weighted mean convergence of orthogonal series, since the error terms can almost always be majorized by some version of the Hardy–Littlewood maximal function. In this context see Rosenblum [1962] and Muckenhoupt [1972]. They are also important in the pointwise convergence of Fourier series as well: let $s^*(f, x) = \sup_n |s_n(f, x)|$, then $\|s^*(f)\|_{L^p_\mu} \leqslant c\|f\|_{L^p_\mu}$, $1 < p < \infty$, provided $w \in A_p$ (cf. Hunt and Young [1974]). A_p weights and their basic properties have been studied extensively; for instance, Fefferman and Muckenhoupt [1974] showed there are doubling measures which are not A_p weights for any $p \geqslant 1$, and Strömberg [1979b] constructed examples to show that aside from the obvious implications, conditions such as doubling and reverse Hölder as well as others we discuss in this section are independent of each other. The proof of Muckenhoupt's Theorem 4.1 we present here is essentially due to Sawyer [1982] and Jawerth [1984] and it does not rely on the (difficult) implication "A_p implies $A_{p-\varepsilon}$" as did the original proof. Sawyer's idea is somehow related to the notion of Carleson measure which will be discussed in Chapter XV. The reader will note, however, that once the elements for the proof are set up, it very much looks like a Calderón–Zygmund decomposition argument, especially relation (4.1). In fact Christ and R. Fefferman [1983] have shown that this is precisely the case; we prefer to give the more abstract proof since it applies to the very general context considered by Jawerth. The proof of the Jones A_p decomposition theorem given here is due to Rubio de Francia [1984] and that of Rubio de Francia's extrapolation theorem is due to García-Cuerva [1983].

Further Results and Problems

8.1 Suppose a nonnegative Borel measure μ is doubling. Show that $\int_{R^n} d\mu(y) = \infty$.

8.2 Let μ be a regular, nonnegative Borel measure and let $f \in L_\mu(R^n)$. Show that for some constant c, independent of f, and $\lambda > 0$, $\lambda\mu(\{M_\mu f > \lambda\}) \leq c \int_{\{M_\mu f > \lambda\}} |f(y)|\, d\mu(y)$. (*Hint*: For a fixed λ let $\mathcal{O} = \{M_\mu f > \lambda\}$ and put $f = f\chi_{\mathcal{O}} + f\chi_{R^n \setminus \mathcal{O}} = f_1 + f_2$, say. If we can show that $\mathcal{O} \subseteq \{M_\mu f_1 > \lambda\}$, then by Theorem 1.1 we are done; but this is easy since to each x in $\mathcal{O}$ there corresponds an open cube I containing x such that $(1/\mu(I)) \int_I |f(y)|\, d\mu(y) > \lambda$, which in turn implies that $\inf_I M_\mu f > \lambda$ and $f = f_1$ on I.)

8.3 The proof of Theorem 1.1 relied on a careful selection procedure of cubes out of an arbitrary family; results of this type are known as "covering lemmas." That proof may also be obtained by making use of the following covering lemma, due to Wiener: Let E be a Borel measurable subset of R^n which is covered by the union of a family of open cubes $\{I_\beta\}$ of bounded sidelength; then from this family we can select a disjoint subsequence $\{I_j\}$ so that $\mu(E) \leq c \sum \mu(I_j)$, where c is a constant that depends only on the doubling constant of μ. Prove this lemma. (*Hint*: Choose I_1 essentially as large as possible, i.e., sidelength $I_1 \geq \frac{1}{2} \sup_\beta(\text{sidelength } I_\beta)$, discard any cube which intersects I_1, and so on.)

8.4 Maximal results, in turn, imply covering results; the following is an example: Assume that for a nonnegative, Borel measure μ and some $1 < p < \infty$ the mapping $f \to M_\mu f$ verifies $\|M_\mu f\|_{L^p_\mu} \leq c\|f\|_{L^p_\mu}$, c independent of f; then given any finite collection of open cubes $\{I_\beta\}$ it is possible to select a sequence $\{I_j\}$ so that (i) $\mu(\bigcup I_\beta) \leq c_1\mu(\bigcup I_j)$ (that is the I_j's cover a good portion of $\bigcup I_\beta$) and (ii) $\int_{R^n} (\sum \chi_{I_\beta}(y))^{p'}\, d\mu(y) \leq c_2\mu(\bigcup I_j)$ (that is the overlap of the I_β's is small when measured in $L^{p'}_\mu(R^n)$ norm, $1/p + 1/p' = 1$); the constants c_1, c_2 depend only on the norm of the maximal operator and on p. (*Hint*: Since the I_β's are finitely many they may be ordered and we choose the first cube as I_1. For I_2 we choose the first I_β among the remaining cubes so that $\mu(I_\beta \cap I_1) \leq \frac{1}{2}\mu(I_\beta)$. For I_3 we choose the first I_β among the cubes listed after I_2 so that $\mu(I_\beta \cap (I_1 \cup I_2)) \leq \frac{1}{2}\mu(I_\beta)$ and so on. Note that if an I_β was not selected then we have $\mu(I_\beta \cap (\bigcup I_j)) > \frac{1}{2}\mu(I_\beta)$

and consequently $\mu(\bigcup I_\beta) \le \mu(\{M_\mu(\chi_{\bigcup I_j}) > \frac{1}{2}\}) \le c2^p \|\chi_{\bigcup I_j}\| L^p_\mu = c_1\mu(\bigcup I_j)$ and (i) holds. Next observe that if $E_k = I_k \setminus \bigcup_{j<k} I_j$, then $\mu(E_k) \ge \frac{1}{2}\mu(I_k)$. We define now a linear operator T: $L^p_\mu(R^n) \to L^p_\mu(R^n)$ as follows:

$$Tf(x) = \sum_k \frac{1}{\mu(I_k)} \left(\int_{I_k} f(y)\, d\mu(y) \right) \chi_{E_k}(x).$$

Clearly $|Tf(x)| \le M_\mu f(x)$. Moreover, the adjoint $T^*: L^{p'}_\mu(R^n) \to L^{p'}_\mu(R^n)$ can be explicitly written as

$$T^*g(y) = \sum_k \frac{1}{\mu(I_k)} \left(\int_{E_k} g(x)\, d\mu(x) \right) \chi_{I_k}(y)$$

and consequently

$$T^*(\chi_{\bigcup I_k})(y) = \sum_k \left(\frac{\mu(E_k)}{\mu(I_k)} \right) \chi_{I_k}(y) \ge \frac{1}{2} \chi_{I_k}(y)$$

and by taking p' norms we get (ii). this technique of proof is known as "linearization" and since at no point did we use the fact that the I_β's were cubes the reader is invited to state and prove a general result in this direction. The substitute result for the case when the maximal operator is of weak-type (1, 1) should also be considered. The proof above is from Córdoba's work [1976] and the weak-type result was done independently by Córdoba [1976] and Hayes [1976].)

8.5 Under very general conditions Corollary 1.3 admits the following converse (we only discuss the unweighted version here): a collection $\mathcal{B} = \{B\}$ of open, bounded subsets of R^n is said to be a translation invariant Buseman–Feller, or B–F, basis if for each x in R^n there is a subfamily $\mathcal{B}(x)$ of $\mathcal{B}$ such that (i) if $B \in \mathcal{B}(x)$, then $x \in B$, (ii) each $\mathcal{B}(x)$ contains sets of arbitrarily small diameter; (iii) $\mathcal{B}(x) = x + \mathcal{B}(0)$. Suppose that B differentiates $L(R^n)$, that, is $\lim_{|B|\to 0,\, B\in\mathcal{B}(x)} (1/|B|) \int_B f(y)\, dy = f(x)$ a.e.; then the mapping $f(x) \to M_{\mathcal{B}} f(x) = \sup_{B\in\mathcal{B}(x)} (1/|B|) \int_B |f(y)|\, dy$ is of weak-type (1, 1). (*Hint*: Suppose not and proceed exactly as in Chapter IV; the result is from de Guzmán and Welland [1971].)

8.6 Although A_p is both necessary and sufficient for the (p, p) type and weak-type of the maximal operator, the same is not true for the restricted weak-type: the inequality $\lambda^p \int_{\{M\chi_E > \lambda\}} d\mu(x) \le c\mu(E)$, all $\lambda > 0$ and measurable subsets $E \subset R^n$, is equivalent to the existence of a positive constant K such that for all cubes I and Lebesgue measurable $E \subseteq I$, $|E|/|I| \le K(\mu(E)/\mu(I))^{1/p}$. (*Hint*: Since $M\chi_E(x) \ge (|E|/|I|)\chi_I(x)$ the condition is necessary. Conversely, we readily see that $M\chi_E(x) \le K(M_\mu \chi_E(x))^{1/p}$ and that μ is doubling. The result is Kerman's and the proof appears in Kerman and Torchinsky [1982].)

8.7 For a nonnegative, locally integrable function w and an open cube I put $m_w(I) = (t_1 t_2)^{1/2}$ where $t_1 = \sup\{t > 0 : |\{x \in I : w(x) \leq t\}| \leq |I|/2\}$ and $t_2 = \inf\{t > 0 : |\{x \in I : w(x) > t\}| \leq |I|/2\}$. Show that for any real number a we have $m_{w^a}(I) = (m_w(I))^a$ and $(w^a)_I \geq \frac{1}{2}(m_w(I))^a$. Furthermore, we say that w satisfies condition A if $w_1 \leq c m_w(I)$, for a constant c independent of I. It is clear that, if $0 < a \leq 1$ and w satisfies condition A, then $(w^a)_I \sim (m_w(I))^a$, all I. Moreover, A_p weights have the following interpretation in terms of the condition A: $w \in A_p$ if and only if w and $w^{-1/(p-1)}$ both satisfy condition A, $1 < p < \infty$. (*Hint*: One direction of the last statement is trivial since $(m_w(I))^{-1/(p-1)} = m_{w^{-1/(p-1)}}(I)$. The other follows from the inequalities $w_I \geq \frac{1}{2} m_w(I) = \frac{1}{2}(m_{w^{-1/(p-1)}}(I))^{-(p-1)}$ and $(w^{-1/(p-1)})_I \geq \frac{1}{2}(m_w(I))^{-1/(p-1)}$. These results and those in the next remark are due to Strömberg and appear in Strömberg and Torchinsky [1980]; they should be compared with Proposition 6.3.)

8.8 Let $1 < p, r < \infty$; then $w \in A_p \cap RH_r$ if and only if w^r and $w^{-1/(p-1)}$ satisfy condition A. (The statement is equivalent to w^a satisfies condition A for all $-1/(p-1) \leq a \leq r$.)

8.9 Assume that a nonnegative function w verifies $|\{y \in I; w(y) < w_I/A^k\}| \leq c\eta^k |I|$, all I, where c is independent of I and $0 < \eta < 1 < A < \infty$. Show that there is a $p > 1$ such that $w \in A_p$. (*Hint*: Note that

$$\int_I w(y)^{-1/(p-1)}\, dy = \int_{[0,\infty)} \left|\left\{\frac{1}{w} > \lambda\right\}\right| d(\lambda^{1/(p-1)})$$

$$= \left(\int_{[0,A/w_I)} + \sum_{k=0}^{\infty} \int_{[A^k/w_I, A^{k+1}/w_I)}\right) \left|\left\{w < \frac{1}{\lambda}\right\}\right| d(\lambda^{1/(p-1)})$$

$$\leq c\left(\frac{A}{w_I}\right)^{1/(p-1)} |I| + c\sum_{k=0}^{\infty} (\eta A^{1/(p-1)})^k (w_I)^{-1/(p-1)} |I|,$$

and choose p so large that $\eta A^{1/(p-1)} < 1$.)

8.10 We say that w verifies Muckenhoupt's A_∞ condition, and write $w \in A_\infty$, if to each $0 < \varepsilon < 1$ there corresponds $0 < \delta < 1$ so that for measurable subsets E of I we have $\mu(E) < \varepsilon \mu(I)$ whenever $|E| < \delta |I|$. By Proposition 4.5(iii) each A_p weight is an A_∞ weight. Show that the converse is also true: if $w \in A_\infty$ there is $1 < p < \infty$ so that $w \in A_p$. (*Hint*: It suffices to show that for appropriate A and η the assumptions of 8.9 hold; $A = 8$ and $\eta = (1 - d/2)$, where d is the δ corresponding to $\varepsilon = \frac{1}{2}$ will do in the one dimensional case, the n dimensional case requires minor adjustments. To see this fix I and k, let $E = \{x \in I : w(x) < w_I/8^k\}$, and observe that $\mu(E) < \mu(I)/8^k < \mu(I)/2$ implies $|E| < (1 - d)|I|$. Now since almost every $x \in E$ is a Lebesgue point of χ_E and Lebesgue measure is regular we may assume that E is compact and each point of E is a Lebesgue point of χ_E. To each x in E we may assign an open interval I_x centered at x such that $|I_x \cap I \cap E| = (1-d)|I_x \cap I|$ (this is possible since for I_x large I_x contains I and $|E| < (1-d)|I|$ and for I_x small $I_x \subset I$ and $|I_x \cap E|/|I_x| \to 1$), $I_x \subseteq cI$, c independent of x and I. Let $S = \bigcup_{x \in E} I_x$, since E is compact we may assume that S is finite and choose I_1 as an I_x in S of largest length. Then after $I_1, \ldots, I_k$ have been chosen let S_k be the family of the remaining I_x's so that $x \notin \bigcup_{j=1}^{k} I_j$ and let I_{k+1} be a largest interval in S_k. Observe that each y in $\bigcup I_j$ belongs to, at most, two of the I_j's and put $E_1 = \bigcup_j (I_j \cap I) \subseteq I$. Then $\mu(E_1) \leq \sum_j \int_{I_j \cap I} d\mu(y) \leq 2 \sum_j \int_{E \cap I_j \cap I} d\mu(y)$ (since $|E \cap I_J \cap I| = (1-d)|I_j \cap I|$ implies $\mu(I_j \cap I) \leq 2\mu(E \cap I_j \cap I)) \leq 4 \int_{E \cup I_j \cup I} d\mu(y)$ (since each y belongs to at most two of the I_j's) $\leq 4 \int_E d\mu(y) \leq 4\mu(I)/8^k$. How about the Lebesgue measure of E_1? Well,

$$|E_1| = |E| + |\textstyle\bigcup_j (I_j \cap I \cap (I \backslash E))| \geq |E| + \frac{1}{2} \sum_j |I_j \cap I \cap (I \backslash E)|$$
$$\geq |E| + (d/2) \textstyle\sum_j |I_j \cap I| \geq |E| + (d/2)|E_1|$$

or $|E_1|, \geq |E|/\eta$. Now, if $k \geq 2$, it is possible to start with $\mu(E_1) < \mu(E)/8^{k-1}$ and repeat the above argument with E replaced by E_1. This gives $E_2 \subset I$, $\mu(E_2) \leq \mu(I)/8^{k-2}$ and $|E_2| > |E|/\eta^2$; repeating the process k times we are done. This result is Muckenhoupt's [1974] and insures that $A_\infty = \bigcup_{1 \leq p < \infty} A_p$.)

8.11 We say that a nonnegative, doubling, Borel measure μ is comparable to a (similar) measure ν if there exist constants ε, $\delta \in (0,1)$ such that whenever E is a measurable subset of a cube I, $\nu(E)/\nu(I) < \delta$ implies $\mu(E)/\mu(I) < \varepsilon$. The following four conditions are equivalent: (i) $\nu(E)/\nu(I) \leq c(\mu(E)/\mu(I))^\eta$ for all $E \subset I$, with c and $\eta > 0$ independent of E and I, (ii) ν is comparable to μ, (iii) μ is comparable to ν, (iv) $d\nu(x) = w(x)\, d\mu(x)$ where $w \in RH_{1+\eta}(d\mu)$, i.e.,

$$\left(\frac{1}{\mu(I)} \int_I w(x)^{1+\eta}\, d\mu(x)\right)^{1/(1+\eta)} \leq c \frac{1}{\mu(I)} \int_I w(x)\, d\mu(x),$$

for every cube I. Moreover, comparability is an equivalence relation. (The proof uses ideas similar to the ones discussed in this chapter, (iii) implies (iv) is the hardest implication. This observation is from Coifman-Fefferman's work [1974].)

8.12 A nonnegative weight $w \in A_\infty$ if and only if

$$\left(\frac{1}{|I|} \int_I w(x)\, dx\right) e^{((1/|I|)\int_I \ln(1/w(x))\, dx)} \leq c, \qquad \text{all } I.$$

(*Hint*: Since $\lim_{p\to\infty}((1/|I|)\int_I w(x)^{-1/(p-1)}\,dx)^{p-1} = e^{((1/|I|)\int_I \ln(1/w(x))\,dx)}$, the assertion here is that A_∞ is obtained as the limiting A_p condition, as $p \to \infty$, much like A_1 is obtained as the limiting condition as $p \to 1$. The proof is computational, and the necessity follows at once from 8.11 and Jensen's inequality. As for the sufficiency, let A denote the sup over I of the expression in question, $A < \infty$. Then for each interval I and disjoint subsets of positive measure E, F of I, $E \cup F = I$, we have $\ln A \geq \ln(w_I) - (|E|/|I|)(\ln w)_E - (|F|/|I|)(\ln w)_F$, where $(\ln w)_E = (1/|E|)\int_E \ln w(x)\,dx$, and similarly for $(\ln w)_F$. Since by Jensen's inequality $(\ln w)_E \leq \ln(w_E)$ we also have that

$$\ln A \geq \frac{|E|}{|I|}\ln\left(\frac{\mu(I)}{\mu(E)}\frac{|E|}{|I|}\right) + \frac{|F|}{|I|}\ln\left(\frac{\mu(I)}{\mu(F)}\frac{|F|}{|I|}\right).$$

Putting $t = |F|/|I|$ and $\tau = \mu(F)/\mu(I)$ we finally get $\ln A \geq (1-t)\ln((1-t)/(1-\tau)) + t\ln(t/\tau)$; elementary considerations obtain now that for a constant c, which depends only on A, $\tau \leq 1 - e^{-c/(1-t_0)} = \tau_0 < 1$ provided that $t < t_0$, in other words $w \in A_\infty$. The proof is from Hruščev's paper [1984].)

8.13 There is yet another way of writing the A_p condition, namely, the S_p condition

$$\int_I M(\chi_I w^{1-p'})(x)^p w(x)\,dx \leq c\int_I w(x)^{1-p'}\,dx, \qquad \text{all } I,$$

with c independent of I. (*Hint*: It is not hard to see that S_p implies A_p. Conversely, if $w \in S_p$ and $x \in I \subset I_0$, then by A_p we readily see that

$$\frac{1}{|I}\int_I \chi_I(y)w(y)^{1-p'}\,dy \leq c\left(\int_I \frac{\chi_{I_0}(y)w(y)^{-1}w(y)\,dy}{\mu(I)}\right)^{p'/p}$$
$$\leq cM_\mu(\chi_{I_0}w^{-1})(x)^{p'/p}.$$

Since for each f with $\operatorname{supp} f \subseteq I_0$ and x in I_0 we have $Mf(x) = \sup_{x\in I\subseteq I_0}(1/|I|)\int_I|f(y)|\,dy$, it follows $M(\chi_{I_0}w^{1-p'})(x)^p \leq cM_\mu(\chi_{I_0}w^{-1})(x)^{p'}$, and S_p follows from Corollary 1.2. This proof is due to Hunt and Kurtz and Neugebauer [1983]; an indirect proof follows from 8.14.)

8.14 (Sawyer's Two Weight Maximal Theorem) Assume that $1 < p \leq q < \infty$, and that v, w are nonnegative, locally integrable functions in R^n. Then the following two conditions are equivalent:

(i) $(\int_{R^n} Mf(x)^q w(x)\,dx)^{1/q} \leq c(\int_{R^n}|f(x)|^p v(x)\,dx)^{1/p}$, with c independent of f.

(ii) $\int_I M(\chi_I v^{1-p'})(x)^q w(x)\,dx \leq c(\int_I v(x)^{1-p'}\,dx)^{q/p} < \infty$, all I, with c independent of I.

The proof of this interesting result is due to Sawyer [1982], and it follows along the lines of Theorem 4.1. An identical result holds for M replaced by the maximal function M_η of fractional order introduced in Chapter VI.

8.15 With the same notation and assumptions as in 7.14 of Chapter III, 7.29 of Chapter IV, and 6.17 of chapter VIII, prove the following: For every real sequence $\{r_k\}$, $-1 < r_k < 1$, with $\sum_{k=0}^{\infty} r_k^2 < \infty$, the infinite product $\prod_{k=0}^{\infty}(1 + r_k f_k(n_k x))$ converges for almost every x to a function $w(x)$ in A_p for $1 < p \leq \infty$. Moreover, $w \in L^p(T)$, $p < \infty$ as well. (*Hint*: the convergence of the product is equivalent to that of the sum $s(x) = \sum_{k=0}^{\infty} r_k f_k(n_k x)$; also for some constant c, $|\ln w(x) - s(x)| \leq c$. The convergence of $s(x)$ follows by classical arguments. Observe now that for some constants c_1, $c_2 > 0$, $c_1 e^{s(x)} \leq w(x) \leq c_2 e^{s(x)}$. This result is from Meyer's work [1979].)

8.16 Under the hypothesis of 8.15 and if $p^*(x) = \sup_{N \geq 0} p_N(x)$, $p_N(x) = \prod_{k=0}^{N}(1 + r_k f_k(n_k x))$, denotes the maximal Riesz product, then $p^* \in A_1$. (*Hint*: First, there is a constant $c > 0$ so that $(1/2\pi)\int_T p^*(x)\,dx \leq c$ whenever $\sum_{k=0}^{\infty} r_k^2 \leq \varepsilon^2$ (this follows since $p_N(x) \leq ce^{s_N(x)}$ and $s^* \in BMO$). Let $I \subseteq T$, if $|I| > 2\pi/n_0$ the estimate holds trivially since $p^*(x) \geq 1 - r_0 > 0$ and $(1/|I|)\int_I p^*(x)\,dx \leq (n_0/2\pi)\int_T p^*(x)\,dx$. Suppose next that $2\pi/n_{N+1} < |I| \leq 2\pi/n_N$ and let x = center of I; then there exists a constant $c > 0$ such that $p_j(t) \leq cp_j(x)$, $p_j(x) \leq cp_j(t)$, whenever $0 \leq j \leq N$ and $t \in I$. The constant c only depends on $\sum r_k^2$. To see this we majorize $|\ln p_j(t) - \ln p_j(x)|$ by

$$\sum_{k=0}^{j} |\ln(1 + r_k f_k(n_k t) - \ln(1 + r_k f_k(n_k x)|$$

$$\leq c \sum_{k=0}^{j} |r_k| n_k^{\eta} |x - t| \leq c.$$

Finally put $\gamma_k(t) = (1 + r_{N+1} f_{N+1}(n_{N+1} t) + \cdots + (1 + r_{N+k} f_{N+k}(n_{N+k} t))$ and note that $p^*(t) = \sup(p_1(t), \ldots, p_N(t), p_N(t)\gamma_1(t), \ldots, p_N(t)\gamma_k(t), \ldots) \leq c \sup(p_1(x), \ldots, p_N(x), p_N(x)\gamma_1(t), \ldots, p_N(x)\gamma_k(t), \ldots) \leq c \sup(a, b\gamma^*(t))$, where $\gamma^*(t) = \sup_{k \geq 1} \gamma_k(t)$. Similarly, $p^*(t) \geq c_1 \sup(a, b\gamma^*(t)) \geq c_1 a$. We will be finished once we show that $(1/|I|)\int_I \gamma^*(t)\,dt \leq c$, but this is not hard. The result is Meyer's.)

8.17 Assume w is a nonnegative function defined in a cube I_0 which verifies

$$\left(\frac{1}{|I|}\int_I w(x)^p\,dx\right)^{1/p} \leq c_1 \frac{1}{|I|}\int_I w(x)\,dx$$

for all subcubes $I \subseteq I_0$, some $p > 1$ and some constant c_1 independent of I. Show that there is $\eta > 0$ so that also

$$\left(\frac{1}{|I|}\int_I w(x)^r\,dx\right)^{1/r} \leq c_2 \frac{1}{|I|}\int_I w(x)\,dx,$$

for $p \leq r < p + \eta$, all $I \subseteq I_0$ and $c_2 = c_{c_1,r}$, but independent of I. (Suppose $I = [0, 1]$, $\int_I w(x)^p\,dx = 1$ and put $E_\lambda = \{x \in I: w(x) > \lambda\}$; the inequality follows at once from the estimate

$$\int_{E_\lambda} w(x)^p\,dx \leq c\lambda^{p-1} \int_{E_\lambda} w(x)\,dx, \qquad \lambda \geq 1,$$

which in turn follows from an argument not unlike that of Theorem 3.5. This result, important both in applications and motivation, is due to Gehring [1973].)

8.18 For a locally integrable function ϕ put $p(\phi) = \{\inf p: e^\phi$ and $e^{-\phi}$ belong to $A_p\}$. Note that $p(\phi)$ can equal ∞; also Hölder's inequality shows that $e^\phi \in A_p$ whenever $p > p(\phi)$. Suppose that $p(\phi) = \infty$ and show that $\phi \in BMO$ and $p(\phi) - 1 = \varepsilon(\phi)$, where $\varepsilon(\phi) = \inf\{\varepsilon > 0: \sup_I (1/|I|)|\{x \in I: |\phi(x) - \phi_I| > \lambda\}| \leq e^{-\lambda/\varepsilon}\}$ whenever $\lambda > \lambda_0 = \lambda_0(\varepsilon, \phi)$. (*Hint*: We must have $p(\phi) \leq 2$ (if $p(\phi) > 2$, then also $e^{\pm\phi} \in A_2$, and this cannot be); then the A_p estimates for $e^{\pm\phi}$ yield that

$$\sup_I \left(\frac{1}{|I|}\int_I e^{\phi(x)/(p-1)}\,dx\right)\left(\frac{1}{|I|}\int_I e^{-\phi(x)/(p-1)}\,dx\right) < \infty,$$

which in turn implies $\varepsilon(\phi) \leq p(\phi) - 1$ since $\eta(\phi) = 1/\varepsilon(\phi)$. Conversely, for $p - 1 > \varepsilon(\phi)$, again by the fact that $\eta(\phi) = 1/\varepsilon(\phi)$, we have that

$$\sup_I \left(\frac{1}{|I|}\int_I e^{\phi(x)/(p-1)}\,dx\right)\left(\frac{1}{|I|}\int_I e^{-\phi(x)/(p-1)}\,dx\right) < \infty$$

and when $p - 1 < 1$, Hölder's inequality shows that both e^ϕ and $e^{-\phi} \in A_p$. The result is from Garnett–Jones [1978].)

8.19 Verify the following statements.

(i) $(w, v) \in A_p$ if and only if $(v^{1-p'}, w^{1-p'})\ A_{p'}$, $1 < p < \infty$.

(ii) if $(w, v) \in A_p$, $0 < \delta < 1$ and $(q - 1)/(p - 1) = \delta$, then $(w^\delta, v^\delta) \in A_q$.

(iii) If $(w, v) \in A_p$, $0 < \delta < 1$, and $d\mu_\delta(x) = w^\delta(x)\,dx$, $d\nu_\delta(x) = v^\delta(x)\,dx$, then $\|Mf\|_{L^p_{\mu_\delta}} \leq c\|f\|_{L^p_{\nu_\delta}}$, $c = c_{\delta,p}$. (*Hint*: Since $(w^\delta, v^\delta) \in A_q$ and $p > \delta(p - 1) + 1 = q$, we get that $\lambda^q_{\mu_\delta}(\{Mf > \lambda\}) \leq c\|f\|^q_{L^q_{\nu_\delta}}$ and (iii) follows by the Marcinkiewicz interpolation theorem. These observations and the next three results are from Neugebauer's work [1983].)

8.20 Assume that

$$\|Mf\|_{L^p_\mu} \leq c\|f\|_{L^p_\nu} \qquad \text{and} \qquad \|Mf\|_{L^{p'}_{\nu_{1-p'}}} \leq c\|f\|_{L^{p'}_{\mu_{1-p'}}};$$

then there are nonnegative functions w_1, w_2 such that $w(x)^{1/p}Mw_1(x) \leq c_1 v(x)^{1/p} w_1(x)$; an identical inequality holds for w_2, and $w(x)^{1/p}v(x)^{1/p'} = w_1(x)w_2(x)^{1-p}$. (*Hint*: Cf. Theorem 5.5 and consider $Tf(x) = w(x)^{1/p}M(|f|v^{-1/p})(x) + v(x)^{-1/ps}M(|f|w^{1/ps})(x)^{1/s}$ where $s = p/p'$.)

8.21 *Let* $(w, v) \in A_p$ *and* $0 < \delta < 1$. Then there exists a nonnegative function $u = u_\delta$ such that $c_1 w^\delta(x) \leqslant u(x) \leqslant c_2 v(x)^\delta$ and $u \in A_p$. (*Hint*: Choose $0 < \varepsilon, \eta < 1$, $\delta = \varepsilon\eta$. From 8.19 (iii) we know that

$$\|Mf\|_{L^p_{\mu_\varepsilon}} \leqslant c\|f\|_{L^p_{\nu_\varepsilon}} \qquad \text{and} \qquad \|Mf\|_{L^{p'}_{\nu_\varepsilon(1-p')}} \leqslant c\|f\|_{L^{p'}_{\mu_\varepsilon(1-p')}}.$$

Thus, by 8.20, $w^{\varepsilon/p}v^{\varepsilon/p'} = w_1 w_2^{1-p}$ where $Mw_j \leqslant cw_j(v/w)^{\varepsilon/p}$, $j = 1, 2$. Note that $v^\varepsilon = w_1(v/w)^{\varepsilon/p}w_2^{1-p} \geqslant cMw_1(Mw_2)^{1-p} \geqslant cw_1w_2^{1-p}(v/w)^{\varepsilon(1-p)/p} = cw^\varepsilon$ and thus $c_1 w^\delta \leqslant (Mw_1)^\eta(Mw_2)^{\eta(1-p)} \leqslant c_2 v^\delta$. Put now $u = (Mw_1)^\eta(Mw_2)^{\eta(1-p)}$.)

8.22 Let (w, v) be a pair of nonnegative functions, then there exists $u \in A_p$ with $c_1 w(x) \leqslant u(x) \leqslant c_2 v(x)$, if and only if there is $\tau > 1$ such that $(w^\tau, v^\tau) \in A_p$. (*Hint*: Since u, $u^{1-p'}$ satisfy a reverse Hölder inequality, there is $\tau > 1$ such that $u^\tau \in A_p$ and $(w^\tau, v^\tau) \in A_p$. As for the converse, use 8.21 with $\delta = 1/\tau$.)

8.23 Suppose that w is a nonnegative function and show that $(w, Mw) \in A_p$, $1 < p < \infty$. In particular $\int_{R^n} Mf(x)^p w(x)\,dx \leqslant c\int_{R^n}|f(x)|^p Mw(x)\,dx$ in the same range of p's. (*Hint*: The proof is reminiscent of Proposition 7.2 (ii). If $w_I = 0$ there is nothing to show. Otherwise, note that $\inf_I Mw \geqslant w_I$; thus $(w, Mw) \in A_p$ with A_p constant 1. Consequently, if $v(x) = Mw(x)$, the maximal operator maps L^p_ν into $wk\text{-}L^p_\mu$ for $1 < p < \infty$ and by interpolation also into L^p_μ for the same p's. The result concerning the integral inequality was originally proved by Fefferman and Stein [1971].)

8.24 Given a nonnegative function w and $1 < p < \infty$, the following conditions are equivalent.

(i) There is a nonnegative, finite a.e. function v such that $\int_{R^n} Mf(x)^p w(x)\,dx \leqslant c\int_{R^n}|f(x)|^p v(x)\,dx$, c independent of f.

(ii) $\int_{R^n} w(x)/(1 + |x|^n)^p\,dx < \infty$.

(*Hint*: (i) implies (ii) follows by considering $f = \chi_A$, measurable A. (ii) implies (i) requires a bit of work; by replacing w by $\max(1, w)$ if necessary, we may assume that $w \geqslant 1$. Let now $u(x) = (1 + |x|^n)^{1-p}$, note that $M(uw)(x) < \infty$ a.e., and put $v(x) = M(uw)(x)/w(x)^3$, $d\nu(x) = v(x)\,dx$. For $k = 0, 1, \ldots$, let $f_k(x) = f(x)\chi_{\{2^k \leqslant |x| < 2^{k+1}\}}(x)$; then by 8.23 it follows that

$$\begin{aligned}\int_{\{|x| \leqslant 2^{k+2}\}} Mf_k(x)^p w(x)\,dx &\leqslant c2^{kn(p-1)}\int_{R^n} Mf_k(x)^p u(x)w(x)\,dx\\ &\leqslant c2^{kn(p-1)}\int_{R^n}|f_k(x)|^p M(uw)(x)\,dx\\ &\leqslant c2^{-2kn(p-1)}\int_{R^n}|f(x)|^p v(x)\,dx.\end{aligned}$$

For $|x| \geqslant 2^{k+2}$, we have

$$\int_{\{|x|>2^k\}} Mf_k(x)^p w(x)\, dx \leqslant c\left(\int_{R^n} \frac{w(x)}{(1+|x|^n)^p}\, dx\right)$$
$$\times \|f\|_{L_v^p}^p \left(\int_{\{2^k \leqslant |x| < 2^{k+1}\}} v(x)^{-1/(p-1)}\, dx\right)^{p-1},$$

where the last term can be estimated and summed. The above proof is from Young's paper [1982] and the result was obtained independently by Gatto and Gutierrez [1983]. The result followed this observation of Rubio de Francia [1981]: given a nonnegative function v and $1 < p < \infty$ the following conditions are equivalent.

(i) There is a nonnegative, finite a.e. function w such that $\int_{R^n} Mf(x)^p w(x)\, dx \leqslant c \int_{R^n} |f(x)|^p v(x)\, dx$, c independent of f.
(ii) $v(x)^{-1/(p-1)}$ is locally integrable and $\limsup_{R\to\infty} |Q_R|^{-p'} \times \int_{Q_R} v(x)^{-1/(p-1)}\, dx < \infty$, where $Q_R = \{x \in R^n\colon \max_{1\leqslant j \leqslant n} |x_j| \leqslant R\}$.)

8.26 We say that a locally integrable function b has bounded lower oscillation, and we write $b \in BLO$, if $b_I - \inf_I b \leqslant c$, all I, where c is independent of I; $BLO \subset BMO$. Then $b \in BLO$ if and only if $e^{\eta b} \in A_1$ for some $\eta > 0$. (*Hint*: If $e^{\eta b} \in A_1$, then $(e^{\eta b})_I \leqslant c \inf_I e^{\eta b}$ and the conclusion follows by Jensen's inequality. To prove the converse, note that the John-Nirenberg inequality gives $(e^{\eta b})_I \leqslant c e^{(\eta b)_I}$, for η sufficiently small, and the conclusion follows easily from this. This means, in particular, that each b in BLO may be written as $\eta \ln f + h$, where $f \geqslant 0$ is integrable and h is bounded. These results are Coifman and Rochberg's [1980].)
8.27 Theorem 7.6 has a weak-type version. More precisely, if T is a sublinear operator which verifies the following property; there is $1 \leqslant p_0 < \infty$, such that for every $w \in A_{p_0}$, $d\mu(x) = w(x)\, dx$, and $\lambda > 0$, $\lambda^{p_0} \mu(\{|Tf| > \lambda\}) \leqslant c\|f\|_{L_\mu^{p_0}}^{p_0}$, where c is independent of f and independent in A_{p_0}; then for every p with $1 < p < \infty$ and every $w \in A_p$, T also verifies $\lambda^p \mu(\{|Tf| > \lambda\}) \leqslant c\|f\|_{L_\mu^p}^p$, where c is independent of f and independent in A_p. (*Hint*: In the first place if $p_0 < p$, then for $\lambda > 0$,

$$\lambda^{p_0} \mu(\{|Tf| > \lambda\})^{p_0/p} = \lambda^{p_0} \|\chi_{\{|Tf|>\lambda\}}\|_{L_\mu^{p/p_0}}$$
$$= \lambda^{p_0} \int_{R^n} \chi_{\{|Tf|>\lambda\}}(x) g(x) w(x)\, dx,$$

for some $g \geqslant 0$ with $\|g\|_{L_\mu^{(p/p_0)'}} = 1$. Associate with g a function G as in Proposition 7.4 and apply the weak-type assumption. If, on the other hand, $1 < p < p_0$, use Proposition 7.5 instead. The result is García-Cuerva's [1983].)

8.28 A nonnegative function w is said to satisfy the $A_{p,q}$ condition, or $w \in A_{p,q}$, if

$$\left(\frac{1}{|I|}\int_I w(x)^q\,dx\right)^{1/q}\left(\frac{1}{|I|}\int_I w(x)^{-p'}\,dx\right)^{1/p'} \leq c, \qquad \text{all} \quad I,$$

where c is independent of I. The infimum over the c's above is called the $A_{p,q}$ constant of w and a statement is said to be independent in $A_{p,q}$ if it only depends on the $A_{p,q}$ constant of the weights involved. Show that, if T is a sublinear operator which verifies

$$\left(\int_{R^n}|Tf(x)|^q w(x)^q\,dx\right)^{1/q} \leq c\left(\int_{R^n}|f(x)|^p w(x)^p\,dx\right)^{1/p},$$

for some pair (p_0, q_0), $1 < p_0 \leq q_0 < \infty$, and all $w \in A_{p_0,q_0}$, where c is independent in A_{p_0,q_0}, then it also satisfies the same inequality for any other pair (p, q), $1 < p \leq q < \infty$ with $1/p - 1/q = 1/p_0 - 1/q_0$ for every $w \in A_{p,q}$ with norm independent in $A_{p,q}$. (The proof, which is similar to that of Theorem 7.6, is in Harboure–Macías–Segovia [1984b]).

8.29 Let

$$Mf_\eta(x) = \sup_I \frac{1}{|I|^{1-\eta}}\int_I |f(y)|\,dy, \qquad \text{where} \quad 0 < \eta < 1,$$

and the sup is taken over all open cubes I containing x. Then, if $0 < 1/q = 1/p - \eta < \infty$ and $w \in A_{p,q}$, there exists a constant c, independent of f and independent in $A_{p,q}$, such that

$$\left(\int_{R^n} M_\eta f(x)^q w(x)^q\,dx\right)^{1/q} \leq c\left(\int_{R^n}|f(x)|^p w(x)^p\,dx\right)^{1/p}.$$

Is the converse true? (*Hint*: The proof follows by a combination of the ideas discussed in this chapter; as an illustration we do the (easier) weak-type result. Assume that $\int_{R^n}|f(x)|^p w(x)^p\,dx = 1$ and note that by Hölder's inequality and $A_{p,q}$ we get that

$$\begin{aligned}\frac{1}{|I|^{1-\eta}}\int_I |f(y)|\,dy &\leq \frac{1}{|I|^{1-\eta}}\left(\int_I |f(y)|^p w(y)^p\,dy\right)^{1/p}\left(\int_I w(y)^{-p'}\,dy\right)^{1/p'}\\ &\leq c\left(\int_I w(y)^q\,dy\right)^{-1/q}\left(\int_I |f(y)|^p w(y)^p\,dy\right)^{1/p}\\ &\leq c\left(\left(\int_I w(y)^q\,dy\right)^{-1}\int_I |f(y)|^p w(y)^p\,dy\right)^{1/q};\end{aligned}$$

the weak-type estimate follows without much difficulty from this. (Cf. the remarks in 8.14).)

8.30 For $0 < 1/q = 1/p - \alpha < 1$ the condition $A_{p,q}$ is necessary and sufficient for the mapping

$$f(x) \to I_\alpha f(x) = \int_{R^n} \frac{f(y)}{|x-y|^{n-\alpha}}\, dy,\ 0 < \alpha < n,$$

to verify

$$\left(\int_{R^n} I_\alpha f(x)^q w(x)^q\, dx\right)^{1/q} \leqslant c\left(\int_{R^n} |f(x)|^p w(x)^p\, dx\right)^{1/p}$$

(cf. Muckenhoupt-Wheeden [1974]). Welland [1975] observed that one may prove this inequality by using Theorem 2.4 in Chapter VI and 8.29. Results similar to 8.24 in this context have been established by Rubio de Francia [1981] and Harboure-Macías-Segovia [1984a].)

8.31 Löfstrom [1983] has shown there exist no nontrivial translation invariant operators on $L^p_\mu(R^n)$, if $d\mu(x) = w(x)\, dx$ and w belongs to a class of rapidly varying weight functions, including for instance $w(x) = e^{\pm|x|^\alpha}$, $\alpha > 1$.

CHAPTER

X

More about R^n

1. DISTRIBUTIONS. FOURIER TRANSFORMS

As we saw in Chapter IX the transition from periodic functions to those defined in R^n may be accomplished smoothly. In this chapter we sketch some of the basic properties of distributions, functions, and operators in Euclidean n-dimensional space which will be useful in what follows. Some of the results are straightforward extensions of the corresponding statements in T and some are not; we will be brief in all cases, though. For instance, the Calderón–Zygmund decomposition is available at all levels now (given $\lambda > 0$, partition R^n into a countable grid of cubes I with $|I| > r$, where $\|f\|_1/r < \lambda$, and observe that the "old" Calderón–Zygmund process at level λ applies on each cube I since $|f|_I < \lambda$) and the Riesz potential operators $I_\alpha f(x) = \int_{R^n} f(y)/|x-y|^{n-\alpha}\,dy$, $0 < \alpha < n$, map $L^p(R^n)$ into $L^q(R^n)$ for $1/q = 1/p - \alpha/n$ (same proof as for periodic case).

On the other hand, other results require some adjustment. For instance, to discuss the notion of distribution we introduce the Schwartz class $\mathcal{S}(R^n)$ as follows. Given an n-tuple of nonnegative integers $\alpha = (\alpha_1, \ldots, \alpha_n)$ with length $|\alpha| = \alpha_1 + \cdots + \alpha_n$, and $x = (x_1, \ldots, x_n)$ in R^n, we put $x^\alpha = x_1^{\alpha_1} \cdots x_n^{\alpha_n}$ and define the differential operator $D^\alpha = \partial^{|\alpha|}/\partial x_1^{\alpha_1} \cdots \partial x_n^{\alpha_n}$. The space $\mathcal{S}(R^n)$ consists of those $C^\infty(R^n)$ functions $\phi(x)$ (i.e., all partial derivatives $D^\alpha \phi(x)$ exist and are continuous) such that

$$\sup_x |x^\beta D^\alpha \phi(x)| = c_{\alpha,\beta}(\phi) < \infty \tag{1.1}$$

for all multi-indices α, β. $\mathcal{S}(R^n)$ contains the space $C_0^\infty(R^n)$ consisting of those C^∞ functions with compact support, but $e^{-|x|^2} \in \mathcal{S}(R^n) \setminus C_0^\infty(R^n)$. We say that a sequence $\{\phi_k\} \subset \mathcal{S}(R^n)$ converges to 0 in $\mathcal{S}$, and we write $\lim_{k\to\infty} \phi_k = 0(\mathcal{S})$ if $\lim_{k\to\infty} c_{\alpha,\beta}(\phi_k) = 0$ for all multi-indices α, β. A linear operation T is said to be continuous in $\mathscr{S}$ provided that $\lim_{k\to\infty} T(\phi_k) = 0(\mathscr{S})$ whenever $\lim_{k\to\infty} \phi_k = 0(\mathscr{S})$. The reader may verify that all usual operations, such as addition, multiplication by a polynomial, and differentiation are continuous in $\mathscr{S}$.

An essential operation which we hope to show to be continuous in $\mathscr{S}$ is the Fourier transformation. The Fourier transform $\hat{f}$ of an integrable function f is defined by the absolutely convergent integral

$$\hat{f}(\xi) = \int_{R^n} f(x)e^{-ix\cdot\xi}\,dx, \qquad \xi \in R^n, \tag{1.2}$$

where $x\cdot\xi$ denotes the usual scalar product $x_1\xi_1 + \cdots + x_n\xi_n$. Some properties of the Fourier transform, such as $\|\hat{f}\|_\infty \leq \|f\|_1$ and $\lim_{|\xi|\to\infty}|\hat{f}(\xi)| = 0$, are readily verified. Since, by the Lebesgue dominated convergence theorem, it follows that for f in $\mathscr{S}(R^n)$, $D^\alpha\hat{f}(\xi) = (-i)^{|\alpha|}(x^\alpha f)\hat{\ }(\xi)$, and $\xi^\alpha\hat{f}(\xi) = (i)^{|\alpha|}(D^\alpha f)\hat{\ }(\xi)$, we also see at once that the Fourier transformation is continuous in $\mathscr{S}$. Moreover, since $\hat{f} \in L(R^n)$ for f in $\mathscr{S}$, we consider the possibility of expressing f in terms of $\hat{f}$ by means of a Fourier inversion formula. In the process of establishing this formula we need to have at hand a specific integrable function ϕ so that $\hat{\phi}$ is also integrable. We construct an example as follows: if $n = 1$ we put $\phi(x) = e^{-|x|}$ and observe that

$$\hat{\phi}(\xi) = \int_{(-\infty,0)} e^{x(1-i\xi)}\,dx + \int_{(0,\infty)} e^{-x(1+i\xi)}\,dx = \frac{2}{(1+\xi^2)}, \tag{1.3}$$

and

$$\int_R \hat{\phi}(\xi)\,d\xi = 2\pi. \tag{1.4}$$

As for arbitrary n note that the Fourier transform of $\eta(x) = \phi(x_1)\cdots\phi(x_n)$ is $\hat{\eta}(\xi) = 2^n/(1+\xi_1^2)\cdots(1+\xi_n^2)$ and

$$\int_{R^n} \hat{\eta}(\xi)\,d\xi = (2\pi)^n. \tag{1.5}$$

We are now in a position to show

Proposition 1.1. Suppose $f \in \mathscr{S}(R^n)$. Then the Fourier inversion formula

$$f(x) = (2\pi)^{-n}\int_{R^n} \hat{f}(\xi)e^{ix\cdot\xi}\,d\xi \tag{1.6}$$

holds.

Proof. For η as above, and since all double integrals involved are absolutely convergent, it readily follows that

$$\begin{aligned}\int_{R^n} \eta(\xi)\hat{f}(\xi)e^{ix\cdot\xi}\,d\xi &= \int_{R^n} \eta(\xi)\int_{R^n} f(y)e^{-iy\cdot\xi}\,dy\,e^{ix\cdot\xi}d\xi\\ &= \int_{R^n} f(y)\int_{R^n} \eta(\xi)e^{-i(y-x)\cdot\xi}\,d\xi\,dy\\ &= \int_{R^n} f(y)\hat{\eta}(y-x)\,dy\\ &= \int_{R^n} f(x+y)\hat{\eta}(y)\,dy. \qquad (1.7)\end{aligned}$$

Thus, by replacing $\eta(\xi)$ by $\eta(\varepsilon\xi)$, $\varepsilon > 0$, in (1.7) and observing that $\eta(\varepsilon\cdot)\hat{\ }(y) = \varepsilon^{-n}\hat{\eta}(y/\varepsilon)$, from (1.7) it follows at once that

$$\int_{R^n} \eta(\varepsilon\xi)\hat{f}(\xi)e^{ix\cdot\xi}\,d\xi = \int_{R^n} f(x+\varepsilon y)\hat{\eta}(y)\,dy. \qquad (1.8)$$

We now let $\varepsilon \to 0$ in (1.8) and by the Lebesgue dominated convergence theorem we get

$$\eta(0)\int_{R^n} \hat{f}(\xi)e^{ix\cdot\xi}\,d\xi = f(x)\int_{R^n} \hat{\eta}(y)\,dy,$$

and the proof is complete. ■

Remark 1.2. A simple variant of the proof above shows that, properly interpreted, (1.6) still holds true if $\hat{f}$ is merely integrable.

Proposition 1.3. Let ϕ, $\psi \in \mathcal{S}(R^n)$, then

$$\int_{R^n} \hat{\phi}(x)\psi(x)\,dx = \int_{R^n} \phi(x)\hat{\psi}(x)\,dx,$$

$$\int_{R^n} \phi(x)\bar{\psi}(x)\,dx = (2\pi)^{-n}\int_{R^n} \hat{\phi}(x)\bar{\hat{\psi}}(x)\,dx,$$

$$(\phi * \psi)\hat{\ }(\xi) = \hat{\phi}(\xi)\hat{\psi}(\xi) \text{ and } (\phi\psi)\hat{\ }(\xi) = (2\pi)^{-n}(\hat{\phi} * \hat{\psi})(\xi).$$

Proof. The immediate verification of these properties is left to the reader. ■

Next we consider tempered distributions.

Definition 1.4. A linear functional F on $\mathcal{S}(R^n)$ is said to be continuous if $\lim_{k\to\infty} F(\phi_k) = 0$ whenever $\lim_{k\to\infty} \phi_k = 0(\mathcal{S})$. The collection of all continuous linear functionals on $\mathcal{S}(R^n)$ is called the space $\mathcal{S}'(R^n)$ of tempered distributions.

As usual $L^p(R^n) \subset \mathcal{S}'(R^n)$, $1 \le p \le \infty$, and the action of the functional f corresponding to the function f in $L^p(R^n)$ is given by the absolutely convergent integral $f(\phi) = \int_{R^n} f(x)\phi(x)\,dx$, $\phi \in \mathcal{S}(R^n)$. Finite Borel measures μ, as well as functions of tempered growth, i.e., those functions f such that $f(x) = O((1+|x|)^k)$ for some integer k also generate distributions in the obvious way.

Tempered distributions F are also infinitely differentiable in $\mathcal{S}'$, and $D^\alpha F$ is defined by $D^\alpha F(\phi) = (-1)^{|\alpha|} F(D^\alpha \phi)$, all ϕ in $\mathcal{S}(R^n)$. Clearly, differentiation is continuous in $\mathcal{S}'(R^n)$ (we say that $F_k \to 0(\mathcal{S}')$ if $F_k(\phi) \to 0$ for each ϕ in $\mathcal{S}(R^n)$, differentiation preserves this property).

As in Chapter I, we may prove

Proposition 1.5. A linear functional L on $\mathcal{S}(R^n)$ is a tempered distribution if and only if there exist a constant $c > 0$ and integers k, m, which depend only on L, such that $|L(\phi)| \le c \sum_{\alpha,\beta} c_{\alpha,\beta}(\phi)$, $|\alpha| \le k$, $|\beta| \le m$.

We also have (cf. Theorem 4.12 in Chapter I)

Proposition 1.6. A tempered distribution F is supported at $x_0 \in R^n$ if and only if F is a linear combination of the Dirac δ concentrated at x_0, i.e., the distribution $\delta(\phi) = \phi(x_0)$ and a finite number of its derivatives.

As for the Fourier transformation we have

Definition 1.7. A tempered distribution F has a well-defined distributional Fourier transform $\hat{F}$ given by $\hat{F}(\phi) = F(\hat{\phi})$, all ϕ in $\mathcal{S}(R^n)$. Moreover, $\hat{F}$ verifies the following inversion formula: if $\tilde{\phi}(x) = \phi(-x)$ and $\tilde{F}(\phi) = F(\tilde{\phi})$, then $\hat{\hat{F}} = (2\pi)^n \tilde{F}$.

We also have the following important property

Proposition 1.8. Suppose $f \in L^2(R^n)$. Then the (distributional) Fourier transform $\hat{f}$ of f coincides with an $L^2(R^n)$ function and $\|\hat{f}\|_2 = (2\pi)^{n/2}\|f\|_2$. Furthermore, Proposition 1.3 holds for ϕ, ψ in L^2.

Proof. By the Definition 1.7, Hölder's inequality and Proposition 1.3 we have $|\hat{f}(\phi)| = |f(\hat{\phi})| \le \|f\|_2 \|\hat{\phi}\|_2 = (2\pi)^{n/2}\|f\|_2\|\phi\|_2$, all ϕ in $\mathcal{S}$. Thus $\hat{f} \in L^2(R^n)$ and $\|\hat{f}\|_2 \le (2\pi)^{n/2}\|f\|_2$. Also replacing f by $\hat{f}$ we see that $(2\pi)^n\|\tilde{f}\|_2 = (2\pi)^n\|f\|_2 = \|\hat{\hat{f}}\|_2 \le (2\pi)^{n/2}\|\hat{f}\|_2 \le (2\pi)^n\|f\|_2$, and the proof is complete. ■

Corollary 1.9. Assume $f \in L^p(R^n)$, $1 < p < 2$. Then the Fourier transform $\hat{f}$ of f is in $L^{p'}(R^n)$, $1/p + 1/p' = 1$, it is given by $\hat{f} = \lim_{N\to\infty} \int_{|x|\leq N} f(x)\, e^{-ix\cdot\xi}\, dx$, where the limit is taken in the $L^{p'}(R^n)$ norm and $\|\hat{f}\|_{p'} \leq \|f\|_p$.

Proposition 1.10. Suppose that the distribution F has compact support. Then $\hat{F}(\xi) = F(e^{-ix\cdot\xi})$ is a $C^\infty(R^n)$ function.

Definition 1.11. For $F \in \mathcal{S}'(R^n)$, $\phi \in \mathcal{S}(R^n)$ we define the convolution $F * \phi$ at x as the $C^\infty(R^n)$ function $F(\tau_x\tilde{\phi})$, where τ_x denotes translation by x in R^n. By Proposition 1.5, $F * \phi(x)$ has tempered growth and its Fourier transform is the distribution $\hat{\phi}\hat{F}$, where the product is understood in $\mathcal{S}'$. A similar definition holds for the convolution of distributions, one of which has compact support.

2. TRANSLATION INVARIANT OPERATORS. MULTIPLIERS

A bounded linear operator T from $L^p(R^n)$ into $L^q(R^n)$ is said to be translation invariant if $\tau_x T = T\tau_x$ for all x in R^n. Unlike for the periodic case, such operators do not always exist.

Proposition 2.1 (Hörmander). If T is a bounded, translation invariant operator from $L^p(R^n)$ into $L^q(R^n)$ and $q < p < \infty$, then $T = 0$.

Proof. First note that $\lim_{|x|\to\infty} \|f + \tau_x f\|_p = 2^{1/p}\|f\|_p$; the proof of this is not hard once we write $f = u + v$, where u is compactly supported and $\|v\|_p$ is small, and observe that for $|x|$ sufficiently large the support of u and that of $\tau_x u$ are disjoint. We show next that the smallest constant c for which $\|Tf\|_q \leq c\|f\|_p$ is 0. By the linearity and translation invariance of T we get $\|Tf + \tau_x Tf\|_q = \|T(f + \tau_x f)\|_q \leq c\|f + \tau_x f\|_p$, and consequently, letting $|x| \to \infty$ it follows that $\|Tf\|_q \leq 2^{1/p-1/q} c\|f\|_p$ where the factor $2^{1/p-1/q} < 1$. By repeating this argument we see that indeed the smallest such constant is 0 and the proof is complete. ∎

Next we would like to discuss the relation between translation invariant and convolution operators; this requires the following result, which is a variant of the well-known Sobolev lemma.

Proposition 2.2. Suppose $f \in L^p(R^n)$, $1 \leq p \leq \infty$, and f has distributional derivatives of order $\leq n + 1$ which coincide with $L^p(R^n)$ functions. Then f equals a continuous function a.e. and there is a constant $c = c_{n,p}$, independent of f and x such that $|f(x)| \leq c \sum_{|\alpha|\leq n+1} \|D^\alpha f\|_p$, $x \in R^n$.

Proof. First, note that for ξ in R^n, $(1+|\xi|)^{n+1} \leqslant c\sum_{|\alpha|\leqslant n+1}|\xi^\alpha|$ for some constant c independent of ξ. Suppose first that $p=1$ and observe that

$$\begin{aligned}|\hat{f}(\xi)| &\leqslant c(1+|\xi|)^{-(n+1)} \sum_{|\alpha|\leqslant n+1} |(D^\alpha f)\hat{\ }(\xi)| \\ &\leqslant c(1+|\xi|)^{-(n+1)} \sum_{|\alpha|\leqslant n+1} \|D^\alpha f\|_1. \end{aligned} \tag{2.1}$$

Since the right-hand side of (2.1) is an integrable function, $\|\hat{f}\|_1 < \infty$. Consequently, by Remark 1.2 $f(x) = (2\pi)^{-n}\int_{R^n}\hat{f}(\xi)e^{ix\cdot\xi}\,d\xi$, x a.e. in R^n, and f may be modified on a set of measure 0 so as to coincide with a continuous function in R^n, and the conclusion follows in this case. Suppose now $p>1$ and choose ϕ in $C_0^\infty(R^n)$ so that $\phi(x)=1$ if $|x|\leqslant 1$ and 0 if $|x|>2$. Then $f\phi$ verifies our assumptions for $p=1$, and by the above proof it coincides a.e. with a continuous function h such that

$$|h(x)| \leqslant \sum_{|\alpha|\leqslant n+1} \|D^\alpha(f\phi)\|_1. \tag{2.2}$$

Since by Leibnitz's rule $D^\alpha(f\phi)(x) = \sum_{\alpha_1+\alpha_2=\alpha} c_{\alpha_1,\alpha_2}D^{\alpha_1}f(x)D^{\alpha_2}\phi(x)$, it follows that the right-hand side of (2.2) is bounded by $c\max_{|\alpha_2|\leqslant|\alpha|}\|D^{\alpha_2}\phi\|_\infty\sum_{|\alpha_1|\leqslant n+1}\|D^{\alpha_1}f\|_p$, with a constant c which depends only on n and p. Moreover, since $\phi=1$ in $|x|\leqslant 1$, h actually coincides with f in that neighborhood of the origin, and the same argument applied to $f\tau_x\phi$, $x\in R^n$, gives the general result. ■

We may now prove

Theorem 2.3. Suppose T is a linear, bounded operator from $L^p(R^n)$ into $L^q(R^n)$, $1\leqslant p\leqslant q\leqslant\infty$, which commutes with translations. Then there exists a unique tempered distribution F such that $T\phi(x) = F*\phi(x)$, for each ϕ in $\mathscr{S}(R^n)$.

Proof. Since T is translation invariant, $T\phi(x)$ has derivatives of all orders which coincide with $L^q(R^n)$ functions and $D^\alpha(T\phi) = T(D^\alpha\phi)$, all α; this observation follows at once since the difference quotients of ϕ converge to the corresponding partial derivatives of ϕ in L^p, and, consequently, T of the quotients converges to T of the derivatives in L^q. By Lemma 2.2, it now follows that $T\phi$ coincides with a continuous function, after correction on a set of measure 0, which verifies

$$|T\phi(0)| \leqslant c\sum_{|\alpha|\leqslant n+1}\|D^\alpha\phi\|_p.$$

By Proposition 1.5, $T\phi(0)$ is a tempered distribution, $\tilde{F}(\phi)$ say. We claim F is the distribution we are seeking; indeed, if $\phi \in \mathcal{S}$, then by Definition 1.11 we have that $F * \phi(x) = F(\tau_x\tilde{\phi}) = F((\tau_{-x}\phi)^{\sim}) = \tilde{F}(\tau_x\phi) = T(\tau_x\phi)(0) = \tau_x T\phi(0) = T\phi(x)$, and we have finished. ■

As in Chapters V and VI, we denote by $M_q^p(R^n)$ the set of Fourier transforms $\hat{k} = m$ of those tempered distributions k which verify $\|k * \phi\|_q \leq c\|\phi\|_p$, c independent of $\phi \in \mathcal{S}$. The elements m of $M_q^p(R^n)$ are called multipliers of type (p, q) and the smallest constant c above is called the norm of m. All the usual properties hold in this setting as well; we show, for instance, that $M_2^2(R^n) = L^\infty(R^n)$ and norm of $m = \|m\|_\infty$ for m in $M_2(R^n)$. Suppose first that $m \in M_2(R^n)$. By Proposition 1.8 it follows that $m(\xi)\hat{\phi}(\xi)$ is a locally integrable function for each ϕ in $\mathcal{S}$, and consequently m itself is locally integrable. Suppose now $\hat{\phi}$ is real valued, and note that for $\varepsilon > 0$ and ξ_0 in R^n, $(k * (e^{i\xi_0 \cdot y}\phi(\varepsilon y)))^{\wedge}(\xi) = m(\xi)\varepsilon^{-n}\hat{\phi}((\xi - \xi_0)/\varepsilon)$, and consequently $\varepsilon^{-n}\int_{R^n} |m(\xi)|^2\hat{\phi}((\xi - \xi_0)/\varepsilon)^2\, d\xi \leq c^2\int_{R^n} \hat{\phi}(\xi)^2\, d\xi$. Thus by letting $\varepsilon \to 0$ we see that for almost every ξ_0 in R^n, $|m(\xi_0)|^2\|\hat{\phi}\|_2^2 \leq$ (norm of $m)^2\|\hat{\phi}\|_2^2$ and $\|m\|_\infty \leq$ norm of m. The opposite inequality is trivial, and the claim is proved.

We make an additional comment due to deLeeuw; in fact, we show that multipliers may be restricted in a natural way to lower dimensional Euclidean spaces preserving their basic properties.

Theorem 2.4. Suppose $m(\xi, \eta) \in M_p(R^{n+m})$. Then, for almost every ξ in R^n, $m(\xi, \cdot) \in M_p(R^m)$, with norm not exceeding that of m.

Proof. Following Jodeit [1971], we assume first that m is continuous everywhere and put $\|m\|$ = norm of m. Also for f_1, g_1 in $\mathcal{S}(R^n)$ and f_2, g_2 in $\mathcal{S}(R^m)$ we set $M(\xi) = \int_{R^m} m(\xi, \eta)\hat{f}_2(\eta)\hat{g}_2(\eta)\, d\eta$ and observe that

$$\left|\int_{R^n} M(\xi)\hat{f}_1(\xi)\hat{g}_1(\xi)\, d\xi\right| = \left|\int_{R^{n+m}} m(\xi, \eta)\hat{f}_1(\xi)\hat{f}_2(\eta)\hat{g}_1(\xi)\hat{g}_2(\eta)\, d\xi\, d\eta\right|$$

$$\leq \|m\|\,\|f_1\|_p\|f\|_2\|_p\|g_1\|_{p'}\|g_2\|_{p'}, \quad 1/p + 1/p' = 1.$$

Thus $M(\xi) \in M_p(R^n)$, with norm $\|M\| \leq \|m\|\,\|f_2\|_p\|g_2\|_{p'}$. Now since $\|M\|_\infty \leq \|M\|$, we also have $|\int_{R^n} m(\varepsilon, \eta)\hat{f}_2(\eta)\hat{g}_2(\eta)\, d\eta| \leq \|m\|\,\|f_2\|_p\|g_2\|_{p'}$ and $m(\xi, \cdot) \in M_p(R^n)$ with norm $\leq \|m\|$. To show the same conclusion for arbitrary m, assume (ξ, η) is a Lebesgue point of m for almost every η in R^m and put $m_\varepsilon(\xi, \eta) = m * \psi_\varepsilon(\xi, \eta)$, where $\psi_\varepsilon(\xi, \eta) = \varepsilon^{-(n+m)}\chi_I(\xi/\varepsilon, \eta/\varepsilon)$ and I is the unit cube in R^{n+m} centered at the origin. Since $m_\varepsilon \in M_p(R^{n+m})$ with norm $\|m_\varepsilon\| \leq \|m\|$ it readily follows that

$$\left|\int_{R^m} m_\varepsilon(\xi, \eta)\hat{f}_2(\eta)\hat{g}_2(\eta)\, d\eta\right| \leq \|m\|\,\|f_2\|_p\|g_2\|_{p'}$$

and the conclusion follows by the Lebesgue dominated convergence theorem. ∎

3. THE HILBERT AND RIESZ TRANSFORM

A classical result asserts that, given a continuous, bounded function f in R, there is a function $F(z)$ analytic in $\operatorname{Im} z \neq 0$, such that

$$\lim_{y\to 0^+} F(x+iy) - F(x-iy) = f(x), \qquad \text{all } x \text{ in } R. \tag{3.1}$$

In fact it is not hard to see that the Cauchy integral of f

$$F(z) = \frac{1}{2\pi i}\int_R \frac{f(s)}{s-z}\,ds, \qquad \operatorname{Im} z \neq 0 \tag{3.2}$$

does the job. Indeed, since for $z = x+iy$ we have that $1/(s-z) - 1/(s-\bar{z}) = 2iy/((s-x)^2+y^2)$, (3.1) holds provided that for $x \in R$

$$\lim_{y\to 0^+} \frac{y}{\pi}\int_R \frac{f(s)}{(s-x)^2+y^2}\,ds = f(x). \tag{3.3}$$

The integral in (3.3) is the convolution $f * P_y(x)$, where $P(x) = (1/\pi)(1/(1+x^2))$ and $P_y(x) = y^{-2}P(x/y)$. Actually, $P_y(x)$ is the Poisson kernel for $R^2_+ = \{(x,y) \in R^2 : y > 0\}$ and (3.3) holds, as we have seen in Chapter VII, even nontangentially and in $L^p(R)$ when f is in $L^p(R)$.

It is also of interest to determine the boundary values $\lim_{y\to 0^+} F(x+iy)$ of F. Suppose that for almost every x in R

$$\lim_{y\to 0^+} F(x+iy) + F(x-iy) = ig(x). \tag{3.4}$$

Then adding (3.1) and (3.4) we have $\lim_{y\to 0^+} F(x+iy) = \frac{1}{2}(f(x)+ig(x))$, x a.e. in R. To find the function $g(x)$ in (3.4) we observe that for $z = x+iy$, $1/(s-z) + 1/(s-\bar{z}) = 2(s-x)/((s-x)^2+y^2)$, and consequently

$$F(x+iy) + F(x-iy) = \frac{i}{\pi}\int_R \frac{f(s)(x-s)}{(x-s)^2+y^2}\,ds. \tag{3.5}$$

The right-hand side of (3.5) is $if * Q_y(x)$, where $Q(x) = (1/\pi)(x/1+x^2)$ and $Q_y(x) = y^{-2}Q(x/y)$ is the conjugate Poisson kernel. Therefore,

$$g(x) = \text{p.v.}\,\frac{1}{\pi}\int_R \frac{f(t)}{x-t}\,dt = \lim_{\varepsilon\to 0}\frac{1}{2}\int_{|x-t|>\varepsilon} \frac{f(t)}{x-t}\,dt, \tag{3.6}$$

in other words $g(x)$ is the Hilbert transform $Hf(x)$ of f. The results of Chapter V will hold in this context, with identical proof, provided we show that $\|Hf\|_2 \leqslant c\|f\|_2$ with c independent of f. This estimate is equivalent to $m(\xi) = ((1/\pi)\text{ p.v. } 1/x)\hat{}(\xi) \in L^\infty(R)$, which is an easy computation since

$$\lim_{\varepsilon\to 0} \int_{|x|>\varepsilon} \frac{e^{-ix\xi}}{x}\, dx = -2i \operatorname{sgn} \xi \int_{[0,\infty)} \frac{\sin t}{t}\, dt.$$

The reader will have no difficulty in establishing the corresponding results of Chapter VII as well.

Next we consider the n-dimensional setting. There are many possible extensions and we only discuss here the one obtained in the direction of the theory of conjugate harmonic functions. Other results will be covered in the next chapters.

Let u be a harmonic function in an open set Ω of R^m, i.e., $u \in C^2(\Omega)$ and $(\partial^2/\partial x_1^2)u(x) + \cdots + (\partial^2/\partial x_m^2)u(x) = 0$ for x in Ω. Motivated by the results in the disk, we consider the gradient $\nabla u(x) = ((\partial/\partial x_1)u(x), \ldots, (\partial u/\partial x_m)u(x)) = (u_1(x), \ldots, u_m(x))$ and observe that in Ω it satisfies the relations

$$\frac{\partial}{\partial x_j} u_i(x) = \frac{\partial}{\partial x_i} u_j(x), \qquad 1 \leqslant i,\ \ j \leqslant m \tag{3.7}$$

and

$$\frac{\partial}{\partial x_1} u_1(x) + \cdots + \frac{\partial}{\partial x_m} u_m(x) = 0. \tag{3.8}$$

In other words, if $W(x)$ is the vector field $(u_1(x), \ldots, u_m(x))$, then curl $W = 0$ and div $W = 0$ in Ω. Conversely, it is well-known that (3.7) implies (in a simply connected region) that there exists $h(x)$ such that $\nabla h = (u_1, \ldots, u_m)$, whereas (3.8) implies that h indeed is harmonic.

We adopt these relations as the extension we have in mind; more precisely, we call (3.7) and (3.8) the generalized Cauchy–Riemann equations in the sense of Stein–Weiss and a solution vector field W to the Cauchy–Riemann equations is called a system of conjugate functions (in the sense of M. Riesz).

With this definition applied to $R^{n+1}_+ = \{(x, t)\colon x \in R^n, t > 0\}$, we pass to discuss the generalization of the Hilbert transform to R^n. First, we find the Poisson kernel $P(x, t)$ in R^{n+1}_+, that is, a function which verifies

(i) $((\partial^2/\partial t^2) + \Delta)P(x, t) = 0$,
(ii) $\int_{R^n} P(x, t)\, dx = 1$, $t > 0$, and
(iii) $\lim_{t\to 0} P(x, t) = \delta$ (in $\mathscr{S}'$).

Taking the Fourier transform in the space x-variables, by (i) we note that $(\partial^2/\partial t^2)\hat{P}(\xi, t) - |\xi|^2\hat{P}(\xi, t) = 0$ and by (iii) that $\lim_{t\to 0} \hat{P}(\xi, t) = 1$. Thus solving the differential equation we have that $\hat{P}(\xi, t) = c_1(\xi)e^{-t|\xi|} + c_2(\xi)e^{t|\xi|}$, where by the tempered nature of P we have $c_2(\xi) = 0$

and by the above observations $c_1(\xi) = 1$. In other words, $\hat{P}(\xi, t) = e^{-t|\xi|}$ and

$$P(x, t) = (2\pi)^{-n} \int_{R^n} e^{-t|\xi|} e^{ix\cdot\xi}\, d\xi. \tag{3.9}$$

It is not hard to find the explicit expressions for $P(x, t)$. Indeed, by (3.9) we see that it is of the form $t^{-n}\phi(|x|/t)$ and from (i) above it follows that ϕ satisfies the differential equation

$$s(1+s^2)\phi''(s) + (2(n+1)s^2 + (n-1))\phi'(s) + sn(n-1)\phi(s) = 0, \; s > 0.$$

Thus $\phi(s) = (1+s^2)^{-(n+1)/2}$ and $P(x, t) = P_t(x) = c_n t^{-n} 1/(1+(|x|/t)^2)^{(n+1)/2}$, $(x, t) \in R_+^{n+1}$, where $c_n = (2\pi)^{-n} \int_{R^n} e^{-|\xi|}\, d\xi$. By passing to polar coordinates, i.e., by putting $d\xi = r^{n-1}\, dr\, d\xi'$, where $d\xi'$ is the surface area element on $\Sigma = \{\xi' \in R^n : |\xi'| = 1\}$, we get that $c_n = (2\pi)^{-n}(n-1)!\, w_n$, w_n = surface area of Σ. Polar coordinates also make it possible to compute w_n; indeed observe that

$$\int_{R^2} e^{-|x|^2}\, dx = \left(\int_R e^{-s^2}\, ds\right)^2 = \int_{[0,\infty)} re^{-r^2} \int_T dx'\, dr = \pi.$$

Thus

$$\int_{R^n} e^{-|x|^2}\, dx = \pi^{n/2} = w_n \int_{[0,\infty)} r^{n-1} e^{-r^2}\, dr = \frac{1}{2}\Gamma\left(\frac{n}{2}\right) w_n,$$

and our computation is complete.

Now for f in $L^p(R^n)$, $1 \le p < \infty$, $u(x, t) = f * P(x)$ is the unique harmonic function in R_+^{n+1} such that $\lim_{t\to 0} u(x, t) = f(x)$ in L^p and $\sup_{t>0} \int_{R^n} |u(x, t)|^p\, dx = \|f\|_p^p < \infty$. It is natural to consider whether u is a (first) component of a system of conjugate functions R_+^{n+1}. Disregarding the question of convergence of the integrals involved, we note that

$$u(x, t) = \frac{\partial}{\partial t}\left(\frac{1}{(1-n)} \frac{c_n}{(2\pi)^n} \int_{R^n} f(y) \frac{1}{(t^2 + |x-y|^2)^{(n-1)/2}}\, dy\right)$$

and by (3.7) the other (possible) components of the system of conjugate functions satisfy

$$\begin{aligned} v_k(x, t) &= \frac{\partial}{\partial x_k}\left(\frac{1}{(1-n)} \frac{c_n}{(2\pi)^n} \int_{R^n} f(y) \frac{1}{(t^2 + |x-y|^2)^{(n-1)/2}}\, dy\right) \\ &= f * Q_k(x, t), \end{aligned} \tag{3.10}$$

where $Q_k(x, t) = (2\pi)^{-n} c_n x_k / (t^2 + |x|^2)^{(n+1)/2}$ denote the conjugate Poisson kernels, $1 \le k \le n$. Since as is readily verified $(\partial/\partial t)u + (\partial/\partial x_1)v_1 + \cdots + (\partial/\partial x_n)v_n = 0$, the vector is indeed a system of conjugate functions. What are the limits of the $v_k(x, t)$'s as $t \to 0$? From (3.10) it is readily seen that they are the Riesz transforms R_k given by

$$R_k f(x) = \text{p.v.} \frac{c_n}{(2\pi)^n} \int_{R^n} f(y) \frac{x_k - y_k}{|x - y|^{n+1}}\, dy, \qquad 1 \leqslant k \leqslant n. \tag{3.11}$$

It is not hard to see that the Riesz operators are bounded in $L^2(R^n)$, since from the relation $Q_k(x, t) = (x_k/t)P(x, t) \in L^2(R^n)$ it follows that $\hat{Q}_k(\xi, t) = (i\xi_k/|\xi|)e^{-t|\xi|}$ and by letting $t \to 0$, $\hat{R}_k(\xi) = (i\xi_k/|\xi|) \in L^\infty(R^n)$. As for the L^p continuity we may either apply the techniques for the Hilbert transform or the following interesting result of Calderón–Zygmund.

Theorem 3.1 (Method of Rotations). Suppose $k(x) = \Omega(x')/|x|^n$, where Ω is odd, homogeneous of degree 0, and $\int_\Sigma |\Omega(x')|\, dx' < \infty$. Then the truncated operator

$$K_\varepsilon f(x) = \int_{|y| > \varepsilon} f(x - y)k(y)\, dy$$

is bounded in $L^p(R^n)$, $1 < p < \infty$, and there is a constant $c_p = c$ independent of ε and f such that $\|K_\varepsilon f\|_p \leqslant c\|f\|_p$.

Proof. By passing to polar coordinates and on account of the fact that Ω is odd we write

$$\begin{aligned} K_\varepsilon f(x) &= \int_\Sigma \int_{[\varepsilon,\infty)} f(x - ry')\Omega(y')\frac{1}{r^n} r^{n-1}\, dr\, dy' \\ &= \int_\Sigma \Omega(y') \int_{[\varepsilon,\infty)} f(x - ry')\frac{1}{r}\, dr\, dy' \\ &= \frac{1}{2}\int_\Sigma \Omega(y') \int_{[\varepsilon,\infty)} \frac{f(x - ry') - f(x + ry')}{r}\, dr\, dy' \\ &= \frac{1}{2}\int_\Sigma \Omega(y') \int_{|r| > \varepsilon} \frac{f(x - ry')}{r}\, dr\, dy'. \end{aligned}$$

For each fixed y' in Σ let Y denote the hyperplane orthogonal to y' which passes through the origin and observe that each x in R^n can be written uniquely as $x = z + sy'$, where $s \in R$ and $z \in Y$. Thus

$$K_\varepsilon f(x) = \frac{1}{2}\int_\Sigma \Omega(y') \int_{|r| > \varepsilon} \frac{f(z + (s - r)y')}{r}\, dr\, dy', \tag{3.12}$$

where the innermost integral in (3.12) is the one-dimensional truncated Hilbert transform H_ε of the function $F(r, z, y') = f(z + ry')$, where z and

y' are now parameters. Therefore by Minkowski's integral inequality and Fubini's theorem it follows that

$$\begin{aligned}\|K_\varepsilon f\|_p &\leq \frac{1}{2}\int_\Sigma |\Omega(y)|\left(\int_{R^n}|H_\varepsilon F(s,z,y')|^p\,dx\right)^{1/p} dy' \\ &= \frac{1}{2}\int_\Sigma |\Omega(y')|\left(\int_Y\int_R |H_\varepsilon F(s,z,y')|^p\,ds\,dz\right)^{1/p} dy' \\ &\leq c\int_\Sigma |\Omega(y')|\left(\int_Y\int_R |f(z+sy')|^p\,ds\,dz\right)^{1/p} dy' \\ &= c\left(\int_\Sigma |\Omega(y')|\,dy'\right)\|f\|_p,\end{aligned}$$

and the proof is complete. ■

4. SOBOLEV AND POINCARÉ INEQUALITIES

We conclude this chapter with a glimpse at some applications of the results covered thus far; first we need some observations of general interest.

In what follows I_0 denotes a fixed, open cube in R^n and I an arbitrary, open subcube of I_0.

Proposition 4.1. Suppose f is supported in I_0 and is C^1 there. Then there is a constant c such that

$$|f(x)| \leq c\int_{I_0}\frac{|\nabla f(y)|}{|x-y|^{n-1}}\,dy, \qquad x\in I_0,$$

c independent of x.

Proof. Following Stein we observe that for y' in Σ, $(d/dt)f(x-ty') = -\nabla f(x-ty')\cdot y'$ and consequently

$$f(x) = \int_{[0,\infty)}\frac{\nabla f(x-ty')\cdot ty'}{t^n}\,t^{n-1}\,dt.$$

Whence integrating this expression over Σ we see that

$$f(x) = \frac{1}{w_n}\int_{R^n}\frac{\nabla f(x-y)\cdot y}{|y|^n}\,dy$$

and the desired result follows at once from this. ■

Proposition 4.2. Assume f is supported in I_0 and is C^1 there and let $I\subseteq I_0$.

Then

$$\frac{1}{|I|}\int_I |f(x) - f(z)|\, dz \leqslant c \int_I \frac{|\nabla f(y)|}{|x - y|^{n-1}}\, dy, \qquad x \in I, \tag{4.1}$$

where c is a dimensional constant.

Proof. As $f(z) - f(x) = \int_{[0,1)} \nabla f(x + t(z - x)) \cdot z - x\, dt$, we see at once that $\int_I |f(x) - f(z)|\, dz \leqslant \int_{[0,1)} \int_I |\nabla f(x + t(z - x))||z - x|\, dz\, dt = A$, say. To bound A we observe that, since the line segment joining x and z is totally contained in I, if we put $y = x + t(z - x)$, then also y is in I, $dy = t^n\, dz$ and $|y - x| = t|z - x| \leqslant ctL$, where L denotes the sidelength of I and c is a dimensional constant. Then

$$\begin{aligned} A &\leqslant c \int_I |\nabla f(y)||x - y| \int_{[|x-y|/cL,\infty)} t^{-(n+1)}\, dt\, dy \\ &\leqslant c \int_I |\nabla f(y)||x - y|(L/|x - y|)^n\, dy \end{aligned}$$

and the conclusion follows at once since $L^n = |I|$. ■

As it is convenient to restate the estimate (4.1) in terms of maximal functions, we introduce the following definitions: for f supported in I_0 and x in I_0 we consider the expressions $\sup(1/|I|) \int_I |f(y) - f_I|\, dy$, where $I \subseteq I_0$ and $x \in I$; in order to keep notations simple we still call this expression the local $M^{\#}f(x)$; we do similarly for the local $M_\eta f(x)$, $0 \leqslant \eta < 1$.

With this notation we have

Corollary 4.3. Let f be as in Proposition 4.2. Then

$$M^{\#}f(x) \leqslant cM_{1/n}(|\nabla f|)(x), \qquad x \in I_0 \tag{4.2}$$

where both maximal functions are local.

Proof. Fix x in I_0 and let I contain x. Then

$$\frac{1}{|I|}\int_I |f(y) - f_I|\, dy \leqslant \frac{1}{|I|^2}\int_I \int_I |f(y) - f(z)|\, dy\, dz \tag{4.3}$$

and by (4.1) the right-hand side of (4.3) does not exceed

$$c\frac{1}{|I|}\int_I \frac{1}{|I|}\int_I \frac{|\nabla f(y)|}{|y - z|^{n-1}}\, dy\, dz.$$

But since as is readily seen

$$\frac{1}{|I|}\int_I \frac{1}{|y - z|^{n-1}}\, dz \leqslant c|I|^{(1-n)/n},$$

(4.2) holds and we are done. ■

We collect now some facts concerning the local maximal functions which will be useful in the applications.

Theorem 4.4. Let $w \in A_\infty(I_0)$ and $f \in L(I_0)$, $f_{I_0} = 0$. Then there is a constant $c = c_q$, independent of f, such that the local maximal function $M^\# f$ verifies

$$\int_{I_0} |f(x)|^q \, d\mu(x) \leq c \int_{I_0} M^\# f(x)^q \, d\mu(x), \qquad 1 < q < \infty. \tag{4.4}$$

We prefer to postpone the proof of this result until Chapter XIII, where the good-λ inequalities are discussed in detail.

The next two facts relate to the continuity properties of the local function $M_\eta f$. We begin with a simple variant of 8.29 in Chapter IX, namely,

Theorem 4.5. Suppose w, v are nonnegative, integrable functions in I_0, and let $d\mu(x) = w(x)\,dx$, $d\nu(x) = v(x)\,dx$. Then for $p, q > 1$, $0 < \eta < 1$, and $1/p - \eta \leq 1/q \leq 1/p$ the local maximal function $M_\eta f$ verifies $\lambda^q \mu(\{x \in I_0 \colon M_\eta f(x) > \lambda\}) \leq c(\|f\|_{L^p_v})^q$ if and only if

$$\sup_{I \subseteq I_0} \left(\int_I w(x)\,dx \right)^{1/q} \left(\int_I v(x)^{-1/(p-1)}\,dx \right)^{1/p'} \leq c|I|^{1-\eta}. \tag{4.5}$$

The condition (4.5) is called $A_{p,q,\eta}(I_0)$ and is indicated by $(w, v) \in A_{p,q,\eta}(I_0)$; the notions of $A_{p,q,\eta}(I_0)$ constant and independent in $A_{p,q,\eta}(I_0)$ are defined as the reader would expect them to be. Also referring to 8.14 in Chapter IX we have

Theorem 4.6. Under the assumption of Theorem 4.5 the inequality

$$\left(\int_{I_0} M_\eta f(x)^q \, d\mu(x) \right)^{1/q} \leq c \left(\int_{I_0} |f(x)|^p \, d\nu(x) \right)^{1/p}$$

holds for the local maximal function $M_\eta f$ if and only if

$$\sup_{I \subseteq I_0} \left(\int_I M_\eta(\chi_I v^{-1/(p-1)})(x)^q \, d\mu(x) \right)^{1/q} \leq c \left(\int_I v(x)^{-1/(p-1)} \, dx \right)^{1/p}. \tag{4.6}$$

The condition (4.6) is called $S_{p,q,\eta}(I_0)$, is indicated by $(w, v) \in S_{p,q,\eta}(I_0)$, and a simple interpolation argument (making use of Kolmogorov's inequality 7.19 in Chapter IV) gives

Proposition 4.7. Suppose $(w, v) \in A_{p,s,\eta}(I_0)$ for some s such that $1/p - \eta \leq 1/s \leq 1/p$. Then for $q < s$, $(w, v) \in S_{p,q,\eta}(I_0)$.

We are now in a position to prove

Theorem 4.8 (Sobolev's embedding theorem). Let $1/p - 1/n \leq 1/s < 1/p$, $w \in A_\infty(I_0)$ and $(w, v) \in A_{p,s,1/n}(I_0)$. Then given any $p \leq q < s$, and for every f supported on I_0 and C^1 there,

$$\left(\int_I |f(x)|^q \, d\mu(x) \right)^{1/q} \leq c \left(\int_I |\nabla f(x)|^p \, d\nu(x) \right)^{1/p}, \qquad I \subseteq I_0, \tag{4.7}$$

where $c \sim \mu(I_0)^\delta$, some $\delta > 0$, is independent of f and independent in $A_\infty(I_0)$ and $A_{p,s,1/n}(I_0)$.

Proof. First, observe that by 8.10 and Proposition 4.4(ii) in Chapter IX there is $r > 1$ such that

$$\left(\frac{1}{|I|}\int_I w(x)^r\,dx\right)^{1/r} \leqslant c\frac{1}{|I|}\int_I w(x)\,dx, \qquad I \subseteq I_0, \tag{4.8}$$

and consequently $(w^r, v) \in A_{p,rs,\eta}(I_0)$, where $\eta = 1/n + 1/s(1 - 1/r)$. That the statement holds for $0 < r < 1$ as well is an easy consequence of Hölder's inequality. Let now $p \leqslant q < s$, and let $r > 1$ be sufficiently close to 1 so that (4.8) holds and the corresponding η verifies $\eta < 2/n \leqslant 1$. By Proposition 4.1 it follows that $\int_{I_0}|f(x)|^q\,d\mu(x) \leqslant c\int_{I_0}(I_{1/n}(|\nabla f|)(x))^q\,d\mu(x) = A$, say, and consequently it suffices to estimate A. Since by Theorem 2.4 in Chapter VI we have that $I_{1/n}(|\nabla f|)(x) \leqslant c(M_\eta(|\nabla f|)(x)M_{\eta_1}(|\nabla f|)(x))^{1/2}$ where $\eta_1 = 1/n - 1/s(1 - 1/r)$, and all maximal functions are local on account of the fact that f is supported in I_0, A may be estimated by

$$\left(\int_{I_0} M_\eta(|\nabla f|)(x)^{rq/(2r-1)}w(x)^{r/(2r-1)}\,dx\right)^{1-1/2r}$$
$$\times\left(\int_{I_0} M_{\eta_1}(|\nabla f|)(x)^{rq}w(x)^r\,dx\right)^{1/2r}. \tag{4.9}$$

Moreover, since $(w^r, v) \in A_{p,rs,\eta}(I_0)$, by Theorem 4.5 M_{η_1} maps $L^p_\nu(I_0)$ into wk-$L^{rs}_\mu(I_0)$ and, similarly, M_η maps $L^p_\nu(I_0)$ into wk-$L^{rs/(2r-1)}_\mu(I_0)$. Thus by Kolmogorov's inequality 7.19 in Chapter IV, and since $q < s$, these operators actually map $L^p_\nu(I_0)$ into $L^{rq}_\mu(I_0)$ (with norm $\leqslant c(\int_{I_0} w(x)^r\,dx)^{1-q/s}$) and into $L^{rq/(2r-1)}_\mu(I_0)$ (with norm $\leqslant c(\int_{I_0} w(x)^{r/(2r-1)}\,dx)^{1-q/s}$), respectively; the desired inequality follows now at once from (4.9). That the constant is of the right order requires a straightforward computation invoking the fact that $w \in RH_r(I_0)$. ■

In the same vein we prove

Theorem 4.9 (Poincaré's inequality). Let $w \in A_\infty(I_0)$ and $(w, v) \in S_{p,p,1/n}(I_0)$ for some $1 < p < \infty$. Then if f is supported in I_0 and is C^1 there, there is a constant c independent of f independent in A_∞ and $S_{p,p,1/n}(I_0)$ so that

$$\int_{I_0}|f(x) - f_{I_0}|^p\,d\mu(x) \leqslant c\int_{I_0}|\nabla f(x)|^p\,d\nu(x). \tag{4.10}$$

Proof. Let $g = f - f_{I_0}$, $g_{I_0} = 0$. By Theorem 4.4 and Proposition 4.1, which apply for the local maximal functions since f is supported in I_0, we get that $\int_{I_0}|g(x)|^p\, d\mu(x) \leqslant c \int_{I_0} M^{\#}g(x)^p\, d\mu(x) \leqslant c \int_{I_0} M_{1/n}(|\nabla f|)(x)^p\, d\mu(x)$ and the conclusion follows at once from Theorem 4.6. ■

Corollary 4.10. Assume w, v are as in Theorem 4.8, then (4.10) holds with a constant $c \sim k\mu(I_0)^\delta$, $\delta > 0$, $k = k_\delta$ is independent in $A_\infty(I_0)$ and $A_{p,s,1/n}(I_0)$.

We are now in a position to consider some applications of Theorems 4.8 and 4.9; we discuss here Harnack's inequality for divergence equations.

More precisely, let Ω be a bounded domain in R^n, and consider (possibly degenerate) elliptic operators L given by $\operatorname{div}(A\nabla)$ where $A(x) = (a_{ij}(x))$ is an $n \times n$ real matrix which verifies

$$0 \leqslant v(x)|\xi|^2 \leqslant \sum_{i,j=1}^{n} a_{ij}(x)\xi_i\xi_j \leqslant w(x)|\xi|^2 \tag{4.11}$$

for all x in Ω and $\xi = (\xi_1, \ldots, \xi_n)$ in $R^n\backslash(0)$. We make the following assumptions on w, v:

(i) $w \in A_\infty(I_0)$, for every $I_0 \subset \Omega$, with A_∞ constant uniform over I_0;

(ii) $(w, v) \in A_{2,s,1/n}(I_0)$ for some $s > 2$, on any $I_0 \subset \Omega$, also with constant uniform over I_0.

Item (ii) implies in particular that $1/v$ is locally integrable, and by (4.10) the same is true of $1/w$. We may then define the Hilbert space $H = L_1^2(\Omega, w, v)$ as the closure of the $C^\infty(\bar{\Omega})$ functions ϕ with respect to the norm $(\int_\Omega|\phi(x)|^2 w(x)\, dx)^{1/2} + (\int_\Omega|\nabla\phi(x)|^2 v(x)\, dx)^{1/2}$ and extend the bilinear form

$$\mathscr{L}(u, v) = \int_\Omega a(u(x), v(x))\, dx = \sum\nolimits_{i,j} \int_\Omega a_{ij}(x)(\partial/\partial x_i)u(x)(\partial/\partial x_j)v(x)\, dx$$

to $H \times H$. Another important, and useful, property of these spaces is the following: if $u \in H$ and $\eta \in C^1(R)$ with η' bounded, then $\eta(u(x)) \in H$ and $(\partial/\partial x_j)\eta(u(x)) = \eta'(u(x))(\partial/\partial x_j)u(x)$ a.e. in Ω, $1 \leqslant j \leqslant n$ (for this and related results the reader may consult Kinderlehrer-Stampacchia [1980] and Gilbarg-Trudinger [1977]).

On account of the assumptions (i) and (ii) on the functions v, w, there is a $q > 2$ for which inequality (4.2) holds with $p = 2$ there and I_0 any cube in Ω; furthermore, by a density argument this inequality can be extended to any f in H with support in I_0. Clearly, similar comments apply to Poincaré's inequality.

A function $u \in L_1^2(\Omega', w, v)$ for each open domain Ω' so that $\bar{\Omega}' \subset \Omega$ is said to be a supersolution (resp. a subsolution) to L if $\mathscr{L}(u, \phi) \geqslant 0$ (resp. $\leqslant 0$) for all $\phi \in H$, supp $\phi \subset \Omega$, and $\phi \geqslant 0$ there. Here, u is a solution to L if it is both a super-and subsolution to L. Our first observation is that nonnegative subsolutions verify a weak form of Harnack's inequality. More precisely, we have

Proposition 4.11. Assume u is a nonnegative subsolution to L. Then there exist constants c, $m>0$, such that if $2I\subset\Omega$, then

$$\sup_{(1/2)I} u \leqslant c\left(\frac{1}{|I|^m}\int_I u(x)^2 w(x)\,dx\right)^{1/2}. \tag{4.12}$$

Proof. Let $\psi\in C_0^\infty(I)$, $\psi=1$ on δI, $0<\delta<1$, $0\leqslant\psi\leqslant 1$ and $|\nabla\psi(x)|\leqslant 2/(1-\delta)|I|^{1/n}$. For $\beta\geqslant 1$ we put $\eta_M(t)=\eta(t)=t^\beta\chi_{[0,M)}(t)+(M^\beta+\beta M^{\beta-1}(t-M))\chi_{[M,\infty)}(t)$ and $\phi(x)=\psi(x)^2\int_{[0,u(x)]}\eta'(t)^2\,dt$; clearly, $\phi\in H$, $\phi(x)\geqslant 0$ and $\operatorname{supp}\phi\subset I$. Since u is a subsolution it follows that $\mathscr{L}(u,\phi)\leqslant 0$, which in turn implies

$$\int_\Omega a(\psi(x)\eta(u(x)),\psi(x)\eta(u(x))\,dx$$
$$\leqslant c\int_\Omega(\eta'(u(x))u(x))^2 a(\psi(x),\eta(x))\,dx. \tag{4.13}$$

Next we note that $\psi\eta(u)\in H$, and has support in I. Thus by Theorem 4.8, (4.11) and (4.13) we get

$$\left(\int_{\delta I}\eta(u(x))^q w(x)\,dx\right)^{2/q}\leqslant\left(\int_I(\psi(x)\eta(u(x)))^q w(x)\,dx\right)^{2/q}$$
$$\leqslant c\int_I|\nabla(\psi\eta(u))(x)|^2 v(x)\,dx$$
$$\leqslant c\int_\Omega a(\psi(x)\eta(u(x)),\psi(x)\eta(u(x)))\,dx$$
$$\leqslant c\int_\Omega(\eta'(u(x))u(x))^2 a(\psi(x),\psi(x))\,dx,$$
$$\leqslant(c/(1-\delta)|I|^{1/n})^{-2}\int_I(\eta'(u(x))u(x))^2 w(x)\,dx. \tag{4.14}$$

Taking limits as $M\to\infty$ in (4.14), we obtain that for any $\beta\geqslant 1$

$$\|u\|_{L_\mu^{2\beta r}(I)}=\left(\frac{c\beta}{(1-\delta)|I|^{1/n}}\right)^{1/\beta}\|u\|_{L_\mu^{2\beta}(I)}, \tag{4.15}$$

where $r=q/2>1$ (cf. 6.33 in Chapter VII). We may iterate (4.15) with $\beta_k=r^k$ and $\delta_k=1-1/2^k$, and, taking limits as $k\to\infty$, the desired conclusion follows at once. ■

Next we use the local boundedness of u to estimate u^p and $\ln u$.

Proposition 4.12. Suppose u is a nonnegative solution to L and $2I \subset \Omega$. Then there are constants c, $m > 0$ such that for any $-1 \leqslant p \leqslant 1$, we have $\sup_{tI} u^p \leqslant (c/(T-t)|I|^{1/n})^m \int_{TI} u(x)^p w(x)\, dx$ provided $\frac{1}{2} \leqslant t < T \leqslant 1$.

Proof. We may suppose $\inf_I u > 0$ (the essential inf that is), and put $\phi(x) = \psi(x)^2 u(x)^{s-1}$, $s \neq 0, 1$, where $\psi \in C_0^\infty(2I)$. Since by Proposition 4.11, u is also bounded above in $2I$, $\phi \in H$ and consequently $\mathscr{L}(u, \phi) = 0$. This gives the inequality

$$\left|1 - \frac{1}{s}\right| \int_\Omega \psi(x)^2 a(u(x)^{s/2}, u(x)^{s/2})\, dx$$
$$\leqslant \left(\frac{1}{|1 - 1/s|}\right) \int_\Omega u(x)^s a(\psi(x), \psi(x))\, dx \tag{4.16}$$

as long as $s \neq 0, 1$. Given now any pair of values $\frac{1}{2} \leqslant t < T \leqslant 1$ we choose a function $\eta \in C_0^\infty(TI)$ such that $\eta = 1$ on tI, $0 \leqslant \eta \leqslant 1$ and $|\nabla \eta(x)| \leqslant 2/|I|^{1/n}(T-t)$. Theorem 4.8 applied to $\eta(x)u(x)^{s/2}$ and combined with (4.11) and (4.16) gives

$$\left(\int_{tI} u(x)^{qs/2} w(x)\, dx\right)^{2/q}$$
$$\leqslant \left(\int_{TI} (\eta(x)u(x)^{s/2})^q w(x)\, dx\right)^{2/q}$$
$$\leqslant c \int_{TI} |\nabla(\eta(x)u(x)^{s/2})|^2 v(x)\, dx$$
$$\leqslant c \int_{TI} u(x)^s a(\eta(x), \eta(x))\, dx$$
$$+ c \int_{TI} \eta(x)^2 a(u(x)^{s/2}, u(x)^{s/2})\, dx$$
$$\leqslant c\left(1 + \frac{1}{|1 - 1/s|^2}\right) \int_{TI} u(x)^s a(\eta(x), \eta(x))\, dx$$
$$\leqslant \frac{c(1 + (1/|1 - 1/s|^2))}{(T-t)|I|^{1/n})^2} \int_{TI} u(x)^s w(x)\, dx.$$

In other words, if $r = q/2 > 1$, $\varepsilon = |1 - 1/s|$, then for any $s \neq 0, 1$ we have

$$\left(\int_{tI} u(x)^{rs} w(x)\, dx\right)^{1/r} \leqslant c\left(\frac{(1 + 1/\varepsilon^2)}{(T-t)|I|^{1/n}}\right)^2 \int_{TI} u(x)^s w(x)\, dx. \tag{4.17}$$

The Moser iteration technique calls for repeated use of this inequality for a sequence of values of s, ε, t. A choice that works in this case is $t_k = \frac{1}{2}(1+2^{-k})$, $s_k = sr^k$ and $\varepsilon_k = \frac{1}{2}|1 - 1/s_k|$, where in order to avoid s_k coming too close to 1 we take s of the form $\frac{1}{2}r^\sigma(r+1)$ for some integer σ. ■

Concerning $\ln u$ we have

Proposition 4.13. Suppose u is a nonnegative solution to L, $2I \subset \Omega$. Then there is a constant c such that

$$\lambda\mu(\{x \in I: |\ln u(x) - (\ln u)_I| > \lambda\}) \leq c\mu(I)/|I|^{1/n}. \tag{4.18}$$

Proof. We may assume $\sup u > 0$. Let $U = \ln u$, by Chebychev's inequality it suffices to prove that $\int_I |U(x) - U_I|\, d\mu(x) \leq c\mu(I)/|I|^{1/n}$. Now since $U \in H$ we may invoke Poincaré's inequality to get

$$\left(\int_I |U(x) - U_I|\, d\mu(x)\right)^2 \leq \mu(I) \int_I |U(x) - U_I|^2\, d\mu(x)$$

$$\leq c\mu(I) \int_I |\nabla U(x)|^2\, d\nu(x). \tag{4.19}$$

To estimate the right-hand side of (4.19), we introduce the function $\phi(x) = \psi^2(x)/u(x)$, where $\psi \in C_0^\infty(2I)$, $\psi \equiv 1$ on I, $0 \leq \psi \leq 1$, and $|\nabla\psi(x)| \leq c/|I|^{1/n}$. Since (by replacing u by $u + \delta$ if necessary and then letting $\delta \to 0$) we may assume that $u(x) \geq \delta > 0$ and by Proposition 4.11 u is bounded on the support of ψ, the function $\phi \in H$ and $\mathscr{L}(u, \phi) = 0$. From this one easily gets that $\int \phi(x)^2 a(U(x), U(x))\, dx \leq 4\int a(\phi(x), \phi(x))\, dx$ and consequently

$$\int_I |\nabla U(x)|^2 v(x)\, dx \leq \int \phi(x)^2 a(U(x), U(x))\, dx$$

$$\leq 4 \int |\nabla\phi(x)|^2 w(x)\, dx \leq \frac{c\mu(I)}{|I|^{2/n}}. \tag{4.20}$$

Whence combining (4.19) and (4.20) our conclusion follows. ■

We still need one more result essentially due to Moser [1971], the proof of which is also achieved by an iteration technique and is left for the reader to attempt.

Proposition 4.14. Assume $\{E(t)\}$ is an increasing family of measurable sets, $\frac{1}{2} \leq t \leq 1$. Let g be a nonnegative, bounded function on $E(1)$ which satisfies the following conditions.

(i) There exist constants c, $m > 0$ such that $\sup_{E(t)} g^p \leqslant c(T-t)^{-m} \int_{E(T)} g(x)^p \, d\mu(x)$, holds for $0 < p < 1$, $\frac{1}{2} \leqslant t \leqslant T \leqslant 1$.

(ii) There is a constant c such that $\lambda\mu(\{x \in E(1): \ln g(x) > \lambda\}) \leqslant c$.

Then there exists a constant k, depending only on the constants c, m of (i) and (ii) above such that $\sup_{E(1/2)} g \leqslant k$.

We are now ready to state and prove Harnack's inequality

Theorem 4.15. Assume u is a nonnegative solution to L. Then for any compact subset K of Ω there exists a constant $c = c_K$ such that

$$\sup_K u \leqslant c \inf_K u. \tag{4.21}$$

Proof. We may, and do, assume that $u(x) \geqslant \delta > 0$ a.e. in Ω. Our conclusion will follow once we show that to each x in Ω there corresponds an open cube I containing x and so that

$$\sup_I u \leqslant c \inf_I u, \tag{4.22}$$

where the constant c in (4.21) depends only on I. I in fact is a cube centered at x so that $4I \subset \Omega$, $|2I| \leqslant 1$ and $\mu(2I) \leqslant 1$. Put $\lambda = (\ln u)_{2I}$ and consider the functions $U_1(x) = e^{-\lambda}u(x)$ and $U_2(x) = e^{\lambda}u(x)$. We claim both of these functions satisfy (i) and (ii) in Proposition 4.14 for $E(t) = 2tI$, $\frac{1}{2} \leqslant t \leqslant 1$. Indeed, U_1, U_2 are bounded because u is bounded above and below. Now, according to Proposition 4.12 we have

$$\sup_{2tI} U_1^p \leqslant c\left(\frac{1}{(T-t)|I|^{1/n}}\right)^m \int_{2TI} U_1(x)^p \, d\mu(x),$$

for $0 < p < 1$ and $\frac{1}{2} \leqslant t < T \leqslant 1$ and (i) holds for U_1; a similar argument shows it also holds for U_2. As for condition (ii), we note that since $\{x \in 2I: \ln U_1(x) > s\} \subseteq \{x \in 2I: |\ln u(x) - \lambda| > s\}$; it holds for U_1 on account of Proposition 4.13, and similarly for U_2. Whence by Proposition 4.14, and with a constant k which may depend on I but not on u, we get

$$\sup_I U_1, \quad \sup_I U_2 \leqslant k, \tag{4.23}$$

and, consequently, since $U_1(x)U_2(x) = 1$, the inequalities in (4.23) can be rewritten as $k^{-1} \leqslant U_1(x) \leqslant k$, all $x \in I$, and $\sup_I U_1 \leqslant k^2 \inf_I U_1$. The desired conclusion follows now by multiplying through by e^{λ}. ∎

It is well known that this result has important consequences, such as the strong maximum principle (a solution to L which attains its maximum or minimum in Ω is constant) and local Hölder continuity,

$$\max_I u - \min_I u \leq c\left(\frac{1}{\mu(I)}\int_I u(x)^q \, d\mu(x)\right)^{1/q},$$

where q depends on the A_p condition of w and I is well inside Ω; for further details the reader should consult Gilbarg and Trudiner [1977]. The exposition of the material presented here follows that of Harboure [1984] and is related to previous work of Fabes, Kenig and Serapioni [1982] and Chanillo and Wheeden [1985].

CHAPTER

XI

Calderón–Zygmund Singular Integral Operators

1. THE BENÉDEK–CALDERÓN–PANZONE PRINCIPLE

In this chapter we extend the continuity results we discussed for the Riesz transforms and more general odd kernels in Chapter X to arbitrary Calderón–Zygmund singular integral kernels. We begin by proving a general principle which summarizes much of what we had to say concerning those kernels.

Theorem 1.1 (Benedek–Calderón–Panzone). Suppose A is a sublinear operation from $C_0^\infty(R^n)$ into measurable functions which satisfies the following two conditions:

(i) A is of weak-type (r, r), $1 < r < \infty$. More precisely, for f in $L^r(R^n)$ and $\lambda > 0$,

$$\lambda^r |\{|Af| > \lambda\}| \leqslant c_1^r \|f\|_r^r,$$

where c_1 is independent of f and λ.

(ii) Let $f \in L(R^n)$, $\operatorname{supp} f \subseteq B(x_0, R) = \{x \in R^n : |x - x_0| < R\}$, $\int_{B(x_0,R)} f(x)\, dx = 0$. Then there are constants $1 < c_2$, c_3, independent of f and $B(x_0, R)$ such that

$$\int_{R^n \setminus B(x_0, c_2 R)} |Af(x)|\, dx \leqslant c_3 \|f\|_1.$$

Then A also is an operator of weak-type $(1, 1)$; in other words, there is a constant $c_4 = c_4(c_1, c_2, c_3, A)$, independent of $f \in C_0^\infty(R^n)$ and $\lambda > 0$, such that

$$\lambda |\{|Af| > \lambda\}| \leqslant c_4 \|f\|_1.$$

Proof. For f in $C_0^\infty(R^n)$ and $\lambda > 0$, let $f = g + b$ denote the Calderón-Zygmund decomposition of f at level λ (cf. Theorem 3.1 in Chapter IV). In particular,

$$b(x) = \sum_j \left(f(x) - \frac{1}{|I_j|} \int_{I_j} f(y)\, dy \right) \chi_{I_j}(x) = \sum b_j(x),$$

say,

$$\int_{I_j} |b(x)|\, dx \leqslant c \int_{I_j} |f(x)|\, dx,$$

$$\sum |I_j| \leqslant \frac{1}{\lambda} \|f\|_1 \qquad \text{and} \qquad \|g\|_r \leqslant c\lambda^{r-1} \|f\|_1 .$$

Also by 7.2 in Chapter IV, $\lim_{N\to\infty} \|\sum_{j=1}^N b_j - b\|_r = 0$.

Now since by the sublinearity of A we have that $|Af(x)| \leqslant |Ag(x)| + |Ab(x)|$, it easily follows that

$$|\{|Af| > \lambda\}| \leqslant |\{|Ag| > \lambda/2\}| + |\{|Ab| > \lambda/2\}|. \tag{1.1}$$

The measure of the set involving g in (2.1) is readily estimated since by Chebychev's inequality we have $(\lambda/2)^r |\{|Ag| > \lambda/2\}| \leqslant c_1^r \|g\|_r^r \leqslant cc_1^r \lambda^{r-1} \|f\|_1$ and consequently $\lambda |\{|Ag| > \lambda/2\}| \leqslant 2^r cc_1^r \|f\|_1$, where c is a dimensional constant. The term involving b is a bit more delicate. First, observe that

$$|Ab(x)| \leqslant \sum_j |Ab_j(x)| \qquad \text{a.e.} \tag{1.2}$$

Indeed, on account of the subadditivity of A we have

$$\begin{aligned} |Ab(x)| &\leqslant \left| A\left(b - \sum_{j=1}^N b_j \right)(x) \right| + \sum_{j=1}^N |Ab_j(x)| \\ &= \phi_N(x) + \sum_{j=1}^N |Ab_j(x)|, \end{aligned} \tag{1.3}$$

say. Now since $\lim_{N\to\infty} \|b - \sum_{j=1}^N b_j\|_r = 0$ and A is of weak-type (r, r), it readily follows that $\phi_N \to 0$ in measure, as $N \to \infty$, and consequently there is a subsequence $N_k \to \infty$ so that $\phi_{N_k}(x) \to 0$ a.e. as $N_k \to \infty$. Using this sequence in (1.3) gives (1.2) at once.

We need one last observation of a geometric nature before we proceed with the proof. Let B_j denote the ball concentric with I_j, with diameter $B_j = c_2$ diameter I_j, and put $\Omega_1 = \bigcup B_j$. Then $|\Omega_1| \leqslant \sum |B_j| \leqslant c \sum |I_j| \leqslant (c/\lambda) \|f\|_1$, where c is a dimensional constant. With this remark out of the way we return to estimate $\lambda |\{|Ab| > \lambda/2\}| \leqslant \lambda |\Omega_1| + \lambda |\{x \in R^n \backslash \Omega_1 : |Ab(x)| > \lambda/2\}| = I + J$, say. We just noted that

$\lambda|\Omega_1| \leq c\|f\|_1$, and I is of the right order. As for J, observe that by Chebychev's inequality and (1.2).

$$J \leq 2\int_{R^n\setminus\Omega_1} |Ab(x)|\,dx \leq 2\int_{R^n\setminus\Omega_1} \sum_j |Ab_j(x)|\,dx$$
$$\leq 2\sum_j \int_{R^n\setminus B_j} |Ab_j(x)|\,dx.$$

Moreover, since each b_j verifies $\int_{I_j} b(x)\,dx = 0$, by (ii) each summand above is bounded by $c_3\int_{I_j}|b(x)|\,dx \leq cc_3\int_{I_j}|f(x)|\,dx$, and $J \leq cc_3\|f\|_1$, which is also of the right order. ■

Remark 1.2. Clearly, the same proof applies to $L_0^\infty(R^n)$, the space of compactly supported, bounded functions on R^n.

Before we proceed with the applications, we discuss the case $r = \infty$, not covered by Theorem 1.1.

2. A THEOREM OF ZÓ

We begin by proving

Proposition 2.1. Suppose A is a sublinear operator which satisfies the following three conditions:

(i) A is of type (∞, ∞). More precisely, $\|Af\|_\infty \leq c_1\|f\|_\infty$, where c_1 is independent of f.

(ii) Same as condition (ii) in Theorem 1.1.

(iii) For every sequence $\{I_j\}$ of pairwise disjoint open cubes and every integrable function h supported in $\bigcup I_j$ and such that $\int_{I_j} h(x)\,dx = 0$, all j,

$$|Ah(x)| \leq \sum_j |A(h\chi_{I_j})(x)| \qquad \text{a.e.}$$

Then A is also of weak-type $(1, 1)$.

Proof. For f in $L(R^n)$ and $\lambda > 0$, consider the Calderón–Zygmund decomposition of f at level λ, $f = g + b$. Since $\|g\|_\infty \leq 2^n\lambda$, $|Ag(x)| \leq c_1 2^n\lambda$ and $\{|Ag| > t\} = \varnothing$ whenever $t \geq c_1 2^n\lambda$. Therefore, $\{|Af| > c_1 2^{n+1}\lambda\} \subseteq \{|Ab| > c_1 2\lambda\}$, and by (ii) and (iii) the measure of this set is estimated exactly as in Theorem 1.1. ■

An interesting application of this result is to maximal operators of the form

$$Tf(x) = \sup_{\alpha\in A}\left|\int_{R^n} k_\alpha(x-y)f(y)\,dy\right|.$$

Our assumptions on the k_α's are those which will insure that (i), (ii), and (iii) of Proposition 2.1 hold. For (i) to hold it is readily seen that we must have

$$\sup_{\alpha \in A} \int_{R^n} |k_\alpha(y)|\, dy \leq c, \tag{2.1}$$

i.e., $\{k_\alpha\}_{\alpha \in A}$ is bounded in $L(R^n)$. As for (iii) note that since

$$\lim_{N \to \infty} \left\| h - \sum_{j=1}^{N} (h\chi_{I_j}) \right\|_1 = 0,$$

then also

$$\lim_{N \to \infty} \left\| h * k_\alpha - \sum_{j=1}^{N} (h\chi_{I_j}) * k_\alpha \right\|_1 = 0 \qquad \text{for each} \quad \alpha \in A.$$

Therefore there is a sequence $N_m \to \infty$ such that

$$\lim_{N_m \to \infty} \left(h - \sum_{j=1}^{N_m} (h\chi_{I_j}) \right) * k_\alpha(x) = 0 \qquad \text{a.e.}$$

and consequently

$$|h * k_\alpha(x)| \leq \sum_j |(h\chi_{I_j}) * k_\alpha(x)| \qquad \text{a.e.}$$

which gives (iii) at once. Finally we consider (ii), this will require a new assumption on the k_α's. We must estimate

$$I = \int_{|x_0 - x| \geq c_2 R} \sup_\alpha |k_\alpha * f(x)|\, dx. \tag{2.2}$$

Now since $\int_{B(x_0,R)} f(y)\, dy = 0$ we have $k_\alpha * f(x) = \int_{|x_0 - y| \leq R} (k_\alpha(x-y) - k_\alpha(x - x_0)) f(y)\, dy$ and

$$\sup_\alpha |k_\alpha * f(x)| \leq \int_{|x_0 - y| \leq R} \sup_\alpha |k_\alpha(x-y) - k_\alpha(x-x_0)| |f(y)|\, dy. \tag{2.3}$$

Therefore, by Tonelli's theorem, putting (2.3) into (2.2) gives

$$I \leq \int_{|x_0 - y| \leq R} |f(y)| \left(\int_{|x_0 - x| \geq c_2 R} \sup_\alpha |k_\alpha(x-y) - k_\alpha(x-x_0)|\, dx \right) dy \tag{2.4}$$

and (ii) will hold provided the innermost integral in (2.4) is finite. More precisely, we have

Theorem 2.2 (Zó). Suppose $\{k_\alpha\}_{\alpha \in A}$ verifies the following two conditions

(i) $\sup_\alpha \int_{R^n} |k_\alpha(x)|\, dx \leq c$

(ii) $\int_{|x| \geq c_2|y|} \sup_\alpha |k_\alpha(x-y) - k_\alpha(x)|\, dx \leq c_3$ where $1 < c_2$, c_3 are constants independent of $y \in R^n$.

Then the mapping $f(x) \to Tf(x) = \sup_\alpha |k_\alpha * f(x)|$ is of weak-type $(1, 1)$.

Corollary 2.3. Suppose a nonnegative, integrable function ϕ verifies $|\nabla\phi(x)| \le c/|x|^{n+1}$. Then if $k \in L(R^n)$ and $|k(x)| \le \phi(x)$, the mapping $f(x) \to \sup_{t>0} |\int_{R^n} t^{-n} k((x-y)/t) f(y)\, dy|$ is of weak-type $(1, 1)$.

Proof. Cf. 8.4 below. ■

Corollary 2.4. Let k be as in Corollary 2.3 and suppose $f \in L(R^n)$. Then

$$\lim_{t\to 0^+} \int_{R^n} t^{-n} k((x-y)/t) f(y)\, dy = \left(\int_{R^n} k(y)\, dy\right) f(x) \text{ a.e.}$$

Proof. Cf. Corollary 2.4 in Chapter IV. ■

3. CONVOLUTION OPERATORS

How does Theorem 1.1 apply to convolution operators? Theorem 2.2 hints that the kernel k in question should verify the following conditions:

$$k \in L(R^n) \tag{3.1}$$

and

$$\int_{|x| \ge 2|y|} |k(x-y) - k(x)|\, dx \le c. \tag{3.2}$$

By Young's convolution theorem, condition (3.1) alone implies that $f \to k * f$ is bounded in $L^p(R^n)$, $1 \le p \le \infty$, with norm $\le \|k\|_1$. But the kernels of interest to us, such as those corresponding to Riesz transforms, fail to be integrable in a neighborhood of the origin and at infinity. Aside from this they are locally integrable in $R^n \backslash (0)$ and they satisfy (3.2), also known as Hörmander's condition.

Our strategy to deal with this more general situation consists of three steps, to wit:

(1) Truncate k both at 0 and ∞, obtain an integrable function, $\operatorname{tr} k$ say, and observe that the mapping $f \to \operatorname{tr} k * f$ is well defined in $L^p(R^n)$, $1 \le p \le \infty$.

(2) Estimate $\|\operatorname{tr} k * f\|_p \le c\|f\|_p$, where c is independent of $\|\operatorname{tr} k\|_1$, $1 < p < \infty$.

(3) Take limits, both in L^p and pointwise senses, to pass from $\operatorname{tr} k$ to k.

In this section we address Step 2 by means of

Theorem 3.1. Assume k satisfies conditions (3.1) and (3.2) and let $Af(x) = k * f(x)$. If in addition to A is of weak-type (r, r) for some $1 < r < \infty$, then it also is of type (p, p), $1 < p < \infty$, with norm independent of $\|k\|_1$.

Proof. We fit our setting into the Benedek–Calderón–Panzone principle. As we saw in the proof of Proposition 2.1, (3.2) implies that hypothesis (ii) of Theorem 1.1 holds, and, consequently, by Remark 1.2, A is of weak-type $(1, 1)$ on $L_0^\infty(R^n)$. Thus by the Marcinkiewicz interpolation theorem A is also of type (p, p), $1 < p < r$, with norm independent of $\|k\|_1$. Let now $h(x) = k(-x)$ and put $A_1 f(x) = h * f(x)$. By a simple duality argument it readily follows that for $1/p + 1/q = 1$, A is bounded in $L^p(R^n)$ if and only if A_1 is bounded in $L^q(R^n)$, with the same norm. Thus in our case A_1 is bounded in $L^q(R^n)$ for $r' < q < \infty$. But since h also satisfies (3.1) and (3.2), A_1 is also of weak-type $(1, 1)$ on $L_0^\infty(R^n)$ and, by the Marcinkiewicz interpolation theorem, of type (q, q) for $1 < q < \infty$ with norm independent of $\|k\|_1$; the same is true for A. ∎

In order to apply Theorem 3.1 we must have an important bit of information, namely, a value of r so that A is of weak-type (or type) (r, r). The results in Section 2 and 3 in Chapter X are quite useful here, especially for $r = 2$. Before we proceed with the other steps we briefly digress to show how Hilbert space techniques apply in this respect.

4. COTLAR'S LEMMA

Cotlar, in the commutative case, and Cotlar and Stein, in the noncommutative case proved the following important result.

Lemma 4.1 (Cotlar's Lemma). Let $\{T_j\}_{j=1}^\infty$ be a sequence of bounded, linear operators from a Hilbert space H into itself which verifies the following "almost orthogonal" property: Suppose there is a positive sequence $\{a(j)\}$ so that $\sum_{j=-\infty}^\infty a(j) = A < \infty$ and

$$\|T_j T_k^*\|, \|T_j^* T_k\| \leq a(j-k)^2, \qquad \text{all} \quad j, k, \tag{4.1}$$

where T^* denotes the (Hilbert space) adjoinnt of T and $\|T\|$ denotes the norm of T as a mapping from H into itself. Then

$$\left\| \sum_{j=1}^N T_j \right\| \leq A, \qquad \text{all} \quad N.$$

Proof. Suppose T is a bounded, linear operator in H. Then $\|T\|^2 = \|TT^*\| = \|(TT^*)^n\|^{1/n}$, all even $n \geqslant 1$; this is a well known property of continuous operators on H. When applied to the T_j's it gives

$$\|T_j\|^2 = \|T_jT_j^*\| \leqslant a(0)^2 \leqslant A^2, \qquad \text{all } j. \tag{4.2}$$

Let now $T = \sum_{j=1}^N T_j$. We want to show that $\|T\| \leqslant A$, independently of N. But this is not hard; indeed, first note that

$$(TT^*)^n = \sum T_{i_1}T_{i_2}^* \cdots T_{i_{2n-1}}T_{i_{2n}}^*, \tag{4.3}$$

where the sum is extended over all possible indices $1 \leqslant i_j \leqslant N$, $1 \leqslant j \leqslant 2n$, and next observe that the norm of each summand in (4.3) may be estimated in one of two ways, that is, either by

$$\|T_{i_1}T_{i_2}^*\| \cdots \|T_{i_{2n-1}}T_{i_{2n}}^*\| \leqslant a(i_1 - i_2)^2 \cdots a(i_{2n-1} - i_{2n})^2 \tag{4.4}$$

or else by

$$\|T_{i_1}\|\,\|T_{i_2}^*T_{i_3}\| \cdots \|T_{i_{2n-2}}^*T_{i_{2n-1}}\|\,\|T_{i_{2n}}^*\|$$
$$\leqslant Aa(i_2 - i_3)^2 \cdots a(i_{2n-2} - i_{2n-1})^2 A. \tag{4.5}$$

Thus combining (4.4) and (4.5), we see that the norm of each summand is bounded by $Aa(i_1 - i_2)a(i_2 - i_3) \cdots a(i_{2n-1} - i_{2n})$ and consequently

$$\|(TT^*)^n\| \leqslant A \sum a(i_1 - i_2) \cdots a(i_{2n-1} - i_{2n}). \tag{4.6}$$

To estimate the right-hand side of (4.6) we sum first over i_{2n} and then successively over $i_{2n-1}, \ldots, i_2$ and note that it does not exceed

$$AA^{2n-1}\sum_{i_1} 1 = A^{2n}N. \tag{4.7}$$

Whence $\|T\|^2 = \|(TT^*)^n\|^{1/n} \leqslant A^2N^{1/n}$, and, by letting $n \to \infty$, we see that $\|T\| \leqslant A$, independently of N, as we wished to show. ■

Cotlar first considered this result to avoid using Fourier transforms in proving the L^2 continuity of convolution operators, and it has become an invaluable tool since.

5. CALDERÓN–ZYGMUND SINGULAR INTEGRAL OPERATORS

We discuss in this section the first step, namely, an appropriate truncation of the kernels. This we achieve for a very general class of kernels introduced by Calderón–Zygmund [1952], [1956].

Definition 5.1. We say that a function k, locally integrable away from the origin, is a Calderón–Zygmund singular kernel, or plainly a CZ kernel, provided it verifies the following three properties

(i) For each $0 < \varepsilon < N$, $|\int_{\varepsilon<|x|<N} k(x)\,dx| \le c_1$, where c_1 is a constant independent of ε, N, and furthermore $\lim_{\varepsilon\to 0}\int_{\varepsilon<|x|<N} k(x)\,dx$ exists for each fixed N.

(ii) $\int_{|x|\le R}|x||k(x)|\,dx \le c_2 R$, all $R > 0$, with c_2 independent of R.

(iii) $\int_{|x|\ge 2|y|}|k(x-y) - k(x)|\,dx \le c_3$, $y \ne 0$, and c_3 independent of y.

The constants c_1, c_2, c_3 are called the CZ constants of k.

Although at a first glance these conditions seem reasonable, since they are verified by the Hilbert and Riesz kernels, even more is true: as we show in 8.3 they are also necessary for the development of a satisfactory theory.

One of the most important observations is

Lemma 5.2. Suppose k is a CZ kernel, and for $0 < \varepsilon < N$ put $k_{\varepsilon,N}(x) = k(x)\chi_{\{\varepsilon<|x|<N\}}(x)$. Then $k_{\varepsilon,N}$ also is a CZ kernel, with constants uniformly bounded, independently of ε and N.

Proof. Fix ε, N and in order to simplify notations put $k_{\varepsilon,N}(x) = h(x)$. Since conditions (i) and (ii) are readily verified for h, with the same constants as k, it suffices to check that h verifies Hörmander's condition with a constant independent of ε, N. In other words, we must estimate

$$\int_{|x|\ge 2|y|} |h(x-y) - h(x)|\,dx, \qquad y \ne 0. \tag{5.1}$$

We consider two cases, namely, $|y| < N$ and $|y| \ge N$. The latter case is immediate since then the integrand in (5.1) vanishes identically. As for the former case there are three possibilities, namely: (1) $|x-y| < \varepsilon$, (2) $\varepsilon \le |x-y| \le N$, and (3) $|x-y| > N$. Since (1) and (3) are dealt in a similar fashion we only discuss (3). We have now $|x| \ge |x-y| - |y| \ge N - |x|/2$, or $N \le 3|x|/2$. Therefore, since $h(x-y) = 0$, the integral in (5.1) is dominated by

$$\begin{aligned}\int_{2N/3\le|x|\le N} |h(x)|\,dx &\le \int_{2N/3\le|x|\le N} \frac{1}{|x|}|x||k(x)|\,dx \\ &\le (3/2N)c_2 N = 3c_2/2,\end{aligned} \tag{5.2}$$

and this estimate is of the right order. To bound (2) we find it convenient to break it up into three subcases according to whether $|x| < \varepsilon$, $\varepsilon \le |x| \le N$, or $|x| > N$. Once again, the first and third subcases are dealt in a similar fashion and we do only the first subcase this time. We have now $|x| < \varepsilon$ and the restrictions in the integral in (5.1) are $|x| > 2|y|$ and $|x-y| > \varepsilon$. This implies in particular that $|y| < |x|/2 < \varepsilon/2$ and therefore also $|x-y| \le |x| + |y| \le 3\varepsilon/2$. Thus the estimate is reduced to

$$\int_{\varepsilon<|x-y|<3\varepsilon/2} |h(x-y)|\,dy \le \int_{\varepsilon<|y|<3\varepsilon/2} |k(y)|\,dy \le \frac{3c_2}{2}.$$

Only one case remains, but now both h terms appear in (5.1) and they equal the k term. Whence (5.1) is bounded by c_3 and the proof is complete. ■

Next we show that the truncated CZ kernels $k_{\varepsilon,N}$ are kernels of convolution operators, uniformly bounded in $L^2(R^n)$.

Lemma 5.3. Suppose k is a CZ kernel and let $k_{\varepsilon,N}$ be as in Lemma 5.2. Then $|\hat{k}_{\varepsilon,N}(\xi)| \leq c$, where c depends only on the CZ constants of k.

Proof. Fix ε, N and put $k_{\varepsilon,N}(x) = h(x)$. First, observe that $\hat{h}(\xi) = e^{iy\cdot\xi} \int_{R^n} h(x-y)e^{-ix\cdot\xi}\,dx$, $y \in R^n$. Since $e^{iy\cdot\xi} = -1$ for $y = \pi\xi/|\xi|^2$, we have that

$$\hat{h}(\xi) = \frac{1}{2}\left(\int_{|x|\leq 2|y|} + \int_{|x|>2|y|}\right)(h(x) - h(x-y))e^{-ix\cdot\xi}\,dx = I + J, \qquad (5.3)$$

say. To estimate J is suffices to invoke the Hörmander condition which, by Lemma 5.2, h verifies independently of ε, N. As for I we rewrite it as

$$\int_{|x|\leq 2|y|} (e^{-ix\cdot\xi} - 1)h(x)\,dx + \int_{|x|\leq 2|y|} h(x)\,dx$$
$$- \int_{|x|\leq 2|y|} e^{-ix\cdot\xi} h(x-y)\,dx = I_1 + I_2 - I_3,$$

say. The bounds for I_1 and I_2 are pretty straightforward. Indeed

$$|I_1| \leq \int_{|x|\leq 2|y|} |e^{-ix\cdot\xi} - 1||h(x)|\,dx$$
$$\leq c\int_{|x|\leq 2|y|} |\xi||x||k(x)|\,dx \leq c|\xi|c_2|y| \leq c.$$

Also

$$|I_2| \leq \left|\int_{|x|\leq 2|y|} k(x)\chi_{\{\varepsilon<|x|<N\}}(x)\,dx\right| \leq c_1.$$

Now, I_3 requires some work. We rewrite it as

$$-\int_{|x|\leq 2|y|} (e^{-ix\cdot\xi} + 1)h(x-y)\,dx + \int_{|x|\leq 2|y|} h(x-y)\,dx = I_4 + I_5,$$

say, and note that since $|e^{-ix\cdot\xi} + 1| = |e^{-ix\cdot\xi} - e^{-iy\cdot\xi}| \leq c|x-y||\xi|$, then

$$|I_4| \leq c|\xi| \int_{|x|\leq 2|y|} |x-y||h(x-y)|\,dx$$
$$\leq c|\xi| \int_{|x|\leq 3|y|} |x||k(x)|\,dx \leq c.$$

Finally to estimate I_5 we observe that since now $|x-y| \leqslant 3|y|$, it equals

$$\int_{|x-y|\leqslant 3|y|} h(x-y)\,dx - \int_{\substack{|x-y|\leqslant 3|y| \\ |x|>2|y|}} h(x-y)\,dx = I_6 - I_7,$$

say. Clearly $|I_6| \leqslant c_1$. As for I_7 note that since $|x-y| \geqslant |x| - |y| > |y|$, then $|I_7| \leqslant \int_{|y|\leqslant|x-y|\leqslant 3|y|} |h(x-y)|\,dx \leqslant 3c_2$. Since we have exhausted all cases the proof is finally complete. ■

We are now in a position to prove

Theorem 5.4 (Calderón–Zygmund). Suppose k is a CZ kernel and for a $C_0^\infty(R^n)$ function f let

$$T_\varepsilon f(x) = \int_{|x-y|>\varepsilon} k(x-y)f(y)\,dy.$$

Then there is a constant $c = c_p$, independent of f, and $\varepsilon > 0$, so that

$$\|T_\varepsilon f\|_p \leqslant c\|f\|_p. \tag{5.4}$$

Moreover, $Tf = \lim_{\varepsilon\to 0} T_\varepsilon f$ exists in $L^p(R^n)$, $1 < p < \infty$, and it also satisfies the estimate (5.4). Furthermore, $T_\varepsilon f$ and Tf are well defined for arbitrary f in $L^p(R^n)$, and the norm inequality (5.4) holds for these functions as well. T_ε and T are called the truncated and the Calderón–Zygmund singular operator associated to k.

Proof. Fix $0 < \varepsilon < N$ and for f in $C_0^\infty(R^n)$ put $T_{\varepsilon,N}f(x) = h * f(x)$, where $h = k_{\varepsilon,N}$ as usual. By Lemma 5.3, $\|T_{\varepsilon,N}f\|_2 \leqslant c\|f\|_2$, c independent of ε, N and, by Lemma 5.2, h satisfies Hörmander's condition also uniformly in ε, N. Therefore, by Theorem 3.1, $T_{\varepsilon,N}f \in L^p(R^n)$, $1 < p < \infty$, and there is a constant $c = c_p$ so that (5.4) holds with $T_\varepsilon f$ replaced by $T_{\varepsilon,N}f$. Now, since $T_\varepsilon f(x) = T_{\varepsilon,N}f(x)$ for every x in R^n for $N \geqslant N_0(\operatorname{supp} f)$, it readily follows by Fatou's lemma that (5.4) holds. In fact, a simple limiting argument shows that (5.4) holds for arbitrary $L^p(R^n)$ functions as well.

Next, and to insure that Tf exists, we verify that $\{T_\varepsilon f\}$ is Cauchy in $L^p(R^n)$ for each f in $C_0^\infty(R^n)$. Let $\varepsilon > \eta > 0$ be given and observe that

$$\begin{aligned} T_\eta f(x) - T_\varepsilon f(x) &= \int_{\eta<|y|\leqslant\varepsilon} k(y)f(x-y)\,dy \\ &= \int_{\eta<|y|\leqslant\varepsilon} k(y)(f(x-y)-f(x))\,dy \\ &\quad + f(x)\int_{\eta<|y|\leqslant\varepsilon} k(y)\,dy \\ &= I + J, \end{aligned} \tag{5.5}$$

say. Moreover, since in our case $\int_{R^n}|f(x-y)-f(x)|^p\,dx \leqslant c|y|^p$, by Minkowski's inequality it readily follows that

$$\|I\|_p \leqslant c\int_{|y|\leqslant\varepsilon}|y||k(y)|\,dy \leqslant cc_2\varepsilon = o(1) \quad \text{as} \quad \varepsilon\to 0.$$

As for the J term, note that by property (i) of CZ kernels $\lim_{\eta\to 0}\int_{\eta<|y|<1}k(y)\,dy$ exists, and consequently $\lim_{\varepsilon,\eta\to 0}|\int_{\eta<|y|\leqslant\varepsilon}k(y)\,dy| = 0$. Whence

$$\|J\|_p \leqslant \left|\int_{\eta<|y|\leqslant\varepsilon}k(y)\,dy\right|\|f\|_p = o(1) \quad \text{as} \quad \eta,\varepsilon\to 0,$$

and our claim is proved. Let $Tf = \lim_{\varepsilon\to 0}T_\varepsilon f$ denote the L^p limit whose existence we have just proved; clearly, $\|Tf\|_p \leqslant c\|f\|_p$ as well. To show that the same holds for an arbitrary $L^p(R^n)$ function f, put $f = g + h$, where g is in $C_0^\infty(R^n)$ and $\|h\|_p \leqslant \delta$ is arbitrarily small. Then $\|T_\eta f - T_\varepsilon f\|_p \leqslant \|T_\eta(f-g)\|_p + \|T_\eta g - T_\varepsilon g\|_p + \|T_\varepsilon(f-g)\|_p \leqslant 2c\delta + o(1)$ as $\varepsilon,\eta\to 0$. Since δ is arbitrary, $\{T_\varepsilon f\}$ is also Cauchy in $L^p(R^n)$, $\lim_{\varepsilon\to 0}T_\varepsilon f = Tf$ exists in L^p and it also verifies (5.4). ■

As for $p = 1$ we have

Theorem 5.5 (Calderón–Zygmund). Suppose the CZ kernel k verifies the following additional property:

(iv) If x_1, x_2, x_3, y are such that $|x_1-x_2|, |x_2-x_3|, |x_1-x_3| \leqslant R/2$ and $|x_1-y|, |x_2-y|, |x_3-y| \geqslant R$, then

$$|k(x_1-y)-k(x_2-y)| \leqslant c_4\frac{|x_1-x_2|}{|x_3-y|^{n+1}}, \tag{5.6}$$

where c_4 is independent of the points involved and R.

Then the mapping $T_\varepsilon f$ defined in Theorem 5.4 is of weak-type (1, 1), with norm independent of ε, $\lim_{\varepsilon\to 0}T_\varepsilon f = Tf$ exists in measure and in the pointwise sense a.e. Furthermore,

$$\lambda|\{|T_\varepsilon f| > \lambda\}| \leqslant c\|f\|_1, \tag{5.7}$$

c independent of λ, ε, f, and

$$\lambda|\{|Tf| > \lambda\}| \leqslant c\|f\|_1, \tag{5.8}$$

with c independent of λ, f.

Proof. Assumption (iv) insures that the proof given in Theorems 1.1 and 2.12 in Chapter V for the Hilbert transform also works in this case with minor adjustments. Another way to go about this is to observe that the estimate (5.6) insures that hypothesis (ii) of Theorem 1.1 holds, and consequently that result also applies. (Also cf. Theorem 3.1 in Chapter XIII.) ∎

There still remains the question of the pointwise convergence of the truncated CZ singular integral operators. This requires the simultaneous control of $T_{\varepsilon,N}f$, and is achieved in the next section with the consideration of the maximal singular integral operators.

6. MAXIMAL CALDERÓN-ZYGMUND SINGULAR INTEGRAL OPERATORS

For a CZ kernel k let

$$T^*f(x) = \sup_{0<\varepsilon<N} |k_{\varepsilon,N} * f(x)|. \tag{6.1}$$

T^* is called the maximal CZ singular integral operator (associated to k) and our aim is to show that under appropriate conditions the statements analogous to Theorems 5.4 and 5.5 hold for T^* as well.

Lemma 6.1 (Cotlar). Suppose the CZ kernel k verifies properties (i) through (iv) of Definition 5.1 and Theorem 5.5. Then for $0 < \eta < 1$ and Tf as in Theorem 5.4,

$$T^*f(x) \leq c(M(|Tf|^\eta)(x)^{1/\eta} + Mf(x)), \tag{6.2}$$

where $c = c_\eta$ is independent of $f \in C_0^\infty(R^n)$ and x.

Proof. As indicated above, for compactly supported functions f the limit as $N \to \infty$ of $T_{\varepsilon,N}f$ can be easily handled, so we may just consider $T_\varepsilon f$. Now given compactly supported functions f, g it is readily seen that there is a sequence $\varepsilon_j \to 0$ so that

$$\lim_{\varepsilon_j \to 0} T_{\varepsilon_j}f(w) = Tf(w) \qquad \text{and} \qquad \lim_{\varepsilon_j \to 0} T_{\varepsilon_j}g(w) = g(w) \tag{6.3}$$

exist simultaneously for almost every w in R^n. Fix x in R^n and $\varepsilon > 0$ and for a given f in $C_0^\infty(R^n)$ let $g(y) = f(y)\chi_{B(x,\varepsilon)}(y)$ be the restriction of f to the ball centered at x of radius ε. Next observe that

$$\begin{aligned} T_\varepsilon f(x) &= \int_{|x-y|>\varepsilon} (k(x-y) - k(w-y))f(y)\,dy \\ &\quad + \left(\int_{|x-y|>\varepsilon} k(w-y)f(y)\,dy - \int_{|w-y|>\varepsilon_j} k(w-y)f(y)\,dy\right) \\ &\quad + \int_{|w-y|>\varepsilon_j} k(w-y)f(y)\,dy = I + J + K, \end{aligned} \tag{6.4}$$

say. In (6.4) we have chosen w to be a point $B(x, \varepsilon/4)$ where (6.3) holds. Clearly, by (the first limit in) (6.3) $\lim_{\varepsilon_j \to 0} K = Tf(w)$. As for the J term, since $w \in B(x, \varepsilon/4)$ and $\varepsilon_j \to 0$, upon choosing ε_j sufficiently small it follows that $B(w, \varepsilon_j) \subset B(x, \varepsilon)$, and, consequently, for those ε_j's $f(y)\chi_{R^n \setminus B(x,\varepsilon)}(y) = f(y)\chi_{R^n \setminus B(x,\varepsilon)}(y)\chi_{B(w,\varepsilon_j)}(y)$. Thus J may be rewritten as $-\int_{|w-y|>\varepsilon_j} k(w-y)g(y)\,dy$ and by (the second limit in) (6.3) $\lim_{\varepsilon_j \to 0} J = -T(f\chi_{B(x,\varepsilon)})(w)$. Since I is independent of ε_j, by letting $\varepsilon_j \to 0$ in (6.4), we obtain that for almost every w in $B(x, \varepsilon/4)$

$$\begin{aligned} T_\varepsilon f(x) = \int_{|x-y|>\varepsilon} &(k(x-y) - k(w-y))f(y)\,dy \\ &- T(f\chi_{B(x,\varepsilon)})(w) + Tf(w) = I + J + K, \end{aligned} \tag{6.5}$$

say. Now, by property (iv) of k, it follows that

$$|I| \le c \int_{|x-y|>\varepsilon} \frac{|x-w|}{|x-y|^{n+1}} |f(y)|\,dy \le c\phi_\varepsilon * |f|(x),$$

where $\phi(x)$ is the radial, decreasing, integrable function $(1+|x|)^{-(n+1)}$, $\phi_\varepsilon(x) = \varepsilon^{-n}\phi(x/\varepsilon)$, and by Proposition 2.3 in Chapter IV, $|I| \le cMf(x)$. Whence by (6.5) we see that

$$|T_\varepsilon f(x)| \le cMf(x) + |T(f\chi_{B(x,\varepsilon)})(w)| + |Tf(w)| \tag{6.6}$$

for almost every w in $B(x, \varepsilon/4)$. Now, if $T_\varepsilon f(x) = 0$, then clearly $|T_\varepsilon f(x)|$ is bounded by the right-hand side of (6.2). Otherwise, let $0 < \lambda < |T_\varepsilon f(x)|$, put $B(x, \varepsilon/4) = B$ and introduce $E_1 = \{w \in B\colon |Tf(w)| > \lambda/3\}$, $E_2 = \{w \in B\colon |T(f\chi_{B(x,\varepsilon)})(w)| > \lambda/3\}$ and $E_3 = \varnothing$ if $cMf(x) \le \lambda/3$ and $= B$ if $cMf(x) > \lambda/3$ (c is the constant in (6.6)). From (6.6) it is clear that $B = E_1 \cup E_2 \cup E_3$, and we pass now to estimate the measure of these sets. In the first place by Chebychev's inequality

$$\left(\frac{\lambda}{3}\right)^\eta |E_1| \le \int_{E_1} |Tf(w)|^\eta\,dw, \qquad 0 < \eta < 1. \tag{6.7}$$

Similarly, and invoking Kolmogorov's inequality 7.19 in Chapter IV and Theorem 5.5, we see that

$$\begin{aligned} \left(\frac{\lambda}{3}\right)^\eta |E_2| &\le \int_{E_2} |T(f\chi_{B(x,\varepsilon)})(w)|^\eta\,dw \\ &\le c|E_2|^{1-\eta} \qquad (\text{wk-}L \text{ norm of } T(f\chi_{B(x,\varepsilon)}))^\eta \\ &\le c|E_2|^{1-\eta} \|f\chi_{B(x,\varepsilon)}\|_1^\eta, \end{aligned}$$

and since $|E_2|$ is finite, we also have

$$\lambda|E_2| \le c \int_{B(x,\varepsilon)} |f(y)|\,dy. \tag{6.8}$$

As for E_3, it either equals B, and in this case $Mf(x)/\lambda > c > 1$, or it is empty. In either case,

$$|E_3| \leq |B| Mf(x)/\lambda. \tag{6.9}$$

Thus combining (6.7), (6.8) and (6.9) it follows that

$$|B| \leq c\lambda^{-\eta} \int_B |Tf(w)|^\eta \, dw + c\lambda^{-1} \int_{B(x,\varepsilon)} |f(y)| \, dy + \frac{|B| Mf(x)}{\lambda},$$

which, multiplying through by $\lambda/|B|$, in turn gives

$$\begin{aligned} \lambda &\leq c\lambda^{1-\eta} \frac{1}{|B|} \int_B |Tf(w)|^\eta \, dw + c\frac{1}{\varepsilon^n} \int_{B(x,\varepsilon)} |f(y)| \, dy + Mf(x) \\ &\leq c(\lambda^{1-\eta} M(|Tf|^\eta)(x) + Mf(x)). \end{aligned} \tag{6.10}$$

Now, since from the inequality $0 < \lambda \leq c_1 \lambda^{1-\eta} + c_2$ we readily get that $\lambda \leq \max((2c_1)^{1/\eta}, 2c_2) \leq (2c_1)^{1/\eta} + 2c_2$, by (6.10) we obtain $\lambda \leq c(M(|Tf|^\eta)(x)^{1/\eta} + Mf(x))$, and (6.2) follows at once from this. ■

Theorem 6.2. Assume the CZ kernel k verifies the assumption of Lemma 6.1. Then

$$\|T^* f\|_p \leq c\|f\|_p, \qquad 1 < p < \infty, \tag{6.11}$$

where $c = c_p$ is independent of f.

Proof. For f in $C_0^\infty(R^n)$, (6.11) follows at once from Lemma 6.1 and the maximal theorem. A by now well-known density argument gives the same result for arbitrary f in $L^p(R^n)$, $1 < p < \infty$. ■

Theorem 6.3. Assume the CZ kernel k verifies the assumptions of Lemma 6.1. Then

$$\lambda |\{T^* f > \lambda\}| \leq c\|f\|_1, \qquad \lambda > 0, \tag{6.12}$$

for a constant c independent of λ and f.

Proof. One way to prove (6.12) is to repeat the argument of Theorem 2.14 in Chapter V. Another way is to note that on account of Lemma 6.1, it is enough to show that for f in $C_0^\infty(R^n)$,

$$\lambda|\mathcal{O}_\lambda| = \lambda |\{M(|Tf|^\eta)(x) > \lambda^\eta\}| \leq c\|f\|_1, \qquad \lambda > 0. \tag{6.13}$$

Now, by 8.2 in Chapter IX, we have that $|\mathcal{O}_\lambda| \leq c\lambda^{-\eta} \int_{\mathcal{O}_\lambda} |Tf(x)|^\eta \, dx < \infty$, and consequently by Kolmogorov's inequality $|\mathcal{O}_\lambda| \le c\lambda^{-\eta}|\mathcal{O}_\lambda|^{1-\eta}(\text{wk-}L$ norm $|Tf|)^\eta \le c\lambda^{-\eta}|\mathcal{O}_\lambda|^{1-\eta}\|f\|_1^\eta$, and we are done. ∎

We are now in a position to prove the existence of the pointwise limit of the truncated Calderón–Zygmund singular integral operators.

Theorem 6.4. Suppose the CZ kernel k verifies the assumptions of Lemma 6.1. Then $\lim_{\varepsilon\to 0} T_\varepsilon f(x)$ exists a.e. for each f in $L^p(R^n)$, $1<p<\infty$, is denoted by p.v. $k * f(x)$, and it coincides with $Tf(x)$, the norm limit.

Proof. For real valued k and f let $\Delta(f)(x) = \limsup_{\varepsilon\to 0} T_\varepsilon f(x) - \liminf_{\varepsilon\to 0} T_\varepsilon f(x) \ge 0$. Since we can write $f = g + h$, $g \in C_0^\infty(R^n)$ and $\|h\|_p \le \delta$, δ arbitrary, we also have $\Delta(f)(x) = \Delta(h)(x)$, since by (5.5) (and making use of (ii) in Definition (5.1)) it readily follows that $\Delta(g)(x) = 0$ everywhere. Thus, $\Delta(f)(x) \le 2T^*h(x)$, and, consequently, for each $\lambda > 0$ $|\{\Delta(f) > \lambda\}| \le |\{T^*h > \lambda/2\}| \le c\lambda^p\delta$. Since δ is arbitrary we immediately get that $|\{\Delta(f) > \lambda\}| = 0$ for each $\lambda > 0$. Thus $\Delta(f)(x) =$ a.e.; in other words $\limsup_{\varepsilon\to 0} T_\varepsilon f(x) = \liminf_{\varepsilon\to 0} T_\varepsilon f(x)$ a.e. and the limit exists for each f in $L^p(R^n)$. Furthermore, since it coincides with $Tf(x)$ for f in $C_0^\infty(R^n)$, a simple argument along the lines of Theorem 2.2 in Chapter IV shows that the same is true for an arbitrary f in $L^p(R^n)$ and we have finished. ∎

7. SINGULAR INTEGRAL OPERATORS IN $L^\infty(R^n)$

As in the periodic case the class *BMO* arises as the image of L^∞ under CZ singular integral operators.

Theorem 7.1. Suppose the CZ kernel k verifies the assumptions of Theorem 5.5, let $k_\varepsilon(x) = k(x)$ if $|x| > \varepsilon$ and 0 otherwise, and for f in $L^\infty(R^n)$ set

$$K_\varepsilon f(x) = \int_{R^n} (k_\varepsilon(x-y) - k_1(-y))f(y)\,dy. \tag{7.1}$$

Then $\lim_{\varepsilon\to 0} K_\varepsilon f(x)$ exists a.e. in x and is a *BMO* function; more precisely, there is a constant c independent of f such that

$$\|Kf\|_* \le c\|f\|_\infty. \tag{7.2}$$

Proof. Observe first that, on account of property (iv) of CZ kernels, the integral in (7.1) converges absolutely for each ε and x; in fact, it is precisely for this reason that the term $-k_1(-y)$ was introduced. Also note that if f is compactly supported, then $K_\varepsilon f(x)$ differs from $T_\varepsilon f(x)$ by a constant. Moreover, well-known arguments by now show that $\lim_{\varepsilon\to 0} K_\varepsilon f(x)$ exists in the L^2 norm on each finite cube as well as pointwise a.e. in R^n.

Next consider $K_\varepsilon f(x) - K_N f(x) = k_{\varepsilon,N} * f(x)$, where by Lemma 5.2 the integrable function $k_{\varepsilon,N}$ is a CZ kernel with constants uniformly bounded, independently of ε and N. It is not hard to see that $\|k_{\varepsilon,N} * f\|_* \leq c\|f\|_\infty$, with c independent of ε, N, and f; in fact, the proof of this estimate is similar to that of Theorem 3.1 in Chapter VIII and is therefore omitted. Thus there is a constant c, independent of ε, N, and f, so that

$$\frac{1}{|I|}\int_I |K_\varepsilon f(x) - K_N f(x) - (K_\varepsilon f)_I + (K_N f)_I|\,dx \leq c\|f\|_\infty, \qquad \text{all} \quad I. \quad (7.3)$$

Next observe that if $c_N = \int_{R^n}(k_N(-y) - k_1(-y))f(y)\,dy$, then $K_N f(x) - c_N = \int_{R^n}(k_N(x-y) - k_N(-y))f(y)\,dy$ tends to 0 uniformly as $N \to \infty$. Moreover, since the inequality in (7.3) remains unchanged if we replace $K_N f$ by $K_N f - c_N$, we get, by first letting $N \to \infty$ and then $\varepsilon \to 0$,

$$\frac{1}{|I|}\int_I |Kf(x) - Kf_I|\,dx \leq c\|f\|_\infty, \qquad \text{all} \quad I. \quad ■$$

8. NOTES; FURTHER RESULTS AND PROBLEMS

Because of the many applications to other branches of analysis and PDE's, the Calderón–Zygmund theory of singular integral operators plays a basic role in harmonic analysis and lies at the heart of much of the work being done in this area nowadays. Although other mathematicians, most notably Giraud and Mikhlin, obtained n-dimensional results, it was only through the techniques introduced by Calderón and Zygmund that the complete picture began to emerge. The classic 1952 *Acta Mathematica* and 1956 *American Journal of Mathematics* papers make inspiring reading and, even though they are quite well understood by now, there is yet much to be learned about the precise meaning of the various conditions discussed there. For instance, only recently Calderón–Zygmund [1979] showed that in the case of kernels of the form $k(x) = \Omega(x')/|x|^n$, Ω homogeneous of degree 0, the Hörmander condition (iii) of Definition 5.1 is actually equivalent to the following one: For a proper rotation ρ of R^n about the origin put $|\rho| = \sup|x' - \rho x'|$. Then if $w_1(t) = \sup_{|\rho|\leq t}\int_\Sigma |k(\rho x') - k(x')|\,dx'$, $\int_{[0,1]} w_1(t)/t\,dt < \infty$. Also Calderón and Capri [1984] have shown that if T is as in Theorem 5.4 and f and Tf are both integrable, then $\lim_{\varepsilon\to 0}\|T_\varepsilon f - Tf\|_1 = 0$.

The versatility of the methods discussed in this chapter is apparent in the consideration of the so-called CZ operators, corresponding to "variable" kernels (also cf. Calderón and Zygmund [1978]), and oscillatory singular integrals.

Further Results and Problems

8.1 De Guzmán [1981] observed that the method of rotations applies to the Hardy–Littlewood maximal operator as well. In other words, assume that the 1-dimensional maximal operator is bounded in L^p, $1 < p < \infty$, and prove that the same is true for the n-dimensional maximal operator. What can be said for the case $p = 1$?

8.2 Suppose $k(x) = \Omega(x')/|x|^n$, Ω homogeneous of degree 0 and $\int_\Sigma \Omega(x')\,dx' = 0$, is a CZ kernel and show that

$$\hat{k}(\xi) = \int_\Sigma \Omega(x')\left(\ln|1/\cos\phi| - i\frac{\pi}{2}\operatorname{sgn}(\cos\phi)\right) dx',$$

where ϕ denotes the angle between ξ and x'. (*Hint*: In polar coordinates we have

$$\hat{k}(\xi) = \lim_{\varepsilon\to 0}\int_\Sigma \Omega(x')\int_{[\varepsilon,\infty)} e^{-ir\cos\phi}\frac{dr}{r}\,dx' = R + iI,$$

say. I is easily computed; as for R note that

$$\int_{[\varepsilon,\infty)} \cos(r\cos\phi)\frac{dr}{r} = \int_{[\varepsilon|\cos\phi|,\varepsilon)} \frac{dr}{r} + \int_{[\varepsilon,1)} \frac{dr}{r}$$
$$+ \int_{[\varepsilon\cos|\phi|,1)} (\cos r - 1)\frac{dr}{r} + \int_{[1,\infty)} \cos r\frac{dr}{r},$$

and take the limit as $\varepsilon \to 0$. The fact that $\int_\Sigma \Omega(x')\,dx' = 0$ is crucial.)

8.3 Suppose the kernel k verifies (ii) in Definition 5.1 and the mappings $T_\varepsilon f(x) = k_\varepsilon * f(x)$ are bounded in some $L^p(R^n)$, $1 < p < \infty$, uniformly in ε. Then for each $0 < \varepsilon < N$, $|\int_{\varepsilon<|x|<N} k(x)\,dx| \le c$, where c is a constant independent of ε and N. If in addition $k(x) = \Omega(x')/|x|^n$, $\Omega(x')$ homogeneous of degree 0, then also $\int_\Sigma \Omega(x')\,dx' = 0$. Furthermore, if $\lim_{\varepsilon\to 0} T_\varepsilon f(x)$ exists a.e. for f in $C_0^\infty(R^n)$ say, then (i) of Definition 5.1 holds. (*Hint*: If $\varepsilon > N/4$, then the conclusion follows from (ii) in Definition 5.1. On the other hand, if $\varepsilon < N/4$ then let $f = \chi_B$, where B is the ball centered at the origin of radius N. Then $T_\varepsilon f(x) - T_\varepsilon f(0) = \int_{R^n}(\chi_B(x+y) - \chi_B(y))k_\varepsilon(-y)\,dy$, and the integrand vanishes if either $|y| \le \varepsilon$ or $\chi_B(x+y) = \chi_B(y)$. For $|x| \le N/8$, this implies that $|T_\varepsilon f(x) - T_\varepsilon f(0)| \le \int_{N/8\le|y|<8N}|k(y)|\,dy \le c$ and consequently $|T_\varepsilon f(0)| \le c + |T_\varepsilon f(x)|$. Averaging the pth power of this estimate over $|x| \le N/8$ gives $|T_\varepsilon f(0)| \le c + c((1/N^n)\|f\|_p)^{1/p} \le c$, which in turn gives the desired conclusion at once.

In case $k(x) = \Omega(x')/|x|^n$, this implies that $|\int_\Sigma \Omega(x')\,dx'|\ln(N/\varepsilon)$ is bounded uniformly in ε and N and consequently $\int_\Sigma \Omega(x')\,dx' = 0$. The last assertion follows upon writing, for f in $C_0^\infty(R^n)$,

$$T_\varepsilon f(x) = \int_{\varepsilon \le |y| \le N} (f(x-y) - f(x))k(y)\,dy + f(x)\int_{\varepsilon \le |y| \le N} k(y)\,dy$$
$$= I + J,$$

say, and noting that I converges since the integrand there is absolutely summable. Results of this nature, and more general ones, are discussed by Jurkat and Sampson [1979].)

8.4 Suppose ϕ is integrable and $|\phi| \le k$, where k is integrable and $|\nabla k(x)| \le c/|x|^{n+1}$. Then the mapping $Tf(x) = \sup_{\varepsilon>0}|\phi_\varepsilon * f(x)|$ is of weak-type (1, 1) and if f is integrable then $\lim_{\varepsilon\to\infty} \phi_\varepsilon * f(x) = (\int_{R^n} \phi(y)\,dy)f(x)$ a.e. (Hint: Since $|\phi_\varepsilon * f(x)| \le k_\varepsilon * f(x)$, it suffices to show the weak-type estimate for ϕ replaced by k. By Theorem 2.2 it is enough to verify that

$$\int_{|x|\ge 2|y|} \sup_{\varepsilon>0}\left(\frac{1}{\varepsilon^n}\left|k\left(\frac{(x-y)}{\varepsilon}\right) - k\left(\frac{x}{\varepsilon}\right)\right|\right) dx \le c.$$

This estimate follows at once from the mean value theorem. Zó [1978] extended this result to the setting $|\phi(x)| \le k_1(|x_1|)k_2(|x_2|)$, where $x = (x_1, x_2) \in R^n$, $x_1 \in R^{n_1}$ denote the first n_1 components of x and $x_2 \in R^{n_2}$ the last n_2.)

8.5 Assume ϕ is integrable in R and $|\phi| \le k$, where supp $k \subseteq [0, 1]$, k is nondecreasing and $k \in L\ln^+ L[0, 1]$. Furthermore, let $\{\lambda_n\}$ be a lacunary sequence, and put $Tf(x) = \sup_n|k_{\lambda_n} * f(x)|$. Then T is of weak-type (1, 1) and if f is locally integrable it also follows that $\lim_{n\to\infty} k_{\lambda_n} * f(x) = (\int_{[0,1]} k(y)\,dy)f(x)$ a.e. By means of Lemma 6.4 in Chapter IV it is possible to construct a nonlacunary sequence $\{\lambda_n\}$ which increases to ∞ and a locally integrable function f so that for k as above $\limsup_{n\to\infty} k_{\lambda_n} * f(x) = +\infty$ a.e. (*Hint*: To show that T is of weak-type (1, 1) it suffices to verify that assumption (ii) in Theorem 2.2 is satisfied. This result is also from Zó's work [1976].)

8.6 Assume $k(x) = \Omega(x')/|x|^n$ is an odd, homogeneous CZ kernel of degree $-n$, so that $\int_\Sigma|\Omega(x')|\,dx' < \infty$, and put $h(x) = b(|x|)k(x)$, where b is a Fourier-Stieltjes transform. Let $T_1 f(x) = \text{p.v. } h * f(x)$ and show that T_1 verifies the same L^p continuity properties as $Tf(x) = \text{p.v. } k * f(x)$ does. (*Hint*: Let $g(t)$ be in $L^p(R)$, $1 < p < \infty$, and $\varepsilon(s)$ be an arbitrary, positive, measurable function in R. Then by the results concerning the Hilbert transform in Section 3 of Chapter X, the expression $e^{-irs}\int_{|s-t|>\varepsilon(s)} e^{irt}g(t)/(s-t)\,dt$ represents a function with L^p norm $\le c\|g\|_p$.

Thus if $\mu(r)$ is a function of bounded variation, from Minkowski's integral inequality it readily follows that the L^p norm of the function (of s) given by

$$\int_R e^{-irs} \int_{|s-t|>\varepsilon(s)} \frac{e^{irt}g(t)}{(s-t)}\,dt\,d\mu(r) \leqslant c(\text{variation of } \mu)\|g\|_p.$$

Now interchange the order of integration in the expression above and observe that since $\varepsilon(s)$ is arbitrary, we may conclude that, if

$$g^*(s) = \sup_\varepsilon \left| \int_{|s-t|>\varepsilon} \frac{b(s-t)}{s-t} g(t)\,dt \right|, \qquad b(r) = \hat{\mu}(r),$$

then also $\|g^*\|_p \leqslant c\|g\|_p$. To pass to the n-dimensional statement it suffices now to invoke the method of rotations, i.e., to use an argument similar to Theorem 3.1 in Chapter X; this result is from Calderón–Zygmund's work [1956]. R. Fefferman [1979] observed that even in the case when b is an arbitrary bounded function the following is true: suppose that for each $r > 0$ we are given a function Ω_r on Σ in such a way that the family $\{\Omega_r\}$ is uniformly in the Dini class; i.e., if $w(t) = \sup\{|\Omega_r(x') - \Omega_r(y')|: |x' - y'| < t,\ r > 0\}$, then $\int_{[0,1]} w(t)/t\,dt < \infty$; and also $\int_\Sigma \Omega_r(x')\,dx' = 0$. Let $h(x) = \Omega_{|x|}(x')/|x|^n$, then $\|\text{p.v.}\,h * f\|_2 \leqslant c\|f\|_2$. If in addition the Dini condition is replaced by a Lipschitz condition of some positive order, then also $\|\text{p.v.}\,h * f\|_p \leqslant c\|f\|_p$, $1 < p < \infty$. The case $p = 2$ amounts to showing that $\hat{h}$ is bounded, but things get complicated when $p \neq 2$ since in general nothing can be said about integrals like $\int_{|x|\geqslant 2|y|} |h(x-y) - h(x)|\,dx$. The proof relies then on the complex method of interpolation. Namazi [1984] relaxed the Lipschitz condition and characterized those b's for which the corresponding mapping is bounded from $L^\infty(R^n)$ into $BMO(R^n)$. Also Shi [1985] proved results for more general CZ operators.)

8.7 Suppose that ϕ is a nonnegative function, supp $\phi \subseteq \{|x| \leqslant 1\}$, such that $\int_{R^n} \phi(x)\,dx = 1$. Let $\varepsilon > 0$ and put $\phi_\varepsilon(x) = \varepsilon^{-n}\phi(x/\varepsilon)$ and define $\delta_\varepsilon(x) = T\phi_\varepsilon(x) - k_\varepsilon(x)$, where k is a CZ singular integral kernel and T is the CZ operator associated to k. Show there exists a constant $c > 0$ such that $\|\delta_\varepsilon\|_1 \leqslant c$, uniformly in ε. (*Hint*: Suppose first $|x| \geqslant 2\varepsilon$ and observe that by the Lebesgue dominated convergence theorem $T\phi_\varepsilon(x) = \int_{|y|\leqslant\varepsilon} k(x-y)\phi_\varepsilon(y)\,dy$. Consequently, $\delta_\varepsilon(x) = \int_{|y|\leqslant\varepsilon}(k(x-y) - k(x))\phi_\varepsilon(y)\,dy$ and the appropriate estimate for $\int_{|x|\geqslant 2\varepsilon} |\delta_\varepsilon(x)|\,dx$ follows from Fubini's theorem and condition (iii) in Definition 5.1. On the other hand

$$\int_{|x|\leqslant 2\varepsilon} |\delta_\varepsilon(x)|\,dx \leqslant \int_{|x|\leqslant 2\varepsilon} |T\phi_\varepsilon(x)|\,dx + \int_{\varepsilon<|x|\leqslant 2\varepsilon} |k(x)|\,dx = I + J,$$

say. To estimate I we invoke the fact that T is bounded in L^2, and to bound J we use condition (ii) in Definition 5.1. This represents the first step in showing that, if f, $Tf \in L(R^n)$, then $\lim_{\varepsilon\to 0}\|T_\varepsilon f - Tf\|_1 = 0$ (cf. Calderón and Capri [1984]).)

8.8 There is yet another way to estimate $T^*f(x)$ under somewhat weaker conditions than those of Lemma 6.1. Rather than assuming condition (iv) of CZ kernels in Theorem 5.5, suppose that k verifies for $|x-y|>\varepsilon$

$$\int_{|x-w|\le\varepsilon/4} |k(x-y)-k(w-y)|\,dw \le \phi\left(\frac{(x-y)}{\varepsilon}\right),$$

where ϕ satisfies the assumptions of Corollary 2.3. Show that under this hypothesis $T^*f(x) \le c(Kf(x)+M(|f|^r)(x)^{1/r}+M(Tf)(x))$, where K is a mapping of weak-type (1, 1) and type (p, p) for $1<p<\infty$, and $1<r<\infty$. (*Hint*: We proceed as Lemma 6.1 and integrate (6.5) over w in $B(x, \varepsilon/4)$ to obtain

$$\begin{aligned}|T_\varepsilon f(x)| &\le c\Big(\int_{|x-y|>\varepsilon} |f(y)|\varepsilon^{-n}\int_{|x-w|\le\varepsilon/4} |k(x-y)-k(w-y)|\,dw\,dy \\ &\quad + \varepsilon^{-n}\int_{|x-w|\le\varepsilon/4} |T(f\chi_{B(x,\varepsilon)})(w)|\,dw + \varepsilon^{-n}\int_{|x-w|\le\varepsilon/4} |Tf(w)|\,dw\Big) \\ &= A+B+C,\end{aligned}$$

say. Clearly, $A \le \sup_\varepsilon |f| * \phi_\varepsilon(x) = Kf(x)$, and by Corollary 2.3, K is as it should be. B is estimated by applying Hölder's inequality and Theorem 5.4 and the bound for C is immediately.)

8.9 Let T_z be a z-weakly measurable and uniformly bounded family of operators in $L^2(R^n)$, $\|T_z\| \le M$ for all z in a measure space (Z, dz). If the inequalities $\|T_zT_{z'}^*\|$, $\|T_z^*T_{z'}\| \le h^2(z, z')$ hold with a function $h(z, z')$ which is the kernel of a bounded integral operator H in $L^2(R^n)$ with norm A, then the operator $T=\int T_z\,dz$ is bounded in $L^2(R^n)$ with norm $\|T\| \le A$. This variant of Cotlar's lemma is crucial in proving the L^2 continuity of pseudo-differential operators (cf. Calderón and Vaillancourt [1971]).

8.10 Cotlar's lemma is quite useful when the Fourier transform is not available. For instance, for $x \ne y$ in R let the kernel $k(x, y)$ verify the following conditions: (i) $D^\alpha((x-y)k(x, y)) \to 0$ at infinity for each α, (ii) $|k(x, y)| \le 1/|x-y|$, (iii) $|(\partial/\partial x)(k(x, y)| \le 1/|x-y|^2$, (iv) $k(x, y) = -k(y, x)$ and (v) p.v. $\int_{|x-y|\le 4^m} k(x, y)\,dy = 0$ for each integer , and $x \in R$. Then the operator $Tf(x) = \lim_{\varepsilon\to 0}\int_{|x-y|>\varepsilon} k(x, y)f(y)\,dy$ is bounded in $L^2(R)$. (*Hint*: For each integer j put χ_j = characteristic function of $\{4^j \le |x-y| \le 4^{j+1}\}$, $k_j(x, y) = k(x, y)\chi_j$ and denote by T_j the integral operator corresponding to k_j. Since for $f \in C_0^\infty(R)\|\sum_{j=-N}^N T_jf - Tf\|_2 \to 0$ as $N \to \infty$ it suffices to show that $\|\sum_{j=-N}^N T_j\|$ is bounded, uniformly in N. By (a variant of) Cotlar's lemma 4.1, and since on account of (iv) $T_j^* = -T_j$ and since the norm of an operator and that of its adjoint coincide, it is enough to verify that for $m \le j$, $\|T_jT_m\| \le a(j-m)^2$, where $\sum a(j) < \infty$. To do this we invoke the well-known and readily verified fact that if $s(x, y)$ is locally integrable in R^2 and $\int_R |s(x, y)|\,dy, \int_R |s(x, y)|\,dx \leq 1$ (a.e.) then the operator with kernel $s(x, y)$ is bounded in $L^2(R)$ with norm ≤ 1. This gives at once that $\|T_jT_j\| = \|T_j\|^2 \leq 100$. On the other hand, if $m < j$, then the kernel $s_{j,m}(x, y)$ of T_jT_m is given by $\int_R k_j(x, t)k_m(y, t)\,dt$ and the following estimates hold:

$$\int_R k_m(y, t)\,dt = 0, \qquad |k_m(y, t)| \leq 4^{-m},$$

$$\left|\frac{\partial}{\partial t}k_j(x, t)\right| \leq 16^{-j}, \qquad \text{and} \qquad |k_j(x, t)| \leq 4^{-j}.$$

This allows us to observe that $|s_{j,m}(x, y)| \leq a_{j,m}(x - y)$, where $a_{j,m}(t) = 0$ if $|t| > 4^{j+1} + 4^{m+1}$, $a_{j,m}(t) \leq 16.4^{-j}$ if $|t \pm 4^{j+1}| \leq 4^{m+1}$, $a_{j,m}(t) \leq 4^{-2j+m}$ if $4^j + 4^{m+1} \leq |t| \leq 4^{j+1} - 4^{m+1}$, finally $a_{j,m}(t) = 0$ if $|t| < 4^j - 4^{m+1}$, $a_{j,m}(t) \leq 16.4^{-j}$ if $|t + 4^j| \leq 4^{m+1}$. Consequently, $\int_R a_{j,m}(t)\,dt \leq 50.4^{-j+m} = a(j - m)$, which is precisely what we wanted to show. This result is from David's work [1982].)

8.11 Motivated by 8.10 and the results in this section we propose the following definitions: we say that a possibly complex-valued, continuous function $k(x, y)$ defined for $x \neq y$ is a Calderón-Zygmund kernel provided the following two conditions are satisfied:

(i) $|k(x, y)| \leq c|x - y|^{-n}$.

(ii) The distributional derivatives $D^\alpha k(x, y)$, $|\alpha| = 1$, coincide with locally bounded functions in $x \neq y$ and verify $|D^\alpha k(x, y)| \leq |x - y|^{-n-1}$.

Associated to k we define the operator T by means of the formula $Tf(x) =$ p.v. $\int_{R^n} k(x, y)f(y)\,dy = \lim_{\varepsilon \to 0} \int_{|x-y|>\varepsilon} k(x, y)f(y)\,dy$, $f \in C_0^\infty(R^n)$, and we say that T is a Calderón-Zygmund operator provided T admits a continuous extension to $L^2(R^n)$. As expected we denote $T_\varepsilon f(x) = \int_{|x-y|>\varepsilon} k(x, y)f(y)\,dy$ and $T^*f(x) = \sup_\varepsilon |T_\varepsilon f(x)|$. Prove that Calderón-Zygmund operators are bounded on $L^p(R^n)$, $1 < p < \infty$, and map $L(R^n)$ into wk-$L(R^n)$. Also prove that the same result holds for T_ε and T^*. (*Hint*: The proofs in Section 5 apply to this setting as well; the notation, as well as the formulation of the next four results, is due to Coifman and Meyer [1978].)

8.12 Show that the conclusion of 8.11 holds under the following assumptions on k in place of (i) and (ii) there: (iii) there exist a constant c and a number $0 < \delta \leq 1$, such that, if $|y - y_0| \leq r$ and $|x - y_0| \geq 2r(x, y, y_0 \in R^n$, $r > 0)$ we have $|k(x, y)| \leq c|x - y|^{-n}$ and $|k(x, y) - k(x, y_0)|$, $|k(y, x) - k(y_0, x)| \leq cr^\delta |x - y_0|^{-n-\delta}$. In fact, Yabuta [1985] has shown that the following assumption on k suffices: $|k(x, y)| \leq c|x - y|^{-n}$ and $|k(x, y) - k(x, y_0)| + |k(y, x) - k(y_0, x)| \leq c|x - y_0|^{-n} w(|y - y_0|/|x - y_0|)$ for all x, y, y_0 with $2|y - y_0| < |x - y_0|$, where w is a nonnegative, nondecreasing function with $\int_{(0,1]} w(t)/t\,dt < \infty$.

8.13 Suppose T is a Calderón-Zygmund operator which verifies the assumptions of either 8.11 or 8.12. Then for each function f in $L^p(R^n)$, $1<p<\infty$, $\lim_{\varepsilon\to 0} T_\varepsilon f(x)$ exists and coincides a.e. with $Tf(x)$ defined as follows: $Tf=$ limit in L^p norm of $T\phi_k$ where $\phi_k \in C_0^\infty(R^n)$ and $\|f-\phi_k\|_p \to 0$.

8.14 If T is a Calderón-Zygmund operator and $f\in C_0^\infty(R^n)$, then for each x in R^n we have $Tf^{\#}(x) \leqslant c\|T\| M(|f|^p)(x)^{1/p}$, where $1<p<\infty$, $c=c_p$ is independent of x and f, and $\|T\|$ denotes the Calderón-Zygmund norm of T, i.e., the norm of T as bounded operator in $L^2(R^n)$, added to the smallest constants for which (i) and (ii), or (iii), hold. On the other hand, there is no constant c so that for every function f in $C_0^\infty(R) Hf^{\#}(x) \leqslant cMf(x)$, $x\in R$ holds.

8.15 Each Calderón-Zygmund operator T induces (canonically) a continuous mapping from $L^\infty(R^n)$ to $BMO(R^n)$. (*Hint*: Since $C_0^\infty(R^n)$ is not dense in $L^\infty(R^n)$, it is necessary to define directly T on $L^\infty(R^n)$. Each cube I in R^n with center and vertices with rational coordinates and rational radius is called a rational cube. Given a rational cube I let $f_1 = f\chi_{2I}$, $f_2 = f - f_1$ and for x, w in I put $F(x,w) = Tf_1(x) - Tf_1(w) + \int_{R^n} (k(x,y) - k(w,y)) f_2(y)\,dy$. It is not hard to see that $F(x,w)$ is well defined and that for almost every w, w' in R^n, $F(x,w) - F(x,w')$ is constant (regarded as a function of x). We then define $(Tf(x)$ as the class in BMO of $x \to F(x,w)$ (since for almost every w, $F(x,w)$ does not depend on w).)

8.16 Suppose the kernel $k(x,y)$ defined on $x\neq y$ verifies

$$\sup_x \sup_\eta \int_{|u|\leqslant 1}\int_{|v|\leqslant 1} |k_\eta(x+u, x+y) - k_\eta(x+v, x+y)|\,du\,dv \leqslant \phi(y),$$

where as usual $k_\eta(x,y) = \eta^{-n} k(x/\eta, y/\eta)$ and for $|y|\geqslant N$, some large value, $\phi(y)$ is a radial, nondecreasing, integrable function. Then if $Tf(x) = \text{p.v.} \int_{R^n} k(x,y) f(y)\,dy$ is of weak-type (1, 1), and $\alpha \leqslant \alpha_0$ is sufficiently small, $M_{0,\alpha}^{\#} Tf(x) \leqslant cMf(x)$, where $c = c_{\alpha,\phi}$ is a constant independent of x and f. (*Hint*: For convenience we assume that all maximal functions are centered, i.e., the sup at x is defined over all open cubes centered at x. Let then I be an open cube centered at x_0. For a fixed locally integrable function f we put $f_1 = f\chi_{n^{1/2}NI}$, $f_2 = f - f_1$. Since T is of weak-type (1, 1) we have $|\{y\in I: |Tf_1(y)| > \lambda\}| \leqslant c\|f_1\|_1/\lambda \leqslant c|n^{1/2}NI| Mf(x_0)/\lambda \leqslant \alpha|I|/2$, provided $\lambda > cMf(x_0)$. As for f_2, by Chebychev's inequality we have

$$\begin{aligned}
&|\{y\in I: |Tf_2(y) - (Tf_2)_I| > \lambda\}| \\
&\leqslant \int_I |Tf_2(y) - (Tf_2)_I|\,dy/\lambda \\
&\leqslant \int_{R^n} |f_2(y)| \left(\frac{1}{|I|}\int_I \int_I |k(x,y) - k(z,y)|\,dx\,dz\right) dy \Big/ \lambda.
\end{aligned}$$

We estimate the innermost integral A in the preceeding inequality. The cube I is obviously contained in a ball with the same center as I and with radius $\delta = \operatorname{diam} Q/2$. Hence by changing variables we see that

$$A \leq \frac{\delta^n}{|I|}\int_{|u|\leq 1}\int_{|v|\leq 1}\left|k_{\delta^{-1}}\left(u+\frac{x_0}{\delta}, y\right) - k_{\delta^{-1}}\left(v+\frac{x_0}{\delta}, y\right)\right| du\, dv$$

$$\leq \frac{\delta^n}{|I|}\phi\left(\frac{(y-x_0)}{\delta}\right) \leq c|I|\phi_\delta(y-x_0).$$

Whence the right-hand side in the above inequality is majorized by $c|I|\int_{R^n}|f_2(y)|\phi_\delta(x_0-y)\,dy/\lambda \leq c|I|Mf(x_0)/\lambda$, and this estimate is also of the right order provided that $\lambda > cMf(x_0)$. Since I is arbitrary our proof is complete. A similar argument shows that, if $\int_{|y|\geq N}\phi(y)\ln|y|\,dy < \infty$ instead, then the stronger inequality $M^{\#}_{0,\alpha}Tf(x) \leq cM^{\#}f(x)$ holds. These results, as well as related ones, are in Jawerth and Torchinsky's work [1985].)

8.17 Let $k \in \mathcal{S}'(R^n)$ have compact support and let $0 < \theta < 1$ be given. Further, suppose that k coincides with a locally integrable function away from the origin, that $\hat{k}$ is a function and that $|\hat{k}(\xi)| \leq A(1+|\xi|)^{-n\theta/2}$ and $\int_{|x|>2|y|^{1-\theta}}|k(x-y)-k(x)|\,dx \leq c$ for $|y|\leq 1$. Then the convolution operator $Tf = \text{p.v. } k*f$, is bounded in $L^p(R^n)$, $1<p<\infty$, and maps $L^\infty(R^n)$ into $BMO(R^n)$. (*Hint*: We may assume that k is integrable by replacing, if necessary, k by $k*\phi_\varepsilon$, where ϕ is a $C_0^\infty(R^n)$ function with integral 1, and observing that the above conditions are satisfied by $k*\phi_\varepsilon$, uniformly in ε. We show first that T maps L^∞ into BMO. Let I be a cube of dianeter δ, which we may assume is centered at the origin. Of the two cases, $\delta \leq 1$ or $\delta > 1$, we only do $\delta \leq 1$. Write $f = f_1 + f_2$, where $f_1 = f$ in the ball $|x| \leq 2\delta^{1-\theta}$, $f_2 = f - f_1$, and $u_1 = Tf_1$, $u_2 = Tf_2$. In terms of Fourier transforms $\hat{u}_1(\xi) = |\xi|^{-n\theta/2}\hat{k}(\xi)|\xi|^{n\theta/2}\hat{f}_1(\xi)$, where according to our assumptions $\hat{k}(\xi)|\xi|^{n\theta/2}$ is bounded. Thus, by (the n-dimensional variant of) Theorem 2.1 in Chapter VI, $\hat{u}_1$ is the Fourier transform of an L^p function with norm $\leq A\|\hat{k}(\xi)|\xi|^{n\theta/2}\hat{f}_1(\xi)\|_2 \leq c\|\hat{f}_1\|_2 \leq c\|f_1\|_2$, with $1/p = 1/2 - \theta/2$. Thus $\int_I |u_1(x)|^p\,dx \leq c\|f_1\|_2^p \leq c\delta^{n(1-\theta)p/2}\|f\|_\infty^p$; in other words $(1/|I|)\int_I|u_1(x)|\,dx \leq c\|f\|_\infty$. Now let $a_I = \int_{R^n} k(-y)f_2(y)\,dy$. Since $u_2(x) - a_I = \int_{R^n}(k(x-y)-k(-y))f_2(y)\,dy$, if $|x|\leq\delta$ (which is the case if $x \in I$), we get that $|u_2(x) - a_I| \leq (\int_{|y|>2|x|^{1-\theta}}|k(x-y)-k(-y)|\,dy)\|f\|_\infty$. Clearly T is bounded in L^2, and by Theorem 4.2 in Chapter VIII, T is also continuous in $L^p(R^n)$ for $2<p<\infty$. The statement for $1<p<2$ follows by duality. These operators were introduced by Fefferman [1970], who also proved that they are of weak-type $(1,1)$. The fact that they map L^∞ continuously into BMO was proved by Fefferman and Stein [1972].)

CHAPTER

XII

The Littlewood–Paley Theory

1. VECTOR-VALUED INEQUALITIES

It is often possible to extend inequalities involving scalar valued functions to functions which take values in a Banach space and thus obtain not only a more general result but also one which can be applied to other situations. The purpose of this chapter is to take systematic advantage of this fact. We begin by giving two important examples, one concerning maximal functions and the other Calderón–Zygmund singular integral operators. In each case the Banach space is $l^p(Z)$, $1 < p < \infty$, and the applications will be discussed later on.

Theorem 1.1. (Fefferman–Stein). Let $f = (f_1, \ldots, f_k, \ldots)$ be a sequence of functions defined on R^n and, corresponding to f, consider the sequence $Mf = (Mf_1, \ldots, Mf_k, \ldots)$ whose kth term is the Hardy-Littlewood maximal function Mf_k of f_k. Then

$$\| \, \|Mf_k\|_{l^r} \|_p \leq c \| \, \|f_k\|_{l^r} \|_p, \qquad 1 < r, \quad p < \infty, \tag{1.1}$$

where $c = c_{r,p}$ is independent of f. Also, and with a constant $c = c_{r,1}$ independent of f,

$$\lambda |\{ \|Mf_k\|_{l^r} > \lambda \}| \leq c \| \, \|f_k\|_{l_r} \|_1, \qquad 1 < r < \infty. \tag{1.2}$$

Proof. To simplify notations put $F(x) = \|f_k(x)\|_{l^r}$ and $mF(x) = \|M_k f(x)\|_{l^r}$. We consider separately the cases $p < r$, $p = r$, and $p > r$. When $p = r$ (1.1) follows at once from the maximal theorem since

$$\|mF\|_r^r = \sum_k \int_{R^n} Mf_k(x)^r \, dx \leq c \sum_k \int_{R^n} |f_k(x)|^r \, dx$$

$$= c \int_{R^n} |F(x)|^r \, dx = c \|F\|_r^r. \tag{1.3}$$

The case $p > r$ is not much harder since

$$\|mF\|_p = \|(mF)^r\|_{p/r}^{1/r} = \sup_{\phi} \left| \int_{R^n} mF(x)^r \phi(x)\, dx \right|^{1/r}, \tag{1.4}$$

where $\phi \in L^{(p/r)'}(R^n)$ and has norm $\leqslant 1$.

To estimate the integral in (1.4) we invoke 8.23 in Chapter IX and note that it is dominated by

$$\begin{aligned}\sum_k \int_{R^n} Mf_k(x)^r |\phi(x)|\, dx &\leqslant c \sum_k \int_{R^n} |f_k(x)|^r M\phi(x)\, dx \\ &= c \int_{R^n} F(x)^r M\phi(x)\, dx \leqslant c\|F^r\|_{p/r} \|M\phi\|_{(p/r)'} \\ &\leqslant c\|F\|_p^r \|\phi\|_{(p/r)'} \leqslant c\|F\|_p^r.\end{aligned}$$

Therefore the right-hand side of (1.4) is bounded by $c\|F\|_p$, which is precisely what we wanted to show.

Finally since the remaining cases of (1.1) follow from (1.2) and the Marcinkiewicz interpolation theorem, we show (1.2). Consider the Calderón–Zygmund decomposition of F at level λ and in particular consider a family of disjoint, open cubes $\{I_j\}$ such that if $\Omega = \bigcup I_j$, then $|\Omega| \leqslant \|F\|_1/\lambda$, $F(x) \leqslant \lambda$ for x in $R^n \backslash \Omega$ and $(1/|I_j|) \int_{I_j} F(x)\, dx \leqslant 2^n \lambda$, all j. Let now $f_k = g_k + h_k$, where $g_k = f_x \chi_{R^n \backslash \Omega}$, $h_k = f_k \chi_\Omega$ and put $G(x) = \|g_k(x)\|_{l^r}$, $mG(x) = \|Mg_k(x)\|_{l^r}$ and similarly for H and mH. Since $Mf_k(x) \leqslant Mg_k(x) + Mh_k(x)$, all x and k, it suffices to show that mG and mH are in $wk - L(R^n)$, with norm $\leqslant c\|F\|_1$. This is immediate for mG since $\|G\|_r^r \leqslant c\lambda^{r-1}\|F\|_1$ and by (1.3) $\|mG\|_r \leqslant c\|G\|_r$; whence $\lambda^r |\{mG > \lambda\}| \leqslant c\lambda^{r-1}\|F\|_1$.

The estimate for H requires some work. In the first place let $\tilde{f}_k(x) = \sum_j ((1/|I_j|) \int_{I_j} f_k(y)\, dy) \chi_{I_j}(x)$ and $\tilde{F}(x)$, $m\tilde{F}(x)$ as usual. Observe that $\operatorname{supp} \tilde{F} \subseteq \Omega$, and for $x \in I_j$, by Minkowski's inequality, we have $\tilde{F}(x) \leqslant (1/|I_j|) \int_{I_j} \|f_k(y)\|_{l^r}\, dy = (1/|I_j|) \int_{I_j} F(y)\, dy \leqslant 2^n \lambda$. Thus $\|\tilde{F}\|_r^r \leqslant c\lambda^r |\Omega| \leqslant c\lambda^{r-1}\|F\|_1$ and as above we see that $\lambda |\{\tilde{F} > \lambda\}| \leqslant c\|F\|_1$. Our proof will thus be complete once we show that, if $\tilde{\Omega} = \bigcup_j (2nI_j)$, then for all k and with a constant c independent of k,

$$Mh_k(x) \leqslant cM\tilde{f}_k(x), \qquad \text{a.e. in} \quad R^n \backslash \tilde{\Omega}. \tag{1.5}$$

To show (1.5) fix a cube I containing x and note that

$$\frac{1}{|I|} \int_I |h_k(y)|\, dy = \frac{1}{|I|} \sum_{j \in J} \int_{I \cap I_j} |h_k(y)|\, dy \tag{1.6}$$

where the sum is only extended over $J = \{\text{those } j\text{'s so that } I \cap I_j \neq \varnothing\}$. In this case, since $x \in I \backslash \tilde{\Omega} \subseteq I \backslash 2nI_j$, by the geometry of the situation, it follows that $I_j \subseteq 2nI$. Therefore, the right-hand side of (1.6) does not exceed

$$\frac{1}{|I|}\sum_{j\in J}\int_{I_j}|h_k(y)|\,dy \leqslant \frac{1}{|I|}\sum_{j\in J}\int_{I_j}|\tilde{f}_k(y)|\,dy$$
$$\leqslant \frac{1}{|I|}\int_I|\tilde{f}_k(y)|\,dy \leqslant \frac{c}{|2nI|}|\tilde{f}_k(y)|\,dy \leqslant cM\tilde{f}_k(x).$$

Since I is arbitrary we conclude that for $x \in R^n\backslash\tilde{\Omega}$, $Mh_k(x) \leqslant cM\tilde{f}_k(x)$ and we are done. ■

A similar result is true for CZ singular integral operators, the key observation being that a statement analogous to 8.23 in Chapter IX holds in this case as well. More precisely, we have

Theorem 1.2 (Córdoba–Fefferman). Assume k is a CZ kernel which verifies the assumptions of Chapter XI. If T denotes the CZ singular integral operator associated to k and w is a nonnegative function so that w^s is locally integrable for some $s > 1$, then

$$\int_{R^n}|Tf(x)|^p w(x)\,dx \leqslant c\int_{R_n}|f(x)|^p M(w^s)(x)^{1/s}\,dx \tag{1.7}$$

for all $f \in L^p(R^n)$, $1 < p < \infty$, and a constant $c = c_{p,s}$ independent of f.

Proof. We begin by pointing out a variant of Theorem 7.1 in Chapter XI, namely,

$$Tf^{\#}(x) \leqslant cM(|f|^q)(x)^{1/q}, \qquad 1 < q < \infty, \tag{1.8}$$

with $c = c_q$ independent of f. To see this fix a cube I containing x and put $f = f\chi_{2I} + f\chi_{R^n\backslash 2I} = f_1 + f_2$, say. Then $Tf = Tf_1 + Tf_2$ and it suffices to show (1.8) with f_1 and f_2 in place of f in the left-hand side of that inequality. In the first place, by Hölder's inequality and Theorem 5.4 in Chapter XI,

$$\frac{1}{|I|}\int_I|Tf_1(y)|\,dy \leqslant \left(\frac{1}{|I|}\int_I|Tf_1(y)|^q\,dy\right)^{1/q}$$
$$\leqslant c\left(\frac{1}{|I|}\int_{2I}|f(y)|^q\,dy\right)^{1/q} \leqslant cM(|f|^q)(x)^{1/q},$$

and this estimate is of the right order. Also for y in I,

$$|Tf_2(y) - (Tf_2)_I| \leqslant \frac{1}{|I|}\int_I\int_{R^n\backslash 2I}|k(y-w) - k(z-w)|\,|f(w)|\,dw\,dz$$
$$\leqslant \frac{c}{|I|}\int_I\int_{R^n\backslash 2I}\frac{|y-z|}{|x-w|^{n+1}}|f(w)|\,dw\,dz \leqslant cMf(x)$$
$$\leqslant cM(|f|^q)(x)^{1/q},$$

and consequently the estimate (1.8) holds.

Now since by Proposition 3.3 in Chapter IX, $M(w^s)(x)^{1/s} \in A_1$, from (a simple variant of) Theorem 4.4 in Chapter X and (1.8) it follows that

$$\begin{aligned}\int_{R^n} |Tf(x)|^p w(x)\, dx &\leq \int_{R^n} |Tf(x)|^p M(w^s)(x)^{1/s}\, dx \\ &\leq c \int_{R^n} Tf^{\#}(x)^p M(w^s)(x)^{1/s}\, dx \\ &\leq c \int_{R^n} M(|f|^q)(x)^{p/q} M(w^s)(x)^{1/s}\, dx. \qquad (1.9)\end{aligned}$$

Suppose now that $1 < q < p$, then $p/q > 1$ and by Theorem 4.1 in Chapter IX we see that the righthand side of (1.9) is dominated by $c \int_{R^n} |f(x)|^p M(w^s)(x)^{1/s}\, dx$, and the proof is complete. ■

We are now in a position to prove

Theorem 1.3 (Córdoba-Fefferman). Let $\{k_j\}$ be a sequence of CZ kernels with unformly bounded CZ constants and let $\{T_j\}$ denote the sequence of CZ singular integral operators which correspond to the k_j's. Then

$$\| \| T_j f_j \|_{l^r} \|_p \leq c \| \| f_j \|_{l^r} \|_p, \qquad 1 < r, p < \infty, \qquad (1.10)$$

where $c = c_{r,p}$ is independent of $f = (f_1, \ldots, f_j, \ldots)$.

Proof. Since a simple duality argument shows that the estimate (1.10) holds with indices p, r if and only if it holds with indices p', r', $1/p + 1/p' = 1/r + 1/r' = 1$, we may assume that $p \geq r$. The case $p = r$ is a simple sequence of Theorem 5.4 in Chapter XI. On the other hand, if $p > r$, then the left-hand side of (1.10) equals

$$\sup_g \left| \int_{R^n} \left(\sum_j |T_j f_j(x)|^r \right) g(x)\, dx \right|^{1/r},$$

where $g \in C_0^\infty(R^n)$ and $\|g\|_{(p/r)'} \leq 1$. Now by Theorem 1.2, and with $1 < s < (p/r)'$ there,

$$\begin{aligned}\sum_j \int_{R^n} |T_j f_j(x)|^r |g(x)|\, dx &\leq c \sum_j \int_{R^n} |f_j(x)|^r M(|g|^s)(x)^{1/s}\, dx \\ &\leq c \| \| f_j \|_{l^r} \|_p^r \| M(|g|^s)^{1/s} \|_{(p/r)'} \\ &\leq c \| \| f_j \|_{l^r} \|_p^r \|g\|_{(p/r)'},\end{aligned}$$

and we are done. ■

We consider next a general result in the direction of Theorem 1.4 and some of its applications.

2. VECTOR-VALUED SINGULAR INTEGRAL OPERATORS

We begin by discussing some preliminary results; we intend to be brief. Given a separable Hilbert space H with inner product $(,)$ and norm $|h|_H = |h| = (h, h)^{1/2}$, we say that a function f defined on R^n and with values in H is measurable (or weakly measurable) if the scalar function $(f(x), h)$ is Lebesgue measurable for every h in H. The class $L^p(R^n, H)$ consists of those measurable f with $\|f\|_p = (\int_{R^n}|f(x)|^p\,dx)^{1/p} < \infty$, $1 \leq p \leq \infty$, and similarly $\|f\|_\infty = \text{ess sup}_{R^n}|f|$ denotes the norm in $L^\infty(R^n, H)$.

For a couple of separable Hilbert spaces H_1, H_2 let $B(H_1, H_2)$ be the (Banach) space of bounded, linear operators T from H_1 into H_2 endowed with the norm $|T|_{B(H_1,H_2)} = |T| = \sup_{h\in H_1}(|Th|_{H_2}/|h|_{H_1})$.

We say that a function f on R^n and with values in $B(H_1, H_2)$ is measurable if $f(x)h$ is an H_2 valued, measurable function or each h in H_1. In this case $|f|_{B(H_1,H_2)}$ is also Lebesgue measurable and the spaces $L^p(R^n, B(H_1, H_2))$ may be defined exactly as above.

The usual facts concerning operations of functions hold in this general setting as well. For instance, suppose that a function k defined on R^n and with values in $B(H_1, H_2)$ is integrable, and for f in $L^p(R^n, H_1)$ put

$$g(x) = \int_{R^n} k(x-y)f(y)\,dy. \tag{2.1}$$

Then the integral in (2.1), as an element in H_2, converges weakly in H_2 for almost every x, and $|g(x)|_{H_2} \leq \int_{R^n}|k(x-y)|_{B(H_1,H_2)}|f(y)|_{H_1}\,dy$. Furthermore, $\|g\|_p \leq \|k\|_1\|f\|_p$, $1 \leq p \leq \infty$.

Another important property concerns the Fourier transformation. For f in $L(R^n, H)$ we define its Fourier transform by $\hat{f}(\xi) = \int_{R^n} e^{-i2\pi x\cdot\xi}f(x)\,dx$. In this case $\hat{f}$ is also H valued and clearly $\|\hat{f}\|_\infty \leq \|f\|_1$. Futhermore, if $f \in L(R^n, H) \cap L^2(R^n, H)$, by means of an appropriate limiting process as was done in Chapter X, also $\hat{f} \in L^2(R^n, H)$ and Plancherel's identity is valid for these functions. This is readily seen by expressing the elements of the Hilbert space in terms of an orthonormal basis and then proceeding as in the scalar case.

To further illustrate the fact that the results we need in this setting are simple extensions of the scalar case we prove the Marcinkiewicz interpolation theorem.

Theorem 2.1. Let A be a sublinear operator defined on $L_0^\infty(R^n, H_1)$, i.e., compactly supported, bounded H_1-valued functions, with values in $M(R^n, H_2)$, i.e., the space of measurable, H_2-valued functions. Suppose in addition that for f in $L_0^\infty(R^n, H_1)\lambda|\{|Af|_{H_2} > \lambda\}| \leq c_1\|f\|_1$, and $\lambda^r|\{|Af|_{H_2} > \lambda\}| \leq c_r^r\|f\|_r^r$, where c_1 and c_r are independent of λ and f. Then for each

$1 < p < r$, we have that $Af \in L^p(R^n, H_2)$ whenever $f \in L^p(R^n, H_1)$ and there is a constant $c = c_{1,r,p}$ independent of f such that $\|Af\|_p \leqslant c\|f\|_p$.

Proof. Let $F(x) = (|f(x)|_{H_1})^{-1} f(x)$ whenever $f(x) \neq 0$ and 0 otherwise. For a scalar valued function g consider $Bg(x) = |A(F(x)g)|_{H_2}$. Clearly B is a sublinear mapping, simultaneously of weak-types $(1, 1)$ and (r, r), with norm $\leqslant c_1, c_r$, respectively. By the Marcinkiewicz interpolation theorem 4.1 in Chapter IV there is a constant c as indicated above so that $\|Bg\|_p \leqslant c\|g\|_p$, $1 < p < r$. Upon setting $g(x) = |f(x)|_{H_1}$ our proof is complete. ∎

An important result for our purposes is the following extension of Theorem 1.1 in Chapter XI,

Theorem 2.2. Suppose a linear operator A defined in $L_0^\infty(R^n, H_1)$ and with values in $M(R^n, H_2)$ verifies

(i) $\lambda^r |\{|Af| > \lambda\}| \leqslant c_1 \|f\|_r$, some $r > 1$.

(ii) If f has support in $B(x_0, R)$ and integral 0, then there are constants $c_2, c_3 > 1$ independent of f so that

$$\int_{R^n \setminus B(x_0, c_2 R)} |Af(x)|\, dx \leqslant c_3 \|f\|_1.$$

Then also $\lambda |\{|Af| > \lambda\}| \leqslant c\|f\|_1$ and by Theorem 2.1 also $\|Af\|_p \leqslant c\|f\|_p$ for $1 < p < r$.

Since the proof is identical to that of the cited result, it is omitted. In the same vein we have

Theorem 2.3. Let k be a function on R^n whose values are bounded linear operators from H_1 to H_2; we assume k to be measurable and integrable on compact sets. For $f \in L_0^\infty(R^n, H_1)$ put

$$Tf(x) = \int_{R^n} k(x - y) f(y)\, dy.$$

If for some $r > 1$ and f in $L^r(R^n, H_1)$ the inequality $\|Tf\|_r \leqslant c_1 \|f\|_r$ holds, and

$$\int_{|x| \geqslant 2|y|} |k(x - y) - k(x)|\, dx \leqslant c_2, \qquad y \in R^n,$$

then $Tf \in L^p(R^n, H_2)$ for all $1 < p < \infty$ and $\|Tf\|_p \leqslant c\|f\|_p$, where c depends on c_1, c_2 and p but is otherwise independent of f.

The proof is identical to that of Theorem 3.1 in Chapter XI and is omitted. As for the vector valued singular integrals we have

Definition 2.4. We say that a function k on R^n whose values are bounded operators from H_1 to H_2 is a vector-valued Calderón-Zygmund integral kernel provided that

(i) k is measurable and integrable on compact sets not containing the origin.

(ii) For $0 < \varepsilon < N$, $|\int_{\varepsilon<|x|<N} k(x)\,dx| \leq c$, and for each h in H_1, $[\int_{\varepsilon<|x|<N} k(x)\,dx]h$ converges as $\varepsilon \to 0$.

(iii) For each h in H_1 with $|h| \leq 1$, $\int_{|x|<R} |x||k(x)h|\,dx \leq cR$, and

(iv) $\int_{|x|\geq 2|y|} |k(x-y) - k(x)|\,dx \leq c$, $y \neq 0$.

For these kernels we have

Theorem 2.5. Given a vector valued Calderón-Zygmund singular integral kernel k let $T_\varepsilon f(x) = \int_{|x-y|>\varepsilon} k(x-y)f(y)\,dy$. Then $\|T_\varepsilon f\|_p \leq c\|f\|_p$, $1 < p < \infty$, with $c = c_p$ depending only on p and the constants in the definition of k. Furthermore, $T_\varepsilon f$ converges in $L^p(R^n, H_2)$ as $\varepsilon \to 0$ to a mapping Tf with similar continuity properties.

The proof follows along the lines of Theorem 5.4 in Chapter XI and is therefore omitted. We also leave to the reader the task of stating and proving the weak-type (1, 1) result as well as the consideration of the maximal singular integral operators and turn our attention to the applications of Theorem 2.5.

3. THE LITTLEWOOD-PALEY g FUNCTION

The first application we consider is to the Littlewood-Paley theory. For this purpose we let $H_1 = C$, the complex numbers, and $H_2 = L^2(R_+, dt/t)$, the Hilbert space of square integrable functions on the positive half-line with respect to the measure dt/t, and norm

$$|h|_{H_2} = \left(\int_{[0,\infty)} |h(t)|^2/t\,dt\right)^{1/2}.$$

Definition 3.1. We say that a scalar valued function ψ on R^n is a Littlewood-Paley function provided it satisfies

(i) $\psi \in L(R^n)$, $\int_{R^n} \psi(x)\,dx = 0$.

(ii) $|\psi(x)| \leq c(1+|x|)^{-(n+\alpha)}$, some $\alpha > 0$.

(iii) $\int_{R^n}|\psi(x+y) - \psi(x)|\,dx \leq c|y|^\gamma$, y in R^n, some $\gamma > 0$.

Clearly any Schwartz function with vanishing integral verifies (i)–(iii). We also have

Proposition 3.2. Suppose ψ is a Littlewood–Paley function and $\xi \in R^n$. Then $|\hat{\psi}(\cdot\,\xi)|_{H_2} \leq c$.

Proof. We begin by showing that

$$|\hat{\psi}(\xi)| \leq c \min(|\xi|^{\alpha/(n+1+\alpha)}, |\xi|^{-\gamma}), \tag{3.1}$$

where c is independent of ξ. Since by identity (5.3) in Chapter XI $\hat{\psi}(\xi) = \frac{1}{2}\int_{R^n}(\psi(x) - \psi(x+y))e^{-ix\cdot\xi}\,dx$, $y = \pi\xi/|\xi|^2$, by (iii) above it follows immediately that $|\hat{\psi}(\xi)| \leq c|\xi|^{-\gamma}$. Furthermore, since ψ has vanishing integral, we also have $\hat{\psi}(\xi) = \int_{R^n}\psi(x)(e^{-ix\cdot\xi} - 1)\,dx$, and consequently

$$\begin{aligned} |\hat{\psi}(\xi)| &\leq 2\int_{R^n}|\psi(x)| \min(|x||\xi|, 1)\,dx \\ &\leq 2|\xi|\int_{|x|\leq\eta}|x||\psi(x)|\,dx + 2\int_{|x|>\eta}|\psi(x)|\,dx \\ &= I + J, \end{aligned}$$

say. Now since $I \leq c|\xi|c_1\eta^{n+1}$ and $J \leq cc_1\int_{|x|>\eta}|x|^{-(n+\alpha)}\,dx = c\eta^{-\alpha}$, we obtain at once that $|\hat{\psi}(\xi)| \leq c(|\xi|\eta^{n+1} + \eta^{-\alpha})$, and (3.1) follows upon minimizing with respect to η. To complete the proof we invoke (3.1) and estimate

$$|\hat{\psi}(\cdot\,\xi)|^2_{H_2} \leq c\int_{[0,\infty)} \min((t|\xi|)^{\alpha/(n+1+\alpha)}, \quad (t|\xi|)^{-\alpha})^2 \frac{dt}{t} \leq c. \quad \blacksquare$$

Let now $k(x) \in L(C, L^2(R_+, dt/t))$ be given by $k(x)a = t^{-n}\psi(x/t)a = \psi_t(x)a$, where $x \in R^n$, a is a complex scalar and ψ is a Littlewood–Paley function. Corresponding to k we consider the singular integral operator

$$\begin{aligned} Tf(x) &= \lim_{\varepsilon\to 0}\int_{|x-y|>\varepsilon} k(x-y)f(y)\,dy \\ &= \lim_{\varepsilon\to 0}\int_{|x-y|>\varepsilon} \psi_t(x-y)f(y)\,dy. \end{aligned} \tag{3.2}$$

We want to show that T falls within the scope of Theorem 2.5 and thus obtain its L^p continuity, $1 < p < \infty$. To get a feeling for the situation we do the L^2 case first.

Proposition 3.3. T is bounded from $L^2(R^n)$ into $L^2(R^n, L^2(R_+, dt/t))$.

Proof. Observe that for f in $L^2(R^n)$ and on account of Tonelli's theorem, Plancherel's identity, and Proposition 3.2 we have

$$\begin{aligned}\int_{R^n} |k * f(x)|^2\, dx &= \int_{R^n}\int_{[0,\infty)} |\psi_t * f(x)|^2 \frac{dt}{t}\, dx\\ &= \int_{[0,\infty)}\int_{R^n} |\psi_t * f(x)|^2\, dx \frac{dt}{t}\\ &= c\int_{[0,\infty)}\int_{R^n} |\hat{\psi}(t\xi)|^2 |\hat{f}(\xi)|^2\, d\xi \frac{dt}{t}\\ &\leqslant c\int_{R^n} |\hat{f}(\xi)|^2 \left(\sup_{\xi}\int_{[0,\infty)} |\hat{\psi}(t\xi)|^2 \frac{dt}{t}\right) d\xi\\ &\leqslant c\|f\|_2^2. \quad \blacksquare\end{aligned}$$

In fact a more precise result holds in the particular case ψ is radial, namely

Proposition 3.4. Suppose ψ is a radial Littlewood–Paley function, then

$$\|Tf\|_2 = (2\pi)^{-n/2}\left(\int_{[0,\infty)} |\hat{\psi}(t)|^2 \frac{dt}{t}\right)^{1/2} \|f\|_2. \tag{3.3}$$

Proof. As above we see that

$$\|Tf\|_2^2 = (2\pi)^{-n}\int_{R^n} |\hat{f}(\xi)|^2 \int_{[0,\infty)} |\hat{\psi}(t\xi)|^2 \frac{dt}{t}\, d\xi \tag{3.4}$$

and since $\hat{\psi}$ is also radial the innermost integral in (3.4) is readily seen to be $\int_{[0,\infty)} |\hat{\psi}(t)|^2/t\, dt$. $\blacksquare$

For the other values of p we have

Theorem 3.5. Suppose T is given by (3.2). Then

$$\|Tf\|_p \leqslant c\|f\|_p, \qquad 1 < p < \infty,$$

whre $c = c_p$ is independent of f.

Proof. We verify that (i)–(iv) in Definition 2.4 are satisfied. (i) is immediate. As for (ii), observe that since $\int_{|x|\leqslant R} \psi(x)\, dx = -\int_{|x|>R} \psi(x)\, dx$, by property (ii) of Definition 3.1

$$\left|\int_{|x|\leqslant R} \psi(x)\, dx\right| \leqslant \frac{cR^n}{(1+R)^{(n+\alpha)}}$$

and consequently $|\int_{|x|\le R}\psi_t(x)\,dx|_{H_2}\le c$, which gives (ii). On the other hand, from (ii) of Definition 3.1 it follows that $|k(x)|\le c|x|^{-n}$ and (iii) also holds. Finally, to show that Hörmander's condition is satisfied let $0<\varepsilon<\min(\alpha,\gamma,n)$ and observe that

$$\int_{|x|>2|y|}|x|^{-(n+\xi)/2}(|x|^{(n+\varepsilon)/2}|k(x-y)-k(x)|)\,dx$$
$$\le c|y|^{-\varepsilon/2}\left(\int_{|x|>2|y|}|x|^{n+\varepsilon}|k(x-y)-k(x)|^2\,dx\right)^{1/2}. \tag{3.5}$$

We want to verify that the integral in (3.5) does not exceed $|y|^{\varepsilon/2}$. In first place note that it is bounded by

$$\left(\int_{|x|>2|y|}|x|^{n+\varepsilon}\int_{[0,\infty)}|\psi((x-y)/t)-\psi(x/t)|^2t^{-2n}\frac{dt}{t}\,dx\right\}^{1/2}. \tag{3.6}$$

Moreover,

$$|\psi((x-y)/t)-\psi(x/t)|\le c(1+(|x-y|/t))^{-(n+\alpha)}+c(1+(|x|/t))^{-(n+\alpha)}$$
$$\le c(1+(|x|/t))^{-(n+\varepsilon)}\le c(t/|x|)^{n+\varepsilon}.$$

Thus the expression in (3.6) is bounded by

$$\left(c\int_{[0,\infty)}t^{-2n}t^{n+\varepsilon}\int_{R^n}\left|\psi\left(\frac{x-y}{t}\right)-\psi\left(\frac{x}{t}\right)\right|\,dx\frac{dt}{t}\right)^{1/2}$$
$$\le c\left(\int_{[0,\infty)}t^{\varepsilon-n}\min\left(t^n2\|\psi\|_1,t^n\left(\frac{|y|}{t}\right)^{\gamma}\right)\frac{dt}{t}\right)^{1/2}$$
$$=c\left(\|\psi\|_1\int_{[0,|y|)}t^{\varepsilon}\frac{dt}{t}+c|y|^{\gamma}\int_{[|y|,\infty)}t^{\varepsilon-\gamma}\frac{dt}{t}\right)^{1/2}$$
$$=c|y|^{\varepsilon/2}.\quad\blacksquare$$

Theorem 2.5 in this context is best expressed as follows: For a Littlewood-Paley function ψ and f in $L^p(R^n)$, $1<p<\infty$, put $F(x,t)=f*\psi_t(x)$ and let

$$g(F)(x)=\left(\int_{[0,\infty)}|F(x,t)|^2\frac{dt}{t}\right)^{1/2} \tag{3.7}$$

denote the Littlewood-Paley g function of F. Then $g(F)$ is in $L^p(R^n)$ and there is a constant $c=c_{p,\psi}$ independent of f so that

$$\|g(F)\|_p\le c\|f\|_p,\qquad 1<p<\infty. \tag{3.8}$$

It is useful to point out that the inequality opposite to (3.8) also holds. This is easiest seen for the particular case when ψ is radial, and normalized so that $\int_{[0,\infty)}|\hat\psi(t)|^2/t\,dt=(2\pi)^n$. In this case by (3.3) we have that $\|g(F)\|_2=\|f\|_2$ and consequently by polarization it follows that

$$(f_1,f_2)=(f_1*\psi_t,f_2*\psi_t)=(F_1,F_2), \tag{3.9}$$

where the first inner product in (3.9) is that of $L^2(R^n)$ and the second one that in $L^2(R^n, L^2(R_+, dt/t))$. Suppose now that $f_1 \in L^2(R^n) \cap L^p(R^n)$, $1 < p < \infty$, and $f_2 \in C_0^\infty(R^n)$, $\|f_2\|_{p'} \leq 1$. Then by (3.9), Hölder's inequality and (3.8) it readily follows that

$$\left|\int_{R^n} f_1(x)\overline{f_2}(x)\,dx\right| \leq \int_{R^n}\int_{[0,\infty)} |F_1(x,t)|\,|F_2(x,t)|\frac{dt}{t}\,dx$$

$$\leq \int_{R^n} g(F_1)(x)g(F_2)(x)\,dx \leq \|g(F_1)\|_p\|g(F_2)\|_{p'}$$

$$\leq c\|g(F_1)\|_p.$$

Whence by the converse to Hölder's inequality we immediately see that

$$\|f_1\|_p \leq c\|g(F_1)\|_p, \qquad f_1 \in L^2(R^n) \cap L^p(R^n), \tag{3.10}$$

and the same inequality holds for general f in $L^p(R^n)$, as a simple limiting argument shows.

In particular our results apply to the function ψ with Fourier transform $\hat{\psi}(\xi) = |\xi|e^{-|\xi|}$; this corresponds to the classical Littlewood–Paley function $\psi(x) = c(\partial/\partial t)(t/(t^2+|x|^2)^{(n+1)/2})]_{t=1}$ obtained by differentiating the Poisson kernel. It is also important to incorporate into the theory the space derivatives $\psi_j(x) = (\partial/\partial x_j)(1+|x|^2)^{-(n+1)/2}$ of the Poisson kernel. There are two difficulties in proving (3.10) for these functions, i.e., ψ_j is not radial, and $\hat{\psi}_j(\xi) = c\xi_j e^{-|\xi|}$ vanishes identically along $\xi_j = 0$; clearly, there is no problem with (3.8). The way to overcome this is to consider instead the gradient of the Poisson kernel whose Fourier transform is the vector $ce^{-|\xi|}\xi$. Indeed, since $\int_{[0,\infty)}(t|\xi|)^2 e^{-2t|\xi|}/t\,dt = \frac{1}{4}$ it readily follows that for sufficiently smooth functions f_1, f_2

$$\int_{R^n} f_1(x)\overline{f_2(x)}\,dx = c\int_{R^n} \hat{f}_1(\xi)\overline{\hat{f}_2}(\xi)\,d\xi$$

$$= c\int_{[0,\infty)}\int_{R^n}\sum_{j=1}^n t\xi_j e^{-t|\xi|}\hat{f}_1(\xi)t\xi_j e^{-t|\xi|}\overline{\hat{f}_2}(\xi)\,d\xi\frac{dt}{t}$$

$$= c\int_{[0,\infty)}\int_{R^n} t\,\nabla(f_1 * P_t)(y)\cdot\overline{t\,\nabla(f_2 * P_t(y)}\,dy\frac{dt}{t}. \tag{3.11}$$

Whence

$$\left|\int_{R^n} f_1(x)\overline{f_2}(x)\,dx\right| \leq c\int_{R^n} g(t|\nabla(f_1 * P_t)|)(y)g(t|\nabla f_2 * P_t|)(y)\,dy,$$

and (3.10) follows as above.

4. THE LUSIN AREA FUNCTION AND THE LITTLEWOOD–PALEY g_λ^* FUNCTION

In order to consider the next application we set $H_1 = C$ and $H_2 = \{h: |h|_{H_2} = (1/va^n)\int_{[0,\infty)}\int_{|y|<a}|h(y,t)|^2/t\,dy\,dt)^{1/2} < \infty\}$, where $a > 0$ and v = volume of the unit ball of R^n.

Let now $k(x) \in L(C, H_2)$ be given by $k(x)a = t^{-n}\psi(x/t - y)a$, where $x \in R^n$, $(y, t) \in R_+^{n+1}$, a is a complex scalar and ψ is a Littlewood–Paley function. Corresponding to k we consider again the singular integral operator

$$\begin{aligned} Tf(x) &= \lim_{\varepsilon\to 0}\int_{|x-w|>\varepsilon} k(x-w)f(w)\,dw \\ &= \lim_{\varepsilon\to 0}\int_{|x-w|>\varepsilon} \psi_t(x - ty - w)f(w)\,dw. \end{aligned} \tag{4.1}$$

We then set $F(x, t) = f * \psi_t(x)$ and $|Tf(x)| = S_a(F)(x)$, the Lusin (or area) function of F (with opening a). If we denote by $\Gamma_a(x) = \{(y, t) \in R_+^{n+1}: |x - y| < at\}$ the cone with vertex at x and opening a, it follows immediately that

$$\begin{aligned} S_a(F)(x) &= \left(\frac{1}{va^n}\int_{[0,\infty)}\int_{|y|<a}|F(x-ty, t)|^2\,dy\frac{dt}{t}\right)^{1/2} \\ &= \left(\frac{1}{v}\int_{\Gamma_a(x)}|F(y,t)|^2(at)^{-n}\,dy\frac{dt}{t}\right)^{1/2}. \end{aligned} \tag{4.2}$$

As we did in case of the g function it may be readily seen that $\|S_a(F)\|_2 \le c\|f\|_2$. Indeed, it suffices to observe that if χ denotes the characteristic function of the unit interval then

$$\begin{aligned} \int_{R^n} S_a(F)(x)^2\,dx &= \int_{R^n}\frac{1}{v}\int_{[0,\infty)}\int_{R^n}\chi\left(\frac{|x-y|}{at}\right)|F(y,t)|^2(at)^{-n}\,dy\frac{dt}{t}\,dx \\ &= \int_{R^n}\int_{[0,\infty)}|F(y,t)|^2\left(\frac{1}{v}\int_{R^n}(at)^{-n}\chi\left(\frac{|x-y|}{at}\right)dx\right)\frac{dt}{t}\,dy \\ &= \int_{R^n} g(F)(x)^2\,dx, \end{aligned} \tag{4.3}$$

and the assertion follows at once from Proposition 3.3. In fact the above argument shows that under the normalization of (3.9), also $\|S_af\|_2 = \|f\|_2$. Also an argument quite similar to that of Theorem 3.5 shows that k verifies (i)–(iv) in Definition 2.4 and consequently

$$\|S_af\|_p \le c\|f\|_p, \qquad 1 < p < \infty, \tag{4.4}$$

where $c = c_{p,a}$ is indepedent of f. To prove the inequality opposite to (4.4) we proceed exactly as in Section 3, so we say no more.

Returning to the constant in (4.4) it is of interest to consider its dependence on a. It is best to approach this question from a geometric point of view and in order to do this we need the following observation

Lemma 4.1. Let $\mathcal{O}$ be an open set in R^n and for $a > 1$ associate to it

$$\mathcal{U} = \{x \in R^n \colon M\chi_{\mathcal{O}}(x) > 1/2a^n\}.$$

Then if $\Gamma_a(R^n \backslash \mathcal{U}) = \bigcup_{x \in R^n \backslash \mathcal{U}} \Gamma_a(x)$, and similarly for $\Gamma_1(R^n \backslash \mathcal{O}) = \Gamma(R^n \backslash \mathcal{O})$, we have

(i) If $(y, t) \in \Gamma_a(R^n \backslash \mathcal{U})$, then $|B(y, t)| \leq 2|B(y, t) \cap (R^n \backslash \mathcal{O})|$.
(ii) $\Gamma_a(R^n \backslash \mathcal{U}) \subseteq \Gamma(R^n \backslash \mathcal{O})$.

Proof. If $(y, t) \in \Gamma_a(R^n \backslash \mathcal{U})$ there is $x \notin \mathcal{U}$ with $|y - x| < at$, or $x \in B(y, at)$. Thus $|B(y, t) \cap \mathcal{O}|/|B(y, t)| \leq a^n|B(y, at) \cap \mathcal{O}|/|B(y, at)| \leq a^n M\chi_{\mathcal{O}}(x) \leq a^n/2a^n = \frac{1}{2}$, and (i) holds. On the other hand if (y, t) is in $\Gamma_a(R^n \backslash \mathcal{U})$, (i) implies in particular that there is $w \in B(y, t) \cap (R^n \backslash \mathcal{O}) \neq \emptyset$. In this case $(y, t) \in \Gamma(w)$, with w in $R^n \backslash \mathcal{O}$, which gives (ii) as well. ■

We are now in a position to show

Lemma 4.2. Suppose $\mathcal{O}$ is an open set of finite measure and let $\mathcal{U}$ be associated to $\mathcal{O}$ as in Lemma 4.1. Then for $a \geq 1$ and with $S_1(F) = S(F)$,

$$\int_{R^n \backslash \mathcal{U}} S_a(F)(x)^2\, dx \leq 2 \int_{R^n \backslash \mathcal{O}} S(F)(x)^2\, dx. \tag{4.5}$$

Proof. From the definition of Lusin function we readily see that

$$\int_{R^n \backslash \mathcal{U}} S_a(F)(x)^2\, dx = \frac{1}{v} \int_{\Gamma_a(R^n \backslash \mathcal{U})} |F(y, t)|^2 |B(y, at) \cap (R^n \backslash \mathcal{U})|(at)^{-n}\, dy \frac{dt}{t} \tag{4.6}$$

and

$$\int_{R^n \backslash \mathcal{O}} S(F)(x)^2\, dx = \frac{1}{v} \int_{\Gamma(R^n \backslash \mathcal{O})} |F(y, t)|^2 |B(y, t) \cap (R^n \backslash \mathcal{O})| t^{-n}\, dy \frac{dt}{t}. \tag{4.7}$$

Now since by Proposition 4.1, $\Gamma_a(R^n \backslash \mathcal{U}) \subseteq \Gamma(R^n \backslash \mathcal{O})$ and $|B(y, at) \cap (R^n \backslash \mathcal{U})|/v(at)^n \leq \frac{1}{2}|B(y, t) \cap (R^n \backslash \mathcal{O})|/vt^n$ for (y, t) in $\Gamma_a(R^n \backslash \mathcal{U})$, the desired conclusion follows by simply comparing (4.6) and (4.7). ■

We distinguish now two cases, in first place we show

Theorem 4.3. Suppose $a \geq 1$ and $0 < p \leq 2$. Then

$$\|S_a(F)\|_p \leq c a^{n(1/p - 1/2)} \|S(F)\|_p, \tag{4.8}$$

where c is an absolute constant.

Proof. The case $p = 2$ follows immediately from (4.3). Otherwise let $\mathcal{O}_\lambda$ be the open set of finite measure $\{S(F) > \lambda\}$ and associate to it $\mathcal{U}_\lambda$ as in Lemma 4.1. If the reader prefers not to show that $\mathcal{O}_\lambda$ is open, the argument given below still works if $\mathcal{O}_\lambda$ is an open set with measure (arbitrarily) close to $\{S(F) > \lambda\}$. Furthermore, let $\mathcal{O}'_\lambda = \{S_a(F) > \lambda\}$ and note that

$$|\mathcal{O}'_{s\lambda}| \leqslant |\mathcal{U}_\lambda| + |\mathcal{O}'_{s\lambda} \cap (R^n \backslash \mathcal{U}_\lambda)| = I + J, \tag{4.9}$$

say. From Chebychev's inequality and Lemma 3.2 it follows that

$$J \leqslant (s\lambda)^{-2} \int_{R^n \backslash \mathcal{U}_\lambda} S_a(F)(x)^2\, dx \leqslant 2(s\lambda)^{-2} \int_{R^n \backslash \mathcal{O}_\lambda} S(F)(x)^2\, dx. \tag{4.10}$$

Also by the maximal theorem we see that

$$I \leqslant ca^n |\mathcal{O}_\lambda|. \tag{4.11}$$

Thus combining (4.9), (4.10) and (4.11) we get

$$\begin{aligned} s^{-p} \|S_a(F)\|_p^p &= \int_{[0,\infty)} |\mathcal{O}'_{s\lambda}|\, d\lambda^p \\ &\leqslant ca^n \int_{[0,\infty)} |\mathcal{O}_\lambda|\, d\lambda^p + 2s^{-2} \int_{[0,\infty)} \lambda^{-2} \int_{R^n \backslash \mathcal{O}_\lambda} S(F)(x)^2\, dx\, d\lambda^p \\ &= L + M, \end{aligned} \tag{4.12}$$

say. Clearly $L = ca^n \|S(F)\|_p^p$. On the other hand,

$$\begin{aligned} M &= 2ps^{-2} \int_{R^n} S(F)(x)^2 \int_{[S(F)(x),\infty)} \lambda^{p-3}\, d\lambda\, dx \\ &= cs^{-2} \int_{R^n} S(F)(x)^{2+(p-2)}\, dx = cs^{-2} \|S(F)\|_p^p. \end{aligned}$$

Whence by (4.12) we obtain $\|S_a(F)\|_p^p \leqslant c(a^n s^p + s^{p-2}) \|S(F)\|_p^p$ and (4.8) follows upon minimizing the right-hand side of the above inequality with respect to s. ■

As for the case $p > 2$ we have

Theorem 4.4. Suppose $a \geqslant 1$ and $2 < p < \infty$. Then

$$\|S_a(F)\|_p \leqslant c \|S(F)\|_p, \tag{4.13}$$

where c is an absolute constant.

Proof. Since $p/2>1$ we may invoke the converse to Hölder's inequality and compare the integrals

$$I=\int_{R^n} S_a(F)(x)^2 g(x)\,dx \qquad \text{and} \qquad J=\int_{R^n} S(F)(x)^2 Mg(x)\,dx,$$

where g is a nonnegative function in $L^{(p/2)'}(R^n)$ with norm $\leqslant 1$. First, observe tha since

$$\begin{aligned}\frac{1}{|B(y,at)|}\int_{B(y,at)} g(x)\,dx &\leqslant \inf_{B(y,t)} Mg\\ &\leqslant \frac{1}{|B(y,t)|}\int_{B(y,t)} Mg(x)\,dx,\end{aligned}$$

it readily follows that

$$\begin{aligned} I &= \frac{1}{v}\int_{R^n}\int_{\Gamma_a(x)} |F(y,t)|^2 (at)^{-n}\,dy\frac{dt}{t}g(x)\,dx\\ &= \int_{R_+^{n+1}} |F(y,t)|^2 \frac{1}{|B(y,at)|}\int_{B(y,at)} g(x)\,dx\,dy\frac{dt}{t}\\ &\leqslant \int_{R_+^{n+1}} |F(y,t)|^2\frac{1}{|B(y,t)|}\int_{B(y,t)} Mg(x)\,dx\,dy\frac{dt}{t}=J.\end{aligned}$$

Thus

$$\begin{aligned} I &\leqslant \|S(F)\|_p^2\|Mg\|_{(p/2)'}\leqslant c\|S(F)\|_p^2,\\ \|S_a(F)^2\|_{p/2} &= \|S_a(F)\|_p^2 \leqslant c\|S(F)\|_p^2,\end{aligned}$$

and we are done. ■

There is yet another important function we consider, namely, the Littlewood–Paley g_λ^* function. It is defined for f in $L^p(R^n)$ and a Littlewood–Paley function ψ by setting $F(y,t)=f*\psi_t(y)$ and

$$g_\lambda^*(F)(x)=\left(\int_{R_+^{n+1}}\frac{|F(y,t)|^2}{(1+(|x-y|/t))^{2\lambda}}t^{-n}\,dy\frac{dt}{t}\right)^{1/2}. \tag{4.14}$$

Since $S(F)(x)\leqslant cg_\lambda^*(F)(x)$, c independent of F, by the known results for the Lusin function we have $\|f\|_p\leqslant c\|S(F)\|_p\leqslant c\|g_\lambda^*(F)\|_p$, $1<p<\infty$. As for the opposite inequality we have

Theorem 4.5. The inequality

$$\|g_\lambda^*(F)\|_p\leqslant c\|S(F)\|_p \tag{4.15}$$

holds, with $c = c_{p,\lambda}$ independent of F, provided either $0 < p \leq 2$ and $\lambda > n/p$ or $2 < p < \infty$ and $\lambda > n/2$.

Proof. We do the case $0 < p \leq 2$ first. Observe that

$$\left(1+\left(\frac{|x-y|}{t}\right)\right)^{-2\lambda} \sim \sum_{k=0}^{\infty} 2^{-2k\lambda}\chi\left(\frac{|x-y|}{2^k t}\right), \tag{4.16}$$

where χ denotes the characteristic function of the interval $[0, 1]$. Thus multiplying (4.16) through by $|F(y, t)|^2 t^{-n}$ and integrating over R_+^{n+1} with respect to $dy\,dt/t$ it readily follows that

$$g_\lambda^*(F)(x)^2 \leq c \sum_{k=0}^{\infty} 2^{-k(2\lambda-n)} S_{2^k}(F)(x)^2. \tag{4.17}$$

Since $p/2 \leq 1$, (4.17) gives at once

$$g_\lambda^*(F)(x)^p \leq c \sum_{k=0}^{\infty} 2^{-k(2\lambda-n)p/2} S_{2^k}(F)(x)^p. \tag{4.18}$$

Whence integrating (4.18) over R^n and invoking Theorem 4.3 we readily obtain

$$\|g_\lambda^*(F)\|_p^p \leq c \sum_{k=0}^{\infty} 2^{-k(2\lambda-n)p/2} \|S_{2^k}(F)\|_p^p \leq c \sum_{k=0}^{\infty} 2^{-k(p\lambda-n)} \|S(F)\|_p^p,$$

where the above series converges, since $p\lambda - n > 0$. On the other hand, if $p > 2$, then $p/2 > 1$, and Minkowski's inequality applied to (4.17) gives

$$\begin{aligned}\|g_\lambda^*(F)\|_p^2 = \|g_\lambda^*(F)^2\|_{p/2} &\leq c \sum_{k=0}^{\infty} 2^{-k(2\lambda-n)} \|S_{2^k}(F)^2\|_{p/2} \\ &= c \sum_{k=0}^{\infty} 2^{-k(2\lambda-n)} \|S_{2^k}(F)\|_p^2 \\ &\leq c \sum_{k=0}^{\infty} 2^{-k(2\lambda-n)} \|S(F)\|_p^2,\end{aligned}$$

where once again the series converges since $2\lambda - n > 0$ now. ■

These results have numerous applications. We discuss multipliers next.

5. HÖRMANDER'S MULTIPLIER THEOREM

A function m defined in $R^n\backslash(0)$ is said to satisfy a Hörmander condition of order k provided that

$$|m(\xi)| \leq c \quad \text{in} \quad R^n\backslash(0) \tag{5.1}$$

and

$$R^{2|\alpha|-n}\int_{R<|\xi|<2R}|D^\alpha m(\xi)|^2\,d\xi\leqslant c \tag{5.2}$$

for all multi-indices α with $|\alpha|\leqslant k$ and c independent of $R>0$.

Associated to m we introduce, as usual, the muliplier operator

$$T\hat{f}(\xi)=m(\xi)\hat{f}(\xi),\qquad f\in\mathcal{S}(R^n)$$

and seek to establish its L^p continuity properties. For this purpose let $\hat{\phi}(\xi)=|\xi|e^{-|\xi|}$, $\hat{\psi}(\xi)=|\xi|^{k+1}e^{-|\xi|}$ and put $F(x,t)=f*\phi_t(x)$, $G(x,t)=Tf*\psi_t(x)$. We then have

Theorem 5.1. Suppose the multiplier m verifies a Hörmander condition of order k. Then

$$S(G)(x)\leqslant cg_k^*(F)(x), \tag{5.3}$$

where $c=c_{k,m}$ is independent of $f\in\mathcal{S}(R^n)$.

Proof. With the notation $\hat{H}(\xi,t)=(t|\xi|)^k e^{-t|\xi|}m(\xi)$ it is readily seen that we may rewrite $\hat{G}(\xi,t)=2^{k+1}\hat{F}(\xi,t/2)\hat{H}(\xi,t/2)$, or

$$G(w,t)=c\int_{R^n}F(w-y,t/2)H(y,t/2)\,dy. \tag{5.4}$$

Thus by a change of variables and Hölder's inequality from (5.4) it readily follows that

$$\begin{aligned}|G(x+w,t)|^2&\leqslant c\int_{R^n}|H(w+y,t/2)|^2(1+(|y|/t))^{2k}\,dy)\\&\quad\times\left(\int_{R^n}|F(x-y,t/2)|^2(1+(|y|/t))^{-2k}\,dt\right)\\&=I\cdot J,\end{aligned} \tag{5.5}$$

say. We show below that $I\leqslant ct^{-n}$. Suppose for the moment this has been done. Then, by multiplying (5.5) through by $1/vt^n$ and integrating the resulting inequality over $(w,t)\in\Gamma(0)$ with respect to $dw\,dt/t$, we see at once that

$$\begin{aligned}S(G)(x)^2&\leqslant c\int_{R^n}\int_{[0,\infty)}\frac{|F(x-y,t/2)|^2}{(1+(|y|/t))^{2k}}t^{-n}\left(\frac{1}{t^n}\int_{|w|\leqslant t}dw\right)dy\frac{dt}{t}\\&\leqslant cg_k^*(F)(x)^2,\end{aligned}$$

which is precisely (5.3). So it only remains to estimate I. Since $(1+(|y|/t))^{2k} \leq c(1+(|w|/t))^{2k} + c(|w+y|/t)^{2k} \leq c + c(|y+w|/t)^{2k}$, we see that

$$I \leq c\int_{R^n} |H(w+y, t/2)|^2\, dy + c\int_{R^n} |H(w+y, t/2)|^2(|w+y|/t)^{2k}\, dy$$
$$= I_1 + I_2,$$

say. To bound I_1 we note it equals

$$c\int_{R^n} (t|\xi|)^{2k} e^{-t|\xi|}|m(\xi)|^2\, d\xi = ct^{-n}\int_{R^n} |\xi|^{2k} e^{-|\xi|}|m(\xi/t)|^2\, d\xi \leq ct^{-n},$$

which is of the right order. As for I_2 note that it does not exceed

$$ct^{-2k}\sum_{j=1}^{n}\int_{R^n} |y_j^k H(y, t/2)|^2\, dy$$
$$= ct^{-2k}\sum_{j=1}^{n}\int_{R^n}\left|\frac{\partial^k}{\partial\xi_j^k}((t|\xi|)^k e^{-t|\xi|/2} m(\xi))\right|^2 d\xi.$$

In other words, it is bounded by expressions involving integrals of the form

$$\int_{R^n} |D^\alpha(|\xi|^k e^{-t|\xi|/2} m(\xi))|^2\, d\xi, \qquad |\alpha| = k. \tag{5.6}$$

Let $\alpha = \alpha_1 + \alpha_2 + \alpha_3$, $|\alpha_1| + |\alpha_2| + |\alpha_3| = k$. Then the derivatives in (5.6) are linear combinations of monomials $D^{\alpha_1}(|\xi|^k) D^{\alpha_2}(e^{-t|\xi|/2}) D^{\alpha_3} m(\xi)$ each of which can be dominated by

$$c|\xi|^{k-|\alpha_1|} t^{|\alpha_2|} e^{-t|\xi|/2} |D^{\alpha_3} m(\xi)|. \tag{5.7}$$

Thus substituting (5.7) in the integral (5.6) we observe it suffices to estimate expressions of the form

$$c\int_{R^n} (t|\xi|)^{2|\alpha_2|} e^{-t|\xi|}(|\xi|^{|\alpha_3|}|D^{\alpha_3} m(\xi)|)^2\, d\xi. \tag{5.8}$$

A quick way of completing the proof at this stage would be to assume $|\xi|^{|\alpha|}|D^\alpha m(\xi)| \leq c$, $|\alpha| \leq k$, for then it follows immediately that the integral in (5.8) is of order ct^{-n}. These assumptions are stricter than (5.2) but sufficient for many of the applications. Returning to the proof, then, by (5.2) it is enough to bound integrals of the form

$$\int_{R^n} (t|\xi|)^j e^{-t|\xi|}\phi(\xi)\, d\xi, \qquad 0 \leq j \leq 2k, \tag{5.9}$$

where ϕ is a nonnegative function which verifies

$$\frac{1}{R^n}\int_{R<|\xi|<2R}\phi(\xi)\,d\xi \leq c, \qquad \text{all} \quad R>0$$

or equivalently

$$\int_{1<|\xi|<2}\phi(\xi/R)\,d\xi \leq c, \qquad \text{all} \quad R>0. \tag{5.10}$$

Now changing variables in (5.9) gives that the integral there is

$$t^{-n}\int_{R^n}|\xi|^j e^{-|\xi|}\phi(\xi/t)\,d\xi \tag{5.11}$$

so we must show that the integral in (5.11) is bounded independently of t. In order to do this break up this integral

$$\sum_{h=-\infty}^{\infty}\int_{2^h<|\xi|<2^{h+1}}|\xi|^j e^{-|\xi|}\phi(\xi/t)\,d\xi$$
$$\leq c\sum_{h=-\infty}^{\infty}2^{h(j+n)}e^{-2h}\sup_h\left(\int_{1<|\xi|<2}\phi\left(\frac{2^h\xi}{t}\right)d\xi\right).$$

By (5.10) this expression does not exceed

$$c\sum_{h=-\infty}^{n}2^{h(j+n)}e^{-2^h}\leq c,$$

whence (5.9) holds and we are done. ■

Corollary 5.2. Suppose the multiplier m verifies a Hörmander condition of order $k>n/2$. Then the mulplier operator associated to m is bounded in $L^p(R^n)$, $1<p<\infty$.

Proof. It is enough to show that m is an $L^p(R^n)$ multiplier for $2\leq p<\infty$. Let $Tf\,\hat{}\,(\xi)=m(\xi)\hat{f}(\xi), f\in\mathcal{S}(R^n)$ and observe that on account of Theorem 5.1, $S(G)(x)\leq cg_k^*(F)(x)$, and consequently by Thorem 4.5, $\|S(G)\|_p\leq c\|g_k^*(F)\|_p\leq c\|S(F)\|_p\leq c\|f\|_p$, provided $2<p<\infty$ and $k>n/2$. In other words, for those functions $\|Tf\|_p\leq c\|f\|_p$ and T admits a bounded extension to $L^p(R^n)$. The case $p=2$ follows at once from assumption 5.1. ■

6. NOTES; FURTHER RESULTS AND PROBLEMS

Marcinkiewicz and Zygmund noted in 1939 that for an arbitrary linear operator T which is bounded in $L^p(R^n)$, with norm $\|T\|$, the inequality

$$\|(\sum|Tf_j|^2)^{1/2}\|_p\leq\|T\|\,\|(\sum|f_j|^2)^{1/2}\|_p \tag{6.1}$$

also holds. This chapter deals with variants and extensions of this estimate. Rubio de Francia [1982] recently observed that some results of Maurey concerning factorization of operators can be used to show that vector-valued inequalities are, to some extent, equivalent to weighted inequalities. In particular, he showed that given a sequence $\{T_j\}$ of sublinear operators bounded from $L^p(R^n)$ into $L^q(R^n)$, and given $\alpha = p/r$, $\beta = q/r < 1$, the estimate $\|(\sum|T_jf_j|^r)^{1/r}\|_q \leqslant c\|(\sum|f_j|^r)^{1/r}\|_p$ holds if and only if for every nonnegative function u in $L^{\beta'}(R^n)$ there exists a nonnegative function U in $L^{\alpha'}(R^n)$, $1/\alpha + 1/\alpha' = 1/\beta + 1/\beta' = 1$, such that $\|U\|_{\alpha'} \leqslant \|u\|_{\beta'}$ and $\int_{R^n}|T_jf(x)|^r u(x)\,dx \leqslant c\int_{R^n}|f_j(x)|^r U(x)\,dx$, all j. A similar statement holds in case $\alpha, \beta < 1$. In the general context of Banach space valued CZ singular integral operators, the theory was developed by Benedek, Calderón, and Panzone [1962] and Rivière [1971]. The proof of Theorem 5.1 is due to Stein.

Further Results and problems

6.1 Prove (6.1). (*Hint*: It suffices to prove the estimate when $f_j = 0, j \geqslant N$, some large N. Let $\sum$ denote the unit sphere in R^N, put $f(x) = (f_1(x), \ldots, f_N(x))$, $Tf(x) = (Tf_1(x), \ldots, Tf_N(x))$ and observe that by the linearity of T we have $T(y' \cdot f(x)) = y' \cdot Tf(x)$. Thus

$$\int_{R^n} |y' \cdot Tf(x)|^p\,dx \leqslant \|T\|^p \int_{R^n} |y' \cdot f(x)|^p\,dx, \qquad \text{all } y' \text{ in } \textstyle\sum.$$

We now invoke the following property: if $w \in R^N \backslash (0)$, then $\int_{\Sigma}|y' \cdot w|^p\,dy' = c\|w\|_{l^2}^p$ with $c \neq 0$ independent of w; the possible dependence on N and p is irrelevant here since c cancels itself out. Thus integrating the above inequality over $\sum$ it follows that

$$c\int_{R^n} \|Tf(x)\|_{l^2}^p\,dx \leqslant c\|T\|^p \int_{R^n} \|f(x)\|_{l^2}^p\,dx.)$$

6.2 Let $\{I_k\}$ be a sequence of disjoint, open cubes in R^n and put $f_k = \chi_{I_k}$. A simple computation gives that

$$\sum_k Mf_k(x)^r \sim \tau_r(x) = \sum_k \frac{|I_k|^r}{(|x - y_k|^n + |I_k|)^r},$$

where y_k denotes the center of I_k and τ_r is a simple modification of the classical Marcinkiewicz integral of order r corresponding to the cubes $\{I_k\}$. More precisely, if d_k = diameter of I_k and $\eta \geqslant n(r-1)$, then

$$\sum_k \frac{d_k^{n+\eta}}{|x - y_k|^{n+\eta} + d_k^{n+\eta}}$$

is bounded by $\tau_r(x)$ (cf. 5.26 in Chapter V). Theorem 1.1 applied to this case gives

(i) For $1/r < q < \infty$, $\|\tau_r\|_q^q \leq c \sum |I_k|$,
(ii) $\lambda^{1/r}|\{\tau_r > \lambda\}| \leq c \sum |I_k|$, and
(iii) If I is a finite cube with $\bigcup I_k \subseteq I$, then τ_r is exponentially integrable over I (extrapolation from (i)).

This interesting application of Theorem 1.1 is also due to Fefferman and Stein [1971].

6.3 There is, of course, a weighted version of Theorem 1.1. More precisely, with the notation of that theorem and $d\mu(x) = w(x)\,dx$ we have

(i) If $1 \leq p < \infty$, there is a constant c such that $\lambda^p \mu(\{\|Mf\|_{l^r} > \lambda\}) \leq c\| \|f\|_{l^r} \|_{L^p_\mu}^p$ if and only if $w \in A_p$.

(ii) If $1 < p < \infty$, there is a constant c such that $\| \|Mf\|_{l^r} \|_{L^p_\mu} \leq c\| \|f\|_{l^r} \|_{L^p_\mu}$ if and only if $w \in A_p$.

(iii) If $w \in A_\infty$ and I is a finite cube then $\|Mf\|_{l^r}^r$ is exponentially integrable over I (with respect to $d\mu$) whenever $\|f(x)\|_{l^r} < \infty$ and supported on I.

These results are included in the work of Anderson and John [1980]; Heinig [1976] considered results in the range $0 < p < 1$ for $w \in A_1$.

6.4 Anderson and John observed that (i) and (ii) in 6.2 hold with the Lebesgue measure replaced by $d\mu(x) = w(x)\,dx$ provided $w \in A_{rq}$, and also (iii) holds provided that $w \in A_\infty$.

6.5 An operator T defined in some L^p space is said to be linearizable if given any f in L^p there is a linear operator $U = U_f$ on L^p and that $|Tf| = |Uf|$ and $|Ug| \leq |Tg|$ for every g in L^p. Maximal operators corresponding to a sequence of linear operators and operators of the form $Tf(x) = (\int_{\mathcal{O}} |T_w f(x)|^r\,dw)^{1/r}$, $1 \leq r < \infty$, where each T_w is linear, are examples of linearizable operators. Suppose now that $\{T_k\}$ is a sequence of linearizable operators which verifies the following property: there is a fixed r, $1 \leq r < \infty$, so that $\|T_k f\|_{L^r_\mu} \leq c\|f\|_{L^r_\mu}$, all k and $d\mu(x) = w(x)\,dx$, $w \in A_r$, with $c = c_{r,\mu}$ independent of k and f. Then, if $f(x) = \{f_k(x)\}$ and $Tf(x) = \{T_k f_k(x)\}$, also $\| \|Tf\|_{l^r} \|_p \leq c\| \|f\|_{l^r} \|_p$, $1 < p < \infty$, with $c = c_{r,p}$ independent of f. (*Hint*: There are three cases, i.e., $p < r$, $p = r$, $p > r$, and we only discuss the case $1 < p < r$ here. We need the following observation: Assume a linear mapping U is bounded on $L^r(R^n)$, $1 < r < \infty$, and let U^* denote its adjoint. Then if $\|Uf\|_{L^r_\mu} \leq c\|f\|_{L^r_\mu}$, with $c = c_{r,\mu}$ whenever $d\mu(x) = w(x)\,dx$, $w \in A_r$, we also have $\|U^*f\|_{L^{r'}_\nu} \leq c\|f\|_{L^{r'}_\nu}$ whenever $d\nu(x) = v(x)\,dx$, $v \in A_{r'}$, $1/r + 1/r' = 1$. With this observation out of the way fix $f = \{f_k\}$ and let $\{U_k\}$ be the sequence of linear operators corresponding to f and Tf. Then for some vector $v'(x) = \{v_k(x)\}$ with $\| \|v'\|_{l^{r'}}\|_{p'} = 1$ we have

$$\int_{R^n} \Big(\sum_k |T_k f_k(x)|^r\Big)^{p/r} dx = \Big(\sum_k \int_{R^n} U_k f_k(x) v_k(x)\, dx\Big)^p$$
$$= \Big(\sum_k \int_{R^n} f_k(x) U_k^* v_k(x)\, dx\Big)^p$$
$$\leq \| \|f\|_{l^r}\|_p^p \| \|U^* v\|_{l^{r'}}\|_{p'}^p = I \cdot J,$$

say. Now $J < \infty$ since the operators $\{U_k^*\}$ verify the hypothesis with r replaced by r' and p by p', and in this case $r' < p' < \infty$. The proof may now be completed by invoking the uniform boundedness principle, Theorem 2.3 in Chapter I. This result is from Rubio de Francia's work [1980].)

6.6 State and prove the vector-valued version of Theorem 7.1 in Chapter XI.

6.7 For an open interval $I \subset R$ (finite or not), let $S_I f$ demote the linear operator corresponding to the multiplier χ_I, i.e., $S_I f\hat{}(\xi) = \chi_I(\xi)\hat{f}(\xi)$, $f \in L^2(R)$, say. Suppose $\{I_k\}$ is a sequence of open, disjoint intervals contained in R and let $Sf = \{S_{I_k} f_k\}$, where $f = \{f_k\}$ is a sequence of $C_0^\infty(R)$ functions. Show that $\| \|Sf\|_{l^2}\|_p \leq c\| \|f\|_{l_2}\|_p$, $1 < p < \infty$. (*Hint*: If $I = (a, \infty)$, then $S_I f(x) = c_1 f(x) + c_2 e^{ix \cdot a} H(e^{-ita} f(t))(x)$, where H = Hilbert transform; and similarly for $I = (-\infty, b)$ and $I = (a, b)$ since then $S_I f = S_{(a,\infty)} S_{(-\infty,b)} f(x)$. Thus the conclusion is a simple consequence of the boundedness of the Hilbert transform and Theorem 3.1. Was the fact that the I_k's are disjoint used?)

6.8 Let $I_k = (2^k, 2^{k+1})$, $J_k = (-2^{k+1}, -2^k)$, $-\infty < k < \infty$ and let Δ denote the collection of disjoint intervals $\{I_k, J_k\}$. This is called dyadic decomposition of R, although strictly speaking a set of measure 0 has been left out. Let now $S_\Delta f(x) = \{S_{I_k} f, S_{J_k} f\}$, and show that $\|S_\Delta f\|_p \sim \|f\|_p$, $1 < p < \infty$. (*Hint*: The results in Section 3 admit the following "discrete" formulation: let $\phi \in \mathcal{S}(R)$, $\hat{\phi}(\xi) = 1$ if $\frac{1}{2} \leq |\xi| \leq 1$, $\phi_k(x) = 2^k \phi(2^k x)$, and put $g(f)(x) = (\sum_k |f * \phi_k(x)|^2)^{1/2}$. It then follows that $\|g(f)\|_p \leq c\|f\|_p$, $1 < p < \infty$. With this result in mind note that $S_{I_k} f(x) = S_{I_k}(f * \phi_{k+1})(x)$, similarly for $S_{J_k} f$, and consequently by 6.7 and the discrete Littlewood–Paley result we immediately have $\|S_\Delta f\|_p \leq c\|f\|_p$. To prove the opposite inequality use the fact that $\sum_k S_{I_k} f + S_{J_k} f = f$, for f smooth and with $\hat{f}(0) = 0$, and use duality.)

6.9 For $n > 1$, the dyadic decomposition of R^n is obtained by taking the product of the dyadic decomposition of R in each direction. In other words, we write R^n as the union of disjoint "rectangles," each of which is a product of intervals which occur in the dyadic decomposition of the coordinate axes. This family of rectangles ρ is denoted by Δ. Show that if S_ρ denotes the operator corresponding to the mutiplier χ_ρ and $f \in L^p(R^n)$, $1 < p < \infty$, then $\|(\sum_{\rho \in \Delta} |S_\rho f|^2)^{1/2}\|_p \sim \|f\|_p$. (*Hint*: As usual it suffices to prove

$\|(\sum_{\rho\in\Delta}|S_\rho f|^2)^{1/2}\|_p \le c\|f\|_p$, $1 < p < \infty$, and then dualize. There are several proofs of this. A way to go about it is by using induction over the number of variables and Theorem 2.5 in its full strength. Let $I_1, I_2, \ldots$ be an arbitrary enumeration of the dyadic intervals of R and observe that each ρ in Δ is of the form $I_{m_1} \times \cdots \times I_{m_n}$; we call such a rectangle ρ_m, $m = (m_1, \ldots, m_n)$. Now as in 6.8 we observe that $(S_{\rho_m} f)^\wedge(\xi) = \chi_{I_m}(\xi)\hat{\phi}(\xi_1/2^{m_1+1}) \cdots \hat{\phi}(\xi_n/2^{m_n+1})\hat{f}(\xi)$. Thus $\|S_\Delta f\|_p \le \|G(f)\|_p$, where $G(f)(x) = (\sum_{m=(m_1,\ldots,m_n)}|f * \psi_m(x)|^2)^{1/2}$ and $\psi_m(x) = 2^{m_1}\phi(2^{m_1}x_1) \cdots 2^{m_n}\phi(2^{m_n}x_n)$. Let $H_1 = l^2(Z^{n-1})$, the space of $(n-1)$-tuples of square summable sequences, $m' = (m_2, \ldots, m_n)$, and apply 6.8, i.e., the one-dimensional result, to $f * \psi_m(x)$ as a sequence of functions of x_1 indexed by m_1 and with values in H_1. This idea is from Rivière's work [1971].)

6.10 The following limiting case of Theorem 4.5 holds: suppose $0 < p < 2$, and $\lambda = n/p$, then $t^p|\{g_\lambda^*(F) > t\}| \le c\|S(F)\|_p^p$. (*Hint*: let $E = \{M(S(F)^p) > t^p/2\}$, $\mathcal{O}_k = \{S(F) > 2^{nk/p}t\}$ and $\mathcal{U}_k = \{M(\chi_{\mathcal{O}_k}) > 1/2^{nk+1}\}$. First notice that $\mathcal{U}_k \subset E$, all k. Indeed, if $x \in \mathcal{U}_k$, then there is a ball $B(y, r)$ which contains x and such that $|B(y, r) \cap \mathcal{O}_k| > |B(y, r)|/2^{nk+1}$. Therefore

$$\int_{B(y,r)} S(F)(x)^p\,dx \ge \int_{B(y,r)\cap\mathcal{O}_k} S(F)(x)^p\,dx \ge t^{p/2} \quad \text{and} \quad x \in E.$$

On the other hand, by Chebychev's inequality we also have that

$$\begin{aligned} I = t^2|\{x \in R^n\backslash E: g_\lambda^*(F)(x) > t\}| &\le \int_{R^n\backslash E} g_\lambda^*(F)(x)^2\,dx \\ &\le c\sum_{k=0}^\infty 2^{-k(\lambda 2-n)}\int_{R^n\backslash\mathcal{U}_k} S_{2^k}(F)(x)^2\,dx \\ &\le c\sum_{k=0}^\infty 2^{-k(\lambda 2-n)}\int_{R^n\backslash\mathcal{O}_k} S(F)(x)^2\,dx \\ &\le \int_{R^n} S(F)(x)^2\left(\sum_{k=0}^\infty 2^{-k(\lambda 2-n)}\chi_{R^n\backslash\mathcal{O}_k}(x)\right)dx. \end{aligned}$$

Let h be the smallest integer k so that $x \in R^n\backslash\mathcal{O}_k$. Then the sum in the above integral is of order $2^{-h(\lambda p-n)}$ and from the definition of h it also follows that $S(F)(x) \le 2^{nh/p}t$. Thus $I \le c\int_{R^n} S(F)(x)^2(S(F)(x)/t)^{-p(\lambda 2-n)/n}\,dx = ct^{p(\lambda 2-n)/n}\int_{R^n} S(F)(x)^p\,dx$. These bounds combine to give the desired conclusion. The proof given here is due to Aguilera and Segovia [1977] and extends some results of Fefferman [1970].)

6.11 Given $f \in \mathcal{S}'(R^n)$ and a Littlewood-Paley function ϕ, let $F(x, t) = f * \phi_t(x)$ and put

$$G(F)(x) = \left(\int_{R_+^{n+1}} |F(y, t)|^2\psi_t(x-y)\,dy\frac{dt}{t}\right)^{1/2}$$

where ψ is integrable. Determine the continuity properties of $G(F)$ as we did for the particular case of the function $g_\lambda^*(F)$. The results of Madych [1974] are relevant here.

6.12 Let m be a measurable function which verifies $\|m\|_\infty \leq B$ and $\int_I |dm(\xi)| \leq B$ for every dyadic interval I of R. Then m is a bounded multiplier in $L^p(R)$, $1 < p < \infty$, with norm which depends only on B and p. (*Hint*: For f in $C_0^\infty(R)$ put $\hat{g}(\xi) = m(\xi)\hat{f}(\xi)$ and for an interval ρ in the dyadic decomposition Δ of R and $w \in \rho$ set $\chi_{\rho,w}(y) = \chi(\{y \in R : y \in \rho, y < w\})$. Furthermore, let $(S_{\rho,w}f)\hat{\ }(\xi) = \chi_{\rho,w}(\xi)\hat{f}(\xi)$ and if $\rho = (2^k, 2^{k+1})$, say, set

$$F(w) = \frac{1}{2\pi}\int_R \chi_{\rho,w}(\xi)\hat{f}(\xi)e^{ix\cdot\xi}\,d\xi = \frac{1}{2\pi}\int_{(2^k,w)} \hat{f}(\xi)e^{ix\xi}\,d\xi.$$

Since $F'(w) = \hat{f}(w)e^{ixw}$ a.e. integrating by parts the expression

$$S_\rho g(x) = \frac{1}{2\pi}\int_\rho m(\xi)\hat{f}(\xi)e^{ix\xi}\,d\xi$$

yields

$$S_\rho g(x) = m(\xi)F(\xi]_{2^k}^{2^{k+1}} - \int_\rho F(\xi)\,dm(\xi) = I - J,$$

say. Clearly, $I = m(2^{k+1})S_\rho f(x)$ and $J = \int_\rho S_{\rho,\xi}f(x)\,dm(\xi)$. Thus

$$|S_\rho g(x)|^2 \leq 2\left(B^2|S_\rho f(x)|^2 + B\left|\int_\rho |S_{\rho,\xi}f(x)|^2\,dm(\xi)\right|\right), \tag{6.2}$$

and

$$\|S(g)\|_p \leq c\left(\|S(f)\|_p + \left(\int_R\left(\sum_{\rho\in\Delta}\int_\rho |S_{\rho,\xi}f(x)|^2\,|dm(\xi)|\right)^{p/2} dx\right)^{1/p}\right).$$

To complete the proof it now only remains to bound the integral above, and this will follow from the estimate

$$\int_R\left(\sum_{j=1}^N \int_{\rho_i} |S_{\rho_i,\xi}f(x)|^2|\,dm(\xi)|\right)^{p/2} dx$$
$$\leq c\int_R\left(\sum_{j=1}^N |S_{\rho_i}f(x)|^2\left(\int_{\rho_i}|dm(\xi)|\right)\right)^{p/2} dx, \tag{6.3}$$

with a constant c independent of N. In first place notice that for $\xi \in \rho$, $(S_{\rho,\xi}f)\hat{\ }(y) = (S_{\rho,\xi}(S_\rho f))\hat{\ }(y)$, so that actually $S_{\rho,\xi}f$ is a partial sum of $S_\rho f$.

Now, divide each ρ_i whch appears in the sums of (6.3) into k equal parts by partitions ξ_j^i, $j = 0, 1, \ldots, k$, $i = 1, \ldots, N$. By 6.8,

$$\int_R \left(\sum_{i=1}^N \left(\sum_{j=1}^k |S_{\rho_i,\xi_j^i} f(x)|^2 \int_{(\xi_{j-1}^i,\xi_j^i)} |dm(y)| \right) \right)^{p/2} dx$$

$$\leqslant c \int_R \left(\sum_{i=1}^N \left(\sum_{j=1}^k |S_{\rho_i} f(x)|^2 \int_{(\xi_{j-1}^i,\xi_j^i)} |dm(y)| \right) \right)^{p/2} dx$$

$$\leqslant c \int_R \sum_{j=1}^N |S_{\rho_i} f(x)|^2 \left(\int_{\rho_i} |dm(y)| \right) \Bigg)^{p/2} dx,$$

which letting $k \to \infty$ gives (6.3) and we are done. For further details and related results the reader may consult Kurtz's work [1980]).

6.13 We think, now of R^n as divided into 2^n "quadrants" by the coordinate axes, the first quadrant being the set $\{x = (x_1, \ldots, x_n) : x_i > 0,\ i = 1, \ldots, n\}$ and so on. Let m be a measurable function which verifies $\|m\|_\infty \leqslant B$, $m \in C^n$ in each quadrant of R^n and so that

$$\sup_{\xi_{k+1},\ldots,\xi_n} \int_\rho \left| \frac{\partial^k}{\partial \xi_1 \cdots \partial \xi_k} m(\xi) \right| d\xi_1 \cdots d\xi_k \leqslant B \qquad \text{for} \quad 0 < k \leqslant n,$$

ρ any dyadic rectangle in R^k, and any permutation of $(\xi_1, \ldots, \xi_n)$. If $1 < p < \infty$, show that m is a bounded multiplier in $L^p(R^n)$, with norm which depends only on B and p. (*Hint*: The proof is similar to that of 6.12. Indeed we decompose $S_\rho g$ into a sum of 2^n pieces now, each of which is handled as before. This result, which has many important applications, is known as the Marcinkiewicz multiplier theorem.)

6.14 Rivière [1971] observed that there is a theory of vector valued multipliers as well. These general results have many interesting applications.

CHAPTER

XIII

The Good λ Principle

1. GOOD λ INEQUALITIES

A good λ inequality is a principle which allows us to derive norm and even local or pointwise estimates of one operator in terms of another provided they satisfy an *a priori* relation of probabilistic or measure theoretic nature. We have already considered some instances of this principle in 4.20 in Chapter VI and Theorem 4.4 in Chapter X.

Definition 1.1. Given a positive, doubling, regular Borel measure μ on R^n we say that the operators T_1, T_2 verify a good λ inequality with respect to μ provided the following three properties hold:

(i) T_1, T_2 are sublinear and positive.

(ii) $\{T_1 f > t\}$ is an open set of finite Lebesgue measure for each f in $C_0^\infty(R^n)$ and $t > 0$.

(iii) If a ball B contains a point x where $T_1 f(x) \leq \lambda$, then to each $0 < \eta < 1$ there corresponds $\gamma = \gamma(T_1, T_2, \eta)$ independent of λ, B and f so that

$$\mu(\{y \in B\colon T_1 f(y) > 3\lambda,\ T_2 f(y) \leq \gamma\lambda\}) \leq \eta\mu(B). \tag{1.1}$$

(1.1) expresses the control in measure alluded to above and the norm estimate is given by

Theorem 1.2. Suppose T_1, T_2 verify a good λ inequality with respect to μ and assume that $\|T_1 f\|_{L^p_\mu} < \infty$, $0 < p < \infty$, for f in $C_0^\infty(R^n)$. Then there is a constant $c = c_{\mu,p}$ independent of $f \in C_0^\infty(R^n)$ so that

$$\|T_1 f\|_{L^p_\mu} \leq c\|T_2 f\|_{L^p_\mu}. \tag{1.2}$$

Proof. Let $f \in C_0^\infty(R^n)$ and assume that $\|T_2 f\|_{L^p_\mu} < \infty$, for otherwise there is nothing to prove. $\mathcal{O}_\lambda = \{T_1 f > \lambda\}$ is an open set of finite Lebesgue measure and consequently to each y in $\mathcal{O}_\lambda$ we may assign a ball $B(y, r_y)$ centered at y and of radius r_y so that

$$B(y, r_y) \subseteq \mathcal{O}_\lambda, \qquad B(y, 3r_y) \cap (R^n \backslash \mathcal{O}_\lambda) \neq \emptyset. \tag{1.3}$$

Let

$$\mathcal{U}_\lambda = \{T_1 f > 3\lambda,\ T_2 f \leq \gamma\lambda\}, \tag{1.4}$$

where γ is a constant yet to be chosen; clearly, $\mathcal{U}_\lambda \subseteq \mathcal{O}_\lambda$. To estimate $\mu(\mathcal{U}_\lambda)$ we consider K, an arbitrary compact subset of $\mathcal{U}_\lambda$. Since $K \subseteq \bigcup_{y \in \mathcal{O}_\lambda} B(y, r_y)$ there exist finitely many balls among the $B(y, r_y)$'s so that actually

$$K \subseteq \bigcup\nolimits_{\text{finitely many}} B(y, r_y). \tag{1.5}$$

Let now $B_1, \ldots, B_m$ be a disjoint subfamily of the family in (1.5) such that

$$K \subseteq \bigcup_{j=1}^{m} 3B_j, \tag{1.6}$$

and let $0 < \eta < 1$ be another constant yet to be selected. Since we will invoke the good λ principle, γ in (1.4) will automatically be determined by property (iii) in Definition 1.1 once η is chosen. Now, since $K \subseteq \mathcal{U}_\lambda$ we can sharpen (1.6) to

$$K \subseteq \bigcup_{j=1}^{m} (\{y \in 3B_j\colon T_1 f(y) > 3\lambda,\ T_2 f(y) \leq \gamma\lambda\}). \tag{1.7}$$

Since it is clear from (1.3) that each ball $3B_j$ contains a point x_j where $T_1 f(x_j) \leq \lambda$, by the good λ principle (still with η to be chosen and with γ to be determined by (1.1)) on account of (1.1), we have that

$$\begin{aligned} \mu(K) &\leq \sum_{j=1}^{m} \mu(\{y \in 3B_j\colon T_1 f(y) > 3\lambda,\ T_2 f(y) \leq \gamma\lambda\}) \\ &\leq \eta \sum_{j=1}^{m} \mu(3B_j) \leq c\eta \sum_{j=1}^{m} \mu(B_j), \end{aligned} \tag{1.8}$$

where the last estimate follows since μ is doubling, and c and η are independent of λ. Furthermore, since the B_j's are disjoint balls totally contained in $\mathcal{O}_\lambda$, and K is an arbitrary compact subset of $\mathcal{U}_\lambda$, from (1.8) we immediately see that

$$\mu(\mathcal{U}_\lambda) \leq c\eta\mu(\mathcal{O}_\lambda), \qquad \lambda > 0. \tag{1.9}$$

This is all we need to complete the proof. Indeed, first observe that by (1.9)

$$\begin{aligned}\|T_1 f\|_{L^p_\mu}^p &= 3^p p \int_{[0,\infty)} \mu(\{T_1 f > 3\lambda\})\lambda^{p-1}\,d\lambda \\ &\leqslant 3^p p \int_{[0,\infty)} \mu(\mathscr{U}_\lambda)\lambda^{p-1}\,d\lambda + 3^p p \int_{[0,\infty)} \mu(\{T_2 f > \gamma\lambda\})\lambda^{p-1}\,d\lambda \\ &\leqslant c\eta 3^p p \int_{[0,\infty)} \mu(\mathcal{O}_\lambda)\lambda^{p-1}\,d\lambda \\ &\quad + \left(\frac{3}{\gamma}\right)^p p \int_{[0,\infty)} \mu(\{T_2 f > \lambda\})\lambda^{p-1}\,d\lambda \\ &= c\eta 3^p \|T_1 f\|_{L^p_\mu}^p + (3/\gamma)^p \|T_2 f\|_{L^p_\mu}^p. \end{aligned} \tag{1.10}$$

We now choose η so that $c\eta 3^p = \frac{1}{2}$ and γ so that (1.1) holds. With this choice, and since by assumption $\|T_1 f\|_{L^p_\mu} < \infty$, we may rewrite (1.10) as $\|T_1 f\|_{L^p_\mu}^p \leqslant 2(3/\gamma)^p \|T_2 f\|_{L^p_\mu}^p$, and we have finished. ■

2. WEIGHTED NORM INEQUALITIES FOR MAXIMAL CZ SINGULAR INTEGRAL OPERATORS

In considering the estimates of interest to us, it is important to decide what restrictions to impose on the weights. In view of the following observation our basic assumption will be the A_p condition.

Proposition 2.1. Let Hf denote the Hilbert transform of f. If μ is a non-negative, regular, Borel measure and H is bounded from $L^p_\mu(R)$ into wk-$L^p_\mu(R)$, $1 \leqslant p < \infty$, then

(i) μ is doubling,
(ii) $Mf(x) \leqslant c(M_\mu(|f|^p)(x))^{1/p}$, and
(iii) μ is absolutely continuous with respect to Lebesgue measure and $d\mu(x) = w(x)\,dx$, where w is an A_p weight.

Proof. Fix an interval I and let I_r and I_l denote the abutting intervals to I, of equal diameter, which lie to the right and to the left of I respectively. Clearly (i) follows from the estimates

$$\mu(I_r),\ \mu(I_l) \leqslant c\mu(I) \tag{2.1}$$

with c independent of I, which we now prove. For a nonidentically 0 function f in $L^p_\mu(R)$ consider the restriction $|f|\chi_I$ of $|f|$ to I. Clearly $|f|\chi_1 \in L^p_\mu(R)$ and for $x \in I_r$ we have $H(|f|\chi_I)(x) \geq ((1/2|I|)\int_I |f(y)|\,dy)\chi_{I_r}(x)$. Thus $I_r \subseteq \{|H(|f|\chi_I)| > (1/2|I|)\int_I |f(y)|\,dy\}$ and

$$\mu(I_r) \leq c\left(2|I| \Big/ \int_I |f(y)|\,dy\right)^p \|f\chi_I\|^p_{L^p_\mu}. \tag{2.2}$$

Obviously a similar estimate holds for $\mu(I_l)$. By putting $f = \chi_I$ in (2.2) we immediately see that $\mu(I_r) \leq c\mu(I)$ and (2.1) holds. Now that we know μ is doubling we may rewrite (2.2) as

$$\left(\frac{1}{|I|}\int_I |f(y)|\,dy\right)^p \leq c\frac{1}{\mu(I)}\int_I |f(y)|^p\,d\mu(y), \qquad \text{all} \quad I,$$

and (ii) follows. Moreover, since μ is doubling and regular, by virtue of (ii) and Theorem 1.1 in Chapter IX the Hardy-Littlewood maximal operator maps $L^p_\mu(R)$ into wk-$L^p_\mu(R)$ and, by Theorem 2.1 in Chapter IX, μ is absolutely continuous with respect to Lebesgue measure and $d\mu(x) = w(x)\,dx$, $w \in A_p$. ■

We pass now to consider the weighted norm inequalities for CZ singular integral operators. Since we will invoke the good λ principle it is convenient to deal with the maximal CZ operators directly.

Theorem 2.2. Suppose the CZ kernel k verifies assumptions (i)-(iv) in Chapter XI and in addition

(v) $|k(x)| \leq c_5|x|^{-n}$, $x \neq 0$. (2.3)

Furthermore, let T^* denote the maximal CZ singular integral operator associated to k and let $d\mu(x) = w(x)\,dx$, $w \in A_p$. Then there is a constant $c = c_p$ such that

$$\|T^*f\|_{L^p_\mu} \leq c\|f\|_{L^p_\mu}, \qquad 1 < p < \infty. \tag{2.4}$$

Proof. We verify that the good λ principle holds for $T_1 = T^*$ and $T_2 = M$; in other words, we check that properties (i)-(iii) in Definition 1.1 hold for these operators.

(i) Suppose $f \in C_0^\infty(R^n)$ and observe that since T^* is of weak-type (1, 1), then $|\mathcal{O}_\lambda| = |\{T^*f > \lambda\}|$ is finite for each $\lambda > 0$. To see that $\mathcal{O}_\lambda$ is open let $\{x_m\}$ be a sequence of points in R^n so that $T^*f(x_m) \leq \lambda$ and $x_m \to x$; we claim that also $T^*f(x) \leq \lambda$. First, put

$$\begin{aligned} T_\varepsilon f(x) &= \int_{|y|>\varepsilon} k(y)(f(x-y) - f(x_m - y))\,dy \\ &\quad + \int_{|y|>\varepsilon} k(y)f(x_m - y)\,dy = I_m + J_m, \end{aligned}$$

say, and note that $I_m \to 0$ as $m \to \infty$ and $|J_m| \le \lambda$, all m. The latter is easy to check since $|J_m| = |T_\varepsilon f(x_m))| \le \lambda$, all m. As for the former observe that for each $\varepsilon > 0$ the function $k_\varepsilon(y)(f(x-y) - f(x_m - y))$ tends to 0 as $m \to \infty$, is integrable, and is pointwise majorized by the integrable function $c|k_\varepsilon(y)|\chi_{B(0,N)}(y)$, $N = N(f)$ sufficiently large. Thus by the Lebesgue dominated convergence theorem it follows that $I_m \to 0$ as $m \to \infty$. This in turn implies that $|T_\varepsilon f(x)| \le \lambda$ for every $\varepsilon > 0$ and consequently $T^*f(x) \le \lambda$ as well.

(ii) Let $B = B(0, R)$ be a ball with large enough radius so that it contains the support of $f \in C_0^\infty(R^n)$ in its interior. Then

$$\|T^*f\|_{L^p_\mu}^p = \left(\int_{10B} + \int_{R^n\setminus 10B}\right) T^*f(x)^p \, d\mu(x) = I + J,$$

say. To estimate I note that by Hölder's inequality and Theorem 6.2 in Chapter XI,

$$I \le \left(\int_{10B} T^*f(x)^{pr'}\,dx\right)^{1/r'} \left(\int_{10B} w(x)^r\,dx\right)^{1/r}$$
$$\le c\|f\|_{pr'}^p \left(\int_{10B} w(x)^r\,dx\right)^{1/r}, \qquad 1/r + 1/r' = 1.$$

Since $w \in RH_r$, for some $r > 1$, with this choice of r it follows that $I < \infty$. As for J, observe that for $x \in R^n\setminus 10B$ and by (v) above $T^*f(x) \le \int_{|y|\le R} |k(x-y)||f(y)|\,dy \le c_5|x|^{-n}\|f\|_1$. Consequently, by Proposition 4.4 in Chapter IX $J \le c\int_{R^n\setminus 10B} |x|^{-np}\,d\mu(x) < \infty$.

(iii) This is the heart of the matter. Suppose B contains a point w so that $T^*f(w) \le \lambda$ and let $E = \{x \in B: T^*f(x) > 3\lambda,\ Mf(x) \le \gamma\lambda\}$, where γ is a constant yet to be chosen. Recall that by Proposition 4.4 in Chapter IX there are constants $c = c_\mu$, r, independent of E, B, so that $\mu(E)/\mu(B) \le c(|E|/|B|)^r$. Thus it suffices to verify (iii) for the Lebesgue measure instead. Write $f = f\chi_{2B} + f\chi_{R^n\setminus 2B} = f_1 + f_2$, say and observe that $E \subseteq \{x \in B: T^*f_1(x) > \lambda\} \cup \{y \in B: T^*f_2(x) > 2\lambda,\ Mf(x) \le \gamma\lambda\} = E_1 \cup E_2$, say. We will show that, given $0 < \eta < 1$, there is $\gamma = \gamma(\eta)$ sufficiently small that simultaneously $|E_1| \le \eta|B|$ and $E_2 = \varnothing$; once this is achieved the proof will be complete.

In the first place observe that since T^* is of weak-type (1, 1)

$$\lambda|E_1| \le c\|f_1\|_1 = \int_{2B} |f(y)|\,dy. \tag{2.5}$$

Let I be a cube concentric with B such that $2B \subset I$ and $|I| \sim |B|$. Then the right-hand side of (2.5) is dominated by

$$c\frac{|I|}{|I|}\int_I |f(y)|\,dy \le c|B| \inf_I Mf \le c|B| \inf_B Mf. \tag{2.6}$$

If $E = \emptyset$ there is nothing to prove. If on the other hand $E \neq \emptyset$, then $\inf_B Mf \leqslant \inf_E Mf \leqslant \gamma\lambda$. Consequently, by (2.5) and (2.6),

$$|E_1| \leqslant c\gamma |B|, \tag{2.7}$$

where c is an absolute constant independent of E, B, and f.

Next we show that E_2 is empty provided γ is judiciously chosen. For this purpose we estimate

$$\begin{aligned} T_\varepsilon f_2(x) &= \left(\int_{|x-y|>\varepsilon} k(x-y) f_2(y)\,dy - \int_{|w-y|>\varepsilon} k(x-y) f_2(y)\,dy\right) \\ &\quad + \left(\int_{|w-y|>\varepsilon} k(x-y) f_2(y)\,dy - \int_{|w-y|>\varepsilon} k(w-y) f_2(y)\,dy\right) \\ &\quad + \int_{|w-y|>\varepsilon} k(w-y) f_2(y)\,dy = I + J + K, \end{aligned}$$

say. We consider J first and observe that

$$\begin{aligned} |J| &\leqslant \int_{|w-y|>\varepsilon} |k(x-y) - k(w-y)|\,|f_2(y)|\,dy \\ &\leqslant c_4 \int_{R^n\setminus 2B} \frac{|x-w|}{|\bar{x}-y|^{n+1}} |f(y)|\,dy, \end{aligned} \tag{2.8}$$

where $\bar{x} \in E$, and in particular $Mf(\bar{x}) \leqslant \gamma\lambda$. Now $|x - w| \leqslant cr$, r = radius of B, and from (2.8) we see at once that

$$|J| \leqslant c \int_{R^n} \frac{r^{-n}}{1 + (|\bar{x}-y|/r)^{n+1}} |f(y)|\,dy \leqslant cMf(\bar{x}) \leqslant c_J \gamma\lambda, \tag{2.9}$$

where c_J is independent of B and f. To bound K we distinguish two cases, namely, (i) $\varepsilon < 2r$ and (ii) $\varepsilon \geqslant 2r$. If (i) holds, then $f_2 = f$ in the integral and $|K| \leqslant T^*f(w) \leqslant \lambda$. If on the other hand we are in case (ii), then we may also assume that $\varepsilon \geqslant r$ for otherwise the part of the integral between ε and r vanishes. Thus

$$\begin{aligned} |K| &\leqslant \left|\int_{|w-y|>r} k(w-y) f_2(y)\,dy - \int_{|w-y|>3r} k(w-y) f_2(y)\,dy\right| \\ &\quad + \left|\int_{|w-y|>3r} k(w-y) f_2(y)\,dy\right| = |K_1| + |K_2|, \end{aligned}$$

say. Clearly $|K_2| \leqslant \lambda$. As for $|K_1|$ it is bounded by

$$\begin{aligned} \int_{r<|w-y|<3r} |k(w-y)|\,|f(y)|\,dy &\leqslant c_5 \int_{r<|w-y|<3r} |w-y|^{-n} |f(y)|\,dy \\ &\leqslant c(3r)^{-n} \int_{|w-y|<3r} |f(y)|\,dy \leqslant c \inf_E Mf \leqslant c_K \gamma\lambda. \end{aligned}$$

Thus collecting the estimates we get

$$|K| \leqslant \lambda + c_K \gamma \lambda, \tag{2.10}$$

again with c_K indepedent of λ, B, and f.

It only remains to estimate I. Since $\chi_{R^n \setminus B(x,\varepsilon)} - \chi_{R^n \setminus B(w,\varepsilon)} = \chi_{B(w,\varepsilon)} - \chi_{B(x,\varepsilon)}$ it readily follows that

$$|I| \leqslant \int_{R^n \setminus 2B} |k(x-y)| |\chi_{B(w,\varepsilon)}(y) - \chi_{B(x,\varepsilon)}(y)| |f(y)| \, dy.$$

Moreover, since $\chi_{R^n \setminus 2B}(y)(\chi_{B(w,\varepsilon)}(y) - \chi_{B(x,\varepsilon)}(y))$ vanishes for $\varepsilon < r$, we may assume $\varepsilon \sim 2^k r$, $k = 0, 1, \ldots$. On the other hand, if this function is not 0, it readily follows that also $|x-y| \sim 2^k r$. Hence

$$\begin{aligned} |I| &\leqslant c \int_{|x-y| \sim 2^k r} |k(x-y)| |f(y)| \, dy \leqslant c(2^k r)^{-n} \int_{|x-y| \sim 2^k r} |f(y)| \, dy \\ &\leqslant c \inf_E Mf \leqslant c_I \gamma \lambda, \end{aligned} \tag{2.11}$$

where c_I is an absolute constant.

Thus combining (2.9), (2.10) and (2.11) we see immediately that $|T_\varepsilon f_2(x)| \leqslant \gamma\lambda(c_I + c_J + c_K) + \lambda$, all $\varepsilon > 0$. It is at this point that we pick γ so that $\gamma(c_I + c_J + c_K) < 1$, in addition γ must be small enough so that $c\gamma < \eta$, where $0 < \eta < 1$ is given and c is the constant in (2.7). In other words, we have proved that for x in E_2, $\sup_\varepsilon |T_\varepsilon f_2(x)| < 2\lambda$, which in turn implies $E_2 = \varnothing$, and that $|E_1| \leqslant \eta |B|$. ■

3. WEIGHTED WEAK-TYPE (1, 1) FOR CZ SINGULAR INTEGRAL OPERATORS

Since the proof follows along the lines of Theorem 2.14 in Chapter V, we shall be brief.

Theorem 3.1. Suppose the CZ kernel k verifies the assumptions of Theorem 2.2 and $d\mu(x) = w(x)\,dx$, $w \in A_1$. If $f \in L_\mu(R^n)$, then $T^*f \in$ wk-$L_\mu(R^n)$, and there is a constant c independent of f such that

$$\lambda\mu(\{T^*f > \lambda\}) \leqslant c\|f\|_{L_\mu}, \qquad \lambda > 0. \tag{3.1}$$

Proof. Let $\{I_j\}$ be the Calderón–Zygmund (interval) decomposition of $|f|$ at level $A\lambda$, where A is yet to be chosen, and set $\Omega = \bigcup I_j$, and $f = g + b$, where g and b are the good and bad parts of f corresponding to the I_j's, respectively. Clearly $T^*f(x) \leqslant T^*g(x) + T^*b(x)$ and it suffices to show (3.1) with f replaced by g and b on the left-hand side there. The estimate for g is straightforward. Observe that since $w \in A_1$,

$$\mu(I_j)\frac{1}{|I_j|}\int_{I_j}|f(y)|\,dy \leqslant c\int_{I_j}|f(y)|\,d\mu(y), \qquad \text{all } j. \tag{3.2}$$

Therefore, since also $w \in A_2$, by Theorem 2.2 and (3.2) it follows that

$$\begin{aligned}\lambda^2\mu(\{T^*g > \lambda\}) &\leqslant c\|g\|_{L^2_\mu}^2\\ &\leqslant cA\lambda\left(\int_{R^n\setminus\Omega}|f(y)|\,d\mu(y) + \sum_j \mu(I_j)\frac{1}{|I_j|}\int_{I_j}|f(y)|\,dy\right)\\ &\leqslant cA\lambda\,\|f\|_{L_\mu},\end{aligned}$$

which leads to an estimate of the right order. (3.2) also implies that if $\Omega_1 = \bigcup_j(2I_j)$, then $\mu(\Omega_1) \leqslant c\|f\|_{L_\mu}/\lambda$, and consequently it is enough to show that

$$\lambda\mu(\{R^n\setminus\Omega_1\colon T^*b > \lambda\}) \leqslant c\|f\|_{L_\mu}. \tag{3.3}$$

For this purpose fix x in $R^n\setminus\Omega_1$, $\varepsilon > 0$, and consider

$$T_\varepsilon b(x) = \sum_j \int_{(R^n\setminus B(x,\varepsilon))\cap I_j} k(x-y)b(y)\,dy. \tag{3.4}$$

We separate the I_j's into three subfamilies, to wit (i) those I_j's contained in $R^n\setminus B(x,\varepsilon)$, (ii) those I_j's contained in $B(x,\varepsilon)$, and (iii) the rest.

Since $\int_{I_j} b(y)\,dy = 0$ and $|b(y)| \leqslant |f(y)| + cA\lambda$, the sum in (3.4) extended over the j's in the first family is readily seen to be dominated, in absolute value, by

$$\begin{aligned}&\sum\int_{I_j}|k(x-y) - k(x-y_j)||b(y)|\,dy\\ &\qquad\leqslant \sum\int_{I_j}|k(x-y) - k(x-y_j)|(|f(y)| + cA\lambda)\,dy,\end{aligned} \tag{3.5}$$

where y_j denotes the center of I_j.

All summands corresponding to the second family vanish.

As for the third family, note that if I_j belongs there, then $I_j \cap B(x,\varepsilon) \neq \varnothing$ and $I_j \cap (R^n\setminus B(x,\varepsilon)) \neq \varnothing$ simultaneously. Moreover, since $\varepsilon > 2L_j = 2$ (sidelength of I_j), there is $R > 0$ so that $\bigcup I_j \subseteq B(x,2R)\setminus B(x,R/2)$. Now let

$$c_j = \frac{1}{|I_j|}\int_{I_j}\chi_{R^n\setminus B(x,\varepsilon)}(y)b(y)\,dy \tag{3.6}$$

and observe that

$$\int_{R^n} (c_j - \chi_{R^n\setminus B(x,\varepsilon)}(y)b(y))\chi_{I_j}(y)\,dy = 0. \tag{3.7}$$

From (3.6) it also readily follows that

$$|c_j| \leqslant cA\lambda. \tag{3.8}$$

To estimate the sum in (3.4) extended over the j's in the third family we note that it is bounded, in absolute value, by

$$\begin{aligned}
&\left|\sum \int_{I_j} k(x-y)\chi_{R^n\setminus B(x,\varepsilon)}(y)b(y)\,dy\right| \\
&\quad\leqslant \left|\sum \int_{I_j} (k(x-y) - k(x-y_j))(\chi_{R^n\setminus B(x,\varepsilon)}(y)b(y) - c_j)\,dy\right| \\
&\qquad + cA\lambda \int_{R^n} \left(\sum \chi_{I_j}(y)\right)|k(x-y)|\,dy \\
&\quad\leqslant \sum_{I_j} |k(x-y) - k(x-y_j)|(|f(y)| + cA\lambda)\,dy \\
&\qquad + cA\lambda \int_{B(x,2R)\setminus B(x;R/2)} |k(x-y)|\,dy.
\end{aligned} \tag{3.9}$$

Furthermore, observe that the last integral in (3.9) is dominated by $c_5 \int_{B(x,2R)\setminus B(x,R/2)} |x-y|^{-n}\,dy \leqslant c$, and consequently the last summand there by

$$cA\lambda. \tag{3.10}$$

Thus choosing A so that $cA < \frac{1}{2}$, where c is the constant in (3.10), and combining (3.5) and (3.9) it readily follows that

$$\begin{aligned}
T^*b(x) &\leqslant \sum_j \int_{I_j} |k(x-y) - k(x-y_j)|(|f(y)| + c\lambda)\,dy + \frac{\lambda}{2} \\
&= \phi(x) + \lambda/2,
\end{aligned}$$

say, and consequently $\{T^*b > \lambda\} \subseteq \{\phi(x) > \lambda/2\}$, $x \in R^n\setminus\Omega_1$. Moreover, by Chebychev's inequality we get that

$$\begin{aligned}
\left(\frac{\lambda}{2}\right)\mu(\{x \in R^n\setminus\Omega_1 : \phi(x) > \lambda\}) &\leqslant \int_{R^n\setminus\Omega_1} \phi(x)\,d\mu(x) \leqslant \sum_j \int_{I_j} (|f(y)| + c\lambda) \\
&\quad \times \left(\int_{R^n\setminus 2I_j} |k(x-y) - k(x-y_j)|\,d\mu(x)\right) dy = M,
\end{aligned} \tag{3.11}$$

say. We take a closer look at the inner integrals in (3.11). If L_j = sidelength of I_j, then $|k(x-y) - k(x-y_j)| \leqslant c_4(L_j/|x-y|^{n+1})$, and we claim that each integral is bounded by

$$cL_jN = cL_j \int_{R^n\setminus 2I_j} \frac{1}{|x-y|^{n+1}}\, d\mu(x) \leq \frac{c\mu(I_j)}{|I_j|}. \tag{3.12}$$

Indeed, first observe that

$$\begin{aligned} N &\leq c\sum_{k=1}^{\infty} \int_{|x-y|\sim 2^kL_j} \frac{1}{|x-y|^{n+1}}\, d\mu(x) \\ &\leq \sum_{k=1}^{\infty} (2^kL_j)^{-(n+1)}\mu(B(y, 2^kL_j)). \end{aligned}$$

Also, since $w \in A_1$, by (2.9) in Chapter IX, $\mu(B(y, 2^kL_j))/(2^kL_j)^n \leq c\mu(B(y, L_j))/L_j^n \leq c\mu(I_j)/|I_j|$, and consequently $N \leq c\mu(I_j)/|I_j|L_j$, as anticipated. Substituting now (3.12) into (3.11), gives that

$$\begin{aligned} M &\leq c\sum_j \int_{I_j} (|f(y)| + c\lambda)\left(\frac{\mu(I_j)}{|I_j|}\right) dy \\ &\leq c\sum_j \int_{I_j} |f(y)|\, d\mu(y) + c\lambda \sum_j \mu(I_j) \\ &\leq c\|f\|_{L^1_\mu} + c\lambda|\Omega| \leq c\|f\|_{L^1_\mu}, \end{aligned}$$

and the proof is complete. ∎

4. NOTES; FURTHER RESULTS AND PROBLEMS

In 1970 Burkholder and Gundy introduced a technique that has since been used quite effectively in establishing the continuity of various kinds of operators, namely the so-called good-λ inequalities. The main application of this idea in this chapter, following Coifman and Fefferman [1974], is to the boundedness of CZ singular integral operators in weighted L^p spaces. In the particular case of the Hilbert transform this result is due to Hunt, Muckenhoupt and Wheeden [1973].

Further Results and Problems

4.1 Suppose $f \in C_0^\infty(R^n)$ and $d\mu(x) = w(x)\, dx$, with $w \in A_\infty$. Show that $\|Mf\|_{L^p_\mu} \leq c\|M^\# f\|_{L^p_\mu}$, $1 < p < \infty$, where $c = c_{p,\mu}$ is independent of f. (*Hint*: It suffices to show that $T_1 = M$ and $T_2 = M^\#$ verify conditions (iii) in Definition 1.1, and this is not hard. Since $w \in A_\infty$, as in Theorem 2.2, we see that in fact we may restrict ourselves to the Lebesgue measure.

We show then that given an open ball B and $\gamma > 0$, $|E| = |\{x \in B: Mf(x) > 3\lambda,\ M^{\#}f(x) \leq \gamma\lambda\}| \leq c\gamma|B|$, where c is a dimensional constant independent of B, λ, γ, f, provided B contains a point w where $Mf(w) \leq \lambda$. Note that this last condition implies that $|f|_I \leq \lambda$ for every open cube I containing B, so if now I denotes a fixed open cube concentric with and containing B, $|I| \sim |B|$, and $x \in E$, then also $M(f\chi_I)(x) > 3\lambda$ and $M((f - f_I)\chi_I)(x) > 2\lambda$. Therefore either E is empty or else $|E| \leq |\{M((f - f_I)\chi_I) > 2\lambda\}| \leq c \int_I |f(y) - f_I|\, dy/\lambda \leq c|B|\gamma\lambda/\lambda = c\gamma|B|$. Since for $f(x) = c$ a.e. we have $Mf(x) = c$ and $M^{\#}f(x) = 0$, it is clear that the result does not hold in general unless some restrictions are placed on f. Some of these restrictions are: $f \in L^q_\mu(R^n)$, some $q < p$, Fefferman-Stein [1972], $|\{Mf > \varepsilon\}| < \infty$ for each $\varepsilon > 0$, Calderón and Scott [1978], and $\inf(1, Mf) \in L^p$, Journé [1983].)

4.2 The assumption $\|T_1 f\|_{L^p_\mu} < \infty$ is essential for the validity of Theorem 1.2. The following pertinent example is due to Miyachi and Yabuta [1984]: Suppose $a > 1$, $\delta, p, \alpha > 0$ and $1 + \delta/p \leq a^2$. Then there exist nonnegative, measurable functions f, g on R^n such that $|\{f > a\lambda, g \leq \gamma\lambda\}| \leq \gamma^\delta|\{f > \lambda\}|$ for all $\gamma, \lambda > 0$, $\|g\|_p < \infty$, $|\{f > \lambda\}| = O(e^{\lambda^{-\alpha}})$ as $\lambda \to 0$, and $\|f\|_p = \infty$. To see this choose a disjoint sequence $\{E_k\}$ of measurable subsets of R^n with $|E_k| = e^{a^{k\alpha}}$ and define $\{\gamma_k\}$ by $\gamma_k^\delta|E_k| = |E_{k-1}|$. Then put $f = a^{-k}$ and $g = a^{-(k+1)}\gamma_{k+1}$ on E_k and $f = g = 0$ outside $\bigcup_{k=0}^\infty E_k$.

4.3 Under the usual assumptions on the vector valued CZ kernel k, Theorem 2.2 and 3.1 remain valid. In other words, the Littlewood-Paley and Lusin functions are bounded in $L^p_\mu(R^n)$, $d\mu(x) = w(x)\,dx$, $w \in A_p$, $1 < p < \infty$, and of weak-type (1, 1) on $L^p_\mu(R^n)$ when $w \in A_1$. As for the g^*_λ function, one must be careful with range of the parameter λ. These and related results are discussed in full detail in the work of Strömberg and Torchinsky [1980]. An immediate corollary to these results and Theorem 5.1 in Chapter XII is a weighted version of the Hörmander multiplier theorem. Another application is to the Marcinkiewicz multiplier theorem 6.13 in Chapter XII. For instance, the proof given in 6.12, Chapter XII, for the case $n = 1$, extends immediately to the weighted setting, provided $w \in A_p$, by simply using the corresponding weighted inequalities in estimates such as (6.2) there (cf. Kurtz [1980]).

4.4 The conclusion of 6.5 in Chapter XII also applies to the Lusin and g^*_λ functions, for instance. We must be careful, however, with the range of values of λ (cf. Rubio de Francia [1980]).

4.5 The following estimates hold for the Calderón-Zygmund operators T introduced in 8.11-8.12 in Chapter XI: $\lambda\mu(\{|Tf| > \lambda\}) \leq c\|f\|_{L_\mu}$, $d\mu(x) = w(x)\,dx$, $w \in A_1$, and $\|Tf\|_{L^p_\mu} \leq c\|f\|_{L^p_\mu}$, $d\mu(x) = w(x)\,dx$, $w \in A_p$, $1 < p < \infty$. Also, and with the same assumptions on T and μ as above, $\lambda\mu(\{\|Tf_k\|_{l^r} > \lambda\}) \leq c\|\,\|f_k\|_{l^r}\|_{L_\mu}$ and $\|\,\|Tf_k\|_{l^r}\|_{L^p_\mu} \leq c\|\,\|f_k\|_{l^r}\|_{L^p_\mu}$, $1 < r < \infty$ (cf. Torrea

Hernandez [1984]). As for the weighted version of Theorem 1.3 in Chapter XII, cf. Andersen and John [1980] and Jawerth and Torchinsky [1985].

4.6 It is convenient to have a way to decompose open sets in R^n so that it will then be easy to set the stage for the good λ procedure to work. A way to go about it is as follows: Let Ω be an open set of finite Lebesgue measure in R^n. Then there exists a collection of closed cubes $\{Q_k\}$ so that (i) $\bigcup_k Q_k = \Omega$, (ii) if I_k = interior of Q_k, then $I_j \cap I_k = \emptyset$, $j \neq k$, and (iii) there are absolute constants c_1, c_2, independent of Ω, such that c_1 diam $Q_k \leq \operatorname{dist}(Q_k, R^n \backslash \Omega) \leq c_2$ diam Q_k. This is the proof: First divide R^n into a mesh of congruent closed cubes, with pairwise disjoint interiors, each of measure $3|\Omega|/2$; in this way none of the cubes may be totally contained in Ω. Next discard any cube in the mesh which does not intersect Ω and subdivide each of the remaining cubes into 2^n congruent closed cubes, again with disjoint interiors, by bisecting the sides. Once again discard any cube in this new mesh which does not intersect Ω and separate those cubes Q of this mesh which are totally contained in Ω and which in addition verify $\operatorname{dist}(Q, R^n \backslash \Omega)/\operatorname{diam} Q \geq 1$, and rename them $Q_1, Q_s, \ldots$, etc. As for those cubes which are left, and these are cubes which either intersect both Ω and $R^n \backslash \Omega$, or they are totally contained in Ω and $\operatorname{dist}(Q, R^n \backslash \Omega)/\operatorname{diam} Q < 1$, subdivide them again into 2^n congruent cubes as before and repeat the selection process. It is not hard to check that (i)-(iii) above hold for the family $\{Q_k\}$ thus selected. It is also possible to be a bit more careful and give a proof that works for arbitrary open sets $\Omega \subset R^n$, whether they have finite measure or not. This result is called the Whitney decomposition and has numerous applications.

CHAPTER

XIV

Hardy Spaces of Several Real Variables

1. ATOMIC DECOMPOSITION

When discussing in Chapter XII the norm relations between the S and g_λ^* functions, we presented our results in the range $0 < p < \infty$. This means, in particular, that the Hörmander multiplier theorem remains true for those functions whose Lusin integral is in $L^p(R^n)$, $0 < p < \infty$. When $1 < p < \infty$ these classes of functions coincide with the usual $L^p(R^n)$ spaces, but what can we say in case $0 < p \leq 1$? More precisely, which classes of functions, or even tempered distributions, are characterized by the fact that their Lusin integral is in $L^p(R^n)$, $0 < p \leq 1$? They are precisely the Hardy $H^p(R^n)$ spaces of several real variables.

In fact, elements in these spaces have the remarkable property that they can be written as sums of elementary components, or atoms. In a sense a very sophisticated Calderón–Zygmund decomposition holds in these spaces. We begin our discussion by giving examples of atoms and then proceed to show how they combine to span the Hardy spaces.

Definition 1.1. Let p, q, N be subject to the following conditions

$$0 < p \leq 1, \qquad 1 < q < \infty, \qquad N = \left[n\left(\frac{1}{p} - 1\right)\right], \tag{1.1}$$

where $[\cdot]$ denotes, as usual, the "greatest integer not exceeding" function. We then say that a function a is a (p, q, N) atom provided that

(i) $a(x) = 0$ off I (some open cube depending on a),
(ii) $\|a\|_q < \infty$, and

(iii) $\int_I x^\alpha a(x)\,dx = 0$, for all multi-indices α with $|\alpha| \le N$. To somehow normalize the atoms, we introduce the quantity $|a|$, or atomic norm of a, by

$$|a| = \inf_I |I|^{1/p-1/q}\|a\|_q, \tag{1.2}$$

where the infimum is taken over those cubes I which verify (i), (iii) above. We then have

Proposition 1.2. Let a be a (p, q, N) atom and suppose that ψ is a Littlewood–Paley function which satisfies, in addition to conditions (i)–(iii) in Definition 3.1 in Chapter XII, the following property: $\psi \in C^{N+1}(R^n)$ and $|D^\alpha\psi(x)| \le c(1+|x|)^{-(n+\varepsilon+N+1)}$ for all multi-indices α with $|\alpha| = N+1$, and some $\varepsilon > 0$. Then, if $A(x, t) = a * \psi_t(x)$, $S(A) \in L^p(R^n)$, and there is a constant $c = c_{p,\psi}$ independent of a so that

$$\|S(A)\|_p \le c|a|. \tag{1.3}$$

Proof. By translating a if necessary we may assume that the cube I, which verifies (i)–(iii) above, is centered at the origin. We estimate

$$\|S(A)\|_p^p = \left(\int_{2I} + \int_{R^n\setminus 2I}\right) S(A)(x)^p\,dx = J + K,$$

say. To bound J, note that, by Hölder's inequality with indices q/p and its conjugate and by (4.4) in Chapter XII,

$$J \le |2I|^{1-p/q}\|S(A)\|_p^p \le c(|I|^{1/p-1/q}\|a\|_q)^p, \tag{1.4}$$

which is an estimate of the right order. To bound K requires some work. By Taylor's expansion formula we have that

$$\psi(x+w) = \sum_{|\alpha|\le N} \frac{w^\alpha}{\alpha!} D^\alpha\psi(x) + R_N(x, w), \tag{1.5}$$

with $|R_N(x, w)| \le c|w|^{N+1}\sup|D^\alpha\psi(x+\eta w)|$, where the sup is taken over $|\alpha| = N+1$ and $0 \le \eta \le 1$. Therefore, in our case

$$|R_N(x, w)| \le c|w|^{N+1} \sup_{0\le\eta\le 1} (1+|x+\eta w|)^{-(n+\varepsilon+N+1)}. \tag{1.6}$$

Now, by (iii) above and (1.5) it is clear that

$$\begin{aligned} A(y, t) &= \int_I a(w)t^{-n}\left(\psi\left(\frac{y-w}{t}\right) - \sum_{|\alpha|\le N}\frac{1}{\alpha!}\left(\frac{(-w)}{t}\right)^\alpha D^\alpha\psi\left(\frac{y}{t}\right)\right)dw \\ &= \int_I a(w)t^{-n}R_N\left(\frac{y}{t}, \frac{-w}{t}\right)dw. \end{aligned} \tag{1.7}$$

We must estimate the integrand in the last integral in (1.7) for $w \in I$ and $(y, t) \in \Gamma(x)$, $x \notin 2I$. Observe first that under these conditions

$$(t + |x|) \leq c(t + |y - \eta w|), \qquad 0 \leq \eta \leq 1, \tag{1.8}$$

for some absolute constant c. Indeed, since there is a constant $0 < c_1 < 1$ so that $|w| \leq c_1|x|$, it follows that $|x| \leq |x - y| + |y - \eta w| + \eta|w| \leq t + |y - \eta w| + c_1|x|$, which gives (1.8) at once. Therefore, combining (1.6) and (1.8) we get

$$t^{-n}|R_N(y/t, -w/t)| \leq ct^{\varepsilon}|w|^{N+1}(t + |x|)^{-(n+\varepsilon+N+1)} \tag{1.9}$$

and this is the estimate we need. Indeed, substituting (1.9) into (1.7) we see that

$$|A(y, t)| \leq ct^{\varepsilon}(t + |x|)^{-(n+\varepsilon+N+1)} \int_I |a(w)||w|^{N+1}\,dw. \tag{1.10}$$

Now, by Hölder's inequality we readily get that the integral in (1.10) does not exceed $(\int_I |w|^{(N+1)q'}\,dw)^{1/q'}\|a\|_q \leq c|I|^{(1+(N+1)/n)-1/q}\|a\|_q$. Consequently, a straightforward estimation gives

$$\begin{aligned} S(A)(x)^2 &\leq c|I|^{2((1+(N+1)/n)-1/q)}\|a\|_q^2 \int_{[0,\infty)} t^{2\varepsilon}(t + |x|)^{-2(n+\varepsilon+N+1)}\frac{dt}{t} \\ &\leq c|I|^{2((1+(N+1)/n-1/q)}\|a\|_q^2|x|^{-2(n+N+1)}, \end{aligned} \tag{1.11}$$

which is all we need to bound K. Indeed, by substituting (1.11) into the definition of K we get

$$K \leq c|I|^{p((1+(N+1)/n)-1/q)}\|a\|_q^p \int_{R^n\setminus 2I} |x|^{-p(n+N+1)}\,dx,$$

and it suffices to bound the last integral above. But since by (1.1) $-p(n + N + 1) < -n$, we readily see that this integral $\leq c|I|^{1-p(1+(N+1)/n)}$, and finally

$$K \leq c(|I|^{(1/p-1/q)}\|a\|_q)^p. \tag{1.12}$$

Thus combining (1.4) and (1.12) and taking the inf over all open cubes I which contain the support of a we get at once that $\|S(A)_p\|_p \leq c|a|^p$, which is precisely what we wanted to show. ■

To discuss the convergence of a sum of atoms in $\mathcal{S}'$ we need a preliminary result.

Lemma 1.3. Suppose a is a (p, q, N) atom and let ψ be integrable and satisfy $|\psi|_{N+1} = \sum_{|\alpha|\leq N+1}\|D^\alpha\psi\|_\infty < \infty$. Then

$$|a(\psi)| \leq c|\psi|_{N+1}|a|, \tag{1.13}$$

where c is a constant independent of a and ψ.

Proof. We may, and do, assume that the cube I which contains the support of a is centered at the origin. We distinguish two cases, namely, $|I| > 1$ and $|I| \leq 1$.

The former case is obvious since by Hölder's inequality and the fact that $|I| \leq |I|^{1/p}$ now,

$$|a(\psi)| \leq \|\psi\|_\infty \|a\|_1 \leq c\|\psi\|_\infty |I|^{1-1/q}\|a\|_q \leq c|\psi|_{N+1}|I|^{1/p-1/q}\|a\|_q.$$

As for the latter case it requires some work. As in (1.7) we see that

$$\begin{aligned}|a(\psi)| &= \left|\int_I a(y)\left(\psi(y) - \sum_{|\alpha|\leq N}\frac{y^\alpha}{\alpha!}D^\alpha\psi(0)\right)dy\right| \\ &\leq c|\psi|_{N+1}\int_I |a(y)||y|^{N+1}\,dy \leq c|\psi|_{N+1}|I|^{(1+(N+1)/n)-1/q}\|a\|_q\end{aligned}$$

and this estimate is also of the right order since by (1.1), $N+1 > n(1/p - 1)$ and now $|I|, p \leq 1$. ■

We are now ready to describe the convergence of sums of atoms.

Theorem 1.4. Suppose $\{a_k\}$ is a sequence of (p, q, N) atoms and that $\sum_k |a_k|^p = A^p < \infty$. Then $\sum a_k$ converges in the sense of distributions to a sum $f \in \mathcal{S}'(R^n)$. Moreover, the convergence is unconditional, i.e., independent of the order of the summands, and f is a distribution of order $N+1$ which satisfies the following two properties.

(i) If ϕ is integrable and $|D^\alpha\phi(x)| \leq c(1+|x|)^{-(n+\varepsilon+|\alpha|)}$, some $\varepsilon > 0$, $|\alpha| \leq N+1$, then $F(x,t) = f * \phi_t(x)$ is well defined and $\lim_{t\to\infty}|F(x,t)| = 0$, uniformly in x.

(ii) If ψ is a Littlewood–Paley function and in addition $|D^\alpha\psi(x)| \leq c(1+|x|)^{-(n+\varepsilon+|\alpha|+1)}$ some $\varepsilon > 0$, $|\alpha| \leq N+1$, and $G(x,t) = f * \psi_t(x)$, $A_k(x,t) = a_k * \psi_t(x)$, then $S(G) \in L^p(R^n)$ and there is a constant c, independent of f, such that $\|S(G)\|_p \leq cA$.

Proof. The unconditional convergence of the sum $\sum a_k$ to a distribution f follows at once from the fact that if ϕ is as in (i) above, then

$$I = \sum_k |a_k(\phi)| < \infty. \tag{1.14}$$

To show (1.14), since $p \leq 1$ Lemma 1.3 gives

$$I \leq \left(\sum_k |a_k(\phi)|^p\right)^{1/p} \leq c|\phi|_{N+1}\left(\sum |a_k|^p\right)^{1/p} \leq c|\phi|_{N+1}A.$$

Moreover, as it also readily follows that

$$f * \phi_t(x) = \sum_k a_k(\phi_t(x - \cdot)), \tag{1.15}$$

we have that $|f * \phi_t(x)| \leq c|\phi|_{N+1}c_p(t)A$, where $c_p(t) = o(1)$ as $t \to \infty$, uniformly in x. This gives (i). As for (ii) note that by (1.15) with ϕ replaced by ψ there and Minkowski's inequality we obtain $S(G)(x) \leq \sum_k S(A_k)(x)$, which in turn gives, since $p \leq 1$,

$$S(G)(x)^p \leq \sum_k S(A_k)(x)^p. \tag{1.16}$$

Whence, by integrating (1.16) over R^n and invoking Proposition 1.2 we get that $\|S(G)\|_p^p \leq cA^p$, which is precisely what we wanted to show. ■

Remark 1.5. Observe that implicit in the above argument is the following fact

$$\left\|S\left(G - \sum_{k=1}^{N} A_k\right)\right\|_p \to 0 \quad \text{as} \quad N \to \infty. \tag{1.17}$$

Indeed, Theorem 1.4 applied to $f - \sum_{k=1}^{N} a_k = \sum_{k=N+1}^{\infty} a_k$ gives immediately that $\|S(G - \sum_{k=1}^{N} A_k)\|_p^p \leq c\sum_{k=N+1}^{\infty}|a_k|^p$, and this last expression, being the tail of a convergent series, goes to 0 as $N \to \infty$.

The following converse to Theorem 1.4 is true: Given $f \in \mathcal{S}'(R^n)$, we set $v(x, t) = t(\partial/\partial t)f * P_t(x)$, where $P_t(x)$ denotes the Poisson kernel. If $S(v) \in L^p(R^n)$, $0 < p \leq 1$, then f can be written as a sum of (p, q, N) atoms which satisfy all the conditions of Theorem 1.4. We choose to work with the Littlewood–Paley function $t(\partial/\partial t)P_t(x)$ here only as a matter of convenience, for then we have the theory of harmonic functions at our disposal. In fact, it is worthwhile to note that the property $S(v) \in L^p(R^n)$ is intrinsic to f.

More precisely, given smooth Littlewood–Paley functions ψ and η, $S(f * \psi_t) \in L^p(R^n)$ if and only if $S(f * \eta_t) \in L^p(R^n)$, $\|S(f * \psi_t)\|_p \sim \|S(f * \eta_t)\|_p$ and the constants involved are independent of f, $0 < p \leq 1$.

Before we proceed with the atomic decomposition we need some preliminary results.

Lemma 1.6. Suppose that for $f \in \mathcal{S}'(R^n)$ we have $S(v) \in L^p(R^n)$, $0 < p \leq 1$. Then

(i) $|v(x, t)| \leq cS(v)(x)$, c independent of x, t, f.
(ii) $|v(x, t)| \leq ct^{-n/p}\|S(v)\|_p$, c independent of x, t, f.
(iii) $\int_{R^n}|v(x, t)|^2\, dx < \infty$.
(iv) $\hat{f}(\xi)$ coincides with a continuous function in R^n and verifies $|\hat{f}(\xi)| \leq c\|S(v)\|_p|\xi|^{n(1/p-1)}$; c is independent of ξ and f.

Proof. Since $(\partial/\partial t)f * P_t(x)$ is harmonic in R_+^{n+1}, we may invoke (an appropriate version of) the Hardy–Littlewood mean-value inequality 5.4 in

Chapter VII, and thus obtain

$$\left|\frac{\partial}{\partial t} f * P_t(x)\right| \leqslant c\left(\frac{1}{|B|}\int_B \left|\frac{\partial}{\partial s} f * P_s(y)\right|^q dy\,ds\right)^{1/q}, \qquad 0 < q < \infty, \quad (1.18)$$

where B is a ball centered at (x, t) totally contained in R_+^{n+1} and $c = c_q$ is independent of B and f. If we actually choose for B the ball with radius $t/2$, then we have $B \subset \Gamma(x)$. Indeed, if $(y, s) \in B$, then $|x - y|^2 + (t - s)^2 \leqslant t^2/4$, which in turn gives $t - s \leqslant t/2$, or $t/2 \leqslant s$. Moreover, since also $|x - t| \leqslant t/2$, then $|x - y| \leqslant s$ and $(y, s) \in \Gamma(x)$. So with $q = 2$ in (1.18) and since for $(y, s) \in B$ also $s \leqslant 3t/2$, we obtain that

$$\begin{aligned} |v(x, t)|^2 &\leqslant c\frac{t^2}{t^{n+1}}\int_B \left|\frac{\partial}{\partial s} f * P_s(y)\right|^2 dy\,ds \\ &\leqslant c\int_{\Gamma(x)} |v(y, s)|^2 s^{-n}\,dy\frac{ds}{s} = cS(v)(x)^2, \end{aligned}$$

which is precisely (i). To show (ii) we invoke (1.18) once again, this time with $q = p$. In this case, and by (i), we have that

$$\begin{aligned} |v(x, t)|^p &\leqslant ct^{-(n+1)}\int_{[0,3t/2]}\int_{R^n} |v(y, s)|^p\,dy\,ds \\ &\leqslant ct^{-(n+1)}t\|S(v)\|_p^p, \end{aligned}$$

which gives (ii) immediately. As for (iii) note that, if $q \geqslant p$, by (i) and (ii) we have

$$\begin{aligned} \int_{R^n} |v(x, t)|^q\,dx &\leqslant (ct^{-n/p}\|S(v)\|_p)^{q-p}\int_{R^n} |v(x, t)|^p\,dx \\ &\leqslant ct^{n(1-q/p)}\|S(v)\|_p^q. \end{aligned}$$

Here, (iii) corresponds to the choice $q = 2$. To prove (iv) we use $q = 1$ instead in conjunction with the obvious facts that $|((\partial/\partial t)f * P_t)\hat{}(\xi)| \leqslant \|(\partial/\partial t)f * P_t(\cdot)\|_1$ and that the Fourier transform of an integrable function is continuous. Thus for $\xi = 0$

$$|\hat{f}(\xi)|t|\xi|e^{-t|\xi|} \leqslant \int_{R^n} |v(x, t)|\,dx \leqslant ct^{n(1-1/p)}\|S(v)\|_p$$

and (iv) follows upon setting $t = 1/|\xi|$. ∎

Lemma 1.7. There is a radial, real-valued, $C_0^\infty(R^n)$ function ψ which satisfies the following properties:

(i) $\operatorname{supp}\psi \subseteq B(0, 1)$, the unit ball centered at 0.
(ii) $D^\alpha\hat{\psi}(\xi)]_{\xi=0} = 0$, $|\alpha| \leqslant N$, N an arbitrary, fixed, integer.
(iii) $\int_{[0,\infty)} e^{-t}\hat{\psi}(t)\,dt = -1$.

Proof. Let η be any real-valued, not identically 0, radial function supported in $B(0, \frac{1}{2})$ and let $\eta_1 = \Delta^M \eta$, for some integer M. Since $\hat{\eta}_1(\xi) = c|\xi|^{2M}\hat{\eta}(\xi)$, η_1 is also radial and $|\hat{\eta}_1(\xi)| \leq c|\xi|^{2M}$. We claim that for an appropriate constant c and M, $\psi(x) = c\eta_1 * \eta_1(x)$ will do. (i) and (ii) are readily verified. Also note that $\int_{[0,\infty)} e^{-t}\hat{\eta}_1(t)^2\, dt \neq 0$, and consequently we may choose c so that (iii) holds as well. ■

Corollary 1.8. Let ψ be the function of Proposition 1.7, then

$$\int_{[0,\infty)} \hat{\psi}(t|\xi|)t|\xi|e^{-t|\xi|}\frac{dt}{t} = -1, \qquad \xi \neq 0.$$

Proposition 1.9. Suppose $f \in \mathcal{S}'(R^n)$ verifies $S(v) \in L^p(R^n), 0 < p \leq 1$, and let η be a Schwartz function with vanishing integral. Then with ψ as in Proposition 1.7 we have

$$f(\eta) = \int_{R_+^{n+1}} v(y, t)\eta * \psi_t(y)\, dy\frac{dt}{t}. \tag{1.19}$$

In other words

$$f = \int_{R_+^{n+1}} v(y, t)\psi_t(y - \cdot)\, dy\frac{dt}{t} \qquad (\mathcal{S}'/\text{polynomials}).$$

Proof. Since by Lemma 1.6(iii) $v(y, t) \in L^2(R^n)$ for almost every $t \in R^+$, by Proposition 1.8 in Chapter X it follows that, if $\tilde{\eta}(y) = \eta(-y)$, then

$$\int_{R^n} v(y, t)\eta * \psi_t(y)\, dy = -(2\pi)^{-n}\int_{R^n} \hat{f}(\xi)\hat{\tilde{\eta}}(\xi)\hat{\psi}(t|\xi|)t|\xi|e^{-t|\xi|}\, d\xi. \tag{1.20}$$

Moreover, since both of the functions $v(y, t)\eta * \psi_t(y)$ and $\hat{f}(\xi)\hat{\tilde{\eta}}(\xi)\hat{\psi}(t|\xi|)t|\xi|e^{-t|\xi|}$ are absolutely integrable with respect to the measures $dy\, dt/t$ and $d\xi\, dt/t$, respectively, we may interchange the order of integration freely in (1.20). Whence, integrating first with respect to t the right-hand side of (1.20) we get

$$\int_{R_+^{n+1}} v(y, t)\eta * \psi_t(y)\, dy\frac{dt}{t}$$
$$= -(2\pi)^{-n}\int_{R^n} \hat{f}(\xi)\hat{\tilde{\eta}}(\xi)\int_{[0,\infty)} \hat{\psi}(t|\xi|)t|\xi|e^{-t|\xi|}\frac{dt}{t}\, d\xi.$$

Since $\hat{\tilde{\eta}}(0) = 0$ we may invoke Corollary 1.8 and thus obtain that the right-hand side above equals $(2\pi)^{-n}\hat{f}(\hat{\tilde{\eta}}) = (2\pi)^{-n}f(\hat{\hat{\tilde{\eta}}}) = f(\eta)$. ■

We are now ready to go ahead with the main result of this section, namely, the atomic decomposition.

Theorem 1.10. Suppose $f \in \mathcal{S}'(R^n)$ verifies that $S(v) \in L^p(R^n)$, $0 < p \leqslant 1$, and that $f * P_t(x) \to 0$ as $t \to \infty$. Then given $1 < q < \infty$ and an integer $N \geqslant [n(1/p - 1)]$ there exists a sequence $\{a_j\}$ of (p, q, N) atoms such that

(i) $\sum |a_j|^p \leqslant c\|S(v)\|_p^p$, where c is independent of f, and
(ii) $\sum a_j = f(\mathcal{S}')$.

Moreover, if $A_j(x, t) = t(\partial/\partial t)a_j * P_t(x)$, then also $\|S(v - \sum_{j=1}^m A_j)\|_p \to 0$ as $m \to \infty$ and $\|S(v)\|_p^p \sim \inf \sum |a_j|^p$, where the inf is taken over all possible decompositions of f into (p, q, N) atoms.

Proof. In order to facilitate the understanding of the proof, which is geometric in nature, we first carry it out in the case $n = 1$ and then indicate the minor changes needed to make it work for arbitrary dimensions. Also since f is fixed we simply put $S(v)(x) = S(x)$ in what follows.

Let $E_k = \{S > 2^k\}$, $k = 0, \pm 1, \ldots$ and $\mathcal{O}_k = \{M(\chi_{E_k}) > \frac{1}{2}\}$. Note that by the Lebesgue differentiation theorem $E_k \subseteq \mathcal{O}_k$ a.e. and by the maximal theorem $|\mathcal{O}_k| \leqslant c|E_k|$, c independent of k. Observe that E_k decreases from R to $\varnothing$ as k increases from $-\infty$ to ∞, and similarly for $\mathcal{O}_k$. For each (y, t) in the upper half plan R_+^2 let $I(y, t) = \{w \in R: |y - w| < t\}$ denote the interval centered at y of diameter $2t$ and let $\phi: R_+^2 \to Z$ be the function given by $\phi(y, t) =$ largest integer k so that

$$|I(y, t) \cap E_k| > |I(y, t)|/2. \tag{1.21}$$

Clearly, by the above observations ϕ is well defined. We list now some properties of the sets $\phi^{-1}(k)$, to wit:

(a) The sets $\phi^{-1}(k)$ are pairwise disjoint and $\bigcup_k \phi^{-1}(k) = R_+^2$.
(b) If $(y, t) \in \phi^{-1}(k)$, then $I(y, t) \subseteq \mathcal{O}_k$.
(c) If $(y, t) \in \phi^{-1}(k)$, then $|I(y, t) \cap (R \backslash E_{k+1})| \geqslant |I(y, t)|/2$.

(a) is obvious. Also, if $(y, t) \in \phi^{-1}(k)$, then (1.21) holds and consequently $\inf_{I(y,t)} M\chi_{E_k} > \frac{1}{2}$, thus proving (b). (c) is equivalent to $|I(y, t) \cap E_{k+1}| \leqslant |I(y, t)|/2$, which also holds on account of the definition of ϕ.

Since $\mathcal{O}_k$ is open it may be written as a countable, disjoint union of open intervals, $\mathcal{O}_k = \bigcup_j I_{j,k}$, say. This allows to localize the situation at hand by setting $T_{j,k} = \{(y, t) \in \phi^{-1}(k): y \in I_{j,k}\}$.

It is clear that the $T_{j,k}$'s are also pairwise disjoint and that

$$\bigcup_j T_{j,k} = \phi^{-1}(k), \bigcup_{j,k} T_{j,k} = R_+^2.$$

These observations are all that is needed to complete the proof. Let ψ be the function constructed in Proposition 1.7 corresponding to the value N in the hypothesis and put

$$a_{j,k}(x) = \int_{T_{j,k}} v(y, t)\psi_t(y - x)\, dy \frac{dt}{t}.$$

We claim that the $a_{j,k}$'s satisfy the following three properties, to wit:

(i) If $a_{j,k}(x) \neq 0$, then $x \in 3I_{j,k}$.
(ii) N moments of $a_{j,k}$ vanish.
(iii) $\|a_{j,k}\|_q \leq c2^k|I_{j,k}|^{1/q}$, $1 < q < \infty$, c independent of j, k.

Assume for the moment that (i)–(iii) have been verified. Then, since $|a_{j,k}| \leq |I_{j,k}|^{1/p-1/q}\|a_{j,k}\|_q \leq c2^k|I_{j,k}|^{1/p}$, we immediately see that

$$\sum_j |a_{j,k}|^p \leq c2^{kp}\sum_j |I_{j,k}| = c2^{kp}|\mathcal{O}_k| \leq c2^{kp}|E_k|.$$

Whence

$$\sum_{j,k}|a_{j,k}|^p \leq c\sum_k 2^{kp}|E_k|. \tag{1.22}$$

To bound the sum in the right-hand side of (1.22) we note that it equals

$$\sum_k 2^{kp}\int_{\{S>2^k\}} dx = \int_R \sum_{\{k:S(x)>2^k\}} 2^{kp}\,dx \leq c\int_{R^n} S(x)^p\,dx.$$

By Theorem 1.7 above there exists a tempered distribution g such that, if $G(y, t) = t(\partial/\partial t)g * P_t(y)$ and $A_{j,k}(y, t) = t(\partial/\partial t)a_{j,k} * P_t(y)$, then

$$\sum_{j,k} a_{j,k} = g(\mathcal{S}'), \qquad \|S(G)\|_p \leq c\sum_{j,k}|a_{j,k}|^p$$

and

$$\left\| S\left(G - \sum_{j\leq N_1, |k|\leq N_2} A_{j,k}\right)\right\|_p \to 0 \quad \text{as} \quad N_1, N_2 \to \infty.$$

Thus we will be finished once we show that $f = g$; this is not hard. Indeed, by the unconditional convergence of $\sum a_{j,k}$ we readily see that its sum actually is $\int_{R^2_+} v(y, t)\psi_t(y - x)/t\,dy\,dt(\mathcal{S}')$. Therefore, by Proposition 1.9 it follows that $f(\eta) = g(\eta)$ for every Schwartz function η with vanishing integral. By Proposition 1.6 in Chapter X this means that $f - g$ is a polynomial $\mathcal{P}$ (that is, the Fourier transform of a distribution supported at the origin). But by assumption and Proposition 1.4 both the Poisson integrals of f and g go to 0 as $t \to \infty$. This implies that for every x, $\mathcal{P} * P_t(x) \to 0$ as $t \to \infty$, and so $\mathcal{P} \equiv 0$. Thus to complete the proof it only remains to verify that properties (i)–(iii) of the $a_{j,k}$'s hold.

(i) If $a_{j,k} \neq 0$, there exists (y, t) in $T_{j,k}$, and y in $I_{j,k}$, so that $|y - x| \leq t$ for otherwise $\psi((y - x)/t)$ vanishes identically. Moreover, since $I(y, t) \subseteq \mathcal{O}_k$ in this case, we actually have that $I(y, t) \subseteq I_{j,k}$ and consequently $t \leq (\text{diam } I_{j,k})/2$. Thus letting $y_{j,k}$ = center of $I_{j,k}$, we see that

$$|x - y_{j,k}| \leq |x - y| + |y - y_{j,k}| \leq t + \text{diam } I_{j,k} \leq 3\,\text{diam}(I_{j,k})/2 \tag{1.23}$$

and we have finished.

(ii) The moments of $a_{j,k}$ coincide with those of ψ.
(iii) We use the expression $\|a_{j,k}\|_q = \sup|\int_R a_{j,k}(x)\eta(x)\,dx|$, where the sup is taken over those functions with $\|\eta\|_{q'} \le 1$, $1/q + 1/q' = 1$. So putting $g(y, t) = \eta * \psi_t(y)$ we estimate $I = |\iint_{T_{j,k}} v(y, t)g(y, t)/t\,dy\,dt|$. By (c) above we immediately get

$$I \le \int_{R_+^2} \chi_{T_{j,k}}(y, t)|v(y, t)||g(y, t)|\,dy\frac{dt}{t}$$
$$\le 2\int_{R_+^2} \chi_{T_{j,k}}(y, t)\frac{|I(y, t) \cap (R\setminus E_{k+1})|}{|I(y, t)|}|v(y, t)||g(y, t)|\,dy\frac{dt}{t}$$
$$\le \int_{R_+^2} \chi_{T_{j,k}}(y, t)\left(\int_{R\setminus E_{k+1}} \chi_{I(y,t)}(x)\,dx\right)|v(y, t)||g(y, t)|\frac{dy}{t}\frac{dt}{t}$$
$$= \int_{R\setminus E_{k+1}} \left(\int_{R_+^2} \chi_{T_{j,k}}(y, t)\chi_{I(y,t)}(x)|v(y, t)||g(y, t)|\frac{dy}{t}\frac{dt}{t}\right) dx.$$

Finally, observe that since $\chi_{T_{j,k}}(y, t)\chi_{I(y,t)}(x) \le \chi_{3I_{j,k}}(x)\chi_{I(x,t)}(y)$, then also

$$I \le 2\int_{(R\setminus E_{k+1})\cap 3I_{j,k}} \left(\int_{\Gamma(x)} |v(y, t)||g(y, t)|\frac{dy}{2t}\frac{dt}{t}\right) dx$$
$$\le 2\int_{(R\setminus E_{k+1})\cap 3I_{j,k}} S(x)S(g)(x)\,dx. \tag{1.24}$$

Now, since $S(x) \le 2^{k+1}$ on $R\setminus E_{k+1}$, and by (4.4) in Chapter XII,

$$\int_{3I_{j,k}} S(g)(x)\,dx \le |3I_{j,k}|^{1/q}\|\eta\|_{q'} \le c|I_{j,k}|^{1/q},$$

(1.24) gives immediately $I \le c2^k|I_{j,k}|^{1/q}$, all j, k. Clearly, the same estimate holds for $\|a_{j,k}\|_q$, and (iii) also holds.

Next we indicate the minor modifications needed for arbitrary dimension n. The first change comes in the definition of ϕ, where we now let $I(y, t)$ be the open cube centered at y of sidelength t; this is a minor change. More important, however, we note that the decomposition of $\mathcal{O}_k = \bigcup I_{j,k}$ given above is no longer valid. This is not a serious obstacle as we can use the Whitney decomposition 4.6 in Chapter XIII instead. Let then $\{Q_{j,k}\}$ be the Whitney decomposition of $\mathcal{O}_k$ and let $I_{j,k}$ = interior of $Q_{j,k}$, all j, k. $T_{j,k}$ and $a_{j,k}$ are defined exactly as before and the only property that has to be checked is (i) concerning the support of $a_{j,k}$. But as in (1.23) observe that if $y_{j,k}$ = center of $I_{j,k}$, then $a_{j,k}(x) \ne 0$ implies there is (y, t) in $T_{j,k}$ such that

$|y-x| \leq t$ and $y \in I_{j,k}$. Then $|x - y_{j,k}| \leq |x-y| + |y - y_{j,k}| \leq t + \operatorname{diam} I_{j,k} \leq c \operatorname{dist}(I_{j,k}, R^n \setminus \mathcal{O}_k) + \operatorname{diam} I_{j,k} \leq c \operatorname{diam} I_{j,k}$, since $I_{j,k}$ corresponds to a Whitney cube in the decomposition of $\mathcal{O}_k$. This completes the proof. ■

2. MAXIMAL FUNCTION CHARACTERIZATION OF HARDY SPACES

In Chapter VII we considered the nontangential maximal function corresponding to a harmonic function in the disk and introduced the Hardy $H^p(T)$ spaces. Similar definitions hold for functions of several real variables. More precisely, given a function $u(y, t)$ defined in R^{n+1}_+ we set

$$N_a(u)(x) = \sup_{(y,t) \in \Gamma_a(x)} |u(y,t)|, \qquad a > 0. \tag{2.1}$$

$N_a(u)$ is called the nontangential maximal function, of opening a, corresponding to u. We are interested in those classes of functions, or more generally, distributions, f which verify the following property: if $u(y, t) = f * P_t(y)$ is the Poisson integral of f, then $N_a(u) \in L^p(R^n)$, $0 < p < \infty$. A straightforward argument shows that this last condition is independent of a (cf. 6.2 below), and for this reason we work with $N_1(u) = N(u)$ in what follows. A deeper fact, due to Fefferman–Stein [1972], is that the Poisson kernel above may be replaced by any sufficiently smooth function ψ with nonvanishing integral and still obtain the same classes.

Now suppose that a is a (p, q, N) atom and ψ is an integrable function, $|D^\alpha \psi(x)| \leq c(1 + |x|)^{-(n+\varepsilon+|\alpha|)}$, some $\varepsilon > 0$, and $|\alpha| \leq N + 1$. If $A(y, t) = a * \psi_t(x)$, then an argument similar to, but simpler than, Proposition 1.2 gives that $\|N(A)\|_p \leq c|a|$, where $c = c_p$ is independent of a. For $f \in \mathcal{S}'(R^n)$ which verifies the assumptions of Theorem 1.10, let $f = \sum a_j(\mathcal{S}')$ be the atomic decomposition given there and put $F(y, t) = f * \psi_t(y)$ and $A_j(y, t) = a_j * \psi_t(y)$. We clearly have that $N(F)(x)^p \leq \sum N(A_j)(x)^p$ and also $\|N(F)\|_p^p \leq c \sum |a_j|^p \leq c\|S(v)\|_p^p$, $c = c_p$ independent of f. In other words, the assumption $S(v) \in L^p(R^n$, $0 < p \leq 1$, implies that the nontangential maximal function associated to extensions of f to R^{n+1}_+ by convolution with the dilations of appropriate functions ψ also belongs to $L^p(R^n)$. What about the converse to this statement: If $N(F) \in L^p(R^n)$, is the same true of $S(v)$? A way to go about this question is to show that, if u is harmonic in R^{n+1}_+ and $N(u) \in L^p(R^n)$, $0 < p \leq 1$, then $u = f * P_t$, where the distribution f admits an atomic decomposition; then apply Theorem 1.4. Although this approach works (cf. 6.5), we prefer to give a direct proof by means of a distribution function inequality we show first.

Lemma 2.1. Suppose $u(y, t) = f * P_t(y)$ is the Poisson integral of an L^2 function f and let $V(y, t) = (t(\partial/\partial t)u(y, t),\ t\nabla u(y, t))$, where $\nabla = (\partial/\partial y_1, \ldots, \partial/\partial y_n)$. Then if $E_\lambda = \{N(y) > \lambda\}$,

$$|\{S(V) > \lambda\}| \leqslant c\lambda^{-2} \int_{R^n \setminus E_\lambda} N(u)(y)^2\, dy + c|E_\lambda|, \tag{2.2}$$

where c is independent of λ and f.

Proof. Let $\mathcal{O}_\lambda = \{N(\chi_{E_\lambda} * \phi_t) > \frac{1}{2}\}$, where the infinitely differentiable function ϕ has support in the unit ball and integral 1 and all its moments of first order vanish. Clearly, $E_\lambda \subseteq \mathcal{O}_\lambda$ a.e., and by a simple extension of Theorem 2.2 in Chapter VII $|\mathcal{O}_\lambda| \leqslant c|E_\lambda|$, with c independent of λ. Let now $(y, t) \in \Gamma(R^n \setminus \mathcal{O}_\lambda) = \bigcup_{x \in R^n \setminus \mathcal{O}_\lambda} \Gamma(x)$. Then $|y - x| < t$ for some x in $R^n \setminus \mathcal{O}_\lambda$ and consequently $\chi_{E_\lambda} * \phi_t(y) \leqslant \frac{1}{2}$, or, equivalently,

$$g(y, t) = \chi_{R^n \setminus E_\lambda} * \phi_t(y) \geqslant \tfrac{1}{2}. \tag{2.3}$$

Next consider

$$\begin{aligned} I &= \int_{R^n \setminus \mathcal{O}_\lambda} S(V)(x)^2\, dx \\ &= c \int_{R^n \setminus \mathcal{O}_\lambda} \int_{R^{n+1}_+} \chi_{\Gamma(x)}(y, t)|V(y, t)|^2 t^{-n}\, dy \frac{dt}{t}\, dx \\ &= c \int_{R^{n+1}_+} |V(y, t)|^2 \left(t^{-n} \int_{R^n \setminus \mathcal{O}_\lambda} \chi_{\Gamma(x)}(y, t)\, dx \right) dy \frac{dt}{t}. \end{aligned} \tag{2.4}$$

Observe that the innermost expression in (2.4) is bounded and that it does not vanish only for those (y, t) in $\Gamma(R^n \setminus \mathcal{O}_\lambda)$. Therefore, by (2.3)

$$I \leqslant c \int_{R^{n+1}_+} |V(y, t)|^2 g(y, t)^2\, dy \frac{dt}{t} = cJ, \tag{2.5}$$

say. Since $g \leqslant 1$, by Proposition 3.4 in Chapter XII, $J \leqslant c\|f\|_2^2 < \infty$. We estimate J by considering separately the integrals corresponding to $t(\partial/\partial t)u(y, t) = v(y, t)$ and $t\nabla u(y, t)$. We begin with the former. First, note that for $0 < \delta < \eta < \infty$ we have

$$\begin{aligned} &\int_{(\delta, \eta)} v(y, t)^2 g(y, t)^2 \frac{dt}{t} \\ &\quad = \int_{(\delta, \eta)} \frac{\partial}{\partial t} u(y, t) v(y, t) g(y, t)^2\, dt \\ &\quad = u(y, t)v(y, t)^2]_\delta^\eta - \int_{(\delta, \eta)} u(y, t) \frac{\partial}{\partial t}(v(y, t)g(y, t)^2)\, dt. \end{aligned}$$

By Corollary 2.4 in Chapter IV, for almost every $y \in R^n$, $\lim_{\delta\to 0} u(y,\delta) = f(y)$, $\lim_{\delta\to 0} v(y,\delta) = 0$, and $\lim_{\delta\to 0} g(y,\delta) = \chi_{R^n\setminus E_\lambda}(y)$. Thus the lower limit of the integrated term above goes to 0 with δ. As for the upper limit we need the estimates

$$|u(y,t)| \leqslant ct^{-n/p}\|N(u)\|_p \tag{2.6}$$

and

$$|v(y,t)| \leqslant cMf(y). \tag{2.7}$$

Inequality (2.7) is a particular case of Proposition 2.3 in Chapter IV and (2.6) follows immediately from the estimate

$$|u(y,t)| \leqslant \inf_{B(y,t)} N(u), \tag{2.8}$$

where $B(y,t)$ denotes the ball centered at y of radius t. Thus also $\lim_{\eta\to\infty} u(y,\eta)v(y,\eta)g(y,\eta)^2 = 0$ and the integrated term vanishes. This gives at once that

$$\begin{aligned}\int_{R_+^{n+1}} v(y,t)^2 g(y,t)^2\,dy\frac{dt}{t} &= \int_{R_+^{n+1}} u(y,t)v(y,t)g(y,t)^2\,dy\frac{dt}{t}\\ &\quad - \int_{R_+^{n+1}} u(y,t)\frac{\partial^2}{\partial t^2}u(y,t)g(y,t)^2\,dy\,dt\\ &\quad - 2\int_{R_+^{n+1}} u(y,t)v(y,t)g(y,t)\frac{\partial}{\partial t}g(y,t)\,dy\,dt.\end{aligned} \tag{2.9}$$

Similarly, but integrating with respect to the space variables first and letting Δ denote the Laplacian in these variables, we get that

$$\begin{aligned}&\int_{R_+^{n+1}} |t\,\nabla u(y,t)|^2 g(y,t)^2\,dy\frac{dt}{t}\\ &\quad = -\int_{R_+^{n+1}} u(y,t)t\,\Delta u(y,t)g(y,t)^2\,dy\,dt\\ &\qquad - 2\int_{R_+^{n+1}} u(y,t)(t\,\nabla u(y,t)\cdot t\,\nabla g(y,t))g(y,t)\,dy\frac{dt}{t}.\end{aligned} \tag{2.10}$$

Thus adding (2.9) and (2.10), and since u is harmonic in R_+^{n+1}, we find that J actually equals

$$-\int_{R_+^{n+1}} u(y,t)v(y,t)g(y,t)^2\,dy\frac{dt}{t}$$

$$-2\int_{R_+^{n+1}} u(y,t)v(y,t)g(y,t)\frac{\partial}{\partial t}g(y,t)\,dy\,dt$$

$$-2\int_{R_+^{n+1}} u(y,t)(t\,\nabla u(y,t)\cdot t\,\nabla g(y,t))g(y,t)\,dy\frac{dt}{t}$$

$$= J_1 + J_2 + J_3, \tag{2.11}$$

say. Since J_2 and J_3 are handled in a similar fashion we only estimate J_2. Let $T(E_\lambda) = \{(y,t) \in R_+^{n+1}: d(y, R^n\backslash E_\lambda) < t\}$. $T(E_\lambda)$ looks roughly like a collection of inverted cones based on the components of the open set E_λ. Now $g(y,t)$ vanishes on $T(E_\lambda)$ and so do all its derivatives. Furthermore, $\{(y,t) \in R_+^{n+1}: |u(y,t)| > \lambda\} \subseteq E_\lambda$, and

$$|u(y,t)| \leqslant \lambda \qquad \text{a.e. on} \quad R^n\backslash E_\lambda. \tag{2.12}$$

From the easily verified estimate $\pm 2ab \leqslant \varepsilon a^2 + \varepsilon^{-1}b^2$, for $\varepsilon > 0$ and a, b arbitrary real numbers, it follows then at once that

$$J \leqslant \varepsilon \int_{R_+^{n+1}} |v(y,t)|^2 g(y,t)^2\,dy\frac{dt}{t}$$

$$+\,\varepsilon^{-1}\int_{R_+^{n+1}} |u(y,t)|^2\left(t\frac{\partial}{\partial t}g(y,t)\right)^2 dy\frac{dt}{t}. \tag{2.13}$$

The first summand in (2.13) does not exceed εJ. As for the second summand, by (2.12) it is bounded by

$$\varepsilon^{-1}\lambda^2\int_{R_+^{n+1}}\left(t\frac{\partial}{\partial t}g(y,t)\right)^2 dy\frac{dt}{t}.$$

Furthermore, since $g(y,t) = 1 - \chi_{E_\lambda} * \phi_t(y)$, by Proposition 3.4 in Chapter XII this last expression is dominated by $c\varepsilon^{-1}\lambda^2\|\chi_{E_\lambda}\|_2^2 = c\varepsilon^{-1}\lambda^2|E_\lambda|$ which is also a bound of the right order. In other words,

$$|J_2| + |J_3| \leqslant 2\varepsilon J + c\varepsilon^{-1}\lambda^2|E_\lambda|, \qquad \varepsilon > 0. \tag{2.14}$$

To complete the proof we only need to estimate J_1; this integral looks like J_2 but there are no derivatives acting on g. In first place note that

$$-2\int_{(\delta,\eta)} u(y,t)\frac{\partial}{\partial t}u(y,t)g(y,t)^2\,dy\,dt$$

$$= -u(y,t)^2 g(y,t)^2]_\delta^\eta + 2\int_{(\delta,\eta)} u(y,t)^2 g(y,t)\frac{\partial}{\partial t}g(y,t)\,dt.$$

On account of (2.6) $\lim_{\eta\to\infty} u(y,\eta)g(y,\eta) = 0$. Also since $\lim_{\delta\to 0} g(y,\delta)^2 = \chi_{R^n\setminus E_\lambda}(y)$, it readily follows that $|\lim_{\delta\to 0} u(y,\delta)^2 g(y,\delta)^2| \leq N(u)(y)^2 \chi_{R^n\setminus E_\lambda}(y)$ and consequently

$$J_1 \leq \int_{R^n\setminus E_\lambda} N(u)(y)^2\,dy + \int_{R^{n+1}_+} u(y,t)^2 g(y,t)\frac{\partial}{\partial t} g(y,t)\,dy\,dt$$
$$= J_4 + J_5, \tag{2.15}$$

say. J_4 is all right. To bound J_5 we need another expression for $(\partial/\partial t)g(y,t) = -(\partial/\partial t)(\chi_{E_\lambda} * \phi_t)(y)$. This is easiest obtained by taking Fourier transforms. Indeed, since $(\partial/\partial t)\hat{g}(\xi,t) = -\hat{\chi}_{E_\lambda}(\xi)(\nabla\hat{\phi}(t\xi)\cdot\xi)$, it readily follows that

$$\frac{\partial}{\partial t} g(y,t) = c\sum_{j=1}^{n} \frac{\partial}{\partial y_j}(\chi_{E_\lambda} * (\eta_j)_t)(y), \tag{2.16}$$

where η_j is the $C_0^\infty(R^n)$ function, supported in the unit ball, such that $\hat{\eta}_j(\xi) = (\partial/\partial\xi_j)\hat{\phi}(\xi)$. By the moment condition on ϕ, clearly, $\int_{R^n} \eta_j(y)\,dy = 0$, $1 \leq j \leq n$. Returning to J_5, by using (2.16) and integrating by parts we obtain that it equals

$$c\int_{[0,\infty)}\int_{R^n} u(y,t)\left(\sum_{j=1}^{n} t\frac{\partial}{\partial y_j} u(y,t)\chi_{E_\lambda} * (\eta_j)_t(y)\right) g(y,t)\,dy\frac{dt}{t}$$
$$+ c\int_{[0,\infty)}\int_{R^n} u(y,t)^2\left(\sum_{j=1}^{n} t\frac{\partial}{\partial y_j} g(y,t)\chi_{E_\lambda} * (\eta_j)_t(y)\right) dy\frac{dt}{t}$$
$$= J_6 + J_7, \tag{2.17}$$

say. A moment's thought suffices to realize that J_6 is a sum of integrals each of which is similar to J_3, and consequently also

$$|J_6| \leq \varepsilon J + c\lambda^2 |E_\lambda|. \tag{2.18}$$

Moreover, since

$$\left|\sum_{j=1}^{n} t\frac{\partial}{\partial y_j} g(y,t)\chi_{E_\lambda} * (\eta_j)_t(y)\right| \leq \left(\sum_{j=1}^{n}\left(t\frac{\partial}{\partial y_j} g(y,t)\right)^2 + \sum_{j=1}^{n} \chi_{E_\lambda} * (\eta_j)_t(y)^2\right)\chi_{R^{n+1}_+\setminus T(E_\lambda)}(y,t),$$

J_7 can be estimated as the second summand in (2.13), that is, by $c\lambda^2|E_\lambda|$. Finally, combining (2.11), (2.14), (2.15), (2.18) and the above observations, and choosing ε sufficiently small, we get that $J \leq \frac{1}{2}J + c\int_{R^n\setminus E_\lambda} N(u)(y)^2\,dy + c\lambda^2|E_\lambda|$. But since $J < \infty$, from (2.5) we obtain at once that

$$I \leq cJ \leq c\int_{R^n\setminus E_\lambda} N(u)(y)^2\,dy + c\lambda^2|E_\lambda|. \tag{2.19}$$

Whence by (2.19) and Chebychev's inequality we see that

$$|\{S(V) > \lambda\}| \leq |\mathcal{O}_\lambda| + |\{x \in R^n \setminus \mathcal{O}_\lambda : S(V)(x) > \lambda\}| \leq c|E_\lambda| + \lambda^{-2} I$$
$$\leq c(|E_\lambda| + \lambda^2 \textstyle\int_{R^n \setminus E_\lambda} N(u)(y)^2\, dy),$$

which is precisely (2.2). ■

We are now ready to prove

Theorem 2.2. Suppose $u(x, t)$ is harmonic in R_+^{n+1} and $N(u) \in L^p(R^n)$, $0 < p < 2$. Then if $v(y, t) = t(\partial/\partial t)u(y, t)$, also $S(v) \in L^p(R^n)$ and there is a constant $c = c_p$ independent of u such that $\|S(v)\|_p \leq c\|N(u)\|_p$.

Proof. Assume first that $u(y, t) = f * P_t(y)$ is the Poisson integral of an L^2 function f. Then by multiplying (2.2) through by λ^{p-1} and integrating, it follows that

$$\|S(v)\|_p^p \leq \|S(V)\|_p^p \leq c \int_{[0,\infty)} \lambda^{p-3} \int_{R^n \setminus E_\lambda} N(u)(y)^2\, dy\, d\lambda$$
$$+ c \int_{[0,\infty)} \lambda^{p-1} |E_\lambda|\, d\lambda$$
$$\leq c \int_{R^n} N(u)(y)^2 \int_{[N(u)(y),\infty)} \lambda^{p-3}\, d\lambda\, dy + c\|N(u)\|_p^p$$
$$= \frac{c}{2-p} \int_{R^n} N(u)(y)^2 N(u)(y)^{p-2}\, dy + c\|N(u)\|_p^p = c\|N(u)\|_p^p,$$

and we have finished in this case.

On the other hand, if u is arbitrary, let $\varepsilon > 0$ and note that by (2.6) and (2.8) $|u(y, t + \varepsilon)| \leq c \min(\varepsilon^{-n/p}\|N(u)\|_p, N(u)(y))$. Thus

$$\int_{R^n} |u(y, t + \varepsilon)|^2\, dy \leq c\varepsilon^{-n(2-p)/p} \|N(u)\|_p^{2-p} \int_{R^n} N(u)(y)^p\, dy < \infty$$

and by Theorem 3.3 in Chapter VII there is an L^2 function f (which depends on ε) so that $u(y, t + \varepsilon) = f * P_t(y)$. Therefore, by the first part of the theorem $S(t, (\partial/\partial t) f * P_t)(x)$ satisfies

$$\left\| S\left(t \frac{\partial}{\partial t} f * P_t \right) \right\|_p^p = \int_{R^n} \left(\int_{[\varepsilon,\infty)} \int_{|x-y|<t-\varepsilon} (t - \varepsilon) \right.$$
$$\left. \times \left| \frac{\partial}{\partial t} u(y, t) \right|^2 (t - \varepsilon)^{-n}\, dy\, dt \right)^{p/2} dx$$
$$\leq c\|N(f * P_t)\|_p^p \leq c\|N(u)\|_p^p,$$

where c is independent of ε. To complete the proof we observe that $\liminf_{\varepsilon\to 0} S(t(\partial/\partial t)f * P_t)(x) = S(v)(x)$ and invoke Fatou's lemma. ∎

Remark 2.3. If u is harmonic in R_+^{n+1} and $\sup_{t>0}\int_{R^n}|u(y, t)|^p\,dy < \infty$ for some $p > 0$, then $\lim_{t\to 0} u(\cdot, t) = f$ exists in the sense of distributions and $u(y, t) = f * P_t(y)(\mathscr{S}')$. Indeed, as in Lemma 1.6 it is readily seen that also $\int_{R^n}|u(y, t)|\,dy \le ct^{n-n/p}$, where c depends on u. Moreover, for any $\varepsilon > 0$ the function $u(y, t+\varepsilon)$ is bounded and harmonic in R_+^{n+1} and continuous in the closure of this set, and, consequently, $u(y, t+\varepsilon)$ is the Poisson integral of $u(\cdot, \varepsilon)$. By taking Fourier transforms we have that $\hat{u}(\xi, t+\varepsilon) = \hat{u}(\xi, \varepsilon)e^{-t|\xi|}$ and $\hat{u}(\xi, t) = g(\xi)e^{-t|\xi|}$ where

$$|g(\xi)| \le e^{t|\xi|}\int_{R^n}|u(y, t)|\,dy \le t^{n-n/p}e^{-t|\xi|}.$$

Taking $t = 1/|\xi|$ it follows that g has tempered growth and consequently $g \in \mathscr{S}'(R^n)$. Let f = inverse Fourier transform of g. Since $\lim_{t\to 0}\hat{u}(\cdot, t) = g(\mathscr{S}')$, we also have $\lim_{t\to 0} u(\cdot, t) = f(\mathscr{S}')$, as we wished to show.

Remark 2.4. Combining the results of this chapter with those of Chapters IV and XII we have proved in particular that the following statements are equivalent for $f \in \mathscr{S}'(R^n)$ and $0 < p < \infty$: Let $u(x, t) = f * P_t(x)$ denote the Poisson integral of f and $v(x, t) = t(\partial/\partial t)u(x, t)$. Then

(i) $\lim_{t\to\infty} u(x, t) = 0$ and $S(v) \in L^p(R^n)$.
(ii) $N(u) \in L^p(R^n)$.

Furthermore, $\|S(v)\|_p \sim \|N(u)\|_p$, and the constants involved in this norm equivalence are independent of f. We introduce the Hardy spaces $H^p(R^n)$ of several real variables to consist of those tempered distributions f for which (ii) and consequently also (i) holds and set $\|f\|_{H^p} = \|N(u)\|_p$, $0 < p < \infty$. It is clear that for $1 < p < \infty$ the $H^p(R^n)$ spaces coincide with the usual Lebesgue $L^p(R^n)$ classes and that the norms are equivalent. The most interesting case occurs then when $0 < p \le 1$ and we pass on to consider some of the natural questions in this setting, including the relation to systems of conjugate harmonic functions, boundedness of multipliers, and interpolation.

3. SYSTEMS OF CONJUGATE FUNCTIONS

Suppose $(n-1)/n < p < \infty$. We say that the $(n+1)$tuple of harmonic functions $U = (u, v_1, \ldots, v_n) \in H^p(R_+^{n+1})$ if it is a system of conjugate functions as defined in Section 3 of Chapter X, i.e., it verifies the generalized Cauchy-Riemann equations, and if in addition

$$|U|_{H^p}^p = \sup_{t>0} \int_{R^n} \left(|u(x,t)|^2 + \sum_{j=1}^n |v_j(x,t)|^2 \right)^{p/2} dx < \infty. \qquad (3.1)$$

We remark first that in this case u essentially determines U. In fact when $p > 1$ we observed that the v_j's could be obtained from u by means of the Riesz transform, that $u(x, t) = f * P_t(x)$ and

$$|U|_{H^p} \sim \|f\|_p + \sum_{j=1}^n \|R_j f\|_p \sim \|f\|_p. \qquad (3.2)$$

In this sense $H^p(R_+^{n+1})$ can be identified with $L^p(R^n)$. What is the situation for $0 < p \leq 1$? The reason we restrict our discussion to the case $p > (n-1)/n$ is that the integrand $|U(x,t)|^p$ of (3.1) is subharmonic when $p \geq (n-1)/n$. For the other values of p we must consider tensor functions of rank >1, satisfying additional conditions. We do not discuss this general case and refer the reader to Fefferman-Stein [1972] where the full picture is presented.

Note that if $U \in H^p(R_+^{n+1})$ a simple extension of Theorem 4.9 in Chapter VII gives that $N(u), N(v_1), \ldots, N(v_n)$ are in $L^p(R^n)$ and

$$\|N(u)\|_p, \|N(v_1)\|_p, \ldots, \|N(v_n)\|_p \leq c|U|_{H^p}, \qquad (3.3)$$

where c is independent of U. Our next observation is along the lines of Theorem 4.12 in Chapter VII.

Lemma 3.1. Suppose u is harmonic in R_+^{n+1} and $N(u) \in L(R^n)$. Then $u(x,t) = f * P_t(x)$, where f and its Riesz transform $R_j f$, $1 \leq j \leq n$, are integrable. Furthermore, there is a constant c independent of u so that

$$\|f\|_1 + \sum_{j=1}^n \|R_j f\|_1 \leq c\|N(u)\|_1. \qquad (3.4)$$

Proof. Since $\int_{R^n} |u(x,t)|\, dx \leq \|N(u)\|_1$, by the analogue of Theorem 3.4 in Chapter VII there is a finite measure μ so that $u(x,t) = \mu * P_t(x)$ and $\lim_{t\to 0} u(x,t) = f(x)$ exists a.e. Moreover, since $|f(x)|, |u(x,t)| \leq N(u)(x)$, also $|u(x,t) - f(x)| \leq 2N(u)(x) \in L(R^n)$, and, by the Lebesgue convergence theorem, $\lim_{t\to 0}\|u(\cdot, t) - f\|_1 = 0$. This gives immediately that $\{u(\cdot, t)\}$ is Cauchy in $L(R^n)$ as $t \to 0$ and consequently μ actually coincides with the integrable function f. But this means that $f \in H^1(R^n)$ and by Theorem 1.10 there is a sequence $\{a_k\}$ of $(1, 2, 0)$ atoms such that $f = \sum a_k(\mathcal{S}')$, $\sum |a_k| \leq c\|f\|_{H^1} = c\|N(u)\|_1$, and $\|\sum_{k=1}^m a_k - f\|_{H^1} \to 0$ as $m \to \infty$.

In order to obtain (3.4), since the Riesz transforms are linear, we are then reduced to showing that for each such atom a,

$$\|R_j a\|_1 \leqslant c|a|, \qquad 1 \leqslant j \leqslant n, \tag{3.5}$$

where c is independent of a. This inequality looks like estimate (1.3) and is proved in a similar fashion. This is not surprising since both results involve estimating a singular integral operator. The details of this verification are therefore left for the reader. A more general result involving multipliers will be discussed in the next section. ■

We are now ready to prove that (3.2) still holds for $p \leqslant 1$ provided we replace the last expression there by $\|f\|_{H^p}$. For simplicity we do only the case $p = 1$, but clearly the argument extends to $(n-1)/n < p \leqslant 1$ as well.

Theorem 3.2. The following statements are equivalent

(a) $U = (u, v_1, \ldots, v_n) \in H^1(R_+^{n+1})$.

(b) There exists an integrable function f, with integrable Riesz transform, so that $u = f * P_t$, $v_j = R_j f * P_t$, $1 \leqslant j \leqslant n$, and $|U|_{H^1} \sim \|f\|_1 + \sum_{j=1}^n \|R_j f\|_1$.

(c) There is a distribution f so that $u = f * P_t, f \in H^1(R^n)$, $v_j = R_j f * P_t$, $1 \leqslant j \leqslant n$ and $|U|_{H^1} \sim \|f\|_{H^1}$.

Proof. (a) *implies* (b). Since by (3.3) $N(u)$, $N(v_1), \ldots, N(v_n)$ are integrable, by Lemma 3.1 there are integrable functions $f, f_1, \ldots, f_n$ such that $u = f * P_t$, $v_j = f_j * P_t$, $1 \leqslant j \leqslant n$, and $R_j f \in L(R^n)$, $1 \leqslant j \leqslant n$. Consider now the system of conjugate functions $V = (f * P_t, R_1 f * P_t, \ldots, R_n f * P_t)$; clearly $V \in H^1(R_+^{n+1})$. So $U - V$ is also a system of conjugate functions in $H^1(R_+^{n+1})$ and its first component is 0. Therefore by the generalized Cauchy-Riemann equations it follows that $f_j * P_t(x) - R_j f * P_t(x) = c_j$, $1 \leqslant j \leqslant n$, where c_j is some constant. Furthermore, since $|f_j * P_t(x) - R_j f * P_t(x)| \leqslant c(\|f_j\|_1 + \|R_j f\|_1)t^{-n} \to 0$ as $t \to \infty$, $c_j = 0$, and $U = V$. The equivalence of the norms follows at once from Lemma 3.1 and an argument similar to Theorem 4.10 in Chapter VII.

(b) *implies* (c). By (3.3), $N(u) \in L(R^n)$ and $\|N(u)\|_1 = \|f\|_{H_1} \leqslant c|U|_{H_1}$. The opposite inequality follows from Lemma 3.1.

(c) *implies* (a). Because of Lemma 3.1 it is obvious. ■

Remark 3.3. Theorem 3.2, in its version for $(n-1)/n < p \leqslant 1$, contains the other half of the Burkholder-Gundy-Silverstein theorem. Indeed, if $f \in H^p(R^n)$, then, by Lemma 3.1, $R_j f \in L^p(R^n)$, and consequently by the implication (c) → (a), $U = (f * P_t, R_1 f * P_t, \ldots, R_n f * P_t) \in H^p(R_+^{n+1})$. For $n = 1$ this covers the whole range $0 < p \leqslant 1$, for $n > 1$ we must also consider Riesz transforms of higher order.

4. MULTIPLIERS

Theorems 5.1 and 4.5 in Chapter XII combine to give that if $0 < p \leqslant 1$ and m satisfies a Hörmander condition of order $L > n/p$ and we denote this by $m \in M(2, L)$, then m is a bounded $H^p(R^n)$ multiplier. In other words, the mapping T given by $Tf\hat{\ }(\xi) = m(\xi)\hat{f}(\xi)$, which by Theorem 1.10 we may think to be originally defined for those $f \in C_0^\infty(R^n)$ which are finite sums of $(p, 2, N)$ atoms, satisfies $\|Tf\|_{H^p} \leqslant c\|f\|_{H^p}$, with c independent of f, and consequently admits a bounded extension to $H^p(R^n)$ with the same bound. But this result is not sharp in the sense that we demand too many derivatives on the multiplier m. The atomic decomposition gives a more precise value of L and we discuss this next.

We begin with some definitions. The notation k is reserved for the kernel obtained as the Fourier transform, in the sense of distributions, of m. First we consider what behavior of k is reflected from the $M(2, L)$ condition on m. For this purpose we say that k verifies the $\tilde{M}(2, L)$ condition, and write $k \in \tilde{M}(2, L)$, if

$$\left(\int_{R \leqslant |x| \leqslant 2R} |D^\beta k(x)|^2\, dx\right)^{1/2} \leqslant cR^{-(n/2+|\beta|)} \tag{4.1}$$

for $R > 0$ and β any multi-index with $|\beta| < L$, and in addition if $\tilde{L}$ is the largest integer strictly less than L and $L = \tilde{L} + \gamma$, then

$$\left(\int_{R \leqslant |x| \leqslant 2R} |D^\beta k(x) - D^\beta k(x-y)|^2\, dx\right)^{1/2}$$

$$\leqslant \begin{cases} c\left(\dfrac{|y|}{R}\right)^{\gamma} R^{-(n/2+\tilde{L})} & \text{when} \quad 0 < \gamma < 1 \\[2ex] c\left(\dfrac{|y|}{R}\right) \ln\left(\dfrac{|y|}{R}\right) R^{-(n/2+\tilde{L})} & \text{when} \quad \gamma = 1 \end{cases} \tag{4.2}$$

for all $|y| < R/2$, $R > 0$ and multi-indices β with $|\beta| = \tilde{L}$. The infimum over the constants c for which (4.1) and (4.2) hold is called the constant of the kernel k; similarly for the constant of m.

A convenient notation is to denote by $|x| \sim R$ the annulus $\{aR \leqslant |x| \leqslant bR\}$, where $0 < a < b < \infty$ are fixed numbers which are unimportant in the conclusions. For instance, $a = 1$, $b = 2$ in (4.1) and (4.2). An important relation is given by

Lemma 4.1. Suppose $m \in M(2, L)$, $L > n/2$. Then $k \in \tilde{M}(2, L - n/2)$.

Proof. We begin with some observations. Note first that, if $\eta \in \mathscr{S}(R^n)$, then multiplication of either m or k by η only increases the constants of these new functions by $c = c_\eta$. This is readily seen by using the product rule for differentiation and it is especially simple when η is supported in an annulus $|x| \sim R$, which is the only case of concern to us. Also the conditions M and $\tilde{M}$ are invariant under dilations of the form $m(\xi) \to m(t\xi)$ and $k(x) \to t^{-n}k(x/t)$, $t > 0$. Hence if m satisfies an M condition, so does $\eta(t\xi)m(\xi)$, with constant bounded uniformly in t.

Now by the dilation invariance we may assume that $R = 1$ and show that the expressions on the right-hand side of (4.1) and (4.2) are finite. Let ϕ be a nonnegative, C^∞ function, supported in $\{\frac{1}{2} < |y| < 2\}$ so that $\sum_{j=-\infty}^{\infty} \phi(2^{-j}y) = 1$ for $y \neq 0$. Such functions are easy to construct (cf. 6.1). Put now $\eta(\xi) = 1 - \sum_{j=1}^{\infty} \phi(2^{-i}\xi)$, and let $m_0(\xi) = \eta(\xi)m(\xi)$ and $m_i(\xi) = \phi(2^{-i}\xi)m(\xi)$, $i = 1, 2, \ldots$. If k_i denotes the distributional Fourier transform of m_i, $i = 0, 1, \ldots$, we estimate first the expressions corresponding to the different k_i's and then add them up.

***Case i* = 0.** We estimate $D^\beta k_0(x)$ and $D^\beta k_0(x) - D^\beta k_0(x-y)$, which have Fourier transform essentially equal to $\xi^\beta \eta(\xi)m(\xi)$ and $\xi^\beta\eta(\xi)(1 - e^{i\xi\cdot y})m(\xi)$, respectively. First, since $0 \leq \eta \leq 1$ and $\operatorname{supp} \eta \subseteq \{|\xi| \leq 2\}$, it readily follows that

$$\left(\int_{|x|\sim 1} |D^\beta k_0(x)|^2\,dx\right)^{1/2} \leq c\left(\int_{R^n} |\xi^\beta \eta(\xi)m(\xi)|^2\,d\xi\right)^{1/2} \leq c\|m\|_\infty \tag{4.3}$$

with $c = c_\beta$. When estimating the difference we also have the factor $(1 - e^{i\xi\cdot y})$ in the Fourier transform side, and since $|1 - e^{i\xi\cdot y}| \leq 2|y|$ on the support of η, we get that

$$\left(\int_{|x|\sim 1} |D^\beta k_0(x) - D^\beta k_0(x-y)|^2\,dx\right)^{1/2} \leq c|y|. \tag{4.4}$$

***Case i* > 0.** Since $\sum_{|\alpha|=L}|x^\alpha| \geq c > 0$ when $|x| \sim 1$, we get that

$$\begin{aligned}\left(\int_{|x|\sim 1} |D^\beta k_i(x)|^2\,dx\right)^{1/2} &\leq c\left(\int_{|x|\sim 1} \sum_{|\alpha|=L} |x^\alpha D^\beta k_i(x)|^2\,dx\right)^{1/2}\\ &\leq c\sum_{|\alpha|=L}\left(\int_{R^n} |x^\alpha D^\beta k_i(x)|^2\,dx\right)^{1/2}\\ &\leq c\sum_{|\alpha|=L}\left(\int_{R^n} |D^\alpha(\xi^\beta m_i(\xi))|^2\,d\xi\right)^{1/2}.\end{aligned}$$

Furthermore, since $\xi^\beta\phi(\xi) \in \mathscr{S}(R^n)$, we see immediately that $2^{-i|\beta|}\xi^\beta m_i(\xi) = (2^{-i}\xi)^\beta\phi(2^{-i}\xi)m(\xi) \in M(2, L)$, with constant bounded uniformly in i, and that this last function is supported $|\xi| \sim 2^i$. Therefore, for each α the corresponding term in the above sum is bounded by $c2^{i(n/2+|\beta|-L)}$ and we conclude that

$$\left(\int_{|x|\sim 1} |D^\beta k_i(x)|^2\, dx\right)^{1/2} \leqslant c2^{i(n/2+|\beta|-L)}. \tag{4.5}$$

The term involving the difference is estimated as before, and we get that

$$\left(\int_{|x|\sim 1} |D^\beta k_i(x) - D^\beta k_i(x-y)|^2\, dx\right)^{1/2} \leqslant c2^{i(n/2+|\beta|-L)}|y|2^i \tag{4.6}$$

when $|y| \leqslant 1$. Notice that when $|y|2^i > 1$, we get a better estimate by using the triangle inequality and (4.5) instead.

Now sum. From (4.3) and (4.5) we see immediately that

$$\left(\int_{|x|\sim 1} |D^\beta k(x)|^2\, dx\right)^{1/2} \leqslant c + c\sum_{i=1}^{\infty} 2^{i(n/2+|\beta|-L)}$$

and this expression is finite provided that $|\beta| < L - n/2$, which is our assumption.

Similarly, by adding (4.4) and (4.6), we observe that when $|\beta| = \tilde{L} =$ largest integer strictly less than $L - n/2$, the difference in (4.2) is bounded by

$$c|y| + c|y| \sum_{2^i \leqslant 1/|y|} 2^{i(1-\gamma)} + \sum_{2^i > 1/|y|} 2^{i\gamma}, \tag{4.7}$$

where $0 < \gamma = L - n/2 - \tilde{L} \leqslant 1$. It is now a simple matter to sum (4.7) and to note that for $|y| \leqslant 1$ it does not exceed $c|y|^\gamma$ when $0 < \gamma < 1$ and $c|y|\ln(2/|y|)$ when $\gamma = 1$. ■

We are now ready to consider the action of multiplier on atoms.

Lemma 4.2. Let $0 < p \leqslant 1$ and $m \in M(2, L)$, where $L > n(1/p - 1/2) \geqslant n/2$. Suppose a is a $(p, 2, N)$ atom, supp $a \subseteq I \subseteq B(x_0, R)$, where R is of order sidelength of I and $|I|^{1/p-1/2}\|a\|_2 \leqslant 2|a|$, and for ϕ as in Lemma 4.1 set $k_i(x) = \phi(x/2^{i+2}R)k(x)$, $i = 1, 2, \ldots$. Then $b_i(x) = k_i * a(x)$ is also a $(p, 2, N)$ atom and $|b_i| \leqslant c2^{-i\varepsilon}|a|$, where c is independent of a, i, and $\varepsilon > 0$.

Proof. As it is readily seen that supp $b_i \subseteq B(x_0, 2^{i+4}R)$ and that the moments of b_i coincide with those of a and thus vanish up to order N, it only remains to bound $\|b_i\|_2$ appropriately. First, note that by Lemma 4.1, $k_i \in \tilde{M}(2, L - n/2)$, with constant bounded uniformly in i and R.

Now let $\tilde{N}$ be the largest integer $\leqslant N$ so that $\tilde{N} < L - n/2$. Furthermore, let $R_i(x, y)$ denote the remainder in the Taylor expansion of $k_i(x - y)$, as a function of y, about x_0 of order $\tilde{N} - 1$. Then we can write $R_i(x, y)$ as

$$\sum_{|\beta| = \tilde{N}} c_\beta (y - x_0)^\beta \int_{[0,1]} (1-s)^{\tilde{N}-1} D^\beta k_i(x - x_0 - s(y - x_0))\, ds$$

and $k_i(x_0 - y) - R_i(x, y)$ is a polynomial of degree at most $\tilde{N} - 1$ when considered as a function of y. Hence, by the moment condition on a, it follows that $b_i(x)$ actually equals

$$\sum_{|\beta|=\tilde{N}} c_\beta \int_{[0,1]} (1-s)^{\tilde{N}-1} \int_{R^n} a(y)(y-x_0)^\beta \times \{D^\beta k_i(x - x_0 - s(y-x_0)) - D^\beta k_i(x-x_0)\}\, dy\, ds. \tag{4.8}$$

Thus, to estimate $\|b_i\|_2$, we may invoke Minkowski's inequality and consider the L^2 norm of the expression in $\{\ldots\}$ in (4.8) above as a function of x for each $0 \leq s \leq 1$ and y in supp a. Since $s|y - x_0| \leq R$ and $|x - x_0| \leq 2^{i+2}R$, we see from the $\tilde{M}(2, L - n/2)$ condition on k_i that $\|\{\ldots\}\|_2$ is bounded by

$$\begin{cases} c(2^iR)^{n/2-n-\tilde{N}}(s|y-x_0|2^iR)^{L-n/2-\tilde{N}} & \text{if } L - n/2 - 1 < \tilde{N} < L - n/2, \\ c(2^iR)^{n/2-n-\tilde{N}}(s|y-x_0|2^iR)\ln(2^iR/s|y-x_0|) & \text{if } \tilde{N} = L - n/2 - 1, \\ c(2^iR)^{n/2-n-\tilde{N}}(s|y-x_0|/2^iR) & \text{if } \tilde{N} < L - n/2 - 1. \end{cases}$$

The first two estimates are immediate and the third follows from the mean value theorem and the bounds for the derivatives of $k_i \in \tilde{M}(2, L - n/2)$. There are then three possible kinds of terms that will appear in estimating (4.8), one corresponding to each of the above expressions. Because all terms are handled in a similar fashion we only do one of them, the first one. In this case the corresponding expression in (4.8) is less than or equal to

$$\begin{aligned} c(2^iR)^{-L} &\int_{[0,1]} (1-s)^{\tilde{N}-1} s^{L-n/2-\tilde{N}} \int_{R^n} |a(y)||y-x_0|^{L-n/2}\, dy\, ds \\ &\leq c(2^iR)^{-L}\|a\|_2 \left(\int_{B(x_0,R)} |y-x_0|^{2(L-n/2)}\, dy \right)^{1/2} \\ &\leq c(2^{i+4}R)^{-n(1/p-1/2)} 2^{-i(L-n(1/p-1/2))} R^{n(1/p-1/2)} \|a\|_2. \end{aligned}$$

In this case $\varepsilon = L - n(1/p - 1/2) > 0$ and

$$R^{n(1/p-1/2)}\|a\|_2 \leq c|I|^{1/p-1/2}\|a\|_2 \leq c|a|.$$

As the other two terms lead to similar expressions and since supp $b_i \subseteq$ open cube Q of sidelength of order $2^{i+4}R$, from the above estimate it follows that $|b_i| \leq |Q|^{1/p-1/2}\|b_i\|_2 \leq c(|Q|/(2^{i+4}R)^n)^{(1/p-1/2)}2^{-i\varepsilon}|a| \leq c2^{-i\varepsilon}|a|$, and we are done. ∎

We also need

Lemma 4.3. For m, a, and ϕ as in Lemma 4.2 let $\phi_0(x) = 1 - \sum_{i=1}^\infty \phi(x/2^{i+2}R)$ and put $k_0(x) = \phi_0(x)k(x)$. Then $b_0(x) = k_0 * a(x)$ is a $(p, 2, N)$ atom and $|b_0| \leq c|a|$, where c is independent of a.

Proof. Observe that $\operatorname{supp} b_0 \subseteq B(x_0, 2^4 R)$ and that the support of the $k_i * a$'s is disjoint with that ball as long as $i \geq 4$. Since the moment condition is not disturbed by convolutions, it only remains to bound $\|b_0\|_2$ appropriately, but this is not hard. First, by the above remarks it readily follows that $|b_0(x)| \leq |Ta(x)| + \sum_{i=1}^{3} |k_i * a(x)|$, where T is the multiplier operator associated to m. Clearly, $\|Ta\|_2 \leq c\|a\|_2$, and consequently $(2^4 R)^{n(1/p-1/2)}\|Ta\|_2 \leq c|a|$. That a similar estimate holds for the other three summands follows at once from Lemma 4.2 and we have finished. ■

We are now ready to prove

Theorem 4.4. Let $0 < p \leq 1$ and $m \in M(2, L)$, where $L > n(1/p - 1/2)$. Then m is a bounded multiplier on $H^p(R^n)$.

Proof. Suppose first that $f \in H^p(R^n)$ is a finite sum of $(p, 2, N)$ atoms, $N > L$, $f(x) = \sum_{j=1}^{h} a_j(x)$, say, so that $\sum_{j=1}^{h} |a_j|^p \leq c\|f\|_{H^p}^p$. Then $T(f) = \sum_{j=1}^{h} T(a_j)$. Moreover, by Lemmas 4.2 and 4.3, and with a different decomposition for the kernel k adjusted to the support of each a_j, we have that

$$Ta_j(x) = \sum_{i=0}^{\infty} k_i * a_j(x) = \sum_{i=0}^{\infty} b_{i,j}(x),$$

say, where the $b_{i,j}$'s are $(2, p, N)$ atoms and

$$\sum_{i=0}^{\infty} |b_{i,j}|^p \leq c \sum_{i=0}^{\infty} 2^{-i\varepsilon p} |a_j|^p \leq c|a_j|^p,$$

with c independent of j. Therefore, by Theorem 1.4, $Ta_j \in H^p(R^n)$, $\|Ta_j\|_{H^p}^p \leq c|a_j|^p$, and the same is true for Tf with

$$\|Tf\|_{H^p}^p \leq c \sum_{j=1}^{h} |a_j|^p \leq c\|f\|_{H^p}^p.$$

To obtain the same result for a general f in $H^p(R^n)$ we invoke Theorem 1.10 and 6.13 below. ■

5. INTERPOLATION

We discuss one more application of the atomic decomposition, this time to a simple interpolation result especially suited to multipliers. We need a preliminary fact.

Proposition 5.1. Let $0 < p_0 \leq 1 < p < 2$ and $f \in L^p(R^n)$. Then given $\lambda > 0$ we may write $f(x) = f_\lambda(x) + f^\lambda(x)$, where $f^\lambda \in H^{p_0}(R^n)$, $f_\lambda \in L^2(R^n)$ and

$$\|f^\lambda\|_{H^{p_0}}^{p_0} \leq c\lambda^{p_0-p}\|f\|_p^p, \qquad \|f_\lambda\|_2^2 \leq c\lambda^{2-p}\|f\|_p^p,$$

with c independent of f and λ.

Proof. The proof is a slight variant of that of Theorem 1.10. Let $v(y, t) = t(\partial/\partial t)f * P_t(y)$ and for the given λ put $E_k = \{S(v) > \lambda 2^k\}$ and $\mathcal{O}_k = \{M(\chi_{E_k}) > \frac{1}{2}\}$. Here M is the Hardy maximal function defined with respect to balls. Let $\mathcal{O}_k = \bigcup_j Q_{j,k}$ be a Whitney decomposition of $\mathcal{O}_k$, and $I_{j,k} =$ interior of $Q_{j,k}$. Then $T_{j,k}$ and $a_{j,k}(x)$ are defined as in Theorem 1.10 with $N > n(1/p_0 - 1)$; we claim that

$$f^\lambda = \sum_{k=1}^{\infty} \sum_j a_{j,k} \quad \text{and} \quad f_\lambda = \sum_{k=-\infty}^{0} \sum_j a_{j,k}$$

will do.

To compute the L^2 norm of f_λ we use duality. If $\|\eta\|_2 = 1$ and, with the notation of Theorem 1.10, $g(y, t) = \eta * \psi_t(y)$, then

$$I = \int_{R^n} f_\lambda(x)\eta(x)\, dx = \int_{R_+^{n+1}} \chi_{\mathcal{U}}(y, t)v(y, t)g(y, t)\, dy\frac{dt}{t},$$

where $\mathcal{U} = \bigcup_{k=-\infty}^{0} \bigcup_j T_{j,k}$. Thus

$$|I| \leq \left(\int_{R_+^{n+1}} \chi_{\mathcal{U}}(y, t)|v(y, t)|^2\, dy\frac{dt}{t}\right)^{1/2} \left(\int_{R_+^{n+1}} |g(y, t)|^2\, dy\frac{dt}{t}\right)^{1/2} = J \cdot K,$$

say. Clearly, $K \leq c\|\eta\|_2 \leq c$. As for J, first, observe that

$$J \leq c \int_{R^n \setminus E_0} S(v)(x)^2\, dx. \tag{5.1}$$

Indeed, if v_n denotes the volume of the unit ball, then

$$\int_{R^n\setminus E_0} S(v)(x)^2\, dx = \int_{R^n\setminus E_0} \int_{\Gamma(x)} |v(y, t)|^2 \frac{t^{-n}}{v_n}\, dy\frac{dt}{t}\, dx$$
$$\geq \int_{R_+^{n+1}} |v(y, t)|^2 |\{x \in R^n\setminus E_0 : (y, t) \in \Gamma(x)\}| \frac{t^{-n}}{v_n}\, dy\frac{dt}{t}.$$

By definition, however, if $(y, t) \in \mathcal{U}$, then $|\{x \in R^n\setminus E_0 : (y, t) \in \Gamma(x)\}|t^{-n}/v_n > \frac{1}{2}$, and the last integral above is greater than or equal to cJ. By the converse to Hölder's inequality and (5.1) we get that

$$\|f_\lambda\|_2^2 \leq cJ \leq c\int_{\{S \leq \lambda\}} S(v)(x)^2\, dx \leq c\lambda^{2-p}\|S(v)\|_p^p \leq c\lambda^{2-p}\|f\|_p^p,$$

as we wanted to show.

To estimate $\|f^\lambda\|_{H^{p_0}}$, observe that if $|a_{j,k}|$ denotes the atomic $(p_0, 2, N)$ norm of $a_{j,k}$, then as in (1.22) of Theorem 1.10 we have that

$$\begin{aligned}\sum_{k=1}^{\infty}\sum_j |a_{j,k}|^{p_0} &\leq c\sum_{k=1}^{\infty}\lambda^{p_0}2^{kp_0}|E_k| \\ &\leq c\lambda^{p_0}\int_{R^n}\sum_{\{k\geq 1: S(v)(x)>\lambda 2^k\}} 2^{kp_0}\,dx \\ &\leq c\int_{\{S(v)>\lambda\}} S(v)(x)^{p_0}\,dx. \qquad (5.2)\end{aligned}$$

Furthermore, since when $S(v)(x)/\lambda > 1$ we also have $S(v)(x)^{p_0} = \lambda^{p_0}(S(v)(x)/\lambda)^{p_0} \leq \lambda^{p_0-p}S(v)(x)^p$, (5.2) and Theorem 1.4 yield $\|f^\lambda\|_{H^{p_0}}^{p_0} \leq c\lambda^{p_0-p}\|S(v)\|_p^p \leq c\lambda^{p_0-p}\|f\|_p^p$. ∎

We are now ready to prove

Theorem 5.2. Suppose T is a sublinear operator of weak-type $(2, 2)$ and bounded from $H^{p_0}(R^n)$ into wk-$L^{p_0}(R^n)$, $0 < p_0 \leq 1$. Then if $p_0 < p < 2$, T maps $H^p(R^n)$ continuously into $L^p(R^n)$.

Proof. Suppose first $p > 1$. Given $f \in L^p(R^n)$ and $\lambda > 0$, write $f = f_\lambda + f^\lambda$ as in Lemma 5.1. Then $\{|Tf| > \lambda\} \subseteq \{|Tf_\lambda| > \lambda/2\} \cup \{|Tf^\lambda| > \lambda/2\}$, and consequently

$$\begin{aligned}|\{|Tf| > \lambda\}| &\leq c\lambda^{-2}\|f_\lambda\|_2^2 + c\lambda^{-p_0}\|f^\lambda\|_{H^{p_0}}^{p_0} \\ &\leq c\lambda^{-2}\lambda^{2-p}\|f\|_p^p + c\lambda^{-p_0}\lambda^{p_0-p}\|f\|_p^p = c\lambda^{-p}\|f\|_p^p.\end{aligned}$$

Thus T is of weak-type (p, p) for $1 < p < 2$ and also of type (p, p) in the same range of p's as a simple application of the Marcinkiewicz interpolation theorem gives.

Consider next the case $p_0 < p \leq 1$. We claim that if a is a $(p, 2, N)$ atom, then $\|Ta\|_p \leq c|a|$, where c is independent of a. Indeed, first observe that there is an open cube I so that supp $a \subseteq I$ and $\|a\|_{H^{p_0}} \leq 2|I|^{1/p_0-1/p}|a|$ and $\|a\|_2 \leq 2|I|^{1/2-1/p}|a|$. Whence

$$\begin{aligned}\|Ta\|_p^p &= p\left(\int_{[0,r)} + \int_{[r,\infty)}\right)\lambda^{p-1}|\{|Ta| > \lambda\}|\,d\lambda \\ &\leq c\int_{[0,r)}\lambda^{p-1}\lambda^{-p_0}\|a\|_{H^{p_0}}^{p_0}\,d\lambda + c\int_{[r,\infty)}\lambda^{p-1}\lambda^{-2}\|a\|_2^2\,d\lambda \\ &\leq c(|I|^{1-p_0/p}r^{p-p_0}|a|^{p_0} + |I|^{1-2/p}r^{p-2}|a|^2). \qquad (5.3)\end{aligned}$$

Setting $r = |a|/|I|^{1/p}$ in (5.3), we see immediately that $\|Ta\|_p^p \leq c|a|^p$, as anticipated. Let now $f \in H^p(R^n)$ be a finite sum $\sum a_j$ of $(p, 2, N)$ atoms so that $\sum|a_j|^p \leq 2\|f\|_{H^p}^p$. Since $p \leq 1$ we see that $|Tf(x)|^p \leq \sum|Ta_j(x)|^p$ and consequently $\|Tf\|_p^p \leq \sum\|Ta_j\|_p^p \leq c\sum|a_j|^p \leq c\|f\|_{H^p}^p$. The general case follows easily by a simple limiting argument. ■

6. NOTES; FURTHER RESULTS AND PROBLEMS

The classical theory of H^p spaces is essentially part of complex analysis with many connections to harmonic functions and Fourier analysis. New methods are therefore needed to rid the theory of one-dimensional techniques such as conformal mapping, and extend the results to several dimensions. The recent n-dimensional real theory was started by Stein and Weiss [1960]; a crucial observation in this context is the fact that, if $F = (u_0, u_1, \ldots, u_n)$ is a (M. Riesz) system of conjugate functions, then $|F|^q$ is subharmonic for $q > n/(n-1)$. In the late 1960s important new developments took place, culminating in the Fefferman–Stein [1972] theory of H^p spaces of several real variables. We single out three such developments here.

(i) Results concerning the boundedness of certain singular integral operators can be extended from $L^p(R^n)$, $1 < p < \infty$, to the $H^p(R^n)$ spaces, $0 < p \leq 1$, and especially $H^1(R^n)$. The methods used involve auxilliary functions such as the Lusin integral, a (vector-valued) singular integral itself.

(ii) The result of Burkholder and Gundy and Silverstein [1971] concerning the characterization of the Hardy space $H_a^p(R_+^2)$ in terms of nontangential maximal functions. This remarkable theorem, proved by probabilistic methods involving Brownian motion, raised many interesting questions, including the possibility of extending these results to R^n and what role the Poisson kernel plays in all of this.

(iii) Fefferman's identification of the dual of $H^1(R^n)$ with $BMO(R^n)$ [1971].

One of the main results of Fefferman and Stein is that the H^p classes can be characterized without any recourse to conjugacy of harmonic functions or Poisson integrals. Elements u in $H^p(R^n)$ can be considered in terms of their boundary values f and have an intrinsic meaning: u is in $H^p(R^n)$ if and only if $N(f * \phi_t)(x) \in L^p(R^n)$ whenever ϕ is a sufficiently smooth function, small at infinity, and has nonvanishing integral; in fact, it suffices to consider the radial maximal function $N_0(f * \phi_t)(x) = \sup_{t>0}|f * \phi_t(x)|$. In a different direction Calderón and Torchinsky [1975] established a similar characterization in terms of the Lusin integral $S(f * \psi_t)(x)$ corresponding to a smooth function ψ, small at infinity and with vanishing integral. The atomic decomposition is due to Coifman [1974] when $n = 1$ and to Latter

[1978] for general n. Both of these results make use of the characterization of $H^p(R^n)$ in terms of maximal functions, and a relatively simple proof along these lines is discussed in 6.5 below. The atomic decomposition given here is based on ideas of Calderón [1977b], Chang and R. Fefferman [1982], and, especially, Cohen [1982]. Fefferman-Stein [1972], Burkholder and Gundy [1972], and Calderón and Torchinsky [1975], considered the distribution function inequalities which allow for the control of the Lusin integral in terms of the nontangential maximal function, and *vice versa*. The proof of Lemma 2.1 is based on some ideas of Merryfield [1985]. The results in Section 3 are due to Stein and Weiss [1960]; the work of Wheeden [1976] is also relevant here. The multiplier results in Section 4, which are due to Calderón and Torchinsky [1977], can be extended in several directions (cf. Taibleson and Weiss [1980], for instance). The proof given here follows along the lines of Strömberg and Torchinsky [1980]; the sharp version of this result is due to Baernstein and Sawyer [1985]. The decomposition in Section 5 is essentially due to Chang and R. Fefferman [1982].

Further Results and Problems

6.1 There is a $C_0^\infty(R^n)$ function ϕ supported in $\{\frac{1}{2} \le |x| \le 2\}$ and such that $\sum_{j=-\infty}^{\infty} \phi(2^{-j}x) = 1$, $x \ne 0$. (*Hint*: If η is nonnegative, nonidentically 0, $C_0^\infty(R^n)$ function supported in $\{\frac{1}{2} \le |x| \le 2\}$, then $\sum_{j=-\infty}^{\infty} \eta(2^{-j}x) \ne 0$ for $x \ne 0$. Look at $\phi(x) = \eta(x)/\sum_{j=-\infty}^{\infty} \eta(2^{-j}x)$.)

6.2 Let f be defined on R_+^{n+1} and suppose $0 < a < b < \infty$. Then $|\{N_b(F) > \lambda\}| \le c(b/a)^n|\{N_a(F) > \lambda\}|$, all $\lambda > 0$, where c is an absolute constant. (*Hint*: Let $\mathcal{O} = \{N_a(F) > \lambda\}$ and put $\mathcal{O}_1 = \{M\chi_{\mathcal{O}} > (a/(a+b))^n\}$. As it is not hard to see that $\{N_b(F) > \lambda\} \subseteq \mathcal{O}_1$ the conclusion follows at once from the maximal theorem. Clearly, this result implies $\|N_b(F)\|_p \le c(b/a)^{n/p}\|N_a(F)\|_p$, $0 < p < \infty$ (cf. Theorem 4.3 in Chapter XII).)

6.3 In case u is harmonic in R_+^{n+1}, then $u \in H^p(R^n)$ if and only if the radial maximal function $N_0(u) \in L^p(R^n)$ and $\|u\|_{H^p} \sim \|N_0(u)\|_p$, $0 < p < \infty$. This statement corresponds to Proposition 5.5 in Chapter VII. The general result is the following: Let $F(x, t)$ be continuously differentiable with respect to the x variables in $t > 0$ and suppose that for some $a, b > 0$, $N_a(F)$ and $N_b(|t\nabla F|)$ are in $L^p(R^n)$, $0 < p < \infty$. Then there is a constant c, depending on a, b, and p such that $\|N_a(F)\|_p \le c\|N_0(F)\|_p^{p/(p+n)}\|N_b(|t\nabla F|)\|^{n/(p+n)}$ if $\|N_0(F)\|_p \le \|N_b(|t\nabla F|)\|_p$ and $\|N_a(F)\|_p \le c\|N_0(F)\|_p$ otherwise. (*Hint*: On account of 6.2 it suffices to prove our assertion for $a = 1$, $b = 2$. The desired inequalities follow without difficulty from the estimate $|\mathcal{U}_1| = |\{N(F) > \lambda,\ N_2(|t\nabla F|) \le r^{-1/p}\lambda\}| \le cr^{-n/p}|\{N_0(F) > \lambda/2\}| = |\mathcal{U}_0|$, where r

is a number between 0 and 1. To show this estimate observe that if $x \in \mathcal{U}_1$, then there exists (y, t) with $|x - y| < t$ and $|F(z, t)| > \lambda/2$ for $|y - z| \leq \frac{1}{2} r^{1/p} t$. Thus, $\mathcal{U}_1 \subseteq \{M\chi_{\mathcal{U}_0} > ((r^{1/p}/2)(1 + r^{1/p}/2))^{-1})^n\}$ and the estimate follows from the maximal theorem. The result is from the work of Calderón and Torchinsky [1975].)

6.4 Suppose $N(F) \in L^p(R^n)$, $0 < p < \infty$, and let $N_\lambda^*(F)(x) = \sup_{(y,t)} |F(y, t)|(1 + |x - y|/t)^{-\lambda}$. Then $\|N_\lambda^*(F)\|_p \leq c\|N(F)\|_p$ if $\lambda > n/p$ and $|\{N_\lambda^*(F) > t\}| \leq ct^{-p}\|N_\lambda^*(F)\|_p^p$ if $\lambda = n/p$. (*Hint*: Note that $N_\lambda^*(F)(x) \leq c \sup_k 2^{-\lambda k} N_{2^k}(F)(x)$, which in turn implies $\{N_\lambda^*(F) > t\} \subseteq \bigcup_{k=0}^\infty \{N_{2^k}(F) > c2^{\lambda k} t\}$ and then use 6.2.)

6.5 Suppose $u(x, t) = f * P_t(x)$ is harmonic in R_+^{n+1} and $N(u) \in L^p(R^n)$, $0 < p \leq 1$. Then f admits an atomic decomposition $f = \sum a_j(\mathcal{S}')$ into (p, q, N) atoms and $\|N(u)\|_p^p \sim \inf \sum |a_j|^p$, where the inf is taken over all possible decompositions. (*Hint*: The proof follows along the lines of Theorem 1.10 and is best understood when $n = 1$. By 6.2 also $N_2(u) \in L^p$, and the open sets $\mathcal{O}_k = \{N_2(u) > 2^k\} = \bigcup_j I_{j,k}$, where the $I_{j,k}$'s are disjoint, open intervals. Let $T(\mathcal{O}_k) = \bigcup_j T(I_{j,k})$, where $T(I) = \{(y, t) \in R_+^2 \colon (y - t, y + t) \subseteq I\}$ and put $T_{j,k} = T(I_{j,k}) \backslash T(\mathcal{O}_{k+1})$. Then

$$f = \sum_{j,k} \int_{T_{j,k}} v(y, t)\psi_t(x - y)\, dy \frac{dt}{t} = \sum_{j,k} a_{j,k}(x),$$

say. As in Theorem 1.10 the proof is reduced to estimate

$$\left(\int_{T_{j,k}} (t|\nabla u(y, t)|^2 + t|v(y, t)|^2)\, dy\, dt \right)^{1/2}.$$

To do this we invoke Green's identity (4.5) in Chapter VII, which applies since the boundary $\partial T_{j,k}$ is smooth enough for Green's theorem to apply, and observe that the term in question is less than or equal to

$$c \int_{\partial T_{j,k}} \left(t|u(y, t)| \left| \frac{\partial}{\partial \nu} u(y, t) \right| + \frac{1}{2} |u(y, t)|^2 \left| \frac{\partial}{\partial \nu} t \right| \right) ds.$$

Since we are working with the level sets for $N_2(u)$, and u is harmonic, it readily follows that $|u(y, t)|$, $t|\nabla u(y, t)|$, $t|v(y, t)| \leq c2^k$ on $\partial T_{j,k}$ and the desired bound follows with no difficulty from this. This proof is due to Wilson [1985].)

6.6 To deal with operators acting on the Hardy spaces it is often necessary to work with sums of atoms, whose supports are stacked one on top of another. More precisely, we say that a function $M(x)$ is a (p, q, N) molecule based at the ball $B(x_0, r)$, $0 < p \leq 1 < q < \infty$, $N > n(q/p - 1)$, provided it satisfies the following three conditions:

(i) $\int_{R^n} |M(x)|^q\, dx \leq cr^{n(1-q/p)}$,

(ii) $\int_{R^n} |M(x)|^q |x - x_0|^N \, dx \le c r^{N+n(1-q/p)}$, and
(iii) $\int_{R^n} M(x)\, dx = 0$ (this condition makes sense since by (i) and (ii), M is integrable).

Show that if (i)-(iii) hold, then $M(x) = \sum_j a_j(x)(L^q(R^n))$, where the a_j's are (p, q, N) atoms supported in $B(x_0, 2^{k+1}r)$ and $\|M\|_{H^p}^p \le c \sum |a_j|^p$, where c depends only on the constants in (i) and (ii). This concept is due to Coifman and Weiss [1977] and is quite useful since it reduces the discussion of the continuity of many operators to showing that map atoms into molecules (cf. Theorem 4.4).

6.7 Show that, if, for $0 < p \le 1$, $f \in H^p(R^n) \cap L(R^n)$, then $\int_{R^n} f(x)\, dx = 0$.

6.8 Show that for each fixed $(x_0, t) \in R_+^{n+1}$, $P_t(x_0 - x) - P_t(x) \in H^1(R^n)$ (as a function of x). Also if f is integrable and $\hat{f}$ vanishes off a compact set K not containing the origin, then $f \in H^1(R^n)$. (*Hint*: $\chi_K(\xi)\xi_j/|\xi|$ is the Fourier transform of an integrable function, $1 \le j \le n$.)

6.9 Suppose ϕ verifies the assumptions of Zó's Theorem 2.2 in Chapter XI and $f \in H^1(R^n)$. Show that $N_0(f * \phi_t) \in L(R^n)$. It is also easy to see that, if $\psi = \chi_I$, I unit cube in R^n, and $N(f * \psi_t) \in L(R^n)$, then $f = 0$, a.e. Along these lines Uchiyama and Wilson [1983] have shown that there is a nonnegative kernel ϕ such that $\{0\} \ne H_\phi^1(R) = \{f \in L(R)\colon N(f * \phi_t) \in L(R)\}$ and $H_\phi^1(R) \ne H^1(R)$.

6.10 There is yet another characterization of the H^p spaces involving maximal functions. Suppose u is harmonic in R_+^{n+1} and let $u_p(x) = \|t^{-(n+1)/p} u(y, t)\|_{\text{wk-}L^p(\Gamma(x),\, dy\, dt)}$, $0 < p < \infty$. Show that $u \in H^p(R^n)$ if and only if $u_p \in L^p(R^n)$ and $\|u\|_{H^p} \sim \|u_p\|_p$, $0 < p < \infty$. (*Hint*: One implication follows at once from the estimate $|\{(y, t) \in \Gamma(x)\colon (N(u)(x)/\lambda)^{p/(n+1)} > t\}| \le c(N(u)(x)/\lambda)^p$. To prove the converse we show that $N_{1/2}(u)(x) \le c u_p(x)$. Indeed, let $(y, t) \in \Gamma_{1/2}(x)$ and note there is a ball B centered at (y, t) and of radius $\sim t$ such that $B \subset \Gamma(x)$. Then by the Hardy-Littlewood Theorem 5.4 in Chapter VII and the estimate 7.5 in Chapter IV, for $0 < q < p$ we have

$$|u(y, t)|^q \le \frac{c}{|B|} \int_B |u(w, s)|^q \, dw\, ds \le \frac{c}{|B|} t^{(n+1)q/p} |B|^{1-q/p} u_p(x)^q \le c u_p(x)^q,$$

and the desired estimate follows easily from this. The result is due to Semmes [1983].)

6.11 (Hardy-Littlewood Imbedding Theorem) Suppose F is defined in R_+^{n+1} and $N(F) \in L^p(R^n)$, $0 < p < \infty$. Then for $p < q < \infty$,

$$\left(\int_{R_+^{n+1}} t^{n(q/p-1)} |F(x, t)|^q \, dx \frac{dt}{t} \right)^{1/q} \le c \|N(F)\|_p.$$

(*Hint*: By (2.6), $|F(x, t)| \le c t^{-n/p} \|N(F)\|_p$. Let $b = c\|N(F)\|_p$ and observe that

$$\int_{R^n} |F(x,t)|^q\, dx = \int_{[0,bt^{-n/p})} |\{|F(\cdot,t)| > \lambda\}|\, d\lambda^q$$
$$\leqslant \int_{[0,bt^{-n/p})} |\{N(F) > \lambda\}|\, d\lambda^q;$$

the desired inequality follows readily from this by multiplying through by $t^{n(q/p-1)}$ and integrating. This proof, as well as some applications, is in the work of Calderón and Torchinsky [1975].)

6.12 The following extension of Paley's inequality, Theorem 1.3 in Chapter VI, holds:

$$\left(\int_{R^n} \frac{|\hat{f}(\xi)|^p}{|\xi|^{n(2-p)}}\, d\xi\right)^{1/p} \leqslant c\|f\|_{H^p} \qquad \text{for} \quad 0 < p \leqslant 1.$$

(*Hint*: From 6.11, applied to $F(x,t) = f * \phi_t(x)$ and $p < q = 1$, it readily follows that $|\hat{f}(\xi)| \leqslant c\|f\|_{H^p}|\xi|^{n(1/p-1)}$. Thus the mapping $f \to |\xi|^n \hat{f}(\xi)$ is bounded from $L^2(R^n)$ into $L^2(R^n, d\xi/|\xi|^{2n})$ and from $H^p(R^n)$ into wk-$L^p(R^n, d\xi/|\xi|^{2n})$. By (a simple variant of) Theorem 5.2 this mapping is also continuous from $H^p(R^n)$ into $L^p(R^n, d\xi/|\xi|^{2n})$, $0 < p \leqslant 1$, which is the desired conclusion. A direct proof using atoms also works.)

6.13 Endowed with the metric $d(f,g) = \|f-g\|_{H^p}^p$, $H^p(R^n)$ is a complete metric space, $0 < p \leqslant 1$. (*Hint*: It suffices to show that, if $\{f_k\} \subset H^p(R^n)$ and $\sum_{k=1}^{\infty}\|f_k\|_{H^p}^p < \infty$, then there exists f in $H^p(R^n)$ such that $\|f - \sum_{k=1}^{m} f_k\|_{H^p} \to 0$ as $m \to \infty$. Observe, first, that, since $p \leqslant 1$, also $\sum_{k=1}^{\infty}\|f_k\|_{H^p} < \infty$ and consequently $|\xi|^{-n(1/p-1)}\sum_{k=1}^{\infty}\hat{f}_k(\xi)$ converges uniformly to a bounded function $|\xi|^{-n(1/p-1)}\hat{f}(\xi)$. This in turn implies that if $\phi \in \mathscr{S}(R^n)$, $F_k(x,t) = f_k * \phi_t(x)$ and $F(x,t) = f * \phi_t(x)$, then the series $\sum_{k=1}^{\infty} F_k(x,t)$ converges pointwise to $F(x,t)$ and $N(F - \sum_{k=1}^{N} F_k)(x) \leqslant \sum_{k=N+1}^{\infty} N(F_k)(x)$. The conclusion follows without difficulty from this.)

6.14 Suppose the multiplier m is bounded from $H^1(R^n)$ into $L(R^n)$. Show that actually m is bounded from $H^1(R^n)$ into itself. (*Hint*: The Riesz transforms are bounded in $H^1(R^n)$).

6.15 Suppose that for some $0 < p \leqslant 1$, m is a bounded multiplier on $H^p(R^n)$. Show that $|m(\xi)| \leqslant c$, where c depends on the norm of m as a bounded multiplier on $H^p(R^n)$. In fact the following result is also true: if $m(\xi)$ is a bounded homogeneous function and the associated operator maps atoms into molecules, then m verifies a Hörmander condition; the precise statement is in Daly [1983].

6.16 Suppose $f \in H^p(R^n)$ and $\phi \in C_0^\infty(R^n)$. Then it is not necessarily the case that $f\phi \in H^p(R^n)$, $0 < p \leqslant 1$. To deal with this and other such inconveniences, Goldberg [1979] introduced the class $h^p(R^n)$ as follows. Let $\phi \in C_0^\infty(R^n)$ be a nonnegative function which equals 1 in a neighborhood of the origin and let $\chi(y, t)$ denote the characteristic function of $R^n \times (0, 1)$. Then $h^p(R^n) = \{f \in \mathcal{S}'(R^n)\colon N(\chi(y, t)f * \phi_t(y))(x) \in L^p(R^n)\}$, $0 < p \leq 1$, $\|f\|_{h^p} = \|N(\chi f * \phi_t)\|_p$. These spaces enjoy many important and useful properties, including the atomic decomposition. (Atoms with small support are the usual atoms; on the other hand, for atoms with large support the moment condition is dropped.)

6.17 Suppose $0 < \alpha < n$, $k(x, y)$ is C^N for $x \neq y$ and verifies

$$\int_{|y-w|\leq\lambda} (|k(w, x)| + |k(x, w|)\, dw \leq c\lambda^\alpha, \qquad \lambda > 0,$$

$$\int_{|x-y|\geq 2|w|, |w|\leq\lambda} |k(x, y+w) - k(x-w, y)|\, dw \leq c\lambda^{n+N}, \qquad \lambda > 0,$$

and

$$\int_{|x-y|\geq 2|w|, |w|\leq\lambda} \left| k(x, y+w) - \sum_{|\beta|\leq m} c_\beta \frac{\partial^{|\beta|}}{\partial v^\beta} k(x, v)]_{v=y} w^\beta \right| dw$$
$$\leq c|x-y|^{-n-N-1+\alpha} \lambda^{n+N+1}, \qquad \lambda > 0, \quad 0 \leq m < N.$$

These estimates are assumed to hold uniformly for x, y in R^n. Furthermore, let $n/(n+N) < p \leq 1$, $1/q = 1/p - \alpha/n$. Then the operator $Tf(x) = \int_{R^n} k(x, y)f(y)\, dy$, defined initially for atoms f, extends to a bounded linear mapping from $H^p(R^n)$ into $h^q(R^n)$. If in addition $k(x, y) = k(x - y)$, then T maps $H^p(R^n)$ into $H^q(R^n)$. The proof follows along the lines of Theorem 4.4 and is due to Krantz [1982].

6.18 Let u be harmonic in R^{n+1}_+, $\lim_{t\to\infty} u(x, t) = 0$, and

$$\left(\int_{[0,\infty)} |\nabla u(x, t)|^2 t\, dt\right)^{1/2} \in L^p(R^n), \qquad 0 < p < \infty.$$

Then $u(x, t) = f * P_t(x)$, $f \in H^p(R^n)$. (*Hint*: It suffices to show that $S(t|\nabla u|)(x) \in L^p(R^n)$; to do this we use the vector-valued version of Proposition 5.5 in Chapter VII. Let $U(x, t) = \nabla u(x, t + s)$, $|U(x, t)| = (\int_{[0,\infty)} |\nabla u(x, t+s)|^2 s\, ds)^{1/2}$. Our assumption is that $N_0(U)(x) = \sup_{t>0} |U(x, t)| = (\int_{[0,\infty)} |\nabla u(x, s)|^2 s\, ds)^{1/2} \in L^p(R^n)$, and by Proposition 5.5 in Chapter VII also $N(U)(x) \in L^p(R^n)$. It is not hard to see now that $N(U)$ dominates the Lusin function of $t\, \nabla u(x, t)$, and we are finished. The result is from Fefferman and Stein [1972]).

6.19 Suppose ϕ verifies the assumptions of Zó's Theorem 2.2 in Chapter XI and let $N_\phi(f)(x) = \sup_{t>0} |f * \phi_t(x)|$. Show that for $1 < r < \infty$, $\| \|N_\phi(f_k)\|_{l^r} \|_1 \leq c \sum_{j=1}^n \| \|R_j f_k\|_{l^r} \|_1$. This, and related results, are in the work of Rubio de Francia, Ruiz and Torrea [1985].

6.20 In addition to vector-valued inequalities, weighted estimates are of interest; the interested reader should consult the work of Strömberg and Torchinsky [1980].

CHAPTER

XV

Carleson Measures

1. CARLESON MEASURES

Suppose $f(y, t)$ is a measurable function on R_+^{n+1} and that $N(f)(x) = \sup_{\Gamma(x)}|f(y, t)|$ is lower semicontinuous. We are interested in finding under what conditions on the nonnegative Borel measure μ on R_+^{n+1} the inequality

$$\left(\int_{R_+^{n+1}}|f(y, t)|^p \, d\mu(y, t)\right)^{1/p} \leq c\|N(f)\|_p, \qquad 0<p<\infty, \tag{1.1}$$

holds. the constant in (1.1) is allowed to depend on p but not on f. A necessary condition for (1.1) is readily obtained as follows. Given an arbitrary subset E of R^n, we consider the tent $T(E)$ over E described by $T(E) = \{(y, t) \in R_+^{n+1}: B(y, t) \subset E\} = \{(y, t) \in R_+^{n+1}: d(y, E^c) > t\}$. Let now B denote an arbitrary open ball and observe that if $f(y, t) = \chi_{T(B)}(y, t)$, then $N(f)(x) = \chi_B(x)$, and, consequently, if (1.1) holds for some p, then also

$$\mu(T(B)) \leq c^p|B|. \tag{1.2}$$

Two quick remarks about (1.2). First, it is easy to see that the family $\{T(B)\}$ may be replaced by any other family which looks roughly like a tent over B. For instance, we may consider instead the collection of cylinders $C(B) = \{(y, t) \in R_+^{n+1}: y \in B(x, r);\ 0<t<r\}$ since given any B there are balls $B_1 \subset B \subset B_2$, with radius B_1, radius $B_2 \sim$ radius B, and so that $C(B_1) \subseteq T(B) \subseteq C(B_2)$. In other words, (1.2) is equivalent to

$$\mu(C(B)) \leq c|B|, \qquad \text{all} \quad B. \tag{1.3}$$

It is also the case that (1.2) is equivalent to the seemingly stronger statement

$$\mu(T(\mathcal{O})) \leq c|\mathcal{O}|, \qquad \text{all open sets } \mathcal{O} \text{ in } R^n. \tag{1.4}$$

Indeed, given an open set $\mathcal{O}$ let $\{Q_j\}$ be a Whitney decomposition of $\mathcal{O}$.

Then there are balls $\{B_j\}$ so that radius $B_j \sim$ sidelength Q_j and $T(Q_j) \subseteq C(B_j)$. Whence, it follows at once that $T(\mathcal{O}) \subseteq \bigcup_j C(B_j)$ and

$$\mu(T(\mathcal{O})) \leqslant \sum_j \mu(C(B_j)) \leqslant c \sum_j |B_j| \leqslant c \sum_j |Q_j| = c|\mathcal{O}|,$$

as claimed.

Those measures which verify (1.2) are called Carleson measures and the infimum over those constants c^p on the right-hand side is denoted by the constant of μ. One of the reasons this concept is important is the following.

Theorem 1.1. Suppose $f(y, t)$ is measurable and $N(f)(x) = \sup_{\Gamma(x)}|f(y, t)|$ is lower semicontinuous. If μ is a Carleson measure on R_+^{n+1}, then (1.1) holds with a constant which depends only on the constant of μ and p. Thus, (1.1) holds if and only if μ is a Carleson measure.

Proof. Given $\lambda > 0$, let $\mathcal{O}_\lambda = \{N(f) > \lambda\}$; by assumption $\mathcal{O}_\lambda$ is open and $\mu(T(\mathcal{O}_\lambda)) \leqslant c|\mathcal{O}_\lambda|$. Furthermore, since $\{|f| > \lambda\} \subseteq T(\mathcal{O}_\lambda)$, then also $\mu(\{|f| > \lambda\}) \leqslant \mu(T(\mathcal{O}_\lambda)) \leqslant c|\mathcal{O}_\lambda|$ and

$$\int_{R_+^{n+1}} |f(y, t)|^p \, d\mu(y, t) = p \int_{[0,\infty)} \mu(\{|f| > \lambda\})\lambda^{p-1} \, d\lambda$$

$$\leqslant cp \int_{[0,\infty)} |\mathcal{O}_\lambda|\lambda^{p-1} \, d\lambda = c\|N(f)\|_p^p, \qquad 0 < p < \infty. \quad \blacksquare$$

Before we go on we discuss some examples of Carleson measures.

Proposition 1.2 (Fefferman). Suppose $f \in BMO$ and ψ is a Littlewood–Paley function. Then $f * \psi_t(y)$ is well defined and $|f * \psi_t(y)|^2(1/t)\, dy\, dt$ is a Carleson measure with constant $\sim \|f\|_*^2$.

Proof. To show that $f * \psi_t(y)$ is given by a convergent integral, write $f = (f - f_I)\chi_I + (f - f_I)\chi_{R^n \setminus I} + f_I$, where I is the open interval centered at y with sidelength t, and note that

$$f * \psi_t(y) = \int_I (f(x) - f_I)\psi_t(y - x)\, dx + \int_{R^n \setminus I} (f(x) - f_I)\psi_t(y - x)\, dx$$

$$= A_1 + A_2,$$

say. Clearly, $|A_1| \leqslant ct^{-n} \int_I |f(x) - f_I|\, dx \leqslant c\|f\|_*$. As for the other term, observe that, by estimate (3.2) in Chapter VIII

$$|A_2| \leqslant \sum_{k=0}^{\infty} \int_{2^{k+1}I \setminus 2^k I} (|f(x) - f_{2^{k+1}I}| + |f_{2^{k+1}I} - f_I|)|\psi_t(y - x)|\, dx$$

$$\leqslant c \sum_{k=0}^{\infty} 2^{-k\alpha}(2^k t)^{-n}\left(\int_{2^{k+1}I} |f(x) - f_{2^{k+1}I}|\, dx + k(2^k t)^n \|f\|_*\right)$$

$$|A_2| \leq c\left(\sum_{k=0}^{\infty} (k+1)2^{-k\alpha}\right)\|f\|_*.$$

Next let B be an open ball in R^n. We must show that, for some constant c independent of B,

$$\int_{T(B)} |f * \psi_t(y)|^2 \, dy \frac{dt}{t} \leq c\|f\|_*^2 |B|. \tag{1.5}$$

Let I denote the smallest open cube concentric with B which contains B, and put $f = (f - f_I)\chi_{2I} + (f - f_I)\chi_{R^n \setminus 2I} + f_I = f_1 + f_2 + f_I$, say. Since $f_I * \psi_t(y)$ vanishes identically, it suffices to show that (1.5) holds with f replaced by f_1 and f_2 there. To bound the term involving f_1 note that, on account of (3.8) in Chapter XII and Corollary 1.5 in Chapter VIII,

$$\int_{T(B)} |f_1 * \psi_t(y)|^2 \, dy \frac{dt}{t} \leq \int_{R^n} \int_{[0,\infty)} |f_1 * \psi_t(y)|^2 \frac{dt}{t} \, dy$$

$$\leq c\|f_1\|_2^2 \leq c \int_{2I} |f(y) - f_I|^2 \, dy \leq c\|f\|_*^2 |B|,$$

which is of the right order. As for the term involving f_2, let r = radius of B and observe that we are interested in estimating $f_2 * \psi_t(y)$ for $0 < t < r$. An argument quite similar to the one used for the term A_2 immediately gives that $|f_2 * \psi_t(y)| \leq c(t/r)^\alpha \|f\|_*$ and consequently

$$\int_{T(B)} |f_2 * \psi_t(y)|^2 \, dy \frac{dt}{t} \leq c \int_B \int_{[0,r)} (t/r)^{2\alpha} \frac{dt}{t} \, dy \, \|f\|_*^2 \leq c|B| \, \|f\|_*^2,$$

which also is of the right order. ■

Corollary 1.3. Suppose f, ψ are as in Proposition 1.2 and $2 \leq q < \infty$. Then $|f * \psi_t(y)|^q (1/t) \, dy \, dt$ is a Carleson measure with constant $\sim \|f\|_*^q$.

Proof. Since $|f * \psi_t(y)| \leq c\|f\|_*$, $|f * \psi_t(y)|^q (1/t) \, dy \, dt \leq c\|f\|_*^{q-2} \times |f * \psi_y(y)|^2 (1/t) \, dy \, dt$, and we have finished. ■

2. DUALS OF HARDY SPACES

An important point we have left open is the determination of the dual space, or space of bounded linear functionals, of the Hardy $H^p(R^n)$ classes. We do the case $p = 1$ first.

Theorem 2.1 (Fefferman). The dual of $H^1(R^n)$ is $BMO(R^n)$ in the following sense. Suppose $f \in BMO(R^n)$. Then the linear functional

$$L(g) = \int_{R^n} g(x)f(x)\,dx \tag{2.1}$$

defined initially for $g \in H^1(R^n) \cap C_0^\infty(R^n)$ has a bounded extension to all of $H^1(R^n)$ with norm $\leq c\|f\|_*$. Conversely, every bounded linear functional L on $H^1(R^n)$ can be written as (2.1) for every $g \in H^1(R^n) \cap C_0^\infty(R^n)$. The function f is uniquely determined and belongs to $BMO(R^n)$, and $\|f\|_* \sim$ norm of L.

Proof. Given $f \in BMO(R^n)$ and $g \in H^1(R^n) \cap C_0^\infty(R^n)$, we show first that the expression $L(g)$ in (2.1) is finite and that it, in fact, verifies $|L(g)| \leq c\|g\|_{H^1}$, with $c \sim \|f\|_*$. Since this collection of g's is dense in $H^1(R^n)$, the first part of our conclusion follows immediately from this.

Let $U(y,t) = t\,\nabla(f * P_t)(y)$, $V(y,t) = t\,\nabla(g * P_t)(y)$ and observe that, on account of (3.11) in Chapter XII,

$$L(g) = \int_{R_+^{n+1}} U(y,t)\cdot V(y,t)\,dy\frac{dt}{t}. \tag{2.2}$$

Now let $E_k = \{S(V) > 2^k\}$, $\mathcal{O}_k = \{N(\chi_{E_k} * \phi_t) > \frac{1}{2}\}$, where ϕ is a $C^\infty(R^n)$ function supported in the unit ball with integral 1, and put $A_k = T(\mathcal{O}_k)\backslash T(E_{k+1})$. The following properties are then readily verified: $\bigcup_{k=-\infty}^{\infty} A_k = R_+^{n+1}$, and

$$\int_{A_k} |V(y,t)|^2\,dy\frac{dt}{t} \leq c\int_{\mathcal{O}_k\backslash E_{k+1}} S(V)(x)^2\,dx. \tag{2.3}$$

Whence from (2.2) and (2.3) it follows that

$$\begin{aligned}
|L(g)| &\leq \sum_{k=-\infty}^{\infty} \int_{A_k} |U(y,t)||V(y,t)|\,dy\frac{dt}{t} \\
&\leq \sum_{k=-\infty}^{\infty} \left(\int_{A_k} |U(y,t)|^2\,dy\frac{dt}{t}\right)^{1/2}\left(\int_{A_k} |V(y,t)|^2\,dy\frac{dt}{t}\right)^{1/2} \\
&\leq c\sum_{k=-\infty}^{\infty} \left(\int_{T(\mathcal{O}_k)} |U(y,t)|^2\,dy\frac{dt}{t}\right)^{1/2}\left(\int_{\mathcal{O}_k\backslash E_{k+1}} S(V)(x)^2\,dx\right)^{1/2} \\
&= c\sum_{k=-\infty}^{\infty} J_k,
\end{aligned} \tag{2.4}$$

say. We examine each of the summands J_k in (2.4). By Proposition 1.2,

$\int_{T(\mathcal{O}_k)}|U(y,t)|^2/t\,dy\,dt \leq c\|f\|_*^2|\mathcal{O}_k|$, and from the definition of the sets involved it follows that $\int_{\mathcal{O}_k\setminus E_{k+1}} S(V)(x)^2\,dx \leq 2^{2(k+1)}|\mathcal{O}_k|$. Consequently, $J_k \leq c\|f\|_*|\mathcal{O}_k|^{1/2}2^k|\mathcal{O}_k|^{1/2} \leq c2^k|E_k|\,\|f\|_*$, and

$$\sum_{k=-\infty}^{\infty} J_k \leq c\|f\|_* \sum_{k=-\infty}^{\infty} 2^k|E_k| \leq c\|f\|_*\|S(V)\|_1 \leq c\|f\|_*\|g\|_{H^1}.$$

Thus, by substituting in (2.4), we get $|L(g)| \leq c\|f\|_*\|g\|_{H^1}$, as anticipated.

To prove the converse, we use the characterization given in Theorem 3.2(b) in Chapter XIV and think of $H^1(R^n)$ as the subspace of $L(R^n)$ consisting of those integrable functions with integrable Riesz transforms. Let B denote the Banach space consisting of the direct sum of $n+1$ copies of $L(R^n)$ normed by $\|G\|_B = \|(g_0, g_1, \ldots, g_n)\|_B = \sum_{j=0}^{n}\|g_j\|_1$. Then $H^1(R^n)$ can be identified with the closed subspace H of B consisting of those G's of the form $(g, R_1g, \ldots, R_ng)$, and by the Hahn–Banach theorem each bounded linear functional L on $H^1(R^n)$, or actually H, can be extended with no increment in its norm to a bounded linear functional, which we also denote by L, on B. Now the dual of B is essentially the direct sum of $n+1$ copies of $L^\infty(R^n)$ and consequently there exist $L^\infty(R^n)$ functions $f_0, \ldots, f_n$ so that

$$L(G) = \sum_{j=0}^{n} \int_{R^n} g_j(x)f_j(x)\,dx, \qquad G \in B, \tag{2.5}$$

and $\sum_{j=0}^{n}\|f_j\|_\infty \leq$ norm of L.

When restricted to those G's in H with g in $C_0^\infty(R^n)$, the identity (2.5) reads

$$\begin{aligned} L(G) &= \int_{R^n} g(x)f_0(x)\,dx + \sum_{j=1}^{n}\int_{R^n} R_jg(x)f_j(x)\,dx \\ &= \int_{R^n} g(x)\left(f_0(x) - \sum_{j=1}^{n} R_jf_j(x)\right)dx = \int_{R^n} g(x)f(x)\,dx, \end{aligned}$$

say. By Theorem 7.1 in Chapter XI, $f \in BMO(R^n)$ and $\|f\|_* \leq c\sum_{j=0}^{\infty}\|f_j\|_\infty \leq c$. norm of L. It thus only remains to show that f is uniquely determined by L. But this is not hard; first, the above representation is readily seen to hold whenever the integral converges and this is the case when g is, for instance, the $H^1(R^n)$ function $P_t(x) - P_t(x-y)$ discussed in 6.8 in Chapter XIV. Whence, if $\int_{R^n} g(x)f(x)\,dx = 0$ for those $g \in H^1(R^n)$, it immediately follows that $\lim_{t\to 0} f * P_t(y) = f(0) = c$ a.e. But constant functions f in BMO are actually (equivalent to) the 0 function and the uniqueness obtains. That the norm of $L \sim \|f\|_*$ requires an easy argument using the first part of the proof. ∎

It seems natural to expect the atomic decomposition to play a role in the consideration of the dual to the Hardy spaces as well. In fact, even the proof of Theorem 2.1 can be simplified by invoking the atomic decomposition. We illustrate this in our next result.

Theorem 2.2. Suppose L is bounded linear functional on $H^p(R^n)$, $0 < p \leq 1$. Then there exists a locally integrable function f such that

$$L(g) = \int_{R^n} g(x)f(x)\,dx \tag{2.6}$$

for every $H^p(R^n)$ function g which is a finite linear combination of $(p, 2, N)$ atoms. Furthermore, f satisfies

$$\left(\frac{1}{|I|}\int_I |f(y) - P_I(f)(y)|^2\,dy\right)^{1/2} \leq c|I|^{1/p-1}, \tag{2.7}$$

where $P_I(f)$ is a polynomial of degree $\leq N = [n(1/p-1)]$, and c is independent of the open cube I. The smallest constant c for which (2.7) holds for every I is $\sim$ norm of L. Conversely, if f verifies (2.7) and g is a finite linear combination of $(p, 2, N)$ atoms, then the integral in (2.6) converges and $|L(g)| \leq M\|g\|_{H^p}$, where $M \sim$ the constant in (2.7) corresponding to f.

Proof. We prove the second statement first. Let $g = \sum_{j=1}^m a_j$ be a finite sum of $(p, 2, N)$ atoms, $\sum_{j=1}^m |a_j|^p \sim \|g\|_{H^p}^p$, and let I_j be open cubes containing the support of a_j with the property that $|I_j|^{1/p-1/2}\|a_j\|_2 \sim |a_j|$. Then

$$\begin{aligned}
\left|\int_{R^n} g(x)f(x)\,dx\right| &= \left|\sum_{j=1}^m \int_{I_j} a_j(x)(f(x) - P_{I_j}(f))\,dx\right| \\
&\leq \sum_{j=1}^m \left(\int_{I_j} |f(x) - P_{I_j}(f)|^2\,dx\right)^{1/2} \|a_j\|_2 \\
&\leq c\sum_{j=1}^m |I_j|^{1/p-1/2}\|a_j\|_2 \leq c\sum_{j=1}^m |a_j| \\
&\leq c\left(\sum_{j=1}^m |a_j|^p\right)^{1/p} \leq c\|g\|_{H^p},
\end{aligned}$$

as we wanted to show.

To discuss the representation of the functionals we start out by fixing an open cube I and consider the subspace H of $L^2(I)$ consisting of those functions with vanishing moments up to order N. Functions a in H belong to $H^p(R^n)$ and $\|a\|_{H^p} \leq c|I|^{1/p-1/2}\|a\|_2$. Thus if $a \in H$, then $|L(a)| \leq c|I|^{1/p-1/2}\|a\|_2$, and, by the Hahn-Banach theorem, L can be extended as a bounded linear functional to $L^2(I)$ with norm not exceeding $c|I|^{1/p-1/2}$. By Proposition 3.2(ii) in Chapter II there is a function $f \in L^2(I)$ such that $\|f\|_2 \leq c|I|^{1/p-1/2}$ and $L(g) = \int_{R^n} f(x)g(x)\,dx$ for $g \in L^2(I)$. Next we estimate $\|(f - P_I(f))\chi_I\|_2$, where $P_I(f)$ is the polynomial of degree $\leq N$ so that $\int_I (f(x) - P_I(f)(x))x^\beta \, dx = 0$, $|\beta| \leqslant N$. For this purpose let $h \in L^2(I)$ be a function in $L^2(I)$ with norm 1 so that $\|(f - P_I(f))\chi_I\|_2 \leqslant 2|\int_I (f(x) - P_I(f)(x))h(x)\, dx|$. Now, if $P_I(h)(x)$ denotes the polynomial of degree $\leqslant N$ so that $\int_I (h(x) - P_I(h)(x))x^\beta \, dx = 0$ for $|\beta| \leqslant N$, it is not hard to see that $a(x) = (h(x) - P_I(h)(x))\chi_I(x)$ is a $(p, 2, N)$ atom with $|a| \leqslant c|I|^{1/p-1/2}$. Therefore,

$$\begin{aligned}\|(f - P_I(f))\chi_I\|_2 &\leqslant 2\left|\int_I (f(x) - P_I(f)(x))h(x)\, dx\right| \\ &= 2\left|\int_I (f(x) - P_I(f)(x))(h(x) - P_I(h)(x))\, dx\right| \\ &= 2\left|\int_I f(x)a(x)\, dx\right| \leqslant c|a| \leqslant c|I|^{1/p-1/2},\end{aligned}$$

which is equivalent to the estimate (2.7) for this particular cube I. It still remains to be shown that the functions which correspond to different cubes I can be thought of as restrictions to I of a single function f which verifies (2.7). But this is not hard to see since any two functions f_1, f_2 which correspond to cubes $I_1 \subset I_2$, say, differ by a polynomial of degree $\leqslant N$ on I_1, and consequently are actually (equivalent to) the 0 function in the space of functions which verify (2.7). ■

Remark 2.3. The reader should compare the description of the dual of the Hardy spaces with the Lipschitz spaces introduced in Chapter VIII. This will be discussed further in 4.10–4.12.

3. TENT SPACES

It is well known that the dual of the space of continuous functions on R_+^{n+1} which vanish at infinity is the space of signed Borel measures on R_+^{n+1}. It is therefore natural to consider whether the space of signed Carleson measures, i.e., those measures μ on R_+^{n+1} which verify

$$\left|\iint_{T(B)} d\mu(y, t)\right| \leqslant c|B|, \qquad \text{all balls } B,$$

can be identified as the dual of some space of continuous functions on R_+^{n+1}. For this purpose, and motivated by Theorem 1.1, let $T = \{f \in C(R_+^{n+1}) : N(f) \in L(R^n)\}$; endowed with the norm $|f| = \|N(f)\|_1$, T becomes a Banach space. The triangle inequality is readily verified and the completeness follows without difficulty from the estimate (2.6) in Chapter

XIV, i.e., $|f(y, t)| \leqslant ct^{-n}\|N(f)\|_1 = ct^{-n}|f|$. The statement we have in mind is

Theorem 3.1. The dual of T is the space of (signed) Carleson measures in the following sense: if μ is a Carleson measure, then

$$L(f) = \int_{R_+^{n+1}} f(y, t)\, d\mu(y, t) \tag{3.1}$$

is a bounded linear mapping on T with norm $\leqslant c\cdot$ the Carleson constant of $|\mu|$. Conversely, to every continuous linear functional L on T there corresponds a unique signed Carleson measure μ so that (3.1) holds and the norm of L is comparable to the Carleson constant of μ.

Proof. The fact that every Carleson measure gives a bounded linear functional on T is contained in the inequality (1.1). The converse is not hard. That L must be given by a Borel measure μ follows from the Riesz representation theorem. To show that μ is actually a Carleson measure put $f(y, t) = \chi_{T(B)}(y, t)$; f does not belong to T but it is the limit, in the T norm, of functions in T. Since $N(f) = \chi_B$, by the continuity of L we get

$$\left|\int_{T(B)} d\mu(y, t)\right| = |L(f)| \leqslant c\|N(f)\|_1 = c|B|. \quad \blacksquare$$

Next, motivated by Proposition 1.2 we introduce the expression

$$C(g)(x) = \sup_B \left(\frac{1}{|B|}\int_{T(B)} |g(y, t)|^2\, dy\frac{dt}{t}\right)^{1/2},$$

where B runs over those balls which contain x, and define T^∞ to be the class of g's for which $C(g) \in L^\infty(R^n)$; the norm in this space is then $|g|_\infty = \|C(g)\|_\infty$. Can we identify T^∞ as a dual Banach space? To answer this question we need one notation. Let $T^p = \{f\colon S(f) \in L^p(R^n)\}$, $0 < p < \infty$; the norm in T^p is $|f|_p = \|S(f)\|_p$. We then have

Theorem 3.2. Suppose $f \in T^1$, $g \in T^\infty$. Then

$$\int_{R_+^{n+1}} |f(y, t)g(y, t)|\, dy\frac{dt}{t} \leqslant c\int_{R^n} S(f)(x)C(g)(x)\, dx. \tag{3.2}$$

Thus each $g \in T^\infty$ induces a continuous linear functional on T^1.

Proof. Let $\Gamma^h(x) = \{(y, t) \in R_+^{n+1}\colon |x - y| < t < h\}$ denote the cone with vertex at x truncated at height h, and set

$$S^h(f)(x) = \left(\int_{\Gamma^h(x)} |f(y, t)|^2 t^{-n}\, dy\frac{dt}{t}\right)^{1/2}.$$

Note that $S^h(f)$ increases with h and $S^\infty(f) = S(f)$. For $g \in T^\infty$ we define the "stopping-time" $h(x)$ as $h(x) = \sup\{h > 0: S^h(g)(x) \leq MC(g)(x)\}$. Here M is a large dimensional constant to be chosen shortly. First, observe that, if $B = B(x, h)$, then for some dimensional constants c_1, $c \geq 1$, we have

$$\int_B S^h(g)(z)^2\,dz \leq c_1 \int_{T(cB)} |g(y,t)|^2\,dy\frac{dt}{t}. \tag{3.3}$$

This is not hard to check; indeed, the left-hand side of (3.3) equals

$$\int_{R^n} \chi_B(z) \int_{R^n}\int_R \chi_{B(y,t)}(z)\chi_{[0,h]}(t)|g(y,t)|^2 t^{-n}\,dy\frac{dt}{t}\,dz$$
$$\leq c_1 \int_{R^n} \chi_{B(x,2h)}(y) \int_{[0,h]} |g(y,t)|^2\,dy\frac{dt}{t} \leq c_1 \int_{T(cB)} |g(y,t)|^2\,dy\frac{dt}{t},$$

as anticipated. Now, from this estimate it readily follows that if M is sufficiently large, then

$$|\{z \in B: h(z) \geq h\}| \geq |B|/2, \qquad \text{all}\quad B. \tag{3.4}$$

To see this let $E_h = \{z \in B: h(z) < h\}$ and observe that for $z \in E_h$ we have automatically that

$$S^h(g)(z) > M \inf_B C(g) \geq M\left(\frac{1}{|cB|}\int_{T(cB)} |g(y,t)|^2\,dy\frac{dt}{t}\right)^{1/2},$$

and consequently by (3.3)

$$M^2\left(\frac{1}{|cB|}\int_{T(cB)} |g(y,t)|^2\,dy\frac{dt}{t}\right)|E_h| \leq \int_{E_h} S^h(g)(z)^2\,dz$$
$$\leq \int_B S^h(g)(z)^2\,dz \leq c_1 \int_{T(cB)} |g(y,t)|^2\,dy\frac{dt}{t}.$$

Thus $|E_h| \leq (c_1/M^2)\,|cB|$ and $|B \backslash E_h| \geq |B|/2$ provided M is sufficiently large, which is precisely (3.4).

Finally, on account of (3.4), Fubini's theorem and Hölder's inequality we get that

$$\int_{R_+^{n+1}} |f(y,t)g(y,t)|\,dy\frac{dt}{t}$$
$$\leq c\int_{R_+^{n+1}} |f(y,t)g(y,t)| \left|\frac{\{x \in B(y,t): h(x) > t\}}{t^n}\right| dy\frac{dt}{t}$$
$$\leq c\int_{R^n}\int_{\Gamma^{h(x)}(x)} |f(y,t)|t^{-n/2}|g(y,t)|t^{-n/2}\,dy\frac{dt}{t}\,dx$$

$$\leqslant c\int_{R^n} S^{h(x)}(f)(x)S^{h(x)}(g)(x)\,dx \leqslant c\int_{R^n} S(f)(x)C(g)(x)\,dx,$$

(3.2) holds, and we have finished. ■

To complete the discussion of the duality of T^1, as in the case of the Hardy spaces, we consider the notion of a T^1 atom. This is a function $a(x, t)$ supported in $T(B)$ for some ball $B \subset R^n$ so that

$$\int_{T(B)} |a(y, t)|^2\, dy\frac{dt}{t} \leqslant c < \infty.$$

If we normalize a by setting

$$|a| = \inf\left\{\left(|B| \int_{T(B)} |a(y, t)|^2\, dy\frac{dt}{t}\right)^{1/2}\right\},$$

where the infimum is taken over all balls B for which supp $a \subseteq T(B)$, then it follows at once that $|a|_1 = \|S(a)\|_1 \leqslant c|a|$, where c is a dimensional constant independent of a. Moreover, the following holds

Theorem 3.3. Every element $f \in T^1$ can be written as $f = \sum_j a_j$, where the a_j's are T^1 atoms, and there is a constant c independent of f so that $\sum_j |a_j| \leqslant c|f|_1$.

Proof. The sketch of the proof follows. Let $E_k = \{S(f) > 2^k\}$ and $\mathcal{O}_k = \{M(\chi_{E_k}) > \gamma\}$, where $0 < \gamma < 1$ is chosen so that supp $f \subseteq \bigcup_k T(\mathcal{O}_k)$. Furthermore, let $\{Q_{k,j}\}$ be a Whitney decomposition of $\mathcal{O}_k$ and for some large constant c let $B_{k,j}$ denote the ball concentric with $Q_{k,j}$ that is c times its diameter. Then we can write $T(\mathcal{O}_{k+1})\backslash T(\mathcal{O}_k)$ as a disjoint union $\bigcup_j A_{k,j}$, where $A_{k,j} = T(B_{k,j}) \cap (Q_{k,j} \times [0, \infty)) \cap (T(\mathcal{O}_k)\backslash T(\mathcal{O}_{k+1}))$, provided the constant c is sufficiently large. Now put $a_{k,j}(y, t) = f(y, t)\chi_{A_{k,j}}(y, t)$. It is clear that the $a_{k,j}$'s are T^1 atoms and that $\sum_{k,j} a_{k,j} = f$. To complete the proof we must then estimate $I = \sum|a_{k,j}|$ and show that it $\leqslant c|f|_1$. But this is not hard since

$$\begin{aligned}
I &\leqslant \sum\left(|B_{k,j}| \int_{T(B_{k,j})\cap A_{k,j}} |f(y, t)|^2\, dy\frac{dt}{t}\right)^{1/2} \\
&\leqslant \sum\left(|B_{k,j}| \int_{T(B_{k,j})\cap(R_+^{n+1}\backslash T(\mathcal{O}_{k+1}))} |f(y, t)|^2\, dy\frac{dt}{t}\right)^{1/2} \\
&\leqslant c\sum\left(|B_{k,j}| \int_{B_{k,j}\cap(R^n\backslash \mathcal{O}_{k+1})} S(f)(x)^2\, dx\right)^{1/2} \\
&\leqslant c\sum(|B_{k,j}|2^{2(k+1)}|B_{k,j}|)^{1/2} \leqslant c\sum 2^k|Q_{k,j}| \\
&= c\sum 2^k|\mathcal{O}_k| \leqslant c\sum 2^k|E_k| = c\|S(f)\|_1 = c|f|_1,
\end{aligned}$$

and the proof is complete. ■

Theorems 3.2 and 3.3 combine to give

Theorem 3.4. Suppose $f \in T^1$ and $g \in T^\infty$, then the pairing $\langle f, g\rangle \to \int_{R_+^{n+1}} f(y, t)g(y, t)/t\, dy\, dt$ realizes T^∞ as equivalent to the Banach space dual of T^1, with equivalence in norms.

Proof. That every g in T^∞ induces a continuous linear functional on T^1 by the pairing described in the statement of the theorem follows at once from (3.2). Conversely, let L be a bounded linear functional on T^1. Notice that whenever $f \in L^2(R_+^{n+1})$ is supported in a compact $K \subset R_+^{n+1}$, then $|f|_1 \leq c\|f\|_2$, where $c = c_K$. Thus L induces a bounded linear functional on $L^2(K)$ and as such it may be represented by a function $g = g_K \in L^2(K)$. Taking an increasing family of compact subsets K with $\bigcup K = R_+^{n+1}$ we thus obtain a function g defined on R_+^{n+1}, which is locally in $L^2(R_+^{n+1})$, and so that $L(f) = \int_{R_+^{n+1}} f(y, t)g(y, t)/t\, dy\, dt$, whenever $f \in T^1$ and f has compact support. That $g \in T^\infty$ and $|g|_\infty \leq c \cdot$ norm of L follows at once by evaluating L over all possible T^1 atoms. ■

Clearly there is a close connection between the Hardy spaces $H^p(R^n)$ and the tent spaces T^p, $0 < p \leq 1$.

Suppose $g \in H^p(R^n)$, $0 < p \leq 1$, and let $f(x, t) = g * \phi_t(x)$, where ϕ is an appropriate Little-Paley function. Then the mapping $g \to f$ is bounded from $H^p(R^n)$ into T^p. On the other hand, Calderón's representation formula, Proposition 1.9 in Chapter XIV, gives that, if $\phi_t(x) = t(\partial/\partial t)P_t(x)$, then there is a $C_0^\infty(R^n)$ function ψ supported in the unit ball and with an arbitrary number of vanishing moments so that

$$g = \int_{[0,\infty)} f(\cdot, t) * \psi_t \frac{dt}{t}. \tag{3.5}$$

It is therefore of interest to consider the integral in (3.5) for arbitrary f in T^p. More precisely, for ψ as in (3.5) we consider the operator $K(f)$ defined on T^p by

$$K(f) = \int_{[0,\infty)} f(\cdot, t) * \psi_t \frac{dt}{t}. \tag{3.6}$$

This operator is initially defined on the dense subspace of T^p consisting of those functions f with compact support on R_+^{n+1}. It is readily seen that in this case $K(f)$ is well defined and we have

Proposition 3.5. $K(f)$ extends to a bounded linear operator from T^p to $L^p(R^n)$ if $1 < p < \infty$, from T^1 into $H^1(R^n)$ and from T^∞ into $BMO(R^n)$.

Proof. We consider the case $1 < p < \infty$ first. It suffices to bound $I = \int_{R^n} K(f)(x)g(x)\,dx$ for $g \in L^{p'}(R^n)$, $1/p + 1/p' = 1$. Let $\eta(x) = \psi(-x)$. Then I equals $\int_{R_+^{n+1}} f(x,t)g * \eta_t(x)/t\,dx\,dt$, which as we observed in Theorem 3.2 in turn is bounded by $\int_{R^n} S(f)(x)S(g * \eta_t)(x)\,dx$. Thus $|I| \leq \|S(f)\|_p \|S(g * \eta_t)\|_{p'} \leq c|f|_p \|g\|_{p'}$ and $\|K(f)\|_p \leq c|f|_p$ as we wanted to prove. To do the case $p = 1$, it suffices to show that K maps a T^1 atom $a(y,t)$ into an $H^1(R^n)$ atom $K(a)(x)$ and $|K(a)| \leq c|a|$, where c is independent of a. Suppose supp $a \subseteq T(B)$. Since ψ is supported in the unit ball it follows from the definition of $K(a)$ that supp $K(a) \subseteq 2B$ and that $\int_{R^n} K(a)(x)\,dx = 0$ since $\int_{R^n} \psi(x)\,dx = 0$. Moreover, by the first part of the proof,

$$\|K(a)\|_2 \leq c|a|_2 = c\left(\int_{T(B)} |a(y,t)|^2\,dy\frac{dt}{t}\right)^{1/2} \leq c|B|^{-1/2}|a|,$$

which is precisely what we wanted to check. To complete our discussion we consider the case $p = \infty$. In this case it is enough to see that $K(f)$ induces a linear functional on $H^1(R^n)$. Let then $g \in H^1(R^n)$; arguing as in Theorem 3.2 we get this time that $|\int_{R^n} Kf(x)g(x)\,dx| \leq c\int_{R^n} C(f)(x)S(g * \eta_t)(x)\,dx \leq c|f|_\infty \|S(g * \eta_t)\|_1 \leq c|f|_\infty \|g\|_{H^1}$, and we have finished. ∎

We point out an immediate application of Proposition 3.5, to wit, the atomic decomposition of $H^1(R^n)$. For, let $g \in H^1(R^n)$ and note that by (3.5), $g = \int_{[0,\infty)} f(\cdot, t) * \psi_t/t\,dt$, where $f(y,t) = t(\partial/\partial t)g * P_t(y) \in T^1$. Now apply the atomic decomposition of T^1 to f and observe that the atomic decomposition of $K(f) = g$ is obtained as in the proof of the proposition.

4. NOTES; FURTHER RESULTS AND PROBLEMS

The notion of Carleson measure is central to the solution of many important problems in complex analysis; a full description of this can be found in Garnett's book [1981]. Our interest in this concept stems from Fefferman's observation that it provides a link in identifying the dual of $H^1(R^n)$ with the class $BMO(R^n)$. Fefferman's original proof of this fact follows along the lines of Theorem 3.2. The identification of the dual of $H^p(R^n)$, $0 < p < 1$, is due to Duren, Romberg, and Shields [1969]. The tent spaces were introduced by Coifman, Meyer, and Stein [1983] as a class of functions especially adapted to the study of singular integral operators; our presentation follows their work.

Further Results and Problems

4.1 Suppose ϕ is a smooth function, 1 near the origin, with nonvanishing integral. Then Theorem 1.1 also holds for convolutions:

$$\int_{R_+^{n+1}} |f * \phi_t(x)|^p \, d\mu(x, t) \leqslant c\|N(f * \phi_t)\|_p^p, \qquad 0 < p < \infty,$$

if and only if μ is a Carleson measure.

4.2 Suppose $f(t) \sim \sum_{n=0}^{\infty} a_n e^{int}$ is in H^1. Show that Paley's inequality holds; namely, if $\{\lambda_n\}$ is an Hadamard sequence, then $\sum_{n=0}^{\infty} |a_{\lambda_n}|^2 \leqslant c\|f\|_{H^1}^2$. (*Hint*: Use duality; 6.18 in Chapter VIII is relevant here.)

4.3 It is well known that for $f \in BMO(R^n)$, $|\nabla f * P_t(x)| \, dx \, dt$ is not necessarily a Carleson measure; in fact, Rudin gave an example of this even for $f \in L^\infty(R^n)$. Following Amar and Bonami [1979], we construct the following example: let ϕ be a rapidly decreasing, continuous function with $\phi(0) = 1$. Given $f \in BMO(T)$, $f \sim \sum c_n e^{int}$, we define its extension F to $R^+ \times T$ by means of $F(x, t) = \sum c_n \phi(nx) e^{int}$. When $\phi(x) = e^{-|x|}$, F coincides, upon changing x into e^{-x}, with the harmonic extension of f into the unit disk. We show next that, in general, $|\nabla F(x, t)| \, dx \, dt$ is not a Carleson measure in $R^+ \times T$; more precisely there are functions f in $BMO(T)$ such that $\int_{[0,1] \times T} |\nabla F(x, t)| \, dx \, dt = \infty$. It suffices to consider functions $f \sim \sum_{n=0}^{\infty} c_n e^{i3^n t}$ which, by Paley's theorem in 4.2, belong to $BMO(T)$ if and only if $\sum_{n=0}^{\infty} |c_n|^2 < \infty$. Now since the L^p norms of Hadamard series are equivalent,

$$\int_T \int_{[0,1]} \left| \frac{\partial}{\partial t} F(xt) \right| \, dx \, dt \sim \int_{[0,1]} \left(\sum |c_n 3^n \phi(3^n x)|^2\right)^{1/2} dx.$$

If $|\phi(x)| \geqslant \frac{1}{2}$ for $|x| \leqslant \varepsilon$, then $(\sum |c_n 3^n \phi(3^n x)|^2)^{1/2} \geqslant c|c_n| 3^n$ when $3^{-(n+1)}\varepsilon \leqslant x \leqslant 3^{-n}\varepsilon$. Thus

$$\int_T \int_{[0,\varepsilon]} \left| \frac{\partial}{\partial t} F(x, t) \right| \, dx \, dt \geqslant c \sum |c_n|$$

and it suffices to take $\{c_n\}$ so that $\sum |c_n| = \infty$ and $\sum |c_n|^2 < \infty$. On the other hand it is not hard to show that, if $\sum |c_n| < \infty$, then $|\nabla F(x, t)| \, dx \, dt$ is a Carleson measure. Also, by considering the function $f(t) = (\sum c_n e^{i3^n t}) \chi_T(t)$, we have the following: there is a compactly supported function f in $BMO(R)$ such that if $\phi \in \mathcal{S}(R)$, $\int_R \phi(y) \, dy = 1$, the extension $F(x, t) = f * \phi_t(x)$ of f to $R^+ \times R$ verifies $\int_{[0,1] \times [0,1]} |\nabla F(x, t)| \, dx \, dt = \infty$.

4.4 Suppose $g \in BMO(R^n)$ and $f \in H^p(R^n)$, $0 < p < \infty$. If ϕ is a smooth function and ψ is a Littlewood-Paley function, then

$$\int_{R_+^{n+1}} |f * \phi_t(x)|^p |g * \psi_t(x)|^2 \, dx \frac{dt}{t} \leqslant c \|N(f * \phi_t)\|_p^p.$$

How does c depend on g?

4.5 Given a nonnegative Borel measure μ in R_+^{n+1} and $x \in R^n$, let

$$C(\mu)(x) = \sup_B \frac{1}{|B|} \int_{T(B)} d\mu(y, t) = \sup_B \frac{\mu(T(B))}{|B|},$$

where B runs over those open balls in R^n which contain x. Suppose $C(\mu)(x)$ is not identically ∞ and let $0 \leqslant \varepsilon < 1$. Show that $C(\mu)(x)^\varepsilon \in A_1$. (*Hint*: For simplicity we do the case $n = 1$. We begin by showing that $C(\mu)$ is of weak-type $(1, 1)$. To each x in the open set $\mathcal{O}_\lambda = \{C(\mu) > \lambda\}$ associate an open interval I_x containing x such that $(1/|I_x|) \int_{T(I_x)} d\mu(y, t) > \lambda$. Then $\mathcal{O}_\lambda \subseteq \bigcup_{x \in \mathcal{O}_\lambda} I_x$ and by passing to a disjoint sequence of intervals, $\{I_j\}$ say, with the property that $\mathcal{O}_\lambda \subseteq \bigcup 2I_j$, the weak-type assertion follows. Fix now I; we must show that $(1/|I|) \int_I C(\mu)(x)^\varepsilon \, dx \leqslant c \inf_I C(\mu)^\varepsilon$, where c is independent of I. To do this for each x in I divide those open intervals Q containing x into two families by setting $\mathcal{I}_1 = \{Q : |Q| \leqslant |2I|\}$ and $\mathcal{I}_2 = \{Q : |Q| > |2I|\}$ and proceed as in Proposition 3.3 in Chapter IX. This observation is due to Deng [1984].)

4.6 The following extension of Theorem 3.2 holds: given f defined in R_+^{n+1}, let

$$A_p(f)(x) = \left(\int_{\Gamma(x)} |f(y, t)|^p \frac{dy \, dt}{t^{n+1}} \right)^{1/p}, \qquad 1 \leqslant p < \infty, \quad A_\infty(f)(x) = N(f)(x).$$

Then

$$\left| \int_{R_+^{n+1}} f(x, t) g(x, t) \, dx \, dt \right| \leqslant c \int_{R^n} A_p(f)(x) C(|g|^{p'})(x)^{1/p'} \, dx,$$

where $1/p + 1/p' = 1$, $1 < p \leqslant \infty$, and c is independent of f and g. (*Hint*: We do the case $n = 1$, $p < \infty$. Let $E_k = \{A_p(f) > 2^k\} = \bigcup_j Q_{j,k}$, where the Q's are disjoint, open intervals and $\mathcal{O}_k = \{M(\chi_{E_k}) > \frac{1}{2}\} = \bigcup_j I_{j,k}$, where the I's are also disjoint open intervals. Observe first that

$$\int_{\mathcal{O}_k} C(|g|^{p'})(x)^{1/p'} \, dx \leqslant c \int_{E_k} C(|g|^{p'})(x)^{1/p'} \, dx,$$

where c is independent of k. This follows immediately from the chain of inequalities

$$\int_{\mathcal{O}_k} C(|g|^{p'})(x)^{1/p'}\,dx \leqslant 4 \int_R M\chi_{E_k}(x)^2 C(|g|^{p'})(x)^{1/p'}\,dx$$

$$\leqslant c \int_R \chi_{E_k}(x)^2 C(|g|^{p'})(x)^{1/p'}\,dx = c \int_{E_k} C(|g|^{p'})(x)^{1/p'}\,dx;$$

here we used the fact that $C(|g|^{p'})(x)^{1/p'} \in A_1$. Let now $\mathcal{U}_{j,k} = T(I_{j,k})\setminus\bigcup_m T(I_{m,k+1})$, $U_{j,k} = I_{j,k}\setminus\bigcup_m J_{m,k+1}$. Then by Hölder's inequality we have

$$\left|\iint_{R^2_+} f(x,t)g(x,t)\,dx\,dt\right| \leqslant \sum_{k=-\infty}^{\infty} \sum_j \left(\iint_{\mathcal{U}_{j,k}} |f(x,t)|^p\,dx\,dt\right)^{1/p}$$

$$\times\left(\iint_{\mathcal{U}_{j,k}} |g(x,t)|^{p'}\,dx\,dt\right)^{1/p'} = \sum_k \sum_j A_{j,k}B_{j,k},$$

say. Furthermore, since

$$A_{j,k} \leqslant c\left(\int_{U_{j,k}} A_p(f)(x)^p\,dx\right)^{1/p}$$

and

$$B_{j,k} \leqslant \left(\iint_{T(I_{j,k})} |g(x,t)|^{p'}\,dx\,dt\right)^{1/p'}$$

$$\leqslant \frac{1}{|I_{j,k}|^{1/p}} \int_{I_{j,k}} C(|g|^{p'})(x)^{1/p'}\,dx,$$

we also have

$$\left|\iint_{R^2_+} f(x,t)g(x,t)\,dx\,dt\right| \leqslant c \sum_{k=-\infty}^{\infty} 2^k \sum_j \int_{I_{j,k}} C(|g|^{p'})(x)^{1/p'}\,dx$$

$$= c \sum_{k=-\infty}^{\infty} 2^k \int_{\mathcal{O}_k} C(|g|^{p'})(x)^{1/p'}\,dx$$

$$\leqslant c \sum_{k=-\infty}^{\infty} 2^k \int_{E_k} C(|g|^{p'})(x)^{1/p'}\,dx$$

$$\leqslant c \int_{R^n} A_p(f)(x) C(|g|^{p'})(x)^{1/p'}\,dx.$$

This result is also from Deng's work [1984], and it can be used to give the following interesting extension of (1.1). Let ψ be a Littlewood–Paley function and suppose $g \in BMO(R)$. Then by Corollary 1.3, $|g * \psi_t(x)|^q/t\,dx\,dt$

is a Carleson measure for $q \geq 2$, and $A_2(f * \psi_t)(x) = S(f * \psi_t)(x)$ verifies $\|A_2(f * \psi_t)\|_p \leq c\|f\|_p$, $1 < p < \infty$. Thus for $1 < p < 2$ we have

$$\int_{R^2_+} |f * \psi_t(x)|^p |g * \psi_t(x)|^\eta \, dx \frac{dt}{t} \leq c\|A_2(f * \psi_t)\|_p^p \leq c\|f\|_p^p \tag{4.1}$$

provided that $2\eta/(2-p) \geq 2$, i.e., $\eta \geq 2 - p$; clearly, (4.1) holds also for $p \geq 2$, $\eta \geq 0$. In particular if $\eta = 1$ we get

$$\int_{R^2_+} |f * \psi_t(x)|^p |g * \psi_t(x)| \, dx \frac{dt}{t} \leq c\|f\|_p^p, \tag{4.2}$$

which is not a consequence of (1.1), since as we have seen in Problem 4.3 $|g * \psi_t(x)|/t \, dx \, dt$ is not necessarily a Carleson measure. Can (4.2) be extended to the case $0 < p \leq 1$?)

4.7 Coifman and Weiss [1977] observed that $H^1(R^n)$ is a dual space; more precisely, it is the dual of $VMO(R^n)$ in the sense that each continuous linear functional L on $VMO(R^n)$ has the form $L(f) = \int_{R^n} f(x)g(x)\,dx$, $f \in C_0(R^n) \cap VMO(R^n)$, $g \in H^1(R^n)$ and the norm of $L \sim \|g\|_{H^1}$.

4.8 A Carleson measure μ with the property that $\lim_{|B| \to 0} \mu(T(B))/|B| = 0$, is called a vanishing Carleson measure. If the mapping $f \to f * P_t$ is compact from $L^p(R^n)$ into $L^p_\mu(R^{n+1}_+)$, $1 < p < \infty$, then μ is a vanishing Carleson measure. How about a converse? (*Hint*: Suppose not. Then there exist $\varepsilon > 0$ and a sequence $\{B_k\}$ such that $|B_k| \to 0$ and $\mu(T(B_k))/|B_k| \geq \varepsilon$. By Proposition 3.2 in Chapter II we can find a subsequence, which we denote by $\{B_k\}$ again, and $f \in L^p(R^n)$ so that $|B_k|^{-1/p}\chi_{B_k}$ converges weakly to f in $L^p(R^n)$. We claim that $f = 0$ a.e. Indeed, let $E_\lambda = \{f > \lambda\}$ and set $g = \chi_{E_\lambda}$; since $f \in L^p(R^n)$ it is easy to see that $g \in L^{p'}(R^n)$, $1/p + 1/p' = 1$. Thus, on the one hand,

$$\lim_{k \to \infty} \int_{R^n} |B_k|^{-1/p} \chi_{B_k}(x) g(x)\,dx = \int_{R^n} f(x)g(x)\,dx$$

$$= \int_{\{f > \lambda\}} f(x)\,dx$$

and on the other hand

$$0 \leq \int_{R^n} |B_k|^{-1/p} \chi_{B_k}(x) g(x)\,dx \leq |B_k|^{1-1/p} \to 0$$

as $k \to \infty$. Consequently $\int_{\{f > \lambda\}} f(x)\,dx = 0$ for each λ and $f = 0$ a.e. Since $P_t(y) \in L^{p'}(R^n)$, we also get that $\lim_{k \to \infty} |B_k|^{-1/p}\chi_{B_k} * P_t(x) = 0$ everywhere, and by the compactness assumption $\||B_k|^{-1/p}\chi_{B_k} * P_t\|_p \to 0$ as $k \to \infty$; this contradicts the fact that $\mu(T(B_k))/|B_k| \geq \varepsilon$. The result, as well as the answer concerning the converse, is in the work of Power [1980].)

4.9 Theorem 2.1 establishes that, given $f \in BMO(R^n)$, we can find $n+1$ bounded functions $f_0, f_1, \ldots, f_n$ so that $f = f_0 - \sum_{j=1}^n R_j f_j$. In this direction Uchiyama [1983] has proved the following result: Given $\theta_1(\xi'), \ldots, \theta_n(\xi') \in C^\infty(\Sigma)$, let $(K_j f)\hat{}(\xi) = \theta_j(\xi/|\xi|)\hat{f}(\xi)$, $1 \leq j \leq n$. It is well known, and not hard to see, that there are n scalars a_j and smooth functions Ω_j homogeneous of degree 0 such that

$$K_j f(x) = a_j f(x) + \text{p.v.} \int_{R^n} \Omega_j\left(\frac{(x-y)}{|x-y|}\right)|x-y|^{-n} f(y)\, dy$$

whenever $f \in L^p(R^n)$, $1 < p < \infty$, say. For $f \in BMO(R^n)$ the definition must be modified as in Theorem 7.1 in Chapter XI. If

$$\operatorname{rank}\begin{pmatrix} \theta_1(\xi') \cdots \theta_m(\xi') \\ \theta_1(-\xi') \cdots \theta_m(-\xi') \end{pmatrix} = 2, \qquad \xi' \in \Sigma,$$

then for any $f \in BMO(R^n)$ with compact support there exist $g_1, \ldots, g_m \in L^\infty(R^n)$ so that

$$f = \sum_{j=1}^m K_j g_j \quad \text{(modulo constants)}, \qquad \text{and} \qquad \sum_{j=1}^m \|g_j\|_\infty \leq c\|f\|_*,$$

where c depends on the θ's but is independent of f. Since

$$\operatorname{rank}\begin{pmatrix} 1 & \xi_1/|\xi| & \cdots & \xi_n/|\xi| \\ 1 & -\xi_1/|\xi| & \cdots & -\xi_n/|\xi| \end{pmatrix} = 2, \qquad \text{on } \Sigma.$$

Uchiyama's result includes the assertion of Theorem 2.1 alluded to in problem 4.9. By duality arguments one can also show $\|f\|_{H^1} \sim \sum_{j=1}^m \|K_j f\|_1$.

4.10 We say that a measure μ in R_+^{n+1} is a Carleson measure of order $\beta \geq 1$ if

$$\mu(T(\mathcal{O})) \leq c|\mathcal{O}|^\beta, \qquad \text{all open sets } \mathcal{O} \subset R^n. \tag{4.3}$$

A simple argument, using a Whitney decomposition, gives that it is sufficient to check (4.3) for cubes. For $0 < p < 1$, let $m \geq 1$ be the smallest integer $> n[1/p - 1]$. Then for harmonic u, $u \to \lim_{t \to 0} \int_{R^n} u(x,t) g(x)\, dx$ gives a continuous linear functional on $H^p(R^n)$ provided that $d\mu(x,t) = |t^m(\partial/\partial t)^m (g * P_t)(x)|^2/t\, dx\, dt$ is a Carleson measure of order β. (*Hint*: The expression in question equals

$$I = c\int_{R_+^{n+1}} t^m\left(\frac{\partial}{\partial t}\right)^m u(x,t) t^m \left(\frac{\partial}{\partial t}\right)^m (g * P_t)(x)\, dx \frac{dt}{t};$$

from this point on the proof proceeds along the lines of Theorem 2.1. I can also be estimated by making use of 6.11 in Chapter XIV if one assumes instead that $|t^m(\partial/\partial t)^m(g * P_t)(x)| \leq ct^{n(1/p-1)}$.)

4.11 Let $0 \leq \alpha$, $k = [\alpha]$. We say that $f \in \mathscr{E}^{\alpha,q}(R^n) = \mathscr{E}^{\alpha,q}$ if

$$|f|_{\alpha,q} = \sup(t^{-n-\alpha q} \inf_{P \in \mathscr{P}_k} \int_{|x-y| \leq t} |f(y) - P(y, x, t)|^q \, dy)^{1/q} < \infty,$$

where the sup is taken over $x \in R^n$, $t > 0$, and $\mathscr{P}_k$ is the class of polynomials of degree $\leq k$. This expression is a seminorm in $\mathscr{E}^{\alpha,q}$ and $|f|_{\alpha,q} = 0$ when f is a polynomial of degree $\leq k$; $\mathscr{E}^{\alpha,q}$ becomes a Banach space when such polynomials are identified. We also say that a harmonic function is in $H^{\alpha,q}(R^{n+1}_+) = H^{\alpha,q}$ if

$$|u|_{\alpha,q} = \sup\left(t^{-n-\alpha q} \int_{|x-y| \leq t} \left(\int_{[0,t)} \left|s^{k+1}\left(\frac{\partial}{\partial s}\right)^{k+1} u(y, s)\right|^2 \frac{ds}{s}\right)^{q/2} dy\right)^{1/q} < \infty,$$

where the sup is taken over $(x, t) \in R^{n+1}_+$. Prove that if $f \in \mathscr{E}^{\alpha,q}$, then $u(x, t) = f * P_t(x) \in H^{\alpha,q}$ and $|u|_{\alpha,q} \leq c|f|_{\alpha,q}$. Conversely, if $u \in H^{\alpha,q}$, $f = \lim_{t \to 0} u(\cdot, t)$ exists, it is in $\mathscr{E}^{\alpha,q}$ and verifies $|f|_{\alpha,q} \leq c|u|_{\alpha,q}$. When $q = 2$ we are dealing with Carleson measures of order $\beta \geq 0$. These "trace" results where considered by Fabes and Johnson and Neri [1976] and Ortiz and Torchinsky [1977].

4.12 In Theorem 2.2 we may use (p, q, N) atoms, $1 < q < \infty$, instead of $(p, 2, N)$ atoms. This gives automatically an equivalence of norms in the spirit of Theorem 5.1 in Chapter VIII.

4.13 Theorem 3.3 holds for $T^p = \{f: |f|_p = \|S(f)\|_p < \infty\}$, $0 < p < 1$, as well. More precisely, suppose $f \in T^p$; then $f = \sum_{j=1}^{\infty} a_j$, where the a_j's are functions supported in $T(B)$ for some ball $B \subset R^n$ so that $\int_{T(B)} |a(y, t)|^q / t \, dy \, dt < \infty$. If we normalize the a_j's by setting

$$|a| = \inf\left\{\left(|B|^{q/p-1} \int_{T(B)} |a(y, t)|^q / t \, dy \, dt\right)^{1/q}\right\},$$

where the infimum is taken over all balls B containing the support of a, then also $\sum_j |a_j|^p \leq c|f|_p^p$.

4.14 $K(f)$ defined by (3.6) extends to a bounded linear operator from T^p into $H^p(R^n)$, $0 < p < 1$.

4.15 Fefferman and Stein [1971] gave a simple proof of the following result of Carleson [1962]: Given $f \in L_{\text{loc}}(R)$, let $Mf(x, t) = \sup(1/|I|) \int_I |f(y)| \, dy$, where the sup is taken over those intervals I which contain x and have diameter at least t. Then

$$\left(\int_{R^2_+} Mf(x, t)^p \, d\mu(x, t)\right)^{1/p} \leq c\|f\|_p, \qquad 1 < p \leq \infty, \tag{4.4}$$

if and only if μ is a Carleson measure. This result includes the Hardy–Littlewood maximal theorem ($d\mu(x, t) = dx$). The proof given by Fefferman and Stein establishes the weak-type (1, 1) estimate for $Mf(x, t)$ first, and (4.4) follows by interpolation. They also showed that $Mf(x, t)$ is bounded from $L^p_\nu(R)$, $d\nu(x) = v(x)\,dx$, into $L^p_\mu(R^2_+)$ provided that $C(\nu)(x) \leq cv(x)$ a.e. Ruiz [1985] and Ruiz-Torrea [1985a] have proved the following results: we say that μ verifies the $C_p(v)$ condition, and denote this by $\mu \in C_p(v)$ if

$$\sup_I \frac{\mu(T(I))}{|I|}\left(\frac{1}{|I|}\int_I v(x)^{-1/(p-1)}\,dx\right)^{p-1} \leq c,$$

where the sup is taken over all open cubes I of R^n. Similarly, μ satisfies the $C_1(v)$ condition if $C(\mu)(x) \leq cv(x)$ a.e. and the $C_\infty(v)$ condition if $\mu(T(I)) \leq cv(I)$, all I in R^n. The following relations are readily verified: if $1 \leq p \leq q \leq \infty$ and $\mu \in C_p(v)$, then also $\mu \in C_q(v)$. Moreover, if $v \in A_p$, then the classes $C_p(v)$ and $C_q(v)$ coincide for $p \leq q$, and $\mu \in C_p(v)$ implies $\mu \in C_{p-\varepsilon}(v)$, some $\varepsilon > 0$.

We then have that, if $\mu \in C_\infty(v)$, then $\mu(\{Mf(x, t) > \lambda\}) \leq c_1 \int_{\{Mf > c_2\lambda\}} v(x)\,dx$. This result in particular implies that for $v \in A_p$, $1 < p < \infty$, then $\mu \in C_\infty(v)$ if and only if $\int_{R^{n+1}_+} Mf(x, t)^p\,d\mu(x, t) \leq c\int_{R^n} |f(x)|^p\,d\nu(x)$. As for the general weak-type result, they show that $Mf(x, t)$ maps $L^p_\nu(R^n)$ into wk-$L^p_\mu(R^{n+1}_+)$, $1 \leq p < \infty$, if and only if $\mu \in C_p(v)$. When $d\mu(x) = w(x)\,dx$, this statement coincides with statement (7.3) in Chapter IX.

To state the strong-type results we need one more notation: we say that μ satisfies the $F_p(v)$ condition, $1 < p < \infty$, and write $\mu \in F_p(v)$ provided that

$$\int_{T(I)} M(v^{1-p'}\chi_I)(x, t)^p\,d\mu(x, t) \leq c\int_I v(x)^{-1/(p-1)}\,dx < \infty,$$

for all open cubes I. Then for $1 < p < \infty$ the following holds: $Mf(x, t)$ maps $L^p_\nu(R^n)$ continuously into $L^p_\mu(R^{n+1}_+)$ if and only if $\mu \in F_p(v)$. The proof follows along the lines of Theorem 4.1 in Chapter IX.

4.16 It is also possible to establish weighted and vector-valued inequalities in this context. We describe briefly a couple of examples from the work of Ruiz-Torrea [1985b]. Consider then the maximal function $Mf(x, t)$ of 4.15 and the fractional maximal function $M_\eta f(x, t)$ given by

$$M_\eta f(x, t) = \sup_I \frac{1}{|I|^{1-\eta/n}}\int_I |f(y)|\,dy,$$

where the sup is taken over those cubes I centered at x of sidelength at least t. The vector-valued estimates that hold are of the type

$$\|\,\|Tf_j\|_{l^r}\|_{L^q(R^{n+1}_+)} \leq c\|\,\|f_j\|_{l^r}\|_{L^p(R^n)} \tag{4.5}$$

and

$$\mu(\{\|Tf_j\|_{l^r} > \lambda\}) \leq c\lambda^{-q}\| \|f_j\|_{l^r}\|^q_{L(R^n)}. \tag{4.6}$$

In case $Tf(x, t) = Mf(x, t)$, (4.5) holds with $1 < p = q < \infty$, and (4.6) with $q = 1$, provided μ is a Carleson measure. In case $Tf(x, t) = M_\eta f(x, t)$ and $\beta = s(1 - \eta/n) \geq 1$, (4.5) holds for $1/p = \eta/n + s(1 - \eta/n)/q$, $s < q < \infty$, and (4.6) with $q = s$, provided μ is a Carleson measure of order β.

4.17 Suppose that $a \in BMO(R^n)$, ψ is a smooth Littlewood-Paley function, and $w \in A_2$. Then $d\mu(x, t) = |a * \psi_t(x)|^2 w(x)\, dx\, dt/t$ is a Carleson measure in $C_\infty(w)$ with constant $\leq c\|a\|^2_*$, where c depends on ψ and w but is independent in A_2. (*Hint*: Repeat the proof of Proposition 1.2 carefully; this result is due to Journé [1983].)

4.18 Assume that $d\nu(x) = v(x)\, dx$, $v \in A_p$, $1 < p < \infty$, and ϕ, ψ are smooth functions with $\int_{R^n} \psi(x)\, dx = 0$. Show that the operator $Tf = (\int_{[0,\infty)} |f * \phi_t|^2 |a * \psi_t|/t\, dt)^{1/2}$ is bounded in $L^p_\nu(R^n)$. (*Hint*: Suppose $p = 2$ first; 4.17 gives at once that T is bounded in $L^2_\nu(R^n)$ with constant independent in A_2. The general result follows by theorem 7.6 in Chapter IX. This observation is also due to Journé.)

CHAPTER

XVI

Cauchy Integrals on Lipschitz Curves

1. CAUCHY INTEGRALS ON LIPSCHITZ CURVES

Suppose Γ is a curve in the complex plane C given by $z(x) = x + i\phi(x)$, $x \in R$; our only assumption is that $\phi' \in L^\infty(R)$. As in Section 3 of Chapter X we are interested in the following question: given a continuous, bounded function f on Γ, does there exist a function $F(z)$, analytic in $C\backslash\Gamma$, so that $\lim_{\eta\to 0^+} F(z(x)+i\eta) - F(z(x)-i\eta) = f(z(x))$, $x \in R$? Our first approach is to consider the Cauchy integral of f on Γ, that is to say the function

$$F(z) = \frac{1}{2\pi i}\int_\Gamma \frac{f(w)}{w-z}\,dw, \qquad z = z(x) + i\eta, \quad \eta \neq 0.$$

Suppose first $\eta > 0$, we then have

$$\begin{aligned} F(z) &= \frac{1}{2\pi i}\int_R \frac{f(z(y))}{z(y)-z(x)-i\eta}\,dz(y) \\ &= \frac{1}{2\pi i}\int_R \frac{z(y)-z(x)}{(z(y)-z(x))^2+\eta^2} f(z(y))\,dz(y) \\ &\quad + \frac{1}{2\pi}\int_R \frac{\eta}{(z(y)-z(x))^2+\eta^2} f(z(y))\,dz(y) \\ &= I + J, \end{aligned}$$

say. Now a straightforward computation using residues gives

$$\frac{1}{\pi}\int_R \frac{\eta}{(z(y)-z(x))^2+\eta^2}\,dz(y) = 1,$$

and consequently $\lim_{\eta\to 0} J = f(z(x))/2$. On the other hand, formally at least,

$$\lim_{\eta\to 0} I = \frac{1}{2\pi i}\,\text{p.v.}\int_{\mathbf{R}} \frac{f(z(y))}{z(y)-z(x)}\,dz(y).$$

A similar argument works for $\eta < 0$ as well. It becomes thus apparent that it is imporant to study the operator

$$\text{p.v.}\frac{1}{2\pi i}\int_{\mathbf{R}} \frac{1+i\phi'(y)}{(x-y)+i(\phi(x)-\phi(y))} f(y+i\phi(y))\,dy,$$

i.e., the singular integral with kernel $k_\phi(x, y) = (1/((x-y)+i(\phi(x)-\phi(y)))$; the factor $1+i\phi'(y)$ may be omitted since $\phi' \in L^\infty$. So, before discussing the various properties of $F(z)$ we introduce the singular integral operator

$$C_\phi f(x) = \text{p.v.}\int_R k_\phi(x, y) f(y)\,dy, \qquad f \in C_0^\infty(R). \tag{1.1}$$

Then we consider, as we did for the case $\phi(x) = x$, i.e., the Hilbert transform, the questions of pointwise and norm convergence and continuity of $C_\phi f$ in the various $L^p(R)$ spaces. Although these problems are easily formulated, they are rather difficult to solve. A way to go about this is to suppose that ϕ has Lipschitz constant $\leqslant 1$ and consider the expansion of $k_\phi(x, y)$ in terms of kernels of the form

$$(\phi(x)-\phi(y))^n/(x-y)^{n+1}, \qquad n = 0, 1, \ldots. \tag{1.2}$$

When $n = 0$ this is the kernel of the Hilbert transform, an operator we can handle. But examine the case $(\phi(x)-\phi(y))/(x-y)^2$, where $n = 1$, which is known as the kernel of Calderón's first commutator and which corresponds to the operator $-(d/dx)[H, M_\phi] + M_{\phi'}H$, where $[H, M_\phi]$ is the commutator of the Hilbert transform H and the multiplication operator M_ϕ by ϕ. This is already quite difficult to handle.

To deal with these operators, we begin with some preliminary results. Let D denote the closed, self-adjoint, densely defined operator $(-i)\,d/dx$ on the Hilbert space $L^2(R)$. Associated with D and the parameter $t \in R$, we introduce the following families of operators on $L^2(R)$: $Q_t = tD(I+t^2D^2)^{-1}$, $P_t = (I+t^2D^2)^{-1} = I - tQ_tD$, and $R_t = P_t - iQ_t = (I+itD)^{-1}$.

One word about the meaning of these operators. Since D corresponds to multiplication by ξ on the Fourier transform side, another way to define Q_t, P_t, and R_t is to consider the operators with symbol $m_{Q_t}(\xi) = t\xi(1+(t\xi)^2)^{-1}$, $m_{P_t}(\xi) = (1+(t\xi)^2)^{-1}$ and $m_{R_t}(\xi) = (1+it\xi)^{-1}$, respectively. In this way it is immediate that they are bounded in $L^2(R)$ and that $\|Q_t\|$ = norm of Q_t as a bounded mapping in $L^2(R) = \sup_\xi |m_{Q_t}(\xi)| = \frac{1}{2}$, $\|P_t\| = \sup_\xi |m_{P_t}(\xi)| = 1$ and $\|R_t\| = \sup_\xi |m_{R_t}(\xi)| = 1$, for all $t \neq 0$.

Another immediate application of this approach is the Littlewood–Paley type identity

$$\int_{[0,\infty)} \|Q_t f\|_2^2 \frac{dt}{t} = \frac{1}{2}\|f\|_2^2. \tag{1.3}$$

Indeed, the left-hand side in (1.3) equals

$$\int_{[0,\infty)} \int_R |(t\xi)(1+(t\xi)^2)^{-1}\hat{f}(\xi)|^2 \, d\xi \frac{dt}{t}$$
$$= \int_R |\hat{f}(\xi)|^2 \int_{[0,\infty)} (t\xi)^2(1+(t\xi)^2)^{-2} \frac{dt}{t} \, d\xi,$$

and the innermost integral above is $\frac{1}{2}$ as a change of variables readily shows.

To study the continuity properties of these operators in the $L^p(R)$ spaces, $p \neq 2$, it is convenient to have their integral representation at hand. This requires the computation of the corresponding kernels. From the results in Section 3 of Chapter X concerning the Poisson kernel it follows immediately that, if $k(x) = \frac{1}{2}e^{-|x|}$, then the kernel of P_t, given by the Fourier inverse transform of $m_{p_t}(\xi)$, is actually $k_{|t|}(x) = |t|^{-1}k(x/|t|)$, that of $Q_t = tDP_t$ is $(-i)t(d/dx)k_{|t|}(x) = i(\operatorname{sgn} x)(\operatorname{sgn} t)k_{|t|}(x)$, and that of $R_t = P_t - iQ_t$ is $2(1/|t|)k(x/t)$ when $(\operatorname{sgn} x)(\operatorname{sgn} t) = 1$ and 0 otherwise. This implies, in particular, that $P_t f(x) = \int_R k_{|t|}(x-y)f(y)\,dy$, and consequently for $1 \leq p \leq \infty$, $\|P_t f\|_p < \|k_{|t|}\|_1\|f\|_p = \|f\|_p$, $t \neq 0$. A similar estimate holds for Q_t since its kernel $q(x,t)$ verifies $|q(x,t)| = k_{|t|}(x)$. As for R_t, its kernel $r(x,t)$ is nonnegative and has the property that $\int_R r(x,t)\,dx = 1$, $t \neq 0$. In short, for $1 \leq p \leq \infty$ and $t \neq 0$, we have that

$$\|P_t f\|_p, \qquad \|Q_t f\|_p, \qquad \|R_t f\|_p \leq \|f\|_p. \tag{1.4}$$

Also, the following stronger statement is true.

Lemma 1.1. The functions $P_t f$, $Q_t f$, defined for $t > 0$, verify

$$N(P_t f)(x), \qquad N(Q_t f)(x) \leq cMf(x). \tag{1.5}$$

Proof. By the above remarks it suffices to prove (1.5) for $P_t f$, where f is a nonnegative function. Now, since $k_{|t|}(y-w) \leq ek_{|t|}(x-w)$ whenever $|x-y| < t$, it follows at once that $P_t f(y) \leq eP_t f(x)$ for $(y,t) \in \Gamma(x)$. Thus $N(P_t f)(x) \leq e \sup_{t>0} P_t f(x)$, which, by an argument similar to Proposition 2.3 in Chapter IV, in turn is bounded by $cMf(x)$. ■

We pass now to study some of the p.v. integral operators that P_t, Q_t, and R_t generate; integrals of operators are always taken in the strong sense of convergence. Consider for instance one of the simplest cases, namely, p.v. $\int_R R_t/t\,dt$. Since P_t is even, as is its kernel $k_{|t|}(x)$, it follows that p.v. $\int_R P_t/t\,dt = 0$. Also, since Q_t is odd we see that

$$\text{p.v.} \int_R Q_t \frac{dt}{t} = \lim_{\varepsilon \to 0} 2 \int_{\varepsilon < t < 1/\varepsilon} Q_t \frac{dt}{t}.$$

Thus

$$\text{p.v.} \int_R R_t \frac{dt}{t} = \lim_{\varepsilon \to 0} -2i \int_{\varepsilon < t < 1/\varepsilon} tD(I + t^2 D^2)^{-1} \frac{dt}{t} = -2i \operatorname{sgn} D.$$

In other words, the symbol of p.v. $\int_R R_t/t\,dt$ is a multiple of sgn ξ, and this operator is essentially the Hilbert transform. We can also get this fact by looking directly at the kernel

$$\lim_{\varepsilon \to 0} \int_{\varepsilon < |t| < 1/\varepsilon} r(x, t) \frac{dt}{t}. \tag{1.6}$$

To evaluate this expression, suppose first $x > 0$; then we also have $t > 0$, and the integral in (1.6) is

$$2 \int_{\varepsilon < t < 1/\varepsilon} \frac{1}{t} e^{-x/t} \frac{dt}{t} = \frac{2}{x} \int_{\varepsilon x < t < x/\varepsilon} d(e^{-1/t})$$

$$= \frac{2}{x}(e^{-x/\varepsilon} - e^{-\varepsilon x}) \to -\frac{2}{x}, \qquad \text{as} \quad \varepsilon \to 0.$$

A similar argument gives the same conclusion for $x < 0$. Thus the kernel of the operator coincides with a multiple of $1/x$, $x \neq 0$, and the operator itself is a multiple of the Hilbert transform. We will use this representation to show that the Hilbert transform is bounded in $L^2(R)$. We begin by deriving a very useful identity.

Lemma 1.2.

$$Q_t = 8Q_t^3 - t\frac{\partial}{\partial t}(Q_t(I - 2P_t)), \qquad t \neq 0. \tag{1.7}$$

Proof. We show that the multiplier corresponding to the operator on the right-hand side of (1.7) is actually $m_{Q_t}(\xi)$. This reduces to a simple computation using that $t(\partial/\partial t)g(t\xi) = t\xi g'(t\xi)$. Indeed, since the multiplier corresponding to $Q_t(I - 2P_t)$ is $m_{Q_t}(\xi)(1 - 2m_{P_t}(\xi))$, we observe that

$$t\frac{d}{dt}\left(\frac{t}{1+t^2}\left(1 - \frac{2}{1+t^2}\right)\right) = t\,\frac{d}{dt}\left(\frac{t}{1+t^2}\,\frac{(t^2-1)}{1+t^2}\right)$$

$$= -\frac{(t^3+t)}{(1+t^2)^2} + \frac{8t^3}{(1+t^2)^3} = \frac{-t}{1+t^2} + 8\left(\frac{t}{1+t^2}\right)^3,$$

and (1.7) follows immediately. ∎

We can now prove

Proposition 1.3. Suppose $f \in L^2(R)$, then

$$\left\| \text{p.v.} \int_R R_t f \frac{dt}{t} \right\|^2 \leqslant 8\|f\|_2. \tag{1.8}$$

Proof. By the above remarks it suffices to show (1.8) with p.v. $2\int_{[0,\infty)} Q_t f/t\, dt$ in the left-hand side there. By Lemma 1.2 this expression equals

$$16 \text{ p.v.} \int_{[0,\infty)} Q_t^3 f \frac{dt}{t} - \text{p.v. } 2\int_{[0,\infty)} \frac{\partial}{\partial t}(Q_t(I - 2P_t)f)\, dt = A + B,$$

say.

Now, by the Lebesgue dominated convergence theorm it readily follows that for f in $L^2(R)$

$$\lim_{\varepsilon \to 0} \int_R (\varepsilon|\xi|(1 + \varepsilon^2|\xi|^2)^{-1}|\hat{f}(\xi)|)^2\, d\xi = 0,$$

and the same is true with ε replaced by $1/\varepsilon$ in the integral above. Thus $B = 2\lim_{\varepsilon\to 0} Q_t(I - 2P_t)f]_\varepsilon^{1/\varepsilon} = 0$, where the limit is taken in the $L^2(R)$ sense. To estimate A we observe first that a minor variant of (1.3) gives that also $\int_R \int_{[0,\infty)} |Q_t^2 g(x)|^2/t\, dt\, dx \leqslant \frac{1}{2}\|g\|_2^2$. Let now $f \in L^2(R)$ and $0 < t_0 < t_1 < \infty$. Then

$$\begin{aligned}\left\| \int_{[t_0,t_1)} Q_t^3 f \frac{dt}{t} \right\|_2 &= \sup \left| \int_{[t_0,t_1)} \int_R Q_t^3 f(x) g(x)\, dx \frac{dt}{t} \right| \\ &= \sup \left| \int_{[t_0,t_1)} \int_R Q_t f(x) Q_t^2 g(x)\, dx \frac{dt}{t} \right|,\end{aligned}$$

where the sup is taken over those g's with $\|g\|_2 = 1$. Now, by Hölder's inequality we get that for fixed t_0, t_1 the integral above does not exceed

$$\begin{aligned}&\left(\int_{[t_0,t_1)} \|Q_t f\|_2^2 \frac{dt}{t} \right)^{1/2} \left(\int_{[0,\infty)} \|Q_t^2 g\|_2^2 \frac{dt}{t} \right)^{1/2} \\ &\leqslant \frac{1}{\sqrt{2}} \left(\int_{[t_0,t_1)} \|Q_t f\|_2^2 \frac{dt}{t} \right)^{1/2},\end{aligned}$$

which on account of (1.3) tends to 0 as $t_0, t_1 \to 0$ and as $t_0, t_1 \to \infty$. So, $\lim_{\varepsilon \to 0} \int_{[\varepsilon, 1/\varepsilon)} Q_t^3 f/t\, dt$ exists in $L^2(R)$ and has norm less than or equal to $(1/\sqrt{2})(\int_{[0,\infty)} \|Q_t f\|_2^2/t\, dt)^{1/2} = \frac{1}{2}\|f\|_2$. ∎

The next question we consider is what other operators can be represented by means of such p.v. integrals and can be thus handled as in Proposition 1.3. Let $\psi \in L^\infty(R)$ and put $K_\psi = \text{p.v.} \int_R R_t M_\psi R_t / t\, dt$. Since $P_t M_\psi P_t$ and $Q_t M_\psi Q_t$ are even it readily follows that actually

$$K_\psi = -i \,\text{p.v.} \int_R Q_t M_\psi P_t \frac{dt}{t} - i \,\text{p.v.} \int_R P_t M_\psi Q_t \frac{dt}{t}.$$

Furthermore, since these terms are the adjoint of each other, we may restrict our attention to either term, the first one, say. Once again we begin by discussing an identity.

Lemma 1.4. Suppose $\psi \in L^\infty(R)$ and $f \in L^2(R)$. Then

$$Q_t M_\psi P_t f = (Q_t \psi)(P_t f) - Q_t((Q_t \psi)(Q_t f)) + P_t((P_t \psi)(Q_t f)), \quad t \neq 0. \qquad (1.9)$$

Proof. We assume first that ψ and f are smooth, in $\mathcal{S}(R)$ say, and check that (1.9) holds in this case. It is not hard then to pass to the limit and obtain the validity of (1.9) for those ψ's and f's for which both sides of the equality make sense. These functions certainly include the ψ's and f's in the statement of our result. First observe that by scaling it suffices to verify (1.9) for $t = 1$. The Fourier transform of the left-hand side of that formula is readily seen to equal $(\xi/(1+\xi^2)) \int_R \hat{\psi}(\xi - \eta)(1/(1+\eta^2)) \hat{f}(\eta)\, d\eta$. As for the right-hand side, its Fourier transform is

$$\int_R \frac{\xi - \eta}{1 + (\xi - \eta)^2} \hat{\psi}(\xi - \eta) \frac{1}{1 + \eta^2} \hat{f}(\eta)\, d\eta$$
$$- \frac{\xi}{1 + \xi^2} \int_R \frac{\xi - \eta}{1 + (\xi - \eta)^2} \hat{\psi}(\xi - \eta) \frac{\eta}{1 + \eta^2} \hat{f}(\eta)\, d\eta$$
$$+ \frac{1}{1 + \xi^2} \int_R \frac{1}{1 + (\xi - \eta)^2} \hat{\psi}(\xi - \eta) \frac{\eta}{1 + \eta^2} \hat{f}(\eta)\, d\eta.$$

That (1.9) holds follows now immediately from the identity

$$\frac{\xi}{1 + \xi^2} = \frac{(\xi - \eta)}{1 + (\xi - \eta)^2} - \frac{\xi\eta(\xi - \eta)}{(1 + \xi^2)(1 + (\xi - \eta)^2)} + \frac{\eta}{(1 + \xi^2)(1 + (\xi - \eta)^2)},$$

for all ξ, η in R. ■

We are now ready to prove

Proposition 1.5. Suppose $\psi \in L^\infty(R)$ and $f \in L^2(R)$. Then

$$\left\| \text{p.v.} \int_R R_t M_\psi R_t f \frac{dt}{t} \right\|_2 \leq c \|\psi\|_\infty \|f\|_2, \qquad (1.10)$$

where c is a constant independent of f.

Proof. By the above remarks, and since $Q_t M_\psi P_t$ is odd, it suffices to show (1.10) with p.v. $\int_{[0,\infty)} Q_t M_\psi P_t / t\, dt$ in the left-hand side there. Motivated by Proposition 1.3 we begin by showing that

$$\int_{[0,\infty} \|Q_t M_\psi P_t f\|_2^2 \frac{dt}{t} \leqslant c\|\psi\|_\infty^2 \|f\|_2^2. \tag{1.11}$$

This is not hard; indeed, by Lemma 1.4, the integral in (1.11) does not exceed a multiple of

$$\int_{[0,\infty} \|(Q_t\psi)(P_t f)\|_2^2 \frac{dt}{t} + \int_{[0,\infty)} \|Q_t((Q_t\psi)(Q_t f))\|_2^2 \frac{dt}{t}$$
$$+ \int_{[0,\infty)} \|P_t((P_t\psi)(Q_t f))\|_2^2 \frac{dt}{t} = A + B + C,$$

say. C is handled easily; by estimates (1.4) and (1.3) it is bounded by

$$\int_{[0,\infty)} \int_R |P_t((P_t\psi)(Q_t f))(x)|^2\, dx \frac{dt}{t}$$
$$\leqslant \int_{[0,\infty)} \int_R |P_t\psi(x)|^2 |Q_t f(x)|^2\, dx \frac{dt}{t}$$
$$\leqslant \|\psi\|_\infty^2 \int_{[0,\infty)} \|Q_t f\|_2^2 \frac{dt}{t} \leqslant \frac{1}{2}\|\psi\|_\infty^2 \| \, \|f\|_2^2,$$

which is of the right order. The estimate for B follows in an identical fashion. The estimate for A requires a different approach since $P_t f$ is not the convolution of f with a Littlewood–Paley function. First, note that since $\psi \in L^\infty(R)$, by Proposition 1.2 in Chapter XV, it follows that $|Q_t\psi(x)|^2/t\, dx\, dt$ is a Carleson measure with constant $\sim \|\psi\|_*^2 \leqslant c\|\psi\|_\infty^2$. Thus by Theorem 1.1 in Chapter XV,

$$A = \int_{R_+^2} |P_t f(x)|^2 |Q_t\psi(x)|^2\, dx \frac{dt}{t} \leqslant c\|\psi\|_\infty^2 \|N(P_t f)\|_2^2,$$

which in turn by Lemma 1.1 is bounded by $c\|\psi\|_\infty^2 \|Mf\|_2^2 \leqslant c\|\psi\|_\infty^2 \|f\|_2^2$, and (1.11) holds. Now the proof follows along the lines of Proposition 1.3. By Lemma 1.2, $Q_t M_\psi P_t f = 8Q_t^3 M_\psi P_t f - t[(\partial/\partial t)(Q_t(I - 2P_t))](M_\psi P_t f)$ and

$$\text{p.v.} \int_{[0,\infty)} Q_t M_\psi P_t f \frac{dt}{t} = 8 \text{ p.v.} \int_{[0,\infty)} Q_t^3 M_\psi P_t f \frac{dt}{t}$$
$$- \text{p.v.} \int_{[0,\infty)} \left[\frac{\partial}{\partial t}(Q_t(I - 2P_t))\right](M_\psi P_t f)\, dt = E - F,$$

say. Since the estimate (1.11) is available, E is bounded exactly as the A term in Proposition 1.3, with a bound of the right order. On the other hand, F requires a further argument. First, observe that the integrand there equals $(\partial/\partial t)(Q_t(I-2P_t)M_\psi P_t f) - Q_t(I-2P_t)M_\psi(\partial/\partial t)(P_t f)$, and consequently

$$F = \lim_{\varepsilon\to 0} Q_t(I-2P_t)M_\psi P_t f]_\varepsilon^{1/\varepsilon} - \text{p.v.}\int_{[0,\infty)} Q_t(I-2P_t)M_\psi t\frac{\partial}{\partial t}P_t f\frac{dt}{t}$$

$$= F_1 - F_2,$$

say. F_1 is 0. As for F_2 note that $t(\partial/\partial t)(P_t f) = -Q_t^2 f$, and by estimate (1.4),

$$\int_{[0,\infty)} \|Q_t(I-2P_t)M_\psi Q_t^2 f\|_2^2\frac{dt}{t} \leqslant c\|\psi\|_\infty^2\int_{[0,\infty)}\|Q_t f\|_2^2\frac{dt}{t}$$

$$\leqslant c\|\psi\|_\infty^2\|f\|_2^2.$$

From this point on the estimate for F_2 proceeds as that for E; observe that the argument works because the string of operators which appears in the integrand of F_2 contains Q_t at least twice. ■

As for the concrete form of the operator K_ψ, it is obtained as a particular case of the following remarkable representation formula.

Proposition 1.6. (McIntosh). Suppose ϕ is a Lipschitz function on R, $\phi' = \psi \in L^\infty(R)$. Then the (distributional) kernel of the operator $\text{p.v.}\int_R R_t(M_\psi R_t)^n/t\,dt$ is $(\phi(x)-\phi(y))^n/(x-y)^{n+1}$, the kernel of Calderón's nth commutator.

Proof. We find first the kernel of the operator $R_t(M_\psi R_t)^n$; a simple computation shows that it is given by the n-interated integral

$$\int_R\cdots\int_R r(x-x_1,t)\psi(x_1)r(x_1-x_2,t)\psi(x_2)\cdots r(x_n-y,t)\,dx_1\cdots dx_n. \tag{1.12}$$

To evaluate this integral we distinguish the cases $t>0$ and $t<0$. If $t>0$ the domain of integration $D(x,y,t)\subset R^n$ of the integral (1.12) is restricted to $x>x_1>\cdots>x_n>y$ and we have

$$r(x-x_1,t)\cdots r(x_n-y,t) = |t|^{-n}r(x-y,t) \tag{1.13}$$

there. If on the other hand $t<0$, then $D(x,y,t)$ becomes $x<x_1<\cdots<x_n<y$, and the identity (1.13) still holds. Whence, in either case (1.12) equals

$$|t|^{-n}r(x-y,t)\int_{D(x,y,t)}\psi(x_1)\cdots\psi(x_n)\,dx_1\cdots dx_n. \tag{1.14}$$

To compute the integral in (1.14), we observe first that if I is the interval $(\min(x, y), \max(x, y))$, then $D(x, y, t) \subset I \times \cdots \times I$. Furthermore, for $\sigma \in S_n$, the permutation group on the set of n elements, the sets $\sigma(D(x, y, t))$ form a measurable partition of $I \times \cdots \times I$, where σ acts on R^n by exchanging the coordinates. Therefore the integral in question verifies

$$n!\int_{D(x,y,t)} \psi(x_1)\cdots\psi(x_n)\,dx_1\cdots dx_n = \left(\int_I \psi(x_1)\,dx_1\right)\cdots\left(\int_I \psi(x_n)\,dx_n\right)$$
$$= (\operatorname{sgn}(x-y))^n(\phi(x)-\phi(y))^n,$$

and the kernel of $R_t(M_\psi R_t)^n$ is then $(1/n!)(\operatorname{sgn}(x-y))^n(\phi(x)-\phi(y))^n|t|^{-n}r(x, y, t)$. It is now easy to complete our proof, for we must only evaluate

$$\frac{1}{n!}(\operatorname{sgn}(x-y))^n(\phi(x)-\phi(y))^n \lim_{\varepsilon\to 0}\int_{\varepsilon<|t|<1/\varepsilon} |t|^{-n}r(x-y, t)\frac{dt}{t}$$
$$= \frac{1}{n!}(\operatorname{sgn}(x-y))^n(\phi(x)-\phi(y))^n \operatorname{sgn}(x-y)|x-y|^{-(n+1)}\int_{[0,\infty)} t^n e^{-t}\,dt$$
$$= \frac{(\phi(x)-\phi(y))^n}{(x-y)^{n+1}},$$

and we are done. ■

Proposition 1.5 asserts, then, that the mapping

$$f \to \text{p.v.}\int_R \frac{\phi(x)-\phi(y)}{(x-y)^2}f(y)\,dy$$

is bounded in $L^2(R)$. To establish the continuity of Calderón's commutators of higher order, we pass to consider the other ingredients in the proof of Propositions 1.3 and 1.5. We begin by introducing the Littlewood–Paley function

$$G_n f(x) = \left(\int_{[0,\infty)} |Q_t(M_\psi P_t)^n f(x)|^2\frac{dt}{t}\right)^{1/2}, \qquad n \geq 0,$$

and prove a preliminary result that will enable us to estimate the dependence of $\|G_n f\|_2$ on n.

Lemma 1.7. For each $t > 0$, $|(M_\psi P_t)^n f(x)| \leq_c \|\psi\|_\infty^n Mf(x)$, c indep. of n.

Proof. Clearly $|(M_\psi P_t)^n f(x)| \leq \|\psi\|_\infty^n(k_t * \cdots * k_t) * |f|(x)$. Now, k is positive, even, decreasing in $[0, \infty)$, and has integral 1, and the same is true of

$k * \cdots * k$. Thus by Proposition 2.3 in Chapter IV, $(k_t * \cdots * k_t) * |f|(x) = (k * \cdots * k)_t * |f|(x) \leqslant_c Mf(x)$, and we have finished. ■

We can now prove

Proposition 1.8. Suppose $\psi \in L^\infty(R^n)$, $\|\psi\|_\infty \leqslant 1$. Then

$$\|G_n f\|_2 \leqslant c(n+1)\|f\|_2, \tag{1.15}$$

where c is independent of n, ψ, and f.

Proof. The case $n = 0$ of (1.15) is (1.3) and the case $n = 1$ is inequality (1.11), which was proved by combining (1.3) and Lemma 1.4; this suggests that an induction argument along those lines might work. Suppose then that (1.15) holds for $0 \leqslant k \leqslant n$, and consider $G_{n+1}f$. Put $f_{n,t} = (M_\psi P_t)^n f$ and observe that $G_{n+1}f(x) = (\int_{[0,\infty)} |Q_t M_\psi P_t f_{n,t}(x)|^2 / t\, dt)^{1/2}$. We invoke Lemma 1.4 for each fixed t and then Minkowski's inequality and we see at once that

$$\begin{aligned} \|G_{n+1}f\|_2 &\leqslant \left(\int_{R_+^2} |Q_t\psi(x) P_t f_{n,t}(x)|^2\, dx \frac{dt}{t}\right)^{1/2} \\ &+ \left(\int_{R_+^2} |Q_t((Q_t\psi)(Q_t f_{n,t}))(x)|^2\, dx \frac{dt}{t}\right)^{1/2} \\ &+ \left(\int_{R_+^2} |P_t((P_t\psi)(Q_t f_{n,t}))(x)|^2\, dx \frac{dt}{t}\right)^{1/2} = A + B + C, \end{aligned}$$

say. To estimate A we observe first that, if $F(x) = \sup_{t>0} |f_{n,t}(x)|$, then by Lemma 1.7, $F(x) \leqslant_c Mf(x)$ and consequently $\|F_2\| \leqslant c\|f\|_2$, with c independent of n and f. Furthermore, since $|P_t f_{n,t}(y)| \leqslant_c P_t F(x)$ for $(y, t) \in \Gamma(x)$, we also have $N(P_t f_{n,t})(x) \leqslant_c MF(x)$ and by Theorem 1.1 in Chapter XV, $A \leqslant c\|\psi\|_\infty \|N(P_t f_{n,t})\|_2 \leqslant c\|MF\|_2 \leqslant c\|f\|_2$, with c independent of n and f. A similar argument works for B as well. Indeed, by (1.4) and since $|q(x, t)| = k_{|t|}(x)$, it readily follows that

$$\begin{aligned} B &\leqslant \left(\int_{R_+^2} |Q_t\psi(x) P_t(|f_{n,t}|)(x)|^2\, dx \frac{dt}{t}\right)^{1/2} \\ &\leqslant c\|\psi\|_\infty \|N(P_t(|f_{n,t}|))\|_2 \leqslant c\|MF\|_2 \leqslant c\|f\|_2, \end{aligned}$$

again with c independent of n and f. It only remains to estimate C, but this is easy since on account of (1.4), applied twice, and the induction hypothesis it readily follows that

$$C \leqslant \|\psi\|_\infty \left(\int_{R_+^2} |Q_t f_{n,t}(x)|^2\, dx \frac{dt}{t}\right)^{1/2} \leqslant \|G_n f\|_2 \leqslant c(n+1)\|f\|_2,$$

where c is an absolute constant. Whence adding the bounds for A, B, and C we get $\|G_{n+1}f\|_2 \leq c(n+1)\|f\|_2 + c\|f\|_2 \leq c(n+2)\|f\|_2$, and the proof is complete. ■

We are now ready to prove one of the basic results in this chapter, namely, the boundedness of Calderón's commutators of higher order, with an appropriate control on the norms.

Theorem 1.9. Suppose ϕ is a Lipschitz function on R, $\|\phi'\|_\infty \leq 1$, and for $f \in C_0^\infty(R)$ put

$$C_n f(x) = \text{p.v.} \int_R \frac{(\phi(x) - \phi(y))^n}{(x-y)^{n+1}} f(y)\, dy.$$

Then there is a constant c, independent of n, ϕ, and f, so that

$$\|C_n f\|_2 \leq c(n+1)^4 \|f\|_2. \tag{1.16}$$

Proof. Let $\psi = \phi'$. By Proposition 1.6 it suffices to prove (1.16) with p.v. $\int_R R_t(M_\psi R_t)^n / t f\, dt$ in the left-hand side there. To study this operator we expand first $R_t(M_\psi R_t)^n = (P_t - iQ_t)M_\psi(P_t - iQ_t) \cdots M_\psi(P_t - iQ_t)$ into 2^{n+1} terms of the form $T_0 M_\psi T_1 \cdots M_\psi T_n$, where T_j is either P_t or $-iQ_t$, $0 \leq j \leq n$. Among these terms there is exactly one with $T_j = P_t$ for all j, and since P_t is even its p.v. is 0. In addition there are $(n+1)$ terms where Q_t appears exactly once, and in the remaining terms Q_t appears at least twice.

We discuss first those terms where Q_t appears at least twice. There are n terms where $T_0 = -iQ_t$ and there are also those terms where $-iQ_t$ appears for the first time in place of T_k and for the last time in place of T_m, $1 \leq k < m \leq n$. Because both cases are handled in a similar fashion, we only discuss the latter situation here. The typical expression we consider, then, is given by the string

$$P_t(M_\psi P_t)^{k-1} M_\psi Q_t (M_\psi R_t)^{m-k-1} M_\psi Q_t (M_\psi P_t)^{n-m}. \tag{1.17}$$

First observe that the p.v. integral corresponding to (1.17) can be written as $\lim_{\varepsilon\to 0} \int_{\varepsilon<t<1/\varepsilon} + \lim_{\varepsilon \to 0} \int_{-1/\varepsilon<t<-\varepsilon}$, and that both integrals are essentially the same; we discuss only the first one. That the limit exists as $\varepsilon \to 0$ follows as in Proposition 1.3. To estimate the L^2 norm of the limit let $g \in L^2(R)$, $\|g\|_2 \leq 1$, and observe that

$$A = \int_R \int_{(\varepsilon,1/\varepsilon)} P_t(M_\psi P_t)^{k-1} M_\psi Q_t (M_\psi R_t)^{m-k-1} M_\psi Q_t (M_\psi P_t)^{n-m} f(x) \frac{dt}{t} g(x)\, dx$$

$$= \int_{(\varepsilon,1/\varepsilon)} \int_R Q_t M_{\bar\psi}(P_t M_{\bar\psi})^{k-1} P_t g(x) (M_\psi R_t)^{m-k-1} M_\psi Q_t (M_\psi P_t)^{n-m} f(x)\, dx \frac{dt}{t}.$$

Whence

$$|A| \leqslant \left(\int_{[0,\infty)} \int_R |Q_t(M_{\bar{\psi}} P_t)^k g(x)|^2 \, dx \frac{dt}{t} \right)^{1/2}$$
$$\times \left(\int_{[0,\infty)} \int_R |(M_\psi R_t)^{m-k-1} M_\psi Q_t (M_\psi P_t)^{n-m} f(x)|^2 \, dx \frac{dt}{t} \right)^{1/2}$$
$$= A_1 A_2,$$

say. By Proposition 1.8, $A_1 \leqslant c(k+1)\|g\|_2 \leqslant c(k+1)$. Also by estimate (1.4) and Proposition 1.8,

$$A_2 \leqslant \left(\int_R |G_{n-m} f(x)|^2 \, dx \right)^{1/2} \leqslant c(n-m+1)\|f\|_2.$$

Combining these bounds we can dominate the contribution of all terms of this form by

$$c\left(\sum_{1 \leqslant k < m \leqslant n} (k+1)(n-m+1) \right) \|f\|_2 \leqslant c\left(\sum_{m=1}^{n} (m+1)^2 (n-m+1) \right) \|f\|_2$$
$$\leqslant c(n+1)^4 \|f\|_2,$$

which is of the right order.

Finally, we consider those terms where Q_t appears exactly once. The integrand of a typical such term is $P_t(M_\psi P_t)^k M_\psi Q_t (M_\psi P_t)^{n-k-1} f$, for $0 \leqslant k \leqslant n-1$. Now, since Q_t is odd the p.v. in question can be replaced by $\lim_{\varepsilon \to 0} 2 \int_{\varepsilon < t < 1/\varepsilon}$. Also by Lemma 1.2, $Q_t = 8Q_t^3 - t(\partial/\partial t)S_t$, where $S_t = Q_t(I - 2P_t)$ and $I - 2P_t$ is bounded in $L^2(R)$ with norm $\sup_\xi |((t\xi)^2 - 1)/(1 + (t\xi)^2)| \leqslant 1$. Once again the fact that the limit exists as $\varepsilon \to 0$ follows as in Proposition 1.3. To bound the L^2 norm of the limit, fix $0 < \varepsilon < 1$ and rewrite the integral as

$$8 \int_{(\varepsilon, 1/\varepsilon)} P_t (M_\psi P_t)^k M_\psi Q_t^3 (M_\psi P_t)^{n-k-1} f \frac{dt}{t}$$
$$- \int_{(\varepsilon, 1/\varepsilon)} P_t (M_\psi P_t)^k M_\psi \frac{\partial}{\partial t} S_t (M_\psi P_t)^{n-k-1} f \, dt = A - B,$$

say. In A, Q_t appears at least twice and by the estimate obtained above for these terms it follows at once that

$$\|A\|_2 \leqslant c(k+1)(n+1-k)\|f\|_2. \tag{1.18}$$

To bound B we first integrate by parts in the t variable. This gives two terms (corresponding to the integrated part) with operator norm $\leqslant \frac{1}{2}$ each, and n terms where $(\partial/\partial t)S_t$ is replaced by S_t and one of the P_t's is replaced by $(\partial/\partial t)P_t = -2Q_t^2/t$. Therefore, Q_t appears at least twice in the string, once in place of T_{k+1} and once in place of $T_j, j \neq k+1$. Thus by an estimate quite similar to the one obtained above we get

$$\|B\|_2 \leqslant \|f\|_2 + c\left(\sum_{j=1}^{k}(j+1)(n-k+1) + \sum_{j=k+1}^{n}(k+1)(n-j+1)\right)\|f\|_2$$
$$\leqslant c((k+1)^2(n-k+1))\|f\|_2. \tag{1.19}$$

Finally, combining (1.18) and (1.19), and summing over k to cover all cases, we get that the contribution of these terms is of order $c(\sum_{k=1}^{n}(k+1)^2(n-k+1))\|f\|_2 \leqslant c(n+1)^4\|f\|_2$, as anticipated. The proof is thus complete. ∎

An important application of Theorem 1.9 is to the study of the Cauchy integral on a Lipschitz curve. The Cauchy kernel associated to a Lipschitz curve Γ is defined by $(1/2\pi i)1/(z(y)-z(x))$ where $z(x) = x + i\phi(x)$ is an arc length parametrization of Γ and ϕ is a real valued, Lipschitz function. For a function f defined on R we also denote by f the function induced on Γ by $f(z(x)) = f(x)$, and vice versa. The stage is now set for

Theorem 1.10 (Calderón). Let Γ be a curve in the complex plane given by $z(x) = x + i\phi(x)$, where ϕ is a real valued, Lipschitz function such that $|\phi(x)-\phi(y)| \leqslant M|x-y|$, all $x \neq y$ in R, and let

$$C_{\phi,\varepsilon}f(x) = \frac{1}{2\pi i}\int_{|x-y|>\varepsilon}\frac{f(y)}{z(y)-z(x)}\,dz(y), \qquad \varepsilon > 0. \tag{1.20}$$

Then for each function f in L^2, $\lim_{\varepsilon\to 0} C_{\phi,\varepsilon}f = C_\phi f$ also belongs to L^2 and there is a constant c, indpendent of ϕ and f, so that

$$\|C_\phi f\|_2 \leqslant c(1+M)^9\|f\|_2. \tag{1.21}$$

Furthermore, the mapping $f \to C_\phi^* f = \sup_\varepsilon|C_{\phi,\varepsilon}f|$ is of weak-type (1, 1) and bounded in L^p, $1 < p < \infty$, with constants which depend on M, but are otherwise independent of ϕ. Also $\lim_{\varepsilon\to 0} C_{\phi,\varepsilon}f$ exists pointwise a.e. for $f \in L^p(R)$, $1 \leqslant p < \infty$.

Proof. We consider first the case when ϕ is infinitely differentiable and has compact support and show that the estimate (1.21) holds. Once this is established the desired result follows by applying what are by now well-known limiting techniques. We take first f in $C_0^\infty(R)$ and observe that the integral in (1.20) converges at x if $f(x) = 0$ or f is constant near x and $\phi'(x)$ exists. From this we conclude that the limit exists everywhere. A simple change of variables gives that this limit actually is

$$\text{p.v.} -\frac{1}{\pi i}\int_R \frac{(1+i\phi'(y))}{(x-y)+i(\phi(x)-\phi(y))}f(y)\,dy. \tag{1.22}$$

Suppose next that $M = \eta < 1$. By scaling we note that the estimate (1.16) in Theorem 1.9 may be rewritten

$$\|C_n f\|_2 \leq c(n+1)^4 \eta^n \|f\|_2. \tag{1.23}$$

Whence

$$\left\|\sum_{n=0}^{\infty}(-1)^n C_n f\right\|_2 \leq c\left(\sum_{n=0}^{\infty}(n+1)^4\eta^n\right)\|f\|_2 \leq c(1-\eta)^{-5}\|f\|_2,$$

or in other words, the singular integral operator with kernel $1/((x-y)+(\phi(x)-\phi(y)))$ is bounded in L^2 with norm $\leq c(1-\eta)^{-5}$, c independent of ϕ.

With all the preliminaries out of the way we are ready to prove (1.21); for this purpose put $\psi(x) = (-M^2x + i\phi(x))(1+M^2)^{-1}$ and observe that $\|\psi'\|_\infty^2 = \|-M^2 + i\phi'\|_\infty^2(1+M^2)^{-2} \leq (M^4+M^2)(1+M^2)^{-2} = 1-(1+M^2)^{-1} = \eta^2 < 1$. Now, since $x + \psi(x) = (x + i\phi(x))(1+M^2)^{-1}$, the kernel of (1.22) can be written as $(1+i\phi'(y))/(x-y+\psi(x)-\psi(y))(1+M^2)$ and consequently by the first part of the argument, the norm of the function in (1.22), and also that of $C_\phi f$, is bounded by $\|1+i\phi'\|_\infty(1+M^2)^{-1}c(1-\eta)^{-5}\|f\|_2$. Furthermore, since $1-\eta^2 = (1-\eta)(1+\eta) = (1+M^2)^{-1}$ and $\|1+i\phi'\|_\infty \leq (1+M^2)^{1/2}$, this expression $\leq c(1+M^2)^{5-1/2}\|f\|_2 \leq c(1+M)^9\|f\|_2$, and (1.21) holds.

Next we show that C_ϕ is bounded in L^p for $1 < p < \infty$; because the techniques we use are familiar we merely outline the arguments. First recall that $\|f/z'\|_p \leq \|f\|_p \leq (1+M^2)^{1/2}\|f/z'\|_p$, $0 < p < \infty$, and consequently (1.21) is equivalent to $\|C_\phi(f/z')\|_2 \leq c\|f\|_2$, where c is a constant which depends only on M. Let now f be a function with vanishing integral, supported in an interval I centered at x_I. From (1.22) it readily follows that $|C_\phi(f/z')(x)| \leq c|x-x_I|^{-2}\|f\|_1$ whenever $x \in R\backslash 2I$, and $\int_{R\backslash 2I}|C_\phi(f/z')(x)|\,dx \leq c\|f\|_1$, where c depends only on M. By Theorem 1.1 in Chapter XI we conclude that

$$\|C_\phi(f/z')\|_p \leq c\|f\|_p, \qquad 1 < p < 2, \tag{1.24}$$

with a constant c which depends only on M and p; we also obtain that $C_\phi(f/z') \in wk - L(R)$ when f is integrable. By duality (with respect to $dz(y)$), (1.24) also holds for $2 < p < \infty$, and C_ϕ is bounded in L^p, $1 < p < \infty$. To prove the same conclusion for C_ϕ^*, it clearly suffices to show that for an arbitrary, positive, measurable function $\varepsilon(x)$, the mapping

$$C_{\phi,\varepsilon(x)}\left(\frac{f}{z'}\right)(x) = \int_{|x-y|>\varepsilon(x)} \frac{f(y)}{z(y)-z(x)}\,dx,$$

is of weak-type (1, 1) and type (p,p) for $1 < p < \infty$. For this purpose let $\eta(y)$ be an even, nonnegative, $C_0^\infty(R)$ function which is 1 for $|y| \leq \frac{1}{2}$ and vanishes for $|y| > 1$, and put

$$\delta(x) = \int_R \eta\left(\frac{(x-y)}{\varepsilon(x)}\right) dz(y).$$

It is easy to see that $|\delta(x)| \sim \varepsilon(x)$. Furthermore, let

$$T_1 f(x) = \int_R \delta(x)^{-1} \eta\left(\frac{(x-y)}{\varepsilon(x)}\right) C_\phi\left(\frac{f}{z'}\right)(y)\, dz(y).$$

$T_1 f$ is a smoothing of $C_\phi(f/z')$ and verifies the same integrability properties, to wit,

$$\|T_1 f\|_p \leqslant c\|f\|_p, \qquad 1 < p < \infty, \tag{1.25}$$

where c depends only on M and p. To see this note that

$$|T_1 f(x)| \leqslant c \sup_\varepsilon \frac{1}{\varepsilon} \int_{|x-y|<\varepsilon} |C_\phi(f/z')(y)|\, dy \leqslant cM(C_\phi(f/z'))(x),$$

which gives (1.25) immediately. This operator is important in our context since it satisfies

$$|C_{\phi,\varepsilon(x)}(f/z')(x) - T_1 f(x)| \leqslant cMf(x), \qquad x \in R. \tag{1.26}$$

(1.26) implies, in particular, that $C_{\phi,\varepsilon(x)}(f/z')(x)$ is bounded in $L^p(R)$, $1 < p < \infty$. Now, to check (1.26) observe that

$$\begin{aligned} C_{\phi,\varepsilon(x)}(f/z')(x) - T_1 f(x) &= \int_R f(y) \int_R \delta(x)^{-1} \eta\left(\frac{(x-u)}{\varepsilon(x)}\right) \\ &\quad \times \left(\frac{\chi(|x-y| > \varepsilon(x))}{z(y)-z(x)} - \frac{1}{z(y)-z(u)}\right) dz(u)\, dy \\ &= \int_R f(y) k(x,y)\, dy, \end{aligned}$$

say. We claim that

$$|k(x,y)| \leqslant c\varepsilon(x)/(\varepsilon(x)^2 + (x-y)^2), \tag{1.27}$$

and in this case (1.26) follows. To estimate $|(k, x, y)|$ we distinguish three cases, namely,

(i) $|x-y| \geqslant 2\varepsilon(x)$,
(ii) $\varepsilon(x) < |x-y| < 2\varepsilon(x)$, and
(iii) $|x-y| \leqslant \varepsilon(x)$.

In case (i) we have $\chi(|x-y|>\varepsilon(x))=1$ and $|y-u|\geqslant|x-y|/2$ for those u's for which the integrand in the integral defining k does not vanish. Whence

$$|k(x,y)|\leqslant c\frac{1}{\varepsilon(x)}\int_R \eta\left(\frac{(x-u)}{\varepsilon(x)}\right)\frac{|z(x)-z(u)|}{|z(y)-z(x)|\,|z(y)-z(u)|}\,du$$

$$\leqslant c\frac{1}{\varepsilon(x)}\int_R \eta\left(\frac{(x-u)}{\varepsilon(x)}\right)\frac{|x-u|}{|y-x|\,|y-u|}\,du\leqslant\frac{c\varepsilon(x)}{(x-y)^2},$$

which is of the right order. If, on the other hand, case (ii) holds, we still have $\chi(|x-y|>\varepsilon(x))=1$, and

$$\begin{aligned}k(x,y)&=\frac{1}{z(y)-z(x)}\\&\quad-\int_{|x-u|<2\varepsilon(x)}\delta(x)^{-1}\left[\eta\left(\frac{(x-u)}{\varepsilon(x)}\right)-\eta\left(\frac{(x-y)}{\varepsilon(x)}\right)\right]\frac{1}{z(y)-z(u)}\,dz(u)\\&\quad-\delta(x)^{-1}\eta\left(\frac{(x-y)}{\varepsilon(x)}\right)\lim_{a\to0}\int_{a\leqslant|x-u|<2\varepsilon(x)}\frac{1}{z(y)-z(u)}\,dz(u)\\&=A-B-C,\end{aligned}\tag{1.28}$$

say. Clearly $|A|=O(|x-y|^{-1})$, which is of the right order. Also $C=0$ since $\eta((x-y)/\varepsilon(x))=0$. As for $|B|$, it is dominated by

$$\frac{c}{\varepsilon(x)}\int_{|x-u|<2\varepsilon(x)}\frac{|u-y|}{\varepsilon(x)}\|\eta'\|_\infty\frac{1}{|y-u|}\,du\leqslant c\varepsilon(x)^{-1}\leqslant c|x-y|^{-1},$$

and this term is also of the right order.

Finally, we consider case (iii). Since $|x-y|\leqslant\varepsilon(x)$ we have, with the notation of (1.28), that $A=0$. The estimate for B is carried out as in case (ii) except that now we dominate $c\varepsilon(x)^{-1}\leqslant c\varepsilon(x)/((x-y)^2+\varepsilon(x)^2)$, which is of the right order. Only C is left. Since the factor in front of the p.v. integral is $\leqslant c\varepsilon(x)^{-1}$, it suffices to show that the p.v. integral itself is bounded. Let $\ln w$ denote the logarithm function, analytic in the complex w plane with the slit $\{w=iy\colon y\leqslant 0\}$ removed. Then for $0<a<2\varepsilon(x)$,

$$\begin{aligned}&-\int_{a\leqslant|x-u|<2\varepsilon(x)}\frac{1}{z(y)-z(u)}\,dz(u)\\&\quad=\int_{[x+a,x+2\varepsilon(x))}d\ln(z(y)-z(u))\\&\quad\quad+\int_{(x-2\varepsilon(x),x-a]}d\ln(z(y)-z(u))\\&\quad=\ln(z(y)-z(u))]_{x+a}^{x+2\varepsilon(x)}+\ln(z(y)-z(u))]_{x-2\varepsilon(x)}^{x-a}\\&\quad=\ln\left(\frac{z(y)-z(x+2\varepsilon(x))}{z(y)-z(x-2\varepsilon(x))}\right)+\ln\left(\frac{z(y)-z(x-a)}{z(y)-z(x+a)}\right).\end{aligned}$$

Whence the p.v. integral in question is bounded by

$$\left|\ln\left(\frac{z(y) - z(x + 2\varepsilon(x))}{z(y) - z(x - 2\varepsilon(x))}\right)\right| \leq c,$$

since $|x - y| \leq \varepsilon(x)$. Thus (1.27), and consequently also (1.26), hold, and the discussion of the boundedness in $L^p(R)$, $1 < p < \infty$, is complete. The *wk*-type (1, 1) estimate requires a further argument. For this purpose let

$$T_2 f(x) = \int_R \left(1 - \eta\left(\frac{(x - y)}{\varepsilon(x)}\right)\frac{f(y)}{z(y) - z(x)}\right) dy.$$

Then the following inequalities are readily verified:

$$\int_{R\setminus 2I} |T_2 f(x)|\, dx \leq c\|f\|_1 \tag{1.29}$$

provided that f is a function with vanishing integral supported in the interval I, and

$$|C_{\phi,\varepsilon(x)}(f/z')(x) - T_2 f(x)| \leq cMf(x). \tag{1.30}$$

On account of (1.30) and (1.26) we obtain at once that T_2 is bounded in $L^p(R)$, $1 < p < \infty$. This in turn, together with the estimate (1.29), implies that the assumptions of Theorem 1.1 in Chapter XI hold, and consquently T_2 is of wk-type (1, 1). That $C_{\phi,\varepsilon(x)}(f/z')(x)$ is also of wk-type (1, 1) follows immediately from estimate (1.30).

So far we have assumed that $\phi \in C_0^\infty(R)$, but since all preceding estimates depend on M only and are otherwise independent of ϕ, a passage to the limit shows that the same results and estimates hold for operators involving general functions ϕ with $\|\phi'\|_\infty < \infty$. Finally, to prove the pointwise existence of $\lim_{\varepsilon\to 0} C_{\phi,\varepsilon}f$, we recall that this limit exists a.e. for $f \in C_0^\infty(R)$; for general f the same conclusion follows from the fact that C_ϕ^* is of wk-type (1, 1) and type (p, p), $1 < p < \infty$. ■

2. RELATED OPERATORS

As deep and interesting as the results discussed in the previous section are, their importance lies in their extensions and applications. We discuss the application to the solution of the Dirichlet and Neumann problems in C^1 domains in the next chapter. As for the extensions an appropriate place to start is

Theorem 2.1 (Coifman-McIntosh-Meyer). Suppose $A_1, \ldots, A_n$ are n

Lipschitz functions on R, and let $k_n(x, y)$ denote the singular kernel

$$k_n(x, y) = \frac{(A_1(x) - A_1(y)) \cdots (A_n(x) - A_n(y))}{(x - y)^{n+1}}, \qquad x \neq y \quad \text{in} \quad R.$$

Then the operator T_n defined by $T_n f(x) = \text{p.v.} \int_R k_n(x, y) f(y)\, dy$ is bounded in $L^2(R)$ and it verifies

$$\| T_n f \|_2 \leq c \| A'_\infty \| \cdots \| A'_n \| (n + 1)^4 \| f \|_2, \tag{2.1}$$

where c is an absolute constant independent of n, the A_j's, and f. The proof follows along the lines of Theorem 1.9 with minor changes. First there is the representation formula, it now reads: suppose $a_j = A'_j$, $1 \leq j \leq n$, and put

$$L_n(a_1, \ldots, a_n) = \text{p.v.} \int_R R_t M_{a_1} R_t \cdots M_{a_n} R_t \frac{dt}{t}.$$

Then

$$T_n = \frac{1}{n!} \sum L_n(a_{\sigma(1)}, \ldots, a_{\sigma(n)}),$$

where the sum is extended over all $\sigma \in S_n$, the permutation group on the set of n elements. Next there are the relevant Littlewood-Paley functions, namely,

$$G_n f(x) = \left(\int_{[0,\infty)} |Q_t M_{a_1} P_t \cdots M_{a_n} P_t f(x)|^2 \frac{dt}{t} \right)^{1/2}.$$

The reader should have no difficulty in completing the details. An interesting application of this result is

Theorem 2.2. Let $F(z) = \sum_{n=0}^{\infty} c_n z^n$ be an analytic function in the disk $|z| < \eta$ and let A be a (complex valued) Lipschitz function on R with $\|A'\|_\infty < \eta$. Let B denote another Lipschitz function on R and let $N(x, y)$ denote the singular kernel

$$N(x, y) = \frac{B(x) - B(y)}{(x - y)^2} F\left(\frac{A(x) - A(y)}{x - y} \right). \tag{2.2}$$

Then the operator T defined by $Tf(x) = \text{p.v.} \int_R N(x, y) f(y)\, dy$ is well defined a.e., is bounded in $L^2(R)$, and it satisfies

$$\| Tf \|_2 \leq c \| B' \|_\infty \| f \|_2, \tag{2.3}$$

where c is a constant which depends on $\|A'\|_\infty$ and F but is otherwise independent of A, B, and f.

Proof. First observe that by the previous theorem the operator

$$\tau_n f(x) = \text{p.v.} \int_R \frac{B(x) - B(y)}{(x-y)^2} \left(\frac{A(x) - A(y)}{x - y} \right)^n f(y)\, dy$$

is bounded in L^2 with norm $\leq c\|B'\|_\infty \|A'\|_\infty^n (n+1)^4$. Furthermore, since $\|A'\|_\infty < \eta$ also $\sum_{n=0}^{\infty} \|A'\|_\infty^n (n+1)^4 |c_n|$ converges, and consequently $Tf(x) = \sum_{n=0}^{\infty} c_n \tau_n f(x)$ is given by a convergent sum in L^2 whenever $f \in L^2$, and verifies (2.3). From the fact that the above sum converges in L^2 it follows that for some sequence $n_k \to \infty$,

$$\lim_{k \to \infty} \sum_{n=0}^{n_k} c_n \tau_n f(x) = Tf(x) \quad \text{exists a.e.,}$$

and the proof is complete. ■

We also have

Theorem 2.3. With $N(x, y)$ as in (2.2), let

$$T^* f(x) = \sup_{\varepsilon > 0} \left| \int_{|x-y| > \varepsilon} N(x, y) f(y)\, dy \right|.$$

Then T^* is bounded in $L^p(R)$, $1 < p < \infty$, and its norm depends only on $\|A'\|_\infty$ and $\|B'\|_\infty$.

The proof follows essentially along the lines of Theorem 1.10 and is therefore omitted.

Remark 2.4. The dependence of the norm of T^* on $\|A'\|_\infty$ and $\|B'\|_\infty$ is important in applications. From estimate (2.3) and the proof of the Benedek-Calderón-Panzone principle, it follows that the norm goes to 0 with $\|B'\|_\infty$. Also, if $F(0) = 0$ it is not hard to see that the norm goes to 0 with $\|A'\|_\infty$.

We also point out an n-dimensional result.

Theorem 2.5. Let A, B be (complex valued) Lipschitz functions on R^n, and let $M(x, y)$ denote the singular kernel

$$M(x, y) = \frac{B(x) - B(y)}{(|x-y|^2 + (A(x) - A(y))^2)^{(n+1)/2}}, \qquad x \neq y \quad \text{in} \quad R^n.$$

Furthermore, let $T_\varepsilon f(x) = \int_{|x-y|>\varepsilon} M(x, y) f(y)\, dy$, and put $T^* f(x) = \sup_{\varepsilon>0} |T_\varepsilon f(x)|$. Then T^* is bounded in $L^p(R^n)$, $1 < p < \infty$, and satisfies $\|T^* f\|_p \leq c\|f\|_p$, where c depends on the Lipschitz constants of A and B and p but is otherwise independent of A, B, and f.

Proof. We invoke the method of rotations, Theorem 3.1 in Chapter X. First observe that

$$T_\varepsilon f(x) = \frac{1}{2}\int_{|y|>\varepsilon} (M(x, x+y)f(x+y) + M(x, x-y)f(x-y))\, dy,$$

which in polar coordinates becomes

$$\begin{aligned}\frac{1}{2}\int_\Sigma \int_{(\varepsilon,\infty)} &(M(x, x+uy')f(x+uy')\\ &+ M(x, x-uy')f(x-uy'))u^{n-1}\, du\, dy'\\ &= \frac{1}{2}\int_\Sigma T_{\varepsilon,y'}f(x)\, dy',\end{aligned}$$

say. Write now x in R^n (uniquely) as $x = w + ty'$, where y' is a fixed vector in Σ, $t \in R$ and $w \in Y$, the hyperplane orthogonal to y' which passes through the origin. With this notation we have

$$\begin{aligned}T_{\varepsilon,y'}f(x) = T_{\varepsilon,y'}f(w+ty') = \int_{|u-t|>\varepsilon} &M(w+ty', w+uy')|t\\ &- u|^{n-1}f(w+uy')\, du.\end{aligned}$$

It is clear that in our case

$$\begin{aligned}&M(w+ty', w+uy')|t-u|^{n-1}\\ &\quad= \frac{B(w+ty') - B(w+uy')}{(t-u)^2}\left(1 + \left(\frac{A(w+ty') - A(w+uy')}{t-u}\right)^2\right)^{-(n+1)/2},\end{aligned}$$

which for each fixed $y' \in \Sigma$ and $w \in R^n$ is one of the operators covered by Theorem 2.3 with $F(z) = 1/(1+z^2)^{(n+1)/2}$ there. By that result it follows that for $1 < p < \infty$,

$$\left(\int_R \left(\sup_{\varepsilon>0}|T_{\varepsilon,y'}f(w+ty')|\right)^p dt\right) \leqslant c\left(\int_R |f(w+ty')|^p\, dt\right)^{1/p},$$

where c is a constant which depends solely on the Lipschitz norm of A and B, and p. Thus, by Minkowski's inequality and Fubini's theorem, we finally get

$$\begin{aligned}\|T^*f\|_p &\leqslant \frac{1}{2}\int_\Sigma \left\|\sup_{\varepsilon>0}|T_{\varepsilon,y'}f(w+ty')|\right\|_p dy'\\ &= \frac{1}{2}\int_\Sigma \left(\int_Y \int_R \left(\sup_{\varepsilon>0}|T_{\varepsilon,y'}f(w+ty')|\right)^p dt\, dw\right)^{1/p} dy'\\ &\leqslant c\int_\Sigma \left(\int_Y \int_R |f(w+ty')|^p\, dt\, dw\right)^{1/p} dy' = c\|f\|_p. \quad \blacksquare\end{aligned}$$

3. THE $T1$ THEOREM

For many operators in analysis an important question is to decide whether they are bounded in L^2. In this section we discuss a simple criteria for this to occur.

Suppose T is a linear operator, which is continuous from the Schwartz class $\mathcal{S}(R^n)$ into $\mathcal{S}'(R^n)$. As in 8.12 in Chapter XI, we assume that there are a kernel $k(x, y)$ defined for $x \neq y$ in R^n and constants c and $0 < \delta \leqslant 1$ such that the following three properties hold:

(i) $|k(x, y)| \leqslant c|x - y|^{-n}$,

(ii) for all x_0, x, y in R^n such that $|x_0 - x| < |x - y|/2$, $|k(x_0, y) - k(x, y)| + |k(y, x_0) - k(y, x)| \leqslant c|x_0 - x|^{\delta}|x - y|^{-(n+\delta)}$,

(iii) for each pair f, ϕ, of disjointly supported, $C_0^\infty(R^n)$ functions, the evaluation of the distribution Tf on the test function ϕ is given by $Tf(\phi) = \int_{R^n}\int_{R^n} k(x, y)f(y)\phi(x)\,dy\,dx$.

As in Chapter XI, T is called a Calderón–Zygmund operator if it can be extended to a bounded operator in $L^2(R^n)$.

It is clear that the adjoint T^* of T, defined by $T^*f(\phi) = T\phi(f)$, is associated with a kernel $h(x, y)$ which verifies the same properties as k; in fact, $h(x, y) = k(y, x)$.

Observe that it is possible to define $T1$, the image of the function identically 1 under T; $T1$ will be a distribution on those test functions in $C_0^\infty(R^n)$ with vanishing integral. In fact, let $f \in L^\infty(R^n) \cap C^\infty(R^n)$ and $\phi \in C_0^\infty(R^n)$ have integral 0; we want to define $Tf(\phi)$. Let f_1 be a $C_0^\infty(R^n)$ function which coincides with f on the support of ϕ, and put $f_2 = f - f_1$. $Tf_2(\phi)$ is well defined, and in analogy with (iii) above we give a meaning to $Tf_1(\phi) = \int_{R^n}\int_{R^n} k(x, y)f_1(y)\phi(x)\,dy\,dx$. Let $x_0 \in \operatorname{supp}\phi$, and note that since ϕ has integral 0, by integrating first with respect to x we obtain

$$\left|\int_{R^n} k(x, y)\phi(x)\,dx\right| \leqslant \int_{R^n} |k(x, y) - k(x_0, y)||\phi(x)|\,dx$$
$$\leqslant c(1 + |x_0 - y|)^{-(n+\delta)},$$

where c depends on ϕ. Whence

$$\int_{R^n}\left|f_1(y)\int_{R^n} k(x, y)\phi(x)\,dx\right|\,dy \leqslant c\|f\|_\infty \int_{R^n} (1 + |x_0 - y|)^{-(n+\delta)}\,dy < \infty,$$

and $Tf_1(\phi)$ makes sense. Since it is clear that $Tf_1(\phi) + Tf_2(\phi)$ is independent of the choice of the decomposition $f_1 + f_2$ of f, we define $Tf(\phi)$ to be this value.

In order to state the desired result we need a couple of observations. $T1 \in BMO$ means that for all $\phi \in C_0^\infty(R^n)$ with integral 0 we have $|T1(\phi)| \leq c\|\phi\|_{H^1}$, where the constant c is independent of ϕ. Also if for a function ψ on R^n we let $\psi_t^z(x) = t^{-n}\psi((x-z)/t)$, then we say that T has the weak boundedness property if for any bounded set $B \subset C_0^\infty(R^n)$ there exists a constant c which depends only on B so that for all ϕ, ψ in B, x in R^n and $t > 0$, $|T\psi_t^x(\phi_t^x)| \leq ct^{-n}$.

We are now ready for

Theorem 3.1 (David–Journé). Let T be a linear operator which is continuous from $\mathcal{S}(R^n)$ into $\mathcal{S}'(R^n)$, and assume that it verifies properties (i)–(iii). Then T can be extended to a Calderón–Zygmund operator (bounded in $L^2(R^n)$) if and only if the following three conditions are satisfied:

$$T1 \in \mathrm{BMO} \tag{3.1}$$

$$T^*1 \in \mathrm{BMO} \tag{3.2}$$

$$T \text{ has the weak boundedness property} \tag{3.3}$$

Proof. That the conditions (3.1) and (3.2) are necessary follows immediately from 8.14 and 8.15 in Chapter XI and (3.3) follows from Hölder's inequality. To show that the conditions are also sufficient we first describe a (weak) representation formula for Tf for smooth f's, and then proceed along the lines of Proposition 1.5.

Let ϕ be an even, nonnegative, $C_0^\infty(R^n)$ function with integral 1, and to conform with the usual notation, let P_t denote the operator convolution with ϕ_t, i.e., $P_tf(x) = f * \phi_t(x)$. Let $f \in C_0^\infty(R^n)$ and observe that since

$$\frac{\partial}{\partial t}(P_t^2 TP_t^2) = \left(\frac{\partial}{\partial t}P_t^2\right) TP_t^2 + P_t^2 T\left(\frac{\partial}{\partial t}P_t^2\right),$$

then

$$\begin{aligned} Tf &= -\lim_{\varepsilon \to 0} P_t^2 TP_t^2 f]_\varepsilon^{1/\varepsilon} = -\lim_{\varepsilon\to 0}\int_{(\varepsilon,1/\varepsilon)} t\frac{\partial}{\partial t}(P_t^2TP_t^2f)\frac{dt}{t} \\ &= -\lim_{\varepsilon\to 0}\int_{(\varepsilon,1/\varepsilon)} t\left(\frac{\partial}{\partial t}P_t^2\right)TP_t^2f\frac{dt}{t} \\ &\quad -\lim_{\varepsilon\to 0}\int_{(\varepsilon,1/\varepsilon)} P_t^2Tt\left(\frac{\partial}{\partial t}P_t^2\right)f\frac{dt}{t} = A - B, \end{aligned}$$

say. This is the representation formula alluded to above. Since the expressions A and B correspond to operators which are the adjoint of one another, it suffices to estimate one of them, A say. We begin by taking a closer look at $t((\partial/\partial t)P_t^2)$. First observe that

$$\left(t\left(\frac{\partial}{\partial t}P_t^2\right)g\right)\hat{}(\xi) = t\frac{\partial}{\partial t}(\hat{\phi}(t\xi)^2)\hat{g}(\xi)$$

$$= 2\sum_{j=1}^{n}\frac{\partial}{\partial \xi_j}\hat{\phi}(t\xi)t\xi_j\hat{\phi}(t\xi)\hat{g}(\xi)$$

$$= 2\left(\sum_{j=1}^{n}\hat{\psi}_j(t\xi)\hat{\eta}_j(t\xi)\right)\hat{g}(\xi),$$

say. Note that according to our assumptions on ϕ, $\hat{\psi}_j(0) = \hat{\eta}_j(0) = 0$, $1 \leq j \leq n$. Thus if ψ and η denote the vectors $(\psi_1, \ldots, \psi_n)$ and $(\eta_1, \ldots, \eta_n)$, respectively, and $Q_{\psi,t}$ and $Q_{\eta,t}$ denote the vector valued operators defined by $(Q_{\psi,t}g)\hat{}(\xi) = \hat{g}(\xi)\hat{\psi}(t\xi)$, $(Q_{\eta,t}g)\hat{}(\xi) = g(\xi)\hat{\eta}(t\xi)$, then $t((\partial/\partial t)P_t^2)g = 2Q_{\psi,t}\cdot Q_{\eta,t}g$. Returning to our representation, and with the notation $M_t = Q_{\eta,t}TP_t$, we study the operator given by

$$\lim_{\varepsilon\to 0}\int_{(\varepsilon,1/\varepsilon)} Q_{\psi,t}\cdot M_tP_tf\frac{dt}{t}. \tag{3.4}$$

To estimate (3.4) it is convenient to have the integral representation of M_tg at hand; as is easily seen it is given by $\int_{R^n} K(x, y, t)g(y)\,dy$, where $K(x, y, t)$ is the n-vector with components $\mu_j(x, y, t) = T^*(\eta_j)_t^x(\phi_t^y)$, $1 \leq j \leq n$. Now, from properties (ii) and (3.3) it readily follows that for some $0 < \delta' < \delta$ and $1 \leq j \leq n$,

$$|\mu_j(x, y, t)| \leq ct^{-n}\left(1 + \frac{|x-y|}{t}\right)^{-(n+\delta')}, \tag{3.5}$$

where c is an absolute constant which depends only on ϕ and η. Estimate (3.5) allows us to compute M_t1; indeed it is the n-vector given by the absolutely convergent integrals

$$\int_{R^n}\mu_j(x, y, t)\,dy = T^*(\eta_j)_t^x(1) = T1((\eta_j)_t^x)$$

$$= (T1) * (\eta_j)_t(x), \qquad 1 \leq j \leq n. \tag{3.6}$$

In particular, since by (3.1) $T1 \in BMO$ and the η_j's are Littlewood–Paley functions, $|M_t1|^2(1/t)\,dx\,dt$ is a Carleson measure with constant $\leq c\|T1\|_*$. This is all we need to know about M_t.

Inspired by Lemma 1.4, we put $M_tP_tf = (P_tf)M_t1 + (M_tP_tf - (P_tf)M_t1)$ and rewrite (3.4) as

$$\lim_{\varepsilon\to 0}\int_{(\varepsilon,1/\varepsilon)} Q_{\psi,t}\cdot (P_tf)M_t1f\frac{dt}{t}$$
$$+\lim_{\varepsilon\to 0}\int_{(\varepsilon,1/\varepsilon)} Q_{\psi,t}\cdot (M_tP_tf-(P_tf)M_t1)f\frac{dt}{t} = A_1 + A_2,$$

say.

To estimate A_1 observe that, as in Proposition 1.3, it is enough to show that

$$\left(\int_{R_+^{n+1}}|Q_{\varepsilon,t}\cdot (P_tf)(x)M_t1(x)|^2\,dx\frac{dt}{t}\right)^{1/2}\leqslant c\|f\|_2,$$

where c is a constant independent of f. Let then $g\in C_0^\infty(R^n)$, $\|g\|_2\leqslant 1$, and bound

$$\int_{[0,\infty)}\left|\int_{R^n} Q_{\psi,t}\cdot (P_tf)(x)M_t1(x)g(x)\,dx\right|\frac{dt}{t}$$
$$=\int_{[0,\infty)}\left|\int_{R^n} P_tf(x)M_t1(x)\cdot Q_{\psi,t}g(x)\,dx\right|\frac{dt}{t}$$
$$\leqslant\left(\int_{R^{n+1}}|P_tf(x)|^2|M_t1(x)|^2\,dx\frac{dt}{t}\right)^{1/2}\left(\int_{R^{n+1}}|Q_{\psi,t}g(x)|^2\,dx\frac{dt}{t}\right)^{1/2}$$
$$= A_3\cdot A_4,$$

say. That $A_4\leqslant c$ follows at once from the Littlewood-Paley theory. To bound A_3 we invoke Theorem 1.1 in Chapter XV and note that it does not exceed $c\|T1\|_*\|N(P_tf)\|_2\leqslant c\|T1\|_*\|f\|_2$. This implies that the limit defining A_1 exists and that its L^2 norm is less than of equal to $c\|f\|_2$, which is an estimate of the right order.

Finally to bound A_2, once again we let $g\in C_0^\infty(R^n)$, $\|g\|_2\leqslant 1$, and observe that

$$\int_{[0,\infty)}\left|\int_{R^n} Q_{\psi,t}\cdot (M_tP_tf-(P_tf)M_t1)(x)g(x)\,dx\right|\frac{dt}{t}$$
$$\leqslant\left(\int_{R_+^{n+1}}|Q_{\psi,t}g(x)|^2\,dx\frac{dt}{t}\right)^{1/2}$$
$$=\left(\int_{R_+^{n+1}}|M_tP_tf(x)-(P_tf)(x)M_t1(x)|^2\,dx\frac{dt}{t}\right)^{1/2} = A_5A_6,$$

say. As before, $A_5\leqslant c$. To estimate A_6 it clearly suffices to bound each of the integrals

$$\left(\int_{R_+^{n+1}}\left|\int_{R^n}\mu_j(x,y,t)(P_tf(y)-P_tf(x))\,dy\right|^2dx\frac{dt}{t}\right)^{1/2},\qquad 1\leqslant j\leqslant n.$$

But this is not hard; indeed, from Hölder's inequality and (3.5) it readily follows that for each j the above expression does not exceed

$$\left(\int_{R_+^{n+1}}\left(\int_{R^n}|\mu_j(x,y,t)|\,dy\right)\left(\int_{R^n}|\mu_j(x,y,t)||P_tf(y)-P_tf(x)|^2\,dy\right)dx\frac{dt}{t}\right)^{1/2}$$
$$\leqslant c\left(\int_{R_+^{n+1}}\int_{R^n}t^{-n}\left(1+\frac{|y|}{t}\right)^{-(n+\delta')}|P_tf(x-y)-P_tf(x)|^2\,dy\,dx\frac{dt}{t}\right).$$

Furthermore, since

$$\int_{R^n}|P_tf(x-y)-P_tf(x)|^2\,dx=\int_{R^n}|\hat{\phi}(t\xi)|^2|e^{-i\xi\cdot y}-1|^2|\hat{f}(\xi)|^2\,d\xi$$
$$\leqslant c\int_{R^n}|\phi(t\xi)|^2(|\xi||y|)^\alpha|\hat{f}(\xi)|^2\,d\xi,\qquad 0<\alpha\leqslant 2,$$

we get that if $0<\alpha<\delta'$,

$$A_6\leqslant c\left(\int_{R^n}|\hat{f}(\xi)|^2\int_{[0,\infty)}|\hat{\phi}(t\xi)|^2(t|\xi|)^\alpha\right.$$
$$\left.\times\int_{R^n}t^{-n}\left(\frac{|y|}{t}\right)^\alpha\left(1+\frac{|y|}{t}\right)^{-(n+\delta')}dy\frac{dt}{t}\,d\xi\right)^{1/2}\leqslant c\|f\|_2.$$

This gives immediately that the limit defining A_2 exists and that its L^2 norm $\leqslant c\|f\|_2$, which is precisely what we wanted to show. ■

4. NOTES; FURTHER RESULTS AND PROBLEMS

The topics discussed in this chapter have their origin in the theory of linear partial differential equations. As Calderón [1978] explains it, the question is one of constructing an algebra, under composition, of differential, or more generally, pseudo-differential, operators. The problem of proving that the composition of two such operators is another operator of the same kind can be reduced to the following problem: let M_a denote, in the one variable case, the operator multiplication by the Lipschitz function a, and show that HM_a is an operator of the same type. Since $HM_a = M_aH + [H, M_a]$, it is sufficient to show that $[H, M_a]D$ is bounded in L^p, $1 < p < \infty$. This Calderón did in 1965 with the aid of the theory of analytic functions. The idea goes as follows: without great difficulty the problem reduces to showing that $[M_a, HD]$ is bounded in L^p, and this operator can be represented by

$$\text{p.v.}\int_R\frac{-1}{(x-y)}\left(\frac{a(x)-a(y)}{x-y}\right)f(y)\,dy,$$

that is, the so-called first commutator. This integral, as well as that representing the higher order commutators, are special cases of

$$\text{p.v.} \int_R \frac{1}{x-y} F\left(\frac{a(x)-a(y)}{x-y}\right) f(y)\, dy,$$

where F is analytic in a neighborhood of $|z| \leqslant \sup(|a(x)-a(y)|/|x-y|)$. Several classical integrals, including the Cauchy integral along a curve Γ, are also special cases of the integral. After a change of variables, and with $z(\lambda) = x + i\lambda a(x)$ and $w(\lambda) = y + i\lambda a(y)$, we are reduced to consider

$$A(\lambda)f = \text{p.v.} \int_R \frac{f(y)}{z(\lambda) - w(\lambda)}\, dy.$$

For $\lambda = 0$ this operator coincides with H. Also differentiating with respect to λ, we get the operator

$$B(\lambda)f = \text{p.v.} \int_R \frac{-i}{(z(\lambda) - w(\lambda))} \left(\frac{a(x)-a(y)}{z(\lambda) - w(\lambda)}\right) f(y)\, dy,$$

whose analogy with Calderón's first commutator is clear. Calderón [1977a] succeeded in using the mehods of the first commutator together with a weighted L^2 estimate for the Lusin function and obtained $(d/d\lambda)\|A(\lambda)\| \leqslant \|B(\lambda)\| \leqslant c(1 + \|A(\lambda)\|)^2$, where the norms are the operator norms in L^2 and c is a constant which depends on the Lipschitz constant of a. From this differential inequality, and the fact that $\|A(0)\|$ equals the norm of the Hilbert transform H, it follows that $\|A(1)\| \leqslant c < \infty$, provided that $\|a'\|_\infty \leqslant M$, some finite constant. David [1982], [1984] removed this unnecessary restiction on M by means of a bootstrap argument. The proof given here, though, is a real variable one and is due to Coifman, McIntosh, and Meyer [1982a]. It is based on some ideas of Coifman and Meyer [1975], [1978], who settled in 1975 the case of the second commutator and soon afterwards extended their results to commutators of arbitrary order; the results of C. P. Calderón [1975], [1979] are relevant here. The proof of Proposition 1.6 is from the work of Coifman, Meyer, and Stein [1983] and that of Theorem 3.1 is from the work of David and Journé [1984] and Coifman and Meyer [1985].

Further Results and Problems

4.1 Assume $k(x, y)$ verifies the assumptions of 8.12 in Chapter XI, and let $Kf(x) = \text{p.v.} \int_{R^n} k(x, y) f(y)\, dy$ denote the Calderón–Zygmund operator associated with it. Furthermore, let $a \in BMO(R^n)$ and consider the commutator $Tf(x) = [M_a, K]f(x)$ of multiplication by a and K. Show that T is bounded in $L^p(R^n)$, $1 < p < \infty$, with norm $\le c\|a\|_*$. (*Hint*: The desired conclusion follows immediately from the pointwise estimate $(Tf)^\#(x) \le c\|a\|_*((M(|Kf|^r)(x))^{1/r} + M(|f|^s)(x))^{1/s}$, $1 < r, s < \infty$. Fix a cube I, then, and note that $Tf(x) = [M_{a-a_I}, K]f(x) = (a(x) - a_I)Kf(x) - K((a - a_I)f\chi_{2I})(x) - K((a - a_I)f\chi_{R^n\setminus 2I})(x) = A + B + C$, say. To estimate the average of $|A|$ over I observe that

$$\frac{1}{|I|}\int_I |A|\,dx \le \left(\frac{1}{|I|}\int_I |a(x) - a_I|^{r'}\,dx\right)^{1/r'}\left(\frac{1}{|I|}\int_I |Kf(x)|^r\right)^{1/r}$$

$$\le c\|a\|_*\left(\inf_I M(|Kf|^r)\right)^{1/r}.$$

The average of $|B|$ over I

$$\le \left(\frac{1}{|I|}\int_I |K((a - a_I)f\chi_{2I})(x)|^q\,dx\right)^{1/q}$$

$$\le c\left(\frac{1}{|I|}\int_{2I} |a(x) - a_I|^q |f(x)|^q\,dx\right)^{1/q}$$

$$\le c\left(\frac{1}{|I|}\int_{2I} |a(x) - a_I|^{qu'}\,dx\right)^{1/qu}\left(\frac{1}{|I|}\int_{2I} |f(x)|^{qu}\,dx\right)^{1/qu}$$

$$\le c\|a\|_*\left(\inf_I M(|f|^s)(x)\right)^{1/s},$$

provided that $qu = s$. To bound C, let x_1 = center of I and note that for x in I,

$$|K((a - a_I)f\chi_{R^n\setminus 2I}(x) - K((a - a_I)f\chi_{R^n\setminus 2I}(x_I)|$$

$$\le \int_{R^n\setminus 2I} |k(x,y) - k(x_I,y)||a(y) - a_I||f(y)|\,dy$$

$$\le \int_{R^n\setminus 2I} \frac{|x - x_I|^\eta}{|x_I - y|^{n+\eta}}|a(y) - a_I||f(y)|\,dy$$

$$\le c\left(\int_{R^n\setminus 2I} \frac{|x - x_I|^\eta}{|x_I - y|^{n+\eta}}|a(y) - a_I|^{s'}\,dy\right)^{1/s'}$$

$$\times\left(\int_{R^n\setminus 2I} \frac{|x - x_I|^\eta}{|x_I - y|^{n+\eta}}|f(y)|^s\,dy\right)^{1/s}$$

$$\le c\|a\|_*\left(\inf_I M(|f|^s)\right)^{1/s}.$$

The result is from Coifman, Rochberg, and Weiss [1976], the above proof is due to Strömberg. In fact the following converse also holds: if $[M_a, R_j]$ is bounded in some $L^p(R^n)$, $1 < p < \infty$, and $1 \leq j \leq n$, then a is in $BMO(R^n)$ and $\|a\|_* \leq c \sum_j(\text{norm in } L^p \text{ of } [M_a, R_j])$. Uchiyama [1978] has shown that T is compact from $L^p(R^n)$ into itself, $1 < p < \infty$, if and only if a is in the $BMO(R^n)$ closure of $C_0^\infty(R^n)$.)

4.2 The following extension of the results of 4.1 is due to Janson [1978]: Let $1 < p < \infty$, and let ϕ and ψ be nondecreasing, positive functions on $[0, \infty)$ connected by the relation $\phi(t) = t^{n/q}\psi^{-1}(t^{-n})$, or equivalently $\psi^{-1}(t) = t^{1/p}\phi(t^{-1/n})$. We assume that ψ is convex, $\psi(0) = 0$ and $\psi(2t) \leq c\psi(t)$. If K is a homogeneous Calderón-Zygmund operator, then a belongs to $BMO_\phi(R^n)$ if and only if $[M_a, K]$ maps $L^p(R^n)$ boundedly into $L_\psi(R^n)$.

4.3 Let

$$I_\alpha f(x) = \int_{R^n} \frac{1}{|x - y|^{n-\alpha}} f(y)\, dy$$

denote the Riesz potential of order α of f, and consider $Tf(x) = [M_a, I_\alpha]f(x)$, the commutator of multiplication by the $BMO(R^n)$ function a and I_α. Show that T maps $L^p(R^n)$ into $L^q(R^n)$, where $1/q = 1/p - \alpha/n$ and $1 < p < n/\alpha$. (*Hint*: The proof follows along the lines of 4.1. The estimate for the A term involves the maximal function M_η introduced in (2.16) in Chapter VI, but is otherwise similar to the A term in 4.1. The bound for the B term requires some care with the indices and uses the fact that the Riesz potentials I_α map $L^p(R^n)$ into $L^q(R^n)$, $1/q = 1/p - \alpha/n$. Finally it is readily seen that the C term is less than or equal to

$$\begin{aligned}
&\int_{R^n\setminus 2I} \frac{|x - x_I|}{|x_I - y|^{n-\alpha+1}} |a(y) - a_I||f(y)|\, dy \\
&\quad\leq \left(\int_{R^n\setminus 2I} \frac{|x - x_I|}{|x_I - y|^{n+1}} |a(y) - a_I|^{r'}\, dy \right)^{1/r'} \left(\int_{R^n\setminus 2I} \frac{|f(y)|^r}{|x - y|^{n+1-\alpha r}}\, dy \right)^{1/r} \\
&\quad\leq c\|a\|_* \left(\inf_I M_{\alpha r/n}(|f|^r) \right)^{1/r}, \qquad 1 < r < p.
\end{aligned}$$

The result is due to Chanillo [1982], who also discusses a converse.)

4.4 The following fact about operators is used repeatedly in Section 1 (cf. Proposition 1.3 for instance): suppose T_t, Z_t, and S_t are bounded, linear operators on a Hilbert space H, depending continuously (in the strong topology) on t. Suppose

$$\|S_{(\cdot)}\| = \sup \left(\int_{[0,\infty)} \|S_t h\|^2\, dt \right)^{1/2} < \infty,$$

where the sup is taken over $h \in H$ with $\|h\| = 1$, and similarly for $\|T_{(\cdot)}\|$. Then $\int_{[0,\infty)} S_t^T Z_t T_t\,dt$ represents a bounded operator on H with norm $\leqslant \sup_{t>0}\|Z_t\|\,\|S_{(\cdot)}\|\,\|T_{(\cdot)}\|$. Furthermore, if $Z_t = I$ and $S_t = T_t$, then $\|\int_{[0,\infty)} S_t^T S_t\,dt\| = \|S_{(\cdot)}\|^2$. This property, as well as an interesting discussion of the "Hilbert space methods" required for the proof of Calderón's theorem, is in the work of Coifman, McIntosh, and Meyer [1982b].

4.5 Let $m \in L^\infty(R)$ be an even function and consider the variant of the Hilbert transform defined by $H_m = \text{p.v.}\int_R R_t m(t)/t\,dt$. The techniques of Section 1 can be used to show that H_m is bounded in $L^2(R)$. The kernel $k_m(x-y)$ of H_m is obtained from the odd function $k_m(x)$ whose restriction to $(0,\infty)$ is $k_m(x) = \int_{[0,\infty)} e^{-xu} m(1/u)\,du$. Given n complex valued, Lipschitz functions, $A_1, \ldots, A_n$, say, let $k_{m,n}(x)$ be a singular integral kernel defined by $n!\,k_{m,n}(x,y) = (A_1(x) - A_1(y)) \cdots (A_n(x) - A_n(y)) D^n k_m(x-y)$. Show that the norm in $L^2(R)$ of the operator with kernel $k_{m,n}(x,y)$ is less than or equal to $c(1+n)^4 \|m\|_\infty \|A_1'\|_\infty \cdots \|A_n'\|_\infty$. A variant of, actually a corollary to, this result states that a similar conclusion holds for the principal value operator with kernel

$$\frac{A_1(x) - A_1(y)}{x-y} \cdots \frac{A_n(x) - A_n(y)}{x-y} k(x-y),$$

where k is now the odd function whose restriction to $(0,\infty)$ is given by $\int_{[0,\infty)} e^{-xu} m(uy)\,du$. An immediate consequence of this observation is the following: let K be a compact, convex subset of the complex plan, F an analytic function on a neighborhood of K, and A a complex-valued, Lipschitz function such that if $x \neq y$, $(A(x) - A(y))/(x-y) \in K$. Then for an odd kernel k as above, the singular kernel $F((A(x) - A(y))/(x-y))k(x-y)$ defines a bounded operator in $L^2(R)$. These results are from the work of Coifman, McIntosh, and Meyer [1982a].

4.6 Coifman and Meyer [1978] developed a method of dealing with commutators by reducing them to certain multilinear operators. In the bilinear case, where the techniques they use are already apparent, the result reads as follows: Let $\hat\phi$, $\hat\psi$ be $C_0^\infty(R^n)$ functions so that at least one of them vanishes in a neighborhood of the origin, and let

$$T(a,f)(x) = \int_{[0,\infty)} f * \phi_t(x) a * \psi_t(x) m(t) \frac{dt}{t}$$

where $m(t)$ is a bounded function. Then

$$\|T(a,f)\|_2 \leqslant c\|m\|_\infty \|a\|_* \|f\|_2, \tag{4.1}$$

where c depends only on ϕ and ψ. (*Hint*: Suppose first both $\hat\phi$ and $\hat\psi$ vanish near the origin, and let $\hat\eta$ be a compactly supported, even function, which is 1 in a neighborhood of the support of $\hat{\phi}$ and $\hat{\psi}$. Then for $h \in C_0^\infty(R^n)$,

$$\int_{R^n} T(a,f)(x)h(x)\,dx$$

$$= \left|\int_{R^n}\int_{[0,\infty)} h * \eta_t(x) f * \phi_t(x) a * \psi_t(x) m(t)\frac{dt}{t}\,dx\right|$$

$$\leq \|m\|_\infty \left(\int_{R_+^{n+1}} |h * \eta_t(x)|^2 |a * \psi_t(x)|^2\,dx\frac{dt}{t}\right)^{1/2}$$

$$\times \left(\int_{R_+^{n+1}} |f * \phi_t(x)|^2\,dx\frac{dt}{t}\right)^{1/2} = AB,$$

say. Since $a \in BMO(R^n)$, $|a * \psi_t(x)|^2(1/t)\,dx\,dt$ is a Carleson measure and $A \leq c\|a\| * \|h\|_2$. Also $B \leq \|f\|_2$, and we have finished in this case. The general result follows readily from this; the above proof is from Calderón [1978]. Since by 4.17 in Chapter XV also $|a * \psi_t(x)|^2 w(x)(1/t)\,dx\,dt$ is a Carleson measure for w in A_2, with constant independent in A_2, the reader is invited to consider the general weighted version of (4.1).)

4.7 There is a weighted version of Theorem 1.10. Corresponding to the operator in (1.22), let

$$C_\varepsilon(\phi,f)(x) = \int_{|x-y|>\varepsilon} \frac{(1+i\phi'(y))}{(x-y)+i(\phi(x)-\phi(y))} f(y)\,dy$$

and $C^*(\phi,f)(x) = \sup_\varepsilon |C_\varepsilon(\phi,f)(x)|$. Then there are constants, k_1, k_2 such that for all $1 < p < \infty$ the following inequality holds:

$$\int_R C^*(\phi,f)(x)^p\,d\mu(x) \leq c\int_R |f(x)|^p\,dx,$$

where c is independent of f, $d\mu(x) = (M((1+M_q^\#(\phi'))^{k_1})(x)^{k_2}$ and $M_q^\# g(x) = \sup((1/|I|)\int_I |g(x)-g_I|^q\,dx)^{1/q}$, where the sup is taken over all open cubes containing x. This result is due to Krikeles [1983].

4.8 These are some examples of operators T which verify the weak boundedness property,

(1) Let k be a standard CZ kernel so that $k(x,y) = -k(-y,x)$, $x \neq y$ in R^n. Then $Tf(g) = \lim_{\varepsilon\to 1}\int_{|x-y|>\varepsilon} k(x,y)f(y)g(x)\,dy\,dx$ defines an operator from $\mathcal{S}(R^n)$ into $\mathcal{S}'(R^n)$ with the weak boundedness property. Indeed, using the antisymmetry of the kernel we have

$$Tf(g) = \lim_{\varepsilon\to 0}\frac{1}{2}\int_{|x-y|>\varepsilon} k(x,y)(f(y)g(x) - f(x)g(y))\,dy\,dx.$$

Thus the smoothness of f and g compensates for the singularity of k and the limit is easily seen to exist and to verify the desired properties.

(2) Suppose ψ is a smooth function with 0 integral and let $Q_t g$ denote the operator convolution with ψ_t, i.e., $Q_t g = g * \psi_t$. If

$$T = \int_{[0,\infty)} Q_t L_t \frac{dt}{t} \quad \text{or} \quad T = \int_{[0,\infty)} L_t Q_t \frac{dt}{t},$$

where the L_t's are uniformly bounded operators on $L^2(R^n)$, then T has the weak boundedness property. Indeed, T is well defined from $\mathscr{S}(R^n)$ into $\mathscr{S}'(R^n)$ by $Tf(g) = \int_{[0,\infty)}(L_t f, Q_t g)\,(1/t)\,dt$ in the first case and by $\int_{[0,\infty)}(Q_t f, L_t^T g)\,(1/t)\,dt$ in the second case. These examples are from the work of David and Journé [1984].

4.9 Suppose k is a CZ kernel which verifies the assumptions of 8.12 in Chapter XI with index $0 < \delta \leq 1$ and let Tf denote the p.v. operator associated to k. If $T^*1 = 0$, then T extends to a continuous operator from $H^p(R^n)$ into itself, $n/(n+\delta) < p \leq 1$. (*Hint*: By 6.6 in Chapter XIV it suffices to show that T maps atoms into appropriate molecules. Let then a be a (p, q, N) atom, supp $a \subseteq B$, where B is a ball centered at x and radius r; we show that Ta is a (p, q, α) molecule based at $B(x, 2r)$. Since T is bounded in $L^q(R^n)$ we have that $\|Ta\|_q^q \leq c\|a\|_q^q \leq c|a|^q r^{n(1-q/p)}$. On the other hand

$$\int_{B(x,2r)} |Ta(y)|^q |x-y|^\alpha \, dy \leq c r^{\alpha + n(1-q/p)}$$

and

$$\int_{R^n \setminus B(x,2r)} |Ta(y)|^q |x-y|^\alpha \, dy$$

$$= \sum_{j=0}^{\infty} \int_{B(x,2^{j+1}r)\setminus B(x,2^j r)} |Ta(y)|^q |x-y|^\alpha \, dy = \sum A_j,$$

say. To estimate each A_j observe that

$$|Ta(y)| = \left| \int_{R^n} (k(y,w) - k(y,x)) a(w)\, dw \right|$$

$$\leq c|x-y|^{-(n+\alpha)} \int_{|w-x|\leq r} |w-x|^\delta |a(w)|\, dw$$

$$\leq c r^{\delta + n/q'} \|a\|_q |x-y|^{-(n+\delta)};$$

this estimate replaced in each A_j allows us to sum that expression and to conclude that it is of the right order as well. Finally, it only remains to check that property (iii) of molecules holds, namely, $\int_{R^n} Ta(y)\,dy = 0$. But observe that $0 = T^*1(a) = Ta(1) = \int_{R^n} Ta(y)\,dy$. This result, which is due to Alvarez and Milman [1985], can be extended in several directions. The above mentioned authors consider, for instance, operators analogous to those intoduced in 8.17 in Chapter XII.)

CHAPTER

XVII

Boundary Value Problems on C^1-Domains

1. THE DOUBLE AND SINGLE LAYER POTENTIALS ON A C^1-DOMAIN

We say that a tempered distribution T is a fundamental solution for the Laplacian Δ in R^n if $\Delta T = \delta$, the Dirac delta at 0. Fundamental solutions are useful for $u = T * f$ solves $\Delta u = f$. It is not hard to find T explicitly. Suppose first $n > 2$ and note that since $\Delta T = \delta$, also $-|\xi|^2 \hat{T}(\xi) = 1$, and consequently, at least formally,

$$T(x) = \frac{-1}{(2\pi)^n} \int_{R^n} \frac{1}{|\xi|^2} e^{ix \cdot \xi} \, d\xi. \tag{1.1}$$

General considerations show that $T(x)$ coincides with a function homogeneous of degree $-(n-2)$ in $R^n \backslash (0)$, but since the constants involved are important we compute them. For this purpose recall that

$$-\int_{[0,\infty)} te^{-t|\xi|} \, dt = -\frac{1}{|\xi|^2}, \qquad \xi \neq 0,$$

which substituted into (1.1) gives

$$T(x) = -\int_{[0,\infty)} t \frac{1}{(2\pi)^n} \int_{R^n} e^{-t|\xi|} e^{ix \cdot \xi} \, d\xi \, dt = -\int_{[0,\infty)} tP(x, t) \, dt,$$

where $P(x, t)$ denotes the Poisson kernel of R_+^{n+1}. Whence by the results in Chapter X,

$$T(x) = -c_n |x|^{-(n-2)} \int_{[0,\infty)} t^2 (1 + t^2)^{-(n+1)/2} \, dt = \frac{1}{(n-2)w_n} |x|^{-(n-2)},$$

where w_n is the surface area of the unit ball in R^n. Also $T(x) = (1/2\pi)\ln|x|$, $n = 2$.

We now turn to the boundary value problems. We restrict ourselves to C^1 domains D in R^n, the precise definition will be given below, and denote points there with capital letters, $X, Y, \ldots$ if they lie in the interior of D and P, Q if they lie on the boundary ∂D of D. Let also dQ denote the surface area element on ∂D.

If u is a harmonic function in a bounded domain D, differentiable up to the boundary, then Green's identity (4.5) in Chapter VII and a limiting argument give

$$u(X) = \int_{\partial D} T(X - Q) \frac{\partial}{\partial N_Q} u(Q)\, dQ - \int_{\partial D} \frac{\partial}{\partial N_Q} T(X - Q) u(Q)\, dQ, \tag{1.2}$$

where $(\partial/\partial N_Q)$ denotes the derivative along the inward normal N_Q into D. A quick verification of (1.2) goes as follows: Since for X, Y in D, $u(Y)\,\Delta T(X - Y) - T(X - Y)\,\Delta u(Y) = u(Y)\delta(X - Y)$, the left-hand side of Green's identity is $\int_D u(Y)\delta(X - Y)\, dY = u(X)$, and the right-hand side is that of (1.2), as claimed.

Identity (1.2) is the starting point for the solution of the Dirichlet and Neumann problems in a bounded domain D. The method we use here, called that of double- and single-layer potentials, involves properties of integral equations on the boundary ∂D, covered by the Fredholm theory. The first term in (1.2) is the single-layer potential of $(\partial/\partial N_Q)u$, and since the singularity of T is of degree lower than the dimension of the boundary this potential is continuous on D and its closure. The second term in (1.2) is the double-layer potential of u and since the singularity of the kernel $(\partial/\partial N_Q)T$ is of order $n - 1$, some care is needed as this singularity is only integrable if the domain is $C^{1+\varepsilon}$, $\varepsilon > 0$.

As in 6.10 in Chapter VII the Dirichlet problem on D is stated as follows: Given $f \in L^p(\partial D)$, find a function u so that $\Delta u = 0$ in D and $u|_{\partial D} = f$. To solve this problem we form the double-layer potential v of f and observe that if f and ∂D are smooth enough, the boundary values of v equal $(\frac{1}{2}I + K)f$, where K is a compact linear mapping when D is C^1. Methods from the Fredholm theory of integral operators yield the invertibility of $\frac{1}{2}I + K$, and the harmonic function u given by the double-layer potential of $(\frac{1}{2}I + K)^{-1}f$ then solves the Dirichlet problem. To do the Neumann problem, i.e., to find a function u so that $\Delta u = 0$ in D and $(\partial/\partial N_Q)u|_{\partial D} = g$, we use the single layer potential instead.

To make all these statements precise we need some preliminary results. Throughout, D denotes an open, connected, bounded subset of R^n such that $R^n \backslash \bar{D}$ is also connected.

Definition 1.1. We say that D is a C^1 domain, and by this we really mean that D is a smooth n-manifold so that its boundary ∂D is a $C^1(n-1)$-manifold without boundary, if the following properties hold: to each Q in ∂D there corresponds a local coordinate system (U_Q, ϕ_Q) such that

(i) U_Q is an open neighborhood of Q and ϕ_Q is a real-valued, compactly supported $C^1(R^{n-1})$ function defined on U_Q.
(ii) In the local Euclidean coordinates we may assume that $Q = (0, \phi(0))$.
(iii) $\phi(0) = 0$ and $(\partial/\partial x_j)\phi(x)]_{x=0} = 0$.
(iv) $D \cap U_Q = \{(x, t) \colon x \in R^{n-1}, t \in R \text{ and } \phi(x) < t\} \cap U_Q$.
(v) $\partial D \cap U_Q = \{(x, t) \colon x \in R^{n-1}, t \in R \text{ and } \phi(x) = t\} \cap U_Q$.

Remark 1.2. On account of the compactness of ∂D and property (iii) above it readily follows that given $\varepsilon > 0$ we can find a finite number of coordinate systems, $\{(U_j, \phi_j)\}_{j=1}^m$ say, so that $\partial D \subseteq \bigcup_{j=1}^n U_j$, (i)–(v) above hold, and $\|\nabla \phi_j\|_\infty \leqslant \varepsilon$, all j.

Remark 1.3. If ϕ is any of the ϕ_Q's above, it is clear that there exists a sequence $\{\phi_j\}$ of $C_0^\infty(R^{n-1})$ functions so that on the support of ϕ, (i) ϕ_j converges uniformly to ϕ, as $j \to \infty$, (ii) $\nabla\phi_j$ converges uniformly to $\nabla\phi$, as $j \to \infty$. Moreover, since ϕ is compactly supported, the ϕ_j's may be explicitly constructed as a sequence of mollifiers of ϕ, and consequently also $\|\nabla\phi_j\|_\infty \leqslant c$, all j.

The first step is to make use of the local coordinates to find the Euclidean expression of the layer potentials and to study them, as well as the traces on the boundary ∂D itself, as operators on L^p.

First observe that since

$$\frac{\partial}{\partial N_Q} T(X-Q) = \frac{-1}{(n-2)w_n}(-(n-2))|X-Q|^{-(n-2)-1}\frac{X-Q}{|X-Q|}\cdot N_Q$$
$$= \frac{1}{w_n}\frac{X-Q\cdot N_Q}{|X-Q|^n},$$

the double-layer potential of f is given by

$$\frac{1}{w_n}\int_{\partial D}\frac{X-Q\cdot N_Q}{|X-Q|^n} f(Q)\,dQ, \qquad X \in D. \tag{1.3}$$

As for the boundary, or trace, double-layer potential it is defined as follows: for $P \in \partial D$ let

$$K_\varepsilon f(P) = \frac{1}{w_n}\int_{\{Q\in\partial D : |P-Q|>\varepsilon\}}\frac{P-Q\cdot N_Q}{|P-Q|^n} f(Q)\,dQ, \tag{1.4}$$

and, when it makes sense,

$$Kf(P) = \text{p.v.} \frac{1}{w_n} \int_{\partial D} \frac{P - Q \cdot N_Q}{|P - Q|^n} f(Q)\, dQ = \lim_{\varepsilon \to 0} K_\varepsilon f(P). \tag{1.5}$$

Similarly, the single-layer potential of f is given by

$$\frac{-1}{(n-2)w_n} \int_{\partial D} \frac{1}{|X - Q|^{n-2}} f(Q)\, dQ, \qquad X \in D, \tag{1.6}$$

and since the singularity of $|P - Q|^{2-n}$ is integrable, the corresponding trace is

$$-\frac{1}{(n-2)w_n} \int_{\partial D} \frac{1}{|P - Q|^{n-2}} f(Q)\, dQ, \qquad P \in \partial D. \tag{1.7}$$

To obtain the Euclidean expression of the operator in (1.4) we localize as follows: Let $\{U_j\}$ be a finite cover of ∂D obtained as in Remark 1.2, say, and let $\{\eta_m\}$ be a nonnegative, smooth, finite partition of unity subordinate to the U_j's. Clearly,

$$K_\varepsilon f(P) = \sum_m \frac{1}{w_n} \int_{\{Q \in \partial D : |P - Q| > \varepsilon\}} \frac{P - Q \cdot N_Q}{|P - Q|^n} \eta_m(Q) f(Q)\, dQ,$$

and it suffices to consider each summand separately. Also by collecting all summands that correspond to each U_j we may assume that we are working with a fixed local coordinate system (U_j, ϕ_j), which we denote simply by (U, ϕ). Next if we change in U the variables Q into $(y, \phi(y))$, and put $P = (x, \phi(x))$ and $\mathcal{U}_\varepsilon = \{y \in R^{n-1} : |x - y|^2 + (\phi(x) - \phi(y))^2 > \varepsilon^2\}$, since $N_Q\, dQ = (-\nabla\phi(y), 1)\, dy$, the resulting expression in (1.4) equals $(1/w_n) \int_{\mathcal{U}_\varepsilon} k(x, y)(\sum_{m'} \eta_{m'}(y, \phi(y))) f(y, \phi(y))\, dy$, where

$$k(x, y) = \frac{\phi(x) - \phi(y) - \nabla\phi(y) \cdot (x - y)}{(|x - y|^2 + (\phi(x) - \phi(y))^2)^{n/2}}. \tag{1.8}$$

Note that if $\phi \in C^2(R^{n-1})$, then $|k(x, y)| \leq c|x - y|^{-n+2}$ and the kernel is locally integrable; in this case most of the complicated arguments we give for the case when ϕ is merely C^1 are unnecessary. At any rate, and with an obvious abuse of notation, it is clear that the L^p continuity properties of the operator in (1.4) follow from those of the Euclidean operator

$$K_\varepsilon f(x) = \frac{1}{w_n} \int_{\mathcal{U}_\varepsilon} k(x, y) f(y)\, dy. \tag{1.9}$$

As a first approximation to study (1.9) we consider

$$\tilde{K}_\varepsilon f(x) = \int_{|x - y| > \varepsilon} k(x, y) f(y)\, dy. \tag{1.10}$$

We then have

Theorem 1.4. Assume ϕ is a Lipschitz function in R^{n-1} such that $\|\nabla\phi\|_\infty = \eta < \infty$. Let k be the kernel in (1.8) and $\tilde{K}_\varepsilon f$ the operator defined in (1.10). Then the mapping $\tilde{K}^*f(x) = \sup_{\varepsilon>0}|\tilde{K}_\varepsilon f(x)|$ is bounded in $L^p(R^{n-1})$, $1 < p < \infty$, with norm which goes to 0 with η. Furthermore $\lim_{\varepsilon\to 0}\tilde{K}_\varepsilon f(x) = \tilde{K}f(x)$ exists pointwise a.e. and in L^p, and $\tilde{K}$ is compact in $L^p(R^{n-1})$.

Proof. Write

$$k(x, y) = \frac{\phi(x) - \phi(y)}{(|x-y|^2 + (\phi(x) - \phi(y))^2)^{n/2}} - \sum_{j=1}^{n} \frac{(\partial/\partial y_j)\phi(y)(x_j - y_j)}{(|x-y|^2 + (\phi(x) - \phi(y))^2)^{n/2}}$$

$$= k_0(x, y) - \sum_{j=1}^{n} \frac{\partial}{\partial y_j}\phi(y)k_j(x, y),$$

say, and put $\tilde{K}^j_\varepsilon f(x) = \int_{|x-y|>\varepsilon} k_j(x, y)f(y)\,dy$, $0 \le j \le n$. Thus

$$\tilde{K}_\varepsilon f(x) = \tilde{K}^0_\varepsilon f(x) - \sum_{j=1}^{n} \tilde{K}^j_\varepsilon\left(f\frac{\partial}{\partial y_j}\phi\right)(x),$$

and consequently, with a self-explanatory notation,

$$\tilde{K}^* f(x) \le \tilde{K}^{0*} f(x) + \sum_{j=1}^{n} \tilde{K}^{j*}\left(f\frac{\partial}{\partial y_j}\phi\right)(x). \tag{1.11}$$

That each of the $\tilde{K}^{j*}$'s is bounded in $L^p(R^{n-1})$, $1 < p < \infty$, follows immediately from Proposition 2.5 in Chapter XVI. When $j = 0$ we apply that result with $A(x) = B(x) = \phi(x)$, and in the remaining cases we put $A(x) = \phi(x)$ and $B(x) = x_j$, $1 \le j \le n$, respectively. We also get that the norm c_0 of $\tilde{K}^{0*}$ goes to 0 with $\|\nabla B\|_\infty = \|\nabla\phi\|_\infty = \eta$ and that the norm of the remaining $\tilde{K}^{j*}$'s depends only on η. Whence from (1.11) it readily follows that

$$\|\tilde{K}^*f\|_p \le \|\tilde{K}^{0*}f\|_p + \sum_{j=1}^{n} \left\|\tilde{K}^{j*}\left(f\frac{\partial}{\partial y_j}\phi\right)\right\|_p$$

$$\le c_0\|f\|_p + c\sum_{j=1}^{n}\left\|f\frac{\partial}{\partial y_j}\phi\right\|_p$$

$$\le (c_0 + n\eta)c\|f\|_p,$$

and the continuity assertion concerning $\tilde{K}^*$ holds.

Next, to prove that $\lim_{\varepsilon\to 0}\tilde{K}_\varepsilon f(x)$ exists pointwise a.e. and in $L^p(R^{n-1})$, consider the sequence $\{\phi_j\}$ introduced in Remark 1.3 above, let

$$k_j(x, y) = \frac{\phi_j(x) - \phi_j(y) - \nabla\phi_j(y)\cdot x - y}{(|x-y|^2 + (\phi(x) - \phi(y))^2)^{n/2}}, \tag{1.12}$$

and put $\tilde{K}_{j,\varepsilon} f(x) = \int_{|x-y|>\varepsilon} k_j(x, y)f(y)\,dy$. As before, it is readily seen that $\tilde{K}^*_j f(x) = \sup_{\varepsilon>0}|\tilde{K}_{j,\varepsilon} f(x)|$ is bounded in L^p, $1 < p < \infty$; we claim that also

$\lim_{\varepsilon\to 0}\tilde{K}_{j,\varepsilon}f(x)$ exists pointwise a.e. for f in $L^p(R^{n-1})$. Indeed, since each k_j verifies $|k_j(x,y)|\leq c|x-y|^{-n+2}$, where c depends on j, we see at once that $\int_{R^{n-1}}|x-y|^{-n+2}|f(y)|\,dy<\infty$ a.e., and our claim is a simple consequence of this.

Suppose now that f is real valued and let

$$L(x)=\limsup_{\varepsilon\to 0}\tilde{K}_\varepsilon f(x)-\liminf_{\varepsilon\to 0}\tilde{K}_\varepsilon f(x). \tag{1.13}$$

We want to show that $L(x)=0$ a.e. First note that

$$\begin{aligned}\tilde{K}_\varepsilon f(x)&=\int_{|x-y|>\varepsilon}(k(x,y)-k_j(x,y))f(y)\,dy+\tilde{K}_{j,\varepsilon}f(x)\\&=H_{j,\varepsilon}f(x)+\tilde{K}_{j,\varepsilon}f(x),\end{aligned} \tag{1.14}$$

say, where the kernel of $H_{j,\varepsilon}$ is given by an expression similar to (1.12) but with ϕ_j replaced by $\phi-\phi_j$ there. Thus $\sup_{\varepsilon>0}|H_{j,\varepsilon}f(x)|$ is bounded in $L^p(R^{n-1})$, $1<p<\infty$, with norm c_j which goes to 0 with $\|\nabla(\phi-\phi_j)\|_\infty$. This is all we need to know; indeed, from (1.13) and (1.14) we get

$$\begin{aligned}L(x)&=\limsup_{\varepsilon\to 0}H_{j,\varepsilon}f(x)-\liminf_{\varepsilon\to 0}H_{j,\varepsilon}f(x)\\&\leq 2\sup_{\varepsilon>0}|H_{j,\varepsilon}f(x)|.\end{aligned}$$

Therefore, for each $\lambda>0$ and j, $\lambda^p|\{L>\lambda\}|\leq\lambda^p|\{\sup_\varepsilon|H_{j,\varepsilon}|>\lambda/2\}|\leq 2^pc_j^p\|f\|_p^p\to 0$ as $j\to\infty$. Thus $|\{L>\lambda\}|=0$ for each $\lambda>0$, $L(x)=0$ a.e., and $\lim_{\varepsilon\to 0}\tilde{K}_\varepsilon f(x)=\tilde{K}f(x)$ exists a.e. That the convergence is also in $L^p(R^{n-1})$, $1<p<\infty$, follows at once from this last result, the boundedness of $\tilde{K}^*$ in $L^p(R^{n-1})$, and the Lebesgue dominated convergence theorem. Finally, we show that for each $1<p<\infty$, $\tilde{K}$ is compact on $L^p(R^{n-1})$. It is well known, and readily verified, that it is sufficient to exhibit a sequence of compact operators which converge to $\tilde{K}$ in norm. As observed in the preceding paragraph $\lim_{j\to\infty}\|\tilde{K}-\tilde{K}_j\|=0$, where $\|T\|$ denotes the norm of T as a mapping on $L^p(R^{n-1})$. Next we show that also $\lim_{\varepsilon\to 0}\|\tilde{K}_j-\tilde{K}_{j,\varepsilon}\|=0$. By a partition of unity argument we may restrict our attention to $L^p(B)$, $B=$ unit ball of R^{n-1}. In this case, and with a constant c that depends on j,

$$\begin{aligned}|\tilde{K}_jf(x)-\tilde{K}_{j,\varepsilon}f(x)|&\leq\int_{|x-y|\leq\varepsilon}|k_j(x,y)||f(y)|\,dy\leq c\int_{|x-y|\leq\varepsilon}\frac{|f(y)|}{|x-y|^{n-2}}\,dy\\&\leq c\left(\int_{|x-y|\leq\varepsilon}\frac{1}{|x-y|^{n-2}}\,dy\right)^{1/p'}\\&\quad\times\left(\int_{|x-y|\leq\varepsilon}\frac{|f(y)|^p}{|x-y|^{n-2}}\,dy\right)^{1/p}\\&=c\varepsilon^{1/p'}\left(\int_{|x-y|\leq\varepsilon}\frac{|f(y)|^p}{|x-y|^{n-2}}\,dy\right)^{1/p},\end{aligned}$$

where $1/p + 1/p' = 1$. Thus, $\int_B |\tilde{K}_j f(x) - \tilde{K}_{j,\varepsilon} f(x)|^p \, dx \leq c\varepsilon^{p/p'} \int_B |f(y)|^p \, dy$, and $\|\tilde{K}_j - \tilde{K}_{j,\varepsilon}\| \leq c\varepsilon^{1/p'} \to 0$ with ε. To check that each $\tilde{K}_{j,\varepsilon}$ is compact, we must show that given a bounded sequence $\{f_m\}$ in $L^p(B)$, i.e., $\|f_m\|_p \leq M$, all m, there exists a subsequence $\{f_{m_k}\}$ such that $\tilde{K}_{j,\varepsilon} f_{m_k}$ converges in $L^p(B)$ as $m_k \to \infty$. But this is not hard; by Proposition 3.2 in Chapter II there exists an L^p function f, $\|f\|_p \leq M$, such that f_{m_k} converges weakly to f. Furthermore, since for each x (in B) $k(x, y)\chi_{|x-y|>\varepsilon}$ is bounded as a function of y and consequently is in $L^{p'}(B)$, it readily follows that $\tilde{K}_{j,\varepsilon} f(x) = \lim_{m_k \to \infty} \tilde{K}_{j,\varepsilon} f_{m_k}(x)$. Moreover, since also $\|\tilde{K}_{j,\varepsilon} f\|_p$, $\|\tilde{K}_{j,\varepsilon} f_{m_k}\|_p \leq cM$, where c depends on ε but is otherwise independent of the functions involved, by the Lebesgue dominated convergence theorem we obtain that $\|K_{j,\varepsilon} f - K_{j,\varepsilon} f_{m_k}\|_p \to 0$ as $m_k \to \infty$, and we are done. ■

We turn now to the study of the trace of the double-layer potential.

Theorem 1.5. Let $K_\varepsilon f(P)$ be the truncated trace double layer potential corresponding to a C^1 domain given by (1.4). Then the mapping $K^* f(P) = \sup_{\varepsilon > 0} |K_\varepsilon f(P)|$ is bounded in $L^p(\partial D)$, $1 < p < \infty$, $\lim_{\varepsilon \to 0} K_\varepsilon f(P) = Kf(P)$ exists pointwise a.e. and in $L^p(\partial D)$, and K is compact in $L^p(\partial D)$.

Proof. By means of a partition of unity argument and by passing to local coordinates, the L^p boundedness of $K^* f(P)$ is readily seen to follow from the corresponding statement for the Euclidean operator $K^* f(x) = \sup_{\varepsilon > 0} |K_\varepsilon f(x)|$, where $K_\varepsilon f(x)$ is defined in (1.9). We begin by showing that

$$\sup_{\varepsilon > 0} |K_\varepsilon f(x) - \tilde{K}_\varepsilon f(x)| \leq cMf(x), \qquad x \in R^{n-1}, \tag{1.15}$$

where $\tilde{K}_\varepsilon f(x)$ is the operator in (1.10) and c is an absolute constant which depends only on the Lipschitz constant η of ϕ. Since we work in local coordinates we may assume that $x = 0$, $\phi(x) = \phi(0) = 0$, and $\nabla\phi(0) = 0$; also the fact that ϕ is C^1 means that $|\phi(y)| = o(|y|)$ and $|\nabla\phi(y)| = o(|y|)$ as $|y| \to 0$. Let then $\mathcal{U}_\varepsilon = \{|y|^2 + \phi(y)^2 > \varepsilon^2\}$ and observe that since $B(0, \varepsilon/(1 + \eta^2)^{1/2}) \subseteq R^n \setminus \mathcal{U}_\varepsilon \subseteq B(0, \varepsilon)$ we have $\chi_{\mathcal{U}_\varepsilon}(y) = \chi_{\mathcal{U}_\varepsilon}(y)\chi_{B(0,\varepsilon)}(y) + \chi_{R^n \setminus B(0,\varepsilon)}(y)$. Thus

$$K_\varepsilon f(0) = \tilde{K}_\varepsilon f(0) + \int_{R^{n-1}} k(0, y)\chi_{\mathcal{U}_\varepsilon}(y)\chi_{B(0,\varepsilon)}(y) f(y) \, dy \tag{1.16}$$

and consequently

$$\sup_{\varepsilon > 0} |K_\varepsilon f(0) - \tilde{K}_\varepsilon f(0)|$$

$$\leq \sup_{\varepsilon > 0} \int_{R^{n-1}} |k(0, y)| \chi_{\mathcal{U}_\varepsilon}(y)\chi_{B(0,\varepsilon)}(y) |f(y)| \, dy. \tag{1.17}$$

Furthermore, since $|k(0, y)| \leq 2\eta|y|^{1-n}$, the expression on the right-hand side of (1.17) does not exceed

$$2\eta(1+\eta^2)^{(n-1)/2}\sup_{\varepsilon>0}\frac{1}{\varepsilon^{n-1}}\int_{|y|\leqslant\varepsilon}|f(y)|\,dy\leqslant cMf(0),$$

and (1.15) follows. In turn, (1.15) gives $K^*f(x)\leqslant\tilde{K}^*f(x)+cMf(x)$, and consequently by Theorem 1.4 $\|K^*f\|_p\leqslant c\|f\|_p$, $1<p<\infty$, as we wanted to show. To prove the existence of the p.v. integral $Kf(P)$ assume first that $f\in C^1(\partial D)$ and observe that

$$K_\varepsilon f(P)=\frac{1}{w_n}\int_{\{|P-Q|>\varepsilon\}}\frac{P-Q\cdot N_Q}{|P-Q|^n}(f(Q)-f(P))\,dQ$$
$$+f(P)\frac{1}{w_n}\int_{\{|P-Q|>\varepsilon\}}\frac{P-Q\cdot N_Q}{|P-Q|^n}\,dQ=A+B,$$

say. Since $|A|\leqslant c\int_{\partial D}|P-Q|^{2-n}\,dQ<\infty$, the limit of this term is readily seen to exist as $\varepsilon\to 0$. As for B, let $D_\varepsilon(P)=\{X\in D:|X-P|>\varepsilon\}$ and note that by 6.11 in Chapter VII, $\int_{\partial D_\varepsilon(P)}(\partial/\partial N_Q)T(P-Q)\,dQ=0$. Whence

$$B=\frac{-f(P)}{w_n}\int_{\{Q\in D:|Q-P|=\varepsilon\}}\frac{P-Q\cdot N_Q}{|P-Q|^n}\,dQ,$$

and this last integral is readily seen to tend to $\frac{1}{2}$ as $\varepsilon\to 0$ at each point P where the plane tangent to D is well defined. This proves the everywhere existence of the p.v. integral $Kf(P)$ when f is smooth. That this p.v. integral exists a.e. and in $L^p(\partial D)$ for an arbitrary f in $L^p(\partial D)$, $1<p<\infty$, follows by a by now familiar argument which is left to the reader.

Finally, we show that for each $1<p<\infty$, K is compact on $L^p(\partial D)$; the assumption that D is a C^1 domain is needed here. Again through the use of a partition of unity argument, and on account of Theorem 1.4, our conclusion will follow from the qualitative version of (1.15), namely,

$$Kf(x)=\tilde{K}f(x)\qquad\text{a.e.}\tag{1.18}$$

In fact, (1.18) holds for those x's for which either side, and consequently the other side also, is well defined. Suppose $x=0$ is such a point and observe that it suffices to show that the integral in (1.16) goes to 0 with ε. But this is not hard; indeed, since $\phi\in C^1$, $|k(0,y)|\leqslant o(|y|)/|y|^n$ as $|y|\to 0$, and the integral does not exceed

$$c\int_{|y|\sim\varepsilon}\frac{o(|y|)}{|y|^n}|f(y)|\,dy\leqslant co(1)\frac{1}{\varepsilon^{n-1}}\int_{|y|\leqslant\varepsilon}|f(y)|\,dy$$
$$\leqslant co(1)Mf(0)=o(1)\text{ as }\varepsilon\to 0.\quad\blacksquare$$

Next we consider the behavior of the double layer potential $Kf(X)$ given by (1.3) for X near the boundary ∂D of D. Since the notion of nontangential convergence is appropriate here we begin by defining cones interior to D. Cones in R^n with vertex at 0 are given by $\{x = (x_1, \ldots, x_n) : |x| < \beta x_n, \beta > 1\}$, and this definition reads in our setting as follows: given $0 < a < 1$ and $P \in \partial D$, the (inner) cone $\Gamma_a(P)$ with vertex at P and opening a is

$$\Gamma_a(P) = \{X \in D : |X - P| < \delta \quad \text{and} \quad a|X - P| < X - P \cdot N_P\}. \tag{1.19}$$

The constant δ in the definition depends on a and D but is independent of P, and N_P denotes as usual the inward normal at P. Similarly, the outer cone $\Gamma_a^e(P)$ with the vertex at P and opening a is defined as

$$\Gamma_a^e(P) = \{X \in R^n \backslash \bar{D} : |X - P| < \delta \quad \text{and} \quad a|X - P| < -(X - P) \cdot N_P\}. \tag{1.20}$$

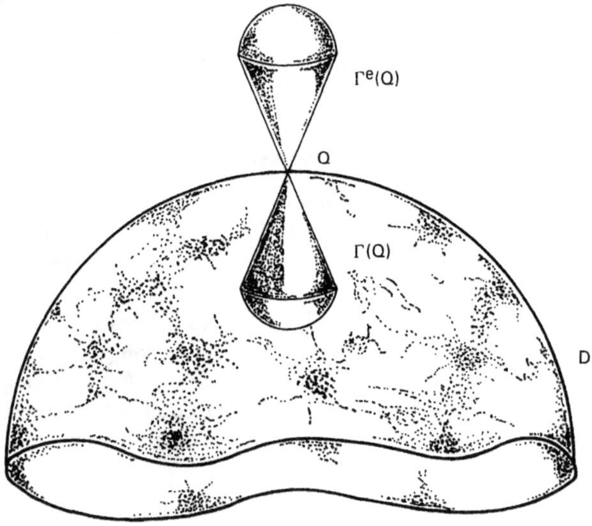

Given $0 < a < 1$, $P \in \partial D$ and a function $u(X)$ in D, we say that the nontangential limit (of order a) of $u(X)$ as X approaches P is L provided that $\lim_{X \to P, X \in \Gamma_a(P)} u(X) = L$. Also the nontangential maximal function $N_a u(P)$ is

$$N_a u(P) = \sup_{X \in \Gamma_a(P)} |u(X)|. \tag{1.21}$$

We then have

Theorem 1.6. Let $Kf(X)$ be the double layer potential corresponding to a

C^1 domain D given by (1.3). Then if $f \in L^p(\partial D)$, $1 < p < \infty$, and $0 < a < 1$, there is a number δ which depends only on a and D, so that for this choice of δ in definition (1.19),

$$\|N_a(Kf)\|_p \leq c\|f\|_p, \qquad 1 < p < \infty, \tag{1.22}$$

where c depends only on p and δ. Furthermore $Kf(X)$ converges nontangentially of order a a.e. on ∂D, and $\lim_{X \to P, X \in \Gamma_a(P)} Kf(X) = \frac{1}{2}f(P) + Kf(P)$, a.e. on ∂D, where $Kf(P)$ is the trace double layer potential in (1.5).

Proof. Let $\partial D \subseteq \bigcup_{j=1}^m B_j$, where each of the balls $B_j = B(P_j, \delta_j)$ corresponds to a coordinate system (B_j, ϕ_j) so that $\sup_{B(P_j, 4\delta_j)}|\nabla \phi_j| \leq A/6$; to obtain this covering apply Remark 1.2. Now let $\delta = \min(\delta_1, \ldots, \delta_m)$ be the value in the definition (1.19) of the cones $\Gamma_a(P)$.

For P in ∂D we want to estimate $N_a(Kf)(P)$. By a partition of unity argument we may assume that $\operatorname{supp} f \subseteq B_j$, some j. We consider two cases, to wit, (i) $P \in B(P_j, 3\delta_j)$ (nearby points) and (ii) $P \notin B(P_j, 3\delta_j)$ (far away points). Case (ii) is easily handled. Since we are interested in estimating $Kf(X)$ for X in $\Gamma_a(P)$ and $\operatorname{supp} f \subseteq B_j$, we must bound the integral in (1.3) when $|X - P| \leq \delta$, $|P_j - Q| \leq \delta_j$ and $|P - P_j| \geq 3\delta_j$. Then also $|X - Q| \geq \delta_j$ and

$$|Kf(X)| \leq \int_{\partial D \cap B_j} |T(X - Q)||f(Q)|\, dQ \leq c \int_{\partial D \cap B_j} \frac{1}{|X - Q|^{n-1}} |f(Q)|\, dQ$$
$$\leq c\left(\frac{1}{\delta_j^{n-1}} \int_{\partial D \cap B_j} |f(Q)|^p\, dQ\right)^{1/p}.$$

In other words

$$N_a(Kf)(P)\chi_{B(P_j, 3\delta_j)}(P) \leq c\|f\|_p. \tag{1.23}$$

Case (i) requires some work. First note that, if $x \in \Gamma_a(P)$, then $|X - P_j| \leq 4\delta_j$, and consequently passing to the local coordinates given by ϕ_j, which we denote by ϕ from now on, we have $\|\nabla \phi\|_\infty \leq a/6$. Thus identifying $Q = (y, \phi(y))$, $X = (x, t)$ and $P = (x_0, \phi(x_0))$, the consideration of $N_a(Kf)(P)$ reduces to the study of

$$\sup_{(x,t)} \left| \int_{R^{n-1}} k(x, t, y) f(y)\, dy \right|, \tag{1.24}$$

where

$$k(x, t, y) = \frac{t - \phi(y) - \nabla\phi(y) \cdot (x - y)}{(|x - y|^2 + (t - \phi(y))^2)^{n/2}}, \tag{1.25}$$

subject to

$$\phi(x) < t, \qquad (X \in D) \tag{1.26}$$

and

$$a(|x - x_0|^2 + (\phi(x) - \phi(x_0))^2)^{1/2} \leqslant \frac{t - \phi(x_0) - \nabla\phi(x_0)\cdot(x - x_0)}{(1 + |\nabla\phi(x_0)|^2)^{1/2}} \qquad (X \in \Gamma_a(P)). \tag{1.27}$$

(1.27) is readily seen to imply

$$t - \phi(x_0) \geqslant 5a|x - x_0|/6. \tag{1.28}$$

To estimate (1.24) we break up the integral there in two parts, $|\int_{|y-x_0|\leqslant M}| + |\int_{|y-x_0|>M}| = I + J$, say where $M = \max(3|x - x_0|, t - \phi(x_0))$. To estimate I, observe that by (1.28) $M \sim t - \phi(x_0)$ and $|k(x, t, y)| \leqslant c(|t - \phi(y)| + |x - y|)^{1-n}$ in the integral. Since as is readily seen (1.28) also implies that $(4/5)(t - \phi(x_0)) \leqslant |t - \phi(y)| + (a/6)|y - x|$, we immediately get

$$I \leqslant c(t - \phi(x_0))^{1-n} \int_{|y-x_0|\leqslant c(t-\phi(x_0))} |f(y)|\, dy \leqslant cMf(x_0). \tag{1.29}$$

Next we estimate

$$\begin{aligned} J &\leqslant \int_{|y-x_0|>M} |k(x, t, y) - k(x_0, t, y)||f(y)|\, dy \\ &\quad + \int_{|y-x_0|>M} |k(x_0, t, y) - k(x_0, \phi(x_0), y)||f(y)|\, dy \\ &\quad + \left|\int_{|y-x_0|>M} k(x_0, \phi(x_0), y) f(y)\, dy\right| \\ &= J_1 + J_2 + J_3, \end{aligned}$$

say. Clearly,

$$J_3 = |\tilde{K}_M f(x_0)| \leqslant \tilde{K}^* f(x_0). \tag{1.30}$$

As for J_1 note that $|k(x, t, y) - k(x_0, t, y)| \leqslant c|x - x_0|/|y - x_0|^n$, whenever $|y - x_0| > 3|x - x_0|$. This estimate follows easily from the mean value theorem. Thus by Proposition 2.3 in Chapter IV,

$$J_1 \leqslant c \int_{|y-x_0|>3|x-x_0|} \frac{|x - x_0|}{|y - x_0|^n} |f(y)|\, dy \leqslant cMf(x_0). \tag{1.31}$$

Similarly, since $|k(x_0, t, y) - k(x_0, \phi(x_0), y)| \leqslant c(t - \phi(x_0))/|y - x_0|^n$ whenever $|y - x_0| > t - \phi(x_0)$, we also have

$$J_2 \leqslant cMf(x_0). \tag{1.32}$$

Whence adding estimates (1.29)-(1.32) we get that $Tf(x_0) \leq c(\tilde{K}^*f(x_0) + Mf(x_0))$, and

$$\|Tf\|_p \leq c\|K^*f\|_p + \|Mf\|_p \leq c\|f\|_p, \qquad 1 < p < \infty. \tag{1.33}$$

The first part of our conclusion, namely, estimate (1.22) follows immediately from (1.23) and (1.33).

To discuss the nontangential boundary values of $Kf(X)$ we consider first the case when $f \in C^1(\partial D)$. We then have

$$Kf(X) = \frac{1}{w_n}\int_{\partial D} \frac{X - Q \cdot N_Q}{|X - Q|^n}(f(Q) - f(P))\, dQ$$

$$+ f(P)\frac{1}{w_n}\int_{\partial D} \frac{X - Q \cdot N_Q}{|X - Q|^n}\, dQ = A + B, \text{ say.}$$

Since $|f(Q) - f(P)| \leq c|Q - P|$, the integrand of A has a summable singularity and the limit exists. An argument along the lines of the B term in Theorem 1.5 gives that the limit actually is $Kf(P) - \frac{1}{2}f(P)$. Also by Green's theorem it readily follows that $B = f(P)$ whenever $X \in D$, and consequently the limit in this case is $Kf(P) - \frac{1}{2}f(P) + f(P) = Kf(P) + \frac{1}{2}f(P)$, as we wanted to show. To show that the same is true for an arbitrary g in $L^p(\partial D)$, assume first that g is real valued and observe that for f in $C^1(\partial D)$

$$\begin{aligned} Lg(P) &= \limsup_{X \to P, X \in \Gamma_a(P)} Kg(X) - \liminf_{X \to P, X \in \Gamma_a(P)} Kg(x) \\ &= \limsup_{X \to P, X \in \Gamma_a(P)} K(g - f)(X) - \liminf_{X \to P, X \in \Gamma_a(P)} K(g - f)(X) \\ &\leq 2N_a(K(g - f))(P). \end{aligned}$$

Thus for each $\lambda > 0$, $\lambda^p|\{Lg > \lambda\}| \leq 2^p\|N_a(K(g - f))\|_p^p \leq c\|K(g - f)\|_p^p \leq c\|g - f\|_p^p$, where the last term above is as small as we want. Consequently, $|\{Lg > \lambda\}| = 0$ for each $\lambda > 0$, and $Lg(P) = 0$ a.e. This is equivalent to the second part of the theorem and we have finished. ■

Similar techniques may be used to study the regularity of the double-layer potential Kf when f is regular; we state the results without proofs. First we need a definition: for $1 < p < \infty$, $L_1^p(\partial D)$ denotes the space of functions f in $L^p(\partial D)$ with the property that for any covering $\{U_j\}$ of ∂D described in Definition 1.1 and for any C^1 function ψ supported in some U_j, the function $\psi(x, \phi_j(x))f(x, \phi_j(x))$ has (distributional) partial derivatives in $L^p(R^{n-1})$. If we fix a covering $\{U_j\}$ and a partition of unity, $\{\psi_j\}$ say, of ∂D subordinate to this cover we can define

$$\|f\|_{L_1^p(\partial D)} = \|f\|_{L^p(\partial D)} + \sum \|\nabla(\psi_j f)\|_{L^p(R^{n-1})},$$

and different coverings give rise to equivalent norms.

Theorem 1.7. For a C^1 domain D and $1 < p < \infty$, the operator Kf given by (1.5) is continuous, and compact on $L_1^p(\partial D)$. Furthermore, given $0 < a < 1$ there is δ which depends only on a and D so that for this δ in definition (1.19), the gradient $\nabla Kf(X)$ of the double layer potential given by (1.3) verifies $\|N_a(|\nabla Kf|)\|_p \leq c\|f\|_{L_1^p(\partial D)}$.

We turn now to the study of the single-layer potential. We begin with some definitions: for $P \in \partial D$ let

$$K_\varepsilon^t f(P) = \frac{-1}{w_n} \int_{\{Q \in \partial D : |Q-P| > \varepsilon\}} \frac{P - Q \cdot N_P}{|P-Q|^n} f(Q)\, dQ \tag{1.34}$$

and, when it makes sense,

$$K^t f(P) = \text{p.v.} \frac{-1}{w_n} \int_{\partial D} \frac{P - Q \cdot N_P}{|P-Q|^n} f(Q)\, dQ = \lim_{\varepsilon \to 0} K_\varepsilon^t f(P). \tag{1.35}$$

The relevant Euclidean integral operator in this case has kernel

$$k^t(x, y) = \frac{\phi(x) - \phi(y) - \nabla\phi(x) \cdot (x - y)}{(|x-y|^2 + (\phi(x) - \phi(y))^2)^{n/2}} \tag{1.36}$$

and represents essentially the adjoint of the operator with kernel $k(x, y)$ defined by (1.8). A statement similar (with almost identical proof) to Theorem 1.4 holds and can be used to prove

Theorem 1.8. Let $K_\varepsilon^t f(P)$ be the potential corresponding to a C^1 domain defined by (1.34). Then the mapping $(K^t)^* f(P) = \sup_{\varepsilon > 0} |K_\varepsilon^t f(P)|$ is bounded in $L^p(\partial D)$, $1 < p < \infty$, $\lim_{\varepsilon \to 0} K_\varepsilon^t f(P) = K^t f(P)$ exists pointwise a.e. and in $L^p(\partial D)$, and K^t is compact in $L^p(\partial D)$.

The proof of this result, being analogous to that of Theorem 1.5 is omitted. In fact, that K^t is compact follows from the fact that its adjoint K is compact on each $L^p(\partial D)$, $1 < p < \infty$.

Theorem 1.9. For f in $L^p(\partial D)$, $1 < p < \infty$, and $X \notin \partial D$, let $u(X)$ be the single-layer potential of f given by (1.6). Then given $0 < a < 1$, there is a number δ which depends only on a and D, so that for this choice of δ in definitions (1.19) and (1.20), $N_a(|\nabla u|)(P)$ and $N_a^e(|\nabla u|)(P) = \sup_{X \in \Gamma_a^e(P)} |\nabla u(X)|$ belong to $L^p(\partial D)$ and there is a constant c which depends only on p and δ so that

$$\|N_a(|\nabla u|)\|_p, \qquad \|N_a^e(|\nabla u|)\|_p \leq c\|f\|_p. \tag{1.37}$$

Furthermore, $\lim_{X \to P, X \in \Gamma_a(P)} (\partial/\partial N_P) u(X) = \lim_{X \to P, X \in \Gamma_a(P)} \nabla u(X) \cdot N_P = \frac{1}{2} f(P) - K^t f(P)$ and $\lim_{X \to P, X \in \Gamma_a^e(P)} (\partial/\partial N_P) u(X) = \frac{1}{2} f(P) + K^t f(P)$, exist pointwise for almost every P in ∂D. Here K^t is the operator in (1.35).

Proof. Observe that since $\nabla((-1/(n-2)w_n)|X-Q|^{2-n}) = (1/w_n)(X - Q/|X-Q|^n)$,

$$\nabla u(X) = \frac{1}{w_n}\int_{\partial D} \frac{X-Q}{|X-Q|^n} f(Q)\,dQ. \tag{1.38}$$

The proof of estimates (1.37) follow along the lines of Theorem 1.6 and is therefore omitted. As for the nontangential convergence it suffices to prove the existence of the pointwise limit for almost every P in ∂D when $f \in C^1(\partial D)$. We consider only the case of the interior nontangential limit, i.e., $X \in D$, the exterior limit being handled analogously. By (1.38)

$$\begin{aligned}
\frac{\partial}{\partial N_P} u(X) &= \frac{1}{w_n}\int_{\partial D} \frac{X-Q\cdot N_P}{|X-Q|^n} f(Q)\,dQ \\
&= \frac{1}{w_n}\int_{\partial D} \frac{X-Q\cdot N_P}{|X-Q|^n} (f(Q)-f(P))\,dQ \\
&\quad + f(P)\frac{1}{w_n}\int_{\partial D} \frac{X-Q\cdot N_Q}{|X-Q|^n}\,dQ \\
&\quad + f(P)\frac{1}{w_n}\int_{\partial D} \frac{X-Q\cdot N_P - N_Q}{|X-Q|^n}\,dQ \\
&= I + J + K,
\end{aligned}$$

say. Observe that $J = f(P)\cdot 1 = f(P)$. Also since $f \in C^1(\partial D)$ it is clear that $\lim_{X\to P, X\in\Gamma_a(P)} I$ exists and equals

$$\text{p.v.}\frac{1}{w_n}\int_{\partial D} \frac{P-Q\cdot N_P}{|P-Q|^n}(f(Q)-f(P))\,dQ.$$

Now consider K. N_P is a continuous function on ∂D and hence there is a sequence of (vector-valued) functions $N_{j,P}$, belonging to $C^1(\partial D)$ such that $N_{j,P} \to N_P$, uniformly in ∂D. The integral in K then equals

$$\begin{aligned}
N_P - N_{j,P}\cdot\frac{1}{w_n}\int_{\partial D}\frac{X-Q}{|X-Q|^n}\,dQ + \frac{1}{w_n}\int_{\partial D}\frac{X-Q\cdot N_{j,P}-N_{j,Q}}{|X-Q|^n}\,dQ \\
+ \frac{1}{w_n}\int_{\partial D}\frac{X-Q\cdot N_{j,Q}-N_Q}{|X-Q|^n}\,dQ = M_1 + M_2 + M_3,
\end{aligned}$$

say. At this time we make use of the following observation: let

$$A_\varepsilon f(P) = \int_{\{Q\in\partial D: |Q-P|>\varepsilon\}} \frac{P-Q}{|P-Q|^n} f(Q)\,dQ,$$

and put $A^*f(P) = \sup_{\varepsilon>0}|A_\varepsilon f(P)|$. Then as we saw in Theorem 1.4, A^* is bounded in $L^p(\partial D)$, $1 < p < \infty$, and $Af(P) = \lim_{\varepsilon\to 0} A_\varepsilon f(P)$ exists pointwise a.e. in ∂D. Moreover, along the lines of Theorem 1.6, we also have that

$$N_a(|Af|)(P) = \sup_{X\in\Gamma_a(P)} \left| \int_{\partial D} \frac{X-Q}{|X-Q|^n} f(Q)\, dQ \right|$$

is bounded in $L^p(\partial D)$, $1 < p < \infty$. From this last remark it follows that for $1 < p < \infty$,

$$\lim_{j\to\infty} \left\| \lim_{X\to P, X\in\Gamma_a(P)} \frac{1}{w_n} \int_{\partial D} \frac{X-Q\cdot N_{j,Q} - N_Q}{|X-Q|^n}\, dQ \right\|_p = 0,$$

and consequently there is a subsequence j_k, which we denote by j again, so that

$$\lim_{X\to P, X\in\Gamma_a(P)} \frac{1}{w_n} \int_{\partial D} \frac{X-Q\cdot N_{j,Q} - N_Q}{|X-Q|^n}\, dQ = 0 \qquad \text{a.e.}$$

In other words, $M_3 = 0$. As for M_2 it is clear that

$$\lim_{j\to\infty} \lim_{X\to P, X\in\Gamma_a(P)} M_2 = \text{p.v.} \frac{1}{w_n} \int_{\partial D} \frac{P-Q\cdot N_P - N_Q}{|P-Q|^n}\, dQ.$$

Clearly $\lim_{j\to\infty} M_1 = 0$. Summing up, we have shown that the nontangential limit exists and it equals

$$\begin{aligned} &\text{p.v.} \frac{1}{w_n} \int_{\partial D} \frac{P-Q\cdot N_P}{|P-Q|^n} (f(Q) - f(P))\, dQ + f(P) \\ &\qquad + f(P)\, \text{p.v.} \frac{1}{w_n} \int_{\partial D} \frac{P-Q\cdot N_P - N_Q}{|P-Q|^n}\, dQ \\ &\quad = \tfrac{1}{2} f(P) - K' f(P). \quad \blacksquare \end{aligned}$$

2. THE DIRICHLET AND NEUMANN PROBLEMS

To solve the Dirichlet and Neumann problems on a C^1 domain we make use of the Fredholm alternative concerning compact operators from a normed space X into itself. We begin with a brief discussion of the Fredholm theory; we are only concerned with the case $X = L^p$, $1 < p < \infty$, here.

Recall that a mapping T from X into itself is said to be compact if for each bounded sequence $\{x_n\} \subset X$ we can find a subsequence $\{x_{n_k}\}$ so that $\{Tx_{n_k}\}$ converges. Also T is compact if and only if its adjoint operator T' is compact.

Proposition 2.1. Let T be a compact, linear operator from X into itself, and let $\lambda \neq 0$ be a complex number. If $\lambda I - T$ is injective, then the range $R(\lambda I - T)$ of $\lambda I - T$ is (strongly) closed.

Proof. Let $y = \lim_{n\to\infty} y_n$, where $y_n = (\lambda I - T)x_n$, $x_n \in X$. If $\{x_n\}$ contains a bounded subsequence, there is yet another subsequence, $\{x_{n_k}\}$ say, so that $\{Tx_{n_k}\}$ converges. Since $x_{n_k} = (y_{n_k} + Tx_{n_k})/\lambda$, then $\{x_{n_k}\}$ itself converges to some element x and $y = (\lambda I - T)x$. If, on the other hand, $\{x_n\}$ contains no bounded subsequence, then $\|x_n\| \to \infty$. Put $z_n = x_n/\|x_n\|$ and note that $\|z_n\| = 1$ and $\lim_{n\to\infty}(\lambda I - T)z_n = 0$. Let $\{z_{n_k}\}$ be a subsequence so that $\{Tz_{n_k}\}$ converges. Since $z_{n_k} - \lambda^{-1}Tz_{n_k} \to 0$ also $\{z_{n_k}\}$ converges, to a limit z, say. Then $\|z\| = 1$ and $(\lambda I - T)z = 0$, contrary to the hypothesis that $(\lambda I - T)$ is injective. ■

To complete our discussion we also need

Proposition 2.2. Let M be a proper, closed subspace of X. Then for $0 < \varepsilon < 1$ we can find an element x_ε which is "nearly orthogonal" to M, i.e.,

$$\|x_\varepsilon\| = 1, \qquad \operatorname{dist}(x_\varepsilon, M) \geqslant \varepsilon.$$

Proof. Let $x \in X \backslash M$; since M is closed $\operatorname{dist}(x, M) = d > 0$. So there exists $y_\varepsilon \in M$ such that $\|x - y_\varepsilon\| \leqslant d/\varepsilon$, and letting $x_\varepsilon = (x - y_\varepsilon)/\|x - y_\varepsilon\|$ we have $\|x_\varepsilon\| = 1$ and for any y in M,

$$\|x_\varepsilon - y\| = \frac{\|x - y_\varepsilon - \|y_\varepsilon - x\|y\|}{\|y_\varepsilon - x\|} \geqslant \frac{d}{\|y_\varepsilon - x\|} \geqslant \varepsilon. \quad ■$$

Proposition 2.3. Let T be a compact, linear operator on X, and suppose that for $\lambda \neq 0$, $\lambda I - T$ is injective. Then $R(\lambda I - T) = X$, $\lambda I - T$ is invertible and

$$\|x\| \leqslant c\|(\lambda I - T)x\|, \tag{2.1}$$

where c is independent of $x \in X$.

Proof. By Proposition 2.1 the sets $R_j = (\lambda I - T)^j X$, $j = 1, 2, \ldots$ form a nonincreasing sequence of closed subspaces of X. Suppose that no two of these spaces coincide, then each is a proper subspace of its predecessor. Hence by Proposition 2.2 there exists a sequence $\{y_n\} \subset X$ such that $y_n \in R_n$, $\|y_n\| = 1$ and $\operatorname{dist}(y_n, R_{n+1}) \geqslant \frac{1}{2}$. Thus if $n > m$, $Ty_m - Ty_n = y_m + (-y_n - (\lambda I - T)y_m + (\lambda I - T)y_n) = y_m - y$ for some $y \in R_{n+1}$. Hence $\|Ty_m - Ty_n\| \geqslant \frac{1}{2}$, contrary to the fact that T is compact. Therefore there is an integer k so that $R_j = R_k$ for $j \geqslant k$. Let $y \in X$, then $(\lambda I - T)^k y \in R_k = R_{k+1}$, and consequently $(\lambda I - T)^k y = (\lambda I - T)^{k+1}x$, for some $x \in X$. In other words, $(\lambda I - T)^k(y - (\lambda I - T)x) = 0$, and since the kernel $N((\lambda I - T)^k)$

of $(\lambda I - T)^k$ is the same as $N(\lambda I - T) = 0$, it follows that $y = (\lambda I - T)x$. Thus $R(\lambda I - T) = R_j = X$, for all j, and $\lambda I - T$ is invertible. Next suppose that estimate (2.1) does not hold. Then we can find a sequence $\{z_n\}$ so that $(\lambda I - T)z_n \to 0$ and $\|z_n\| = 1$. Since T is compact there is a subsequence, $\{z_{n_k}\}$ say, such that $Tz_{n_k} \to x \in X$. Furthermore, since $\lambda z_{n_k} = (\lambda I - T)z_{n_k} + Tz_{n_k}$, then λz_{n_k} also converges to x. It is then readily seen that $x \in N(\lambda I - T)$, and consequently $x = 0$. But this contradicts the fact that $\|z_{n_k}\| = 1$. ∎

We begin discussing the Dirichlet problem.

Theorem 2.4. Assume D is a C^1 domain and $R^n \backslash \bar{D}$ is connected, and let $Kf(P)$ denote the trace double-layer potential defined by (1.5). Then $\frac{1}{2}I + K$ is invertible on $L^p(\partial D)$ for each $1 < p < \infty$.

Proof. We show in fact that the adjoint of $\frac{1}{2}I + K$, namely $\frac{1}{2}I + K^t$, where K^t is given by (1.35), is invertible on each $L^p(\partial D)$, $1 < p < \infty$. Since, by Theorem 1.8, K^t is compact in $L^p(\partial D)$, by Proposition 2.3 it is enough to prove that $\frac{1}{2}I + K^t$ is injective.

First observe that if $f \in L^p(\partial D)$ and $(\frac{1}{2}I + K^t)f = 0$, then actually $f \in L^q(\partial D)$ for every $1 < q < \infty$. To see this let $B = B(P_0, \delta)$ be a ball centered on ∂D and with radius δ sufficiently small so that for the local coordinate system (B, ϕ) we have $\|\nabla \phi\|_\infty \leq \varepsilon$, where ε is a fixed, small, positive number. Let, then, η and ψ be C^∞ functions supported in B so that $\eta = 1$ in $B(P_0, \delta/3)$ and 0 in $R^n \backslash B(P_0, 2\delta/3)$ and ψ is identically 1 in $B(P_0, 3\delta/4)$. Notice that since ψ is 1 on the support of η, we have $\eta\psi = \eta$. Now, since $\eta\psi(\frac{1}{2}I + K^t)f = 0$, we also have $\eta\psi f + 2\eta\psi K^t f - 2\psi(K^t\eta f) + 2\psi K^t \eta f = 0$, or $\eta f + 2\psi K^t \psi\eta f = -2\psi(\eta K^t - K^t\eta)f = g$, say. The function

$$g(P) = 2\psi(P) \text{ p.v. } \frac{1}{w_n} \int_{\partial D} \frac{P - Q \cdot N_P}{|P - Q|^n} (\eta(Q) - \eta(P)) f(Q) \, dQ,$$

is readily seen to verify

$$|g(P)| \leq c \int_{\partial D} \frac{|f(Q)|}{|P - Q|^{n-2}} \, dQ,$$

where c depends on δ. By the Sobolev embedding theorem (cf. Theorem 2.1 in Chapter VI and Theorem 4.8 in Chapter X), we see that $g \in L^q(\partial D)$ where $1/q = 1/p - 1/(n-1) > 0$, or $g \in L^q(\partial D)$, $1 < q < \infty$ if $1/p \leq 1/(n-1)$. In either case, since the norm of $\psi K^t \psi$ is small on L^q, we conclude that ηf, and consequently f itself, belongs to $L^q(\partial D)$, $p < q$. Iterating this process we get that $f \in L^q(\partial D)$, $1 < q < \infty$, as anticipated.

Let now $u(X)$ denote the single layer potential of the function f over ∂D given by (1.7), and consider the integral $I = \int_{R^n \backslash \bar{D}} |\nabla u(X)|^2 \, dX$.

If $\partial^e D$ denotes the boundary of $R^n \backslash \bar{D}$ (it coincides with ∂D except for the orientation), then by Green's theorem

$$I = \int_{R^n \backslash \bar{D}} \operatorname{div}(\nabla u(X))\, dX = \int_{\partial^e D} u(Q) \frac{\partial}{\partial N_Q^e} u(Q)\, dQ, \tag{2.2}$$

where $(\partial/\partial N_Q^e)$ indicates the derivative in the direction of the inward normal $N_Q^e = -N_Q$ into $\partial^e D$. The application of Green's theorem is justified since by Theorem 1.9 the last integral in (2.2) is absolutely convergent. Also by Theorem 1.9 $(\partial/\partial N_Q^e)u(Q) = -(\frac{1}{2}I + K')f(Q) = 0$, Q a.e. in ∂D, and consequently $I = 0$. Therefore $u(X)$ is constant on $R^n \backslash \bar{D}$, and since $\lim_{|X| \to \infty} u(X) = 0$ and $R^n \backslash \bar{D}$ is connected, then $u(X)$ is identically 0 in $R^n \backslash \bar{D}$. Furthermore, since $u(X)$ is a continuous function on R^n and $u|_{\partial D} = 0$, by the uniqueness principle of harmonic functions, Proposition 4.3 in Chapter VII, we obtain that $u(X)$ is identically 0 on R^n. From Theorem 1.9 it now follows that also $(\frac{1}{2}I + K')f(Q) = 0$ a.e. on ∂D, and consequently $f(Q) = (\frac{1}{2}I + K')f(Q) + (\frac{1}{2}I - K')f(Q) = 0$ a.e. on ∂D. In other words, $\frac{1}{2}I + K'$ is injective, and the proof is complete. ■

Corollary 2.5. $\frac{1}{2}I + K$ is invertible on $L_1^p(\partial D)$, $1 < p < \infty$.

We are now ready to prove the existence and uniqueness of the solution to the Dirichlet problem.

Theorem 2.6. Suppose D is a C^1 domain and $R^n \backslash \bar{D}$ is connected. Given $f \in L^p(\partial D)$, $1 < p < \infty$, there exists a unique harmonic function $u(X)$ defined for X in D, such that for each $0 < a < 1$, there exists a $\delta > 0$ which depends only on a and D, so that for this choice of δ in definition (1.19), $N_a u$ belongs to $L^p\,(\partial D)$ and

$$\|N_a u\|_p \leq c\|f\|_p, \tag{2.3}$$

with c independent of f. Moreover, $\lim_{X \to P, X \in \Gamma_a(P)} u(X) = f(P)$ a.e. on ∂D.

Proof. By Theorem 2.4 and Proposition 2.3, $\frac{1}{2}I + K$ has a continuous inverse in $L^p(\partial D)$, $1 < p < \infty$. Let $u(X)$ be the double-layer potential $u(X) = (1/w_n)\int_{\partial D}((X - Q \cdot N_Q)/(|X - Q|^n))(\frac{1}{2}I + K)^{-1}f(Q)\, dQ$. By Theorem 1.6 the nontangential limit of $u(X)$ is $f(P)$ a.e. on ∂D, and $\|N_a u\|_p \leq c\|(\frac{1}{2}I + K)^{-1}f\|_p \leq c\|f\|_p$, which is (2.3).

The proof of the uniqueness requires some work. For X, Y, in D and $Q \in \partial D$, let $F(X, Q) = (\frac{1}{2}I + K)^{-1}(1/|X - \cdot|^{n-2})(Q)$ and consider the Green's function $G(X, Y)$ defined by

$$G(X, Y) = \frac{1}{|X - Y|^{n-2}} - \frac{1}{w_n}\int_{\partial D} \frac{Y - Q \cdot N_Q}{|Y - Q|^n} F(X, Q)\, dQ.$$

Next, for fixed $\varepsilon > 0$ consider the set $D_\varepsilon = \{Y \in D\colon \operatorname{dist}(Y, \partial D) \le \varepsilon\}$ and let $\psi_\varepsilon(Y) \in C_0^\infty(D)$ satisfy $0 \le \psi_\varepsilon \le 1$, $\psi_\varepsilon = 1$ on D_ε and $|\partial^\alpha/\partial Y^\alpha \psi_\varepsilon| \le c\varepsilon^{|\alpha|}$, where c depends only on α. For a fixed X in D, and for small ε, by Green's identity we see that

$$u(X) = u(X)\psi_\varepsilon(X) = \int_D G(X, Y)\, \Delta(u\psi_\varepsilon)(Y)\, dY. \tag{2.4}$$

Moreover, if u is harmonic in D, integrating by parts (2.4) gives

$$\begin{aligned} u(X) &= -2\int_D \nabla_Y G(X, Y) \cdot \nabla\psi_\varepsilon(Y) u(Y)\, dY \\ &\quad - \int_D G(X, Y)\, \Delta\psi_\varepsilon(Y) u(Y)\, dY \\ &= A + B, \end{aligned}$$

say. We will show that under the additional assumption that the nontangential boundary values of u are 0, then A, $B \to 0$ with ε, and consequently u vanishes identically. Since the proofs for A and B are similar we only do A here. For this purpose let $\{\psi_j\}$ be a finite family of nonnegative, $C_0^\infty(R^n)$ functions, such that $\sum \psi_j(Y) = 1$ on $\{Y \in R^n\colon \operatorname{dist}(Y, \partial D) \le \delta\}$ and $\operatorname{supp}\psi_j \subseteq B_j$, where (B_j, ϕ_j) is a local coordinate system for D. It clearly suffices to show that for each j, $I_j = \int_D |\nabla_Y G(X, Y)||\nabla\psi_\varepsilon(Y)||u(Y)|\psi_j(Y)\, dY$, goes to 0 with ε. Fix j, put $I_j = I$, $\psi_j = \psi$, $\phi_j = \phi$, and passing the variable Y to Euclidean coordinates note that

$$\begin{aligned} I &\le \frac{c}{\varepsilon}\int_{|y|\le c}\int_{[0,\varepsilon)} |\nabla_Y G(X, y, t + \phi(y))||u(y, t + \phi(y))|\, dt\, dy \\ &\le c\int_{|y|\le c} \sup_{0\le s\le\varepsilon} |\nabla_Y G(X, y, s + \phi(y))| \frac{1}{\varepsilon}\int_{[0,\varepsilon)} |u(y, t + \phi(y))|\, dt\, dy. \end{aligned}$$

Since $G(X, Q) \in L_1^q(\partial D)$ for each $1 < Q < \infty$, $\sup_{0\le s\le\varepsilon}|\nabla_Y G(X, y, s + \phi(y))| \le N_a(|\nabla_Y G|)(y, \phi(y)) \in L^q(\{y\colon |y| < c\})$. It is also easy to see that there is $0 < a < 1$, so that $\sup_{0<t<\varepsilon}|u(y, t + \phi(y))| \le N_a u(y, \phi(y))$. Therefore, if in addition to being harmonic, u verifies $N_a u(y, \phi(y)) \in L^p(\{y\colon |y| \le c\})$ and $u(y, t + \phi(y)) \to 0$ as $t \to 0$ for almost every $|y| \le c$, then $|A| \to 0$ with ε, and the proof is complete. ∎

Concerning the regularity properties of the solution to the Dirichlet problem we have

Theorem 2.7. Suppose D is as in Theorem 2.6. If $f \in L_1^p(\partial D)$, $1 < p < \infty$, then the solution $u(X)$ of the Dirichlet problem given by Theorem 2.6 has the additional property that $N_a(|\nabla u|) \in L^p(\partial D)$ and there is a constant c, independent of f, so that

$$\|N_a(|\nabla u|)\|_p \leq c\|f\|_{L_1^p(\partial D)}.$$

The proof of this result, being analogous to that of Theorem 2.6 is omitted; Theorem 1.7 is relevant here.

Finally, we consider the Neumann problem; we begin by showing

Theorem 2.8. Suppose D is a bounded, connected, C^1 domain, and let K' be the trace single layer potential defined by (1.35). Then for each $1 < p < \infty$, $\frac{1}{2}I - K'$ is invertible on the subspace of $L^p(\partial D)$ consisting of those functions f with $\int_{\partial D} f(Q)\, dQ = 0$.

Proof. Since by Theorem 1.8, K' is compact, by Proposition 2.3 it is enough to prove that $\frac{1}{2}I - K'$ is injective. So assume that $f = 2K'f$ and $\int_{\partial D} f(Q)\, dQ = 0$. As in Theorem 2.4 we conclude that $f \in L^q(\partial D)$ for every $1 < q < \infty$. Let now $u(X)$ denote the single layer potential of f over ∂D defined by (1.7). Integrating by parts we get

$$\int_D |\nabla u(X)|^2\, dX = \int_{\partial D} u(Q) \frac{\partial u}{\partial N_Q}(Q)\, dQ$$

$$= \int_{\partial D} u(Q)\left(\frac{1}{2} I - K'\right) f(Q)\, dQ = 0.$$

Hence $u(X)$ is constant in $\bar{D}$. In $R^n \setminus \bar{D}$, $u(X)$ is harmonic and $\lim_{|X| \to \infty} u(X) = 0$. As noted $u|_{\partial D} = c$, a constant. Since the maximum or minimum of u in $R^n \setminus D$ are assumed on ∂D, then they both occur at every P in ∂D and the nontangential limit of $(\partial/\partial N_P)u(X)$ as $X \to P$, $X \in R^n \setminus \bar{D}$ is of constant sign. But by Theorem 1.9 the limit in our case is $-\frac{1}{2}f(P) - K'f(P) = -f(P)$. Thus f is of constant sign, and since it has vanishing integral we must have $f(P) = 0$ on ∂D. ■

We are now ready to prove the existence and uniqueness of the solution to the Neumann problem.

Theorem 2.9. Suppose D is a bounded, connected, C^1 domain and $R^n \setminus \bar{D}$ is connected. Given $g \in L^p(\partial D)$, $1 < p < \infty$, with $\int_{\partial D} g(Q)\, dQ = 0$, there exists a unique harmonic function $u(X)$ defined for X in D such that for each $0 < a < 1$, we can find a $\delta > 0$ which depends only on a and D, so that for this choice of δ in definition (1.19), $N_a(|\nabla u|)$ belongs to $L^p(\partial D)$ and

$$\|N_a(|\nabla u|)\|_p \leq c\|g\|_p, \tag{2.5}$$

with c independent of g. Moreover

$$\lim_{X\to P, X\in\Gamma_a(P)} \frac{\partial}{\partial N_P} u(X) = g(P) \qquad \text{a.e. on} \quad \partial D.$$

Proof. By Theorem 2.8 and Proposition 2.3, $\frac{1}{2}I - K'$ has a continuous inverse in the subspace of $L^p(\partial D)$ consisting of those functions with integral 0, $1 < p < \infty$. Let $u(X)$ be the single layer potential

$$u(X) = \frac{-1}{(n-2)w_n} \int_{\partial D} \frac{1}{|X-Q|^{n-2}} \left(\frac{1}{2}I - K'\right)^{-1} g(Q)\, dQ.$$

By Theorem 1.9 the nontangential limit of $(\partial/\partial N_P)u(X)$ is $g(P)$ a.e. on ∂D, and $\|N_a(|\nabla u|)\|_p \leqslant c\|(\frac{1}{2}I - K')^{-1}g\|_p \leqslant c\|g\|_p$, which is (2.5).

As for the uniqueness, for X, Y, in D and $Q \in \partial D$, let

$$G(X, Q) = \left(\frac{1}{2}I - K'\right)^{-1} \left(\frac{(X - \cdot)\cdot N.}{|X - \cdot|^n} - w_n\right)(Q),$$

and consider the Neumann's function $N(X, Y)$ defined by

$$N(X, Y) = \frac{1}{(n-2)|X-Y|^{n-2}} + \frac{1}{(n-2)w_n} \int_{\partial D} \frac{1}{|Y-Q|^{n-2}} G(X, Q)\, dQ.$$

Integrating by parts we get

$$\sum_{j=1}^{n} \int_D \frac{\partial}{\partial Y_j} N(X, Y) \frac{\partial}{\partial Y_j} u(Y)\, dY = u(X) + c,$$

where c is a constant. However, if $N_a(|\nabla u|) \in L^p(\partial D)$ and $(\partial/\partial N_P)u(X) \to 0$ as $X \to P$ nontangentially, then the left-hand side above is 0 and $u(X)$ is constant in D. ■

3. NOTES

The method of layer potentials to solve boundary value problems in smooth domains is classical, and goes back to Giraud and Mikhlin. However, for C^1, and Lipschitz, domains D the techniques needed were not developed until the late 1970s. In 1977 Dahlberg, through a careful study of the Poisson kernel of D, solved the Dirichlet problem for the Laplacian in the case of C^1 domains for data in $L^p(\partial D)$, $1 < p < \infty$, and in the case of Lipschitz domains for data in $L^p(\partial D)$, $2 - \varepsilon \leqslant p < \infty$, where ε depends on D. In 1978 Fabes, Jodeit Jr, and Rivière were able to utilize Calderón's theorem on the boundedness of the Cauchy integral on C^1 curves to extend the classical methods of layer potentials to C^1 domains; their presentation is the one we followed in this chapter.

The situation is different when ∂D is not smooth. For example, Fabes, Jodeit Jr, and Lewis [1977], showed that in the case of the Laplacian the double layer potential in the first quadrant leads to an integral equation which is not solvable for $p = \frac{3}{2}$, but solvable for all other $1 < p < \infty$. Corners and edges in ∂D determine the bad p's; in particular an edge causes an interval of bad p's, whereas isolated vertices yield isolated bad p's.

Fabes and Kenig [1981] have also studied the Hardy $H^1(D)$ and $BMO(D)$ spaces and extended the results of this chapter to that setting.

Bibliography

Adams, D. R.

[1973] A trace inequality for generalized potentials, *Studia Math.* **48**, 99–105.

[1975] A note on Riesz potentials, *Duke Math. J.* **42**, 765–778.

Aguilera, N., and Segovia, C.

[1977] Weighted norm inequalities relating the g_λ^* and the area functions, *Studia Math.* **61**, 293–303.

Alvarez, J., and Milman, M.

[1985] H^p continuity of Calderón–Zygmund type operators, preprint.

Amar, E., and Bonami, A.

[1979] Mesures de Carleson d'ordre α et solutions au bord d l'equation $\bar{\partial}$, *Bull. Soc. Math. France* **107**, 23–48.

Andersen, K. F., and John, R. T.

[1980] Weighted inequalities for vector-valued maximal functions and singular integrals, *Studia Math.* **69**, 19–31.

Axler, S., and Shields, A.

[1982] Extreme points in *VMO* and *BMO*, *Indiana Univ. Math. J.* **31**, 1-6.

Baernstein II, A., and Sawyer, E. T.

[1985] "Embedding and Multiplier Theorems for $H^p(R^n)$," *Mem. Amer. Math. Soc.* **318**, Providence, R.I.

Baishanski, B. M., and Coifman, R. R.

[1978] Pointwise estimates for commutator singular integrals, *Studia Math.* **62**, 1–15.

Benedek, A., Calderón, A. P., and Panzone, R.

[1962] Convolution operators on Banach space valued functions, *Proc. Nat. Acad. Sci. U.S.A.* **48**, 356-365.

Bennett, C., DeVore, R. A., and Sharpley, R.

[1981] Weak-L^∞ and *BMO*, *Ann. of Math.* **113**, 601-611.

Boas, R. P.

[1967] "Integrability Theorems for Trigonometric Transforms," Springer-Verlag, Berlin and New York.

Brown, G.

[1977] Construction of Fourier multipliers, *Bull. Austral. Math. Soc.* **16**, 463–472.

Burkholder, D. L.

[1973] Distribution function inequalities for martingales, *Ann. Prob.* **1**, 19–42.

Burkholder, D. L., and Gundy, R. F.

[1970] Extrapolation and interpolation of quasilinear operators on martingales, *Acta Math.* **124**, 249-304.

[1972] Distribution function inequalities for the area integral, *Studia Math.* **44**, 527-544.

Burkholder, D. L., Gundy, R. F., and Silverstein, M. L.

[1971] A maximal function characterization of the class H^p, *Trans. Amer. Math. Soc.* **157**, 137-153.

Calderón, A. P.

[1960] "Integrales Singulares y sus Aplicaciones a Ecuaciones Diferenciales Hiperbólicas," *Cursos y Seminarios de Matematicas* 3, Univ. de Buenos Aires.

[1965] Commutators of singular integral operators, *Proc. Nat. Acad. Sci. U.S.A.* **53**, 1092-1099.

[1972] Estimates for singular integral operators in terms of maximal functions, *Studia Math.* **44**, 563-582.

[1976] Inequalities for the maximal function relative to a metric, *Studia Math.* **57**, 297-306.

[1977a] Cauchy integrals on Lipschitz curves and related operators, *Proc. Nat. Acad. Sci. U.S.A.* **74**, 1324-1327.

[1977b] An atomic decomposition of distributions in parabolic H^p spaces, *Advances in Math.* **25**, 216-225.

[1978] Commutators, singular integrals on Lipschitz curves and applications, *Proc. I.M.S., Helsinki*, Vol. I, 84-96.

Calderón, A. P., Calderón, C. P., Fabes, E. B., Jodeit Jr., M., Rivière, N.

[1978] Applications of the Cauchy integral on Lipschitz curves, *Bull. Amer. Math. Soc.* **84**, 287-290.

Calderón, A. P., and Capri, O. N.

[1984] On the convergence in L^1 of singular integrals, *Studia Math.* **78**, 321-327.

Calderón, A. P., and Scott, R.

[1978] Sobolev type inequalities for $p > 0$, *Studia Math.* **62**, 75-92.

Calderón, A. P., and Torchinsky, A.

[1975] Parabolic maximal functions associated with a distribution, *Advances in Math.* **16**, 1-64.

[1977] Parabolic maximal functions associated with a distribution, II, *Advances in Math.* **24**, 101-171.

Calderón, A. P., and Vaillancourt, R.

[1971] On the boundedness of pseudo-differential operators, *J. Math. Soc. Japan* **23**, 374-378.

Calderón, A. P., and Zygmund, A.

[1952] On the existence of certain singular integrals, *Acta Math.* **88**, 85-139.

[1956] On singular integrals, *Amer. J. Math.* **78**, 289-309.

[1957] Singular integral operators and differential equations, *Amer. J. Math.* **79**, 801-821.

[1978] On singular integrals with variable kernel, *Applicable Ana.* **7**, 221-238.

[1979] A note on singular integrals, *Studia Math.* **65**, 77-87.

Calderón, C. P.

[1975] On commutators of singular integrals, *Studia Math.* **53**, 139-174.

[1979] On a singular integral, *Studia Math.* **65**, 313-335.

Carleson, L.

[1962] Interpolation by bounded analytic functions and the corona problem, *Ann. of Math.* **76**, 547-559.

[1966] On convergence and growth of partial sums of Fourier series, *Acta Math.* **116**, 135-157.

Chang, S.-Y. A., and Fefferman, R.

[1982] The Calderón-Zygmund decomposition on product domains, *Amer. J. Math.* **104**, 455-468.

Chanillo, S.

[1982] A note on commutators, *Indiana Univ. Math. J.* **31**, 7-16.

Chanillo, S., and Wheeden, R. L.

[1985] Weighted Poincaré and Sobolev inequalities and estimates for weighted Peano maximal functions, *Amer. J. Math.* **107**, 1191-1226.

Chipot, M.

[1984] "Variational Inequalities and Flow in Porous Media," *Applied Math. Sci.* **52**, Springer-Verlag, Berlin and New York.

Christ, M., and Fefferman, R.

[1983] A note on weighted norm inequalities for the Hardy-Littlewood maximal operator, *Proc. Amer. Math. Soc.* **87**, 447-448.

Cohen, G. M.

[1982] Hardy Spaces: Atomic Decomposition, Area Functions and some New Spaces of Distributions, Thesis, The University of Chicago.

Coifman, R. R.

[1974] A real variable characterization of H^p, *Studia Math.* **51**, 269-274.

Coifman, R. R., and Fefferman, C.

[1974] Weighted norm inequalities for maximal functions and singular integrals, *Studia Math.* **51**, 241-250.

Coifman, R. R., Jones, P. W., and Rubio de Francia, J. L.

[1983] Constructive decomposition of *BMO* functions and factorization of A_p weights, *Proc. Amer. Math. Soc.* **87**, 665-666.

Coifman, R. R., McIntosh, A., and Meyer, Y.

[1982a] L'integrale de Cauchy sur les courbes lipschitziennes, *Ann. of Math.* **116**, 361-387.

[1982b] The Hilbert transform on Lipschitz curves, *Proc. Miniconf. on Partial Diff. Eqns., Centre for Math. Analysis, Australian Nat. Univ., Canberra, Australia*, 26-69.

Coifman, R. R., and Meyer, Y.

[1975] On commutators of singular integrals and bilinear singular integrals, *Trans. Amer. Math. Soc.* **212**, 315-331.

[1978] "Au delà des Opérateurs Pseudo-différentiels," *Astérisque* **57**, Société Math. de France, Paris.

[1979] Fourier analysis of multilinear convolutions, Calderón's theorem and analysis on Lipschitz curves, *Lecture Notes in Math.* **779**, 104-122. Springer-Verlag, Berlin and New York.

[1985] A simple proof of a theorem by G. David and J. L. Journé, preprint.

Coifman, R. R., Meyer, Y., and Stein, E.

[1983] Un nouvel espace fonctionnel adapte a l'étude des opératéurs définis par des intégrales singulières, *Proc. Conf. on Harmonic Analysis, Cortona*, Lecture Notes in Math. **992**, 1-15. Springer-Verlag, Berlin and New York.

[1985] Some new function spaces and their applications to harmonic analysis, preprint.

Coifman, R. R., and Rochberg, R.

[1980] Another characterization of *BMO*, *Proc. Amer. Math. Soc.* **79**, 249-254.

Coifman, R. R., Rochberg, R., and Weiss, G.

[1976] Factorization theorems for Hardy spaces in several variables, *Ann. of Math.* **103**, 611-635.

Coifman, R. R., and Weiss, G.

[1977] Extensions of Hardy spaces and their use in analysis, *Bull. Amer. Math. Soc.* **83**, 569-645.

Córdoba, A.

[1976] On the Vitali covering properties of a differentiation basis, *Studia Math.* **57**, 91-95.

Córdoba, A., and Fefferman, C.

[1976] A weighted norm inequality for singular integrals, *Studia Math.* **57**, 97-101.

Cotlar, M., and Cignoli, R.

[1974] "An Introduction to Functional Analysis," North-Holland, Amsterdam and London.

Cowling, M. G., and Fournier, J. F.

[1976] Inclusions and non-inclusion of spaces of convolution operators, *Trans. Amer. Math. Soc.* **221**, 59-95.

Dahlberg, B. E. J.

[1979] On the Poisson integral for Lipschitz and C^1 domains, *Studia Math.* **66**, 13-24.

Daly, J. E.

[1983] On the necessity of the Hörmander condition for multipliers on $H^p(R^n)$, *Proc. Amer. Math. Soc.* **88**, 321-325.

David, G.

[1982] Commutateurs de Calderón et Lemme de Cotlar, *Sém. d'Analyse Harmonique, 1981-1982, Publ. Math., Orsay*, 59-73.

[1984] Opérateurs intégraux singuliers sur certaines courbes du plan complexe, *Ann. Sci. École Norm. Sup.* **17**, 157-189.

David, G., and Journé, J.-L.

[1984] A boundedness criterion for generalized Calderón-Zygmund operators, *Ann. of Math.* **120**, 371-397.

de Guzmán, M.

[1981] "Real Variable Methods in Fourier Analysis," Mathematics Studies 46, North-Holland, Amsterdam, New York and Oxford.

de Guzmán, M., and Welland, G. V.

[1971] On the differentiation of integrals, *Rev. Un. Mat. Argentina* **25**, 253-276.

deLeeuw, K.

[1965] On L^p multipliers, *Ann. of Math.* **81**, 364-370.

Deng, D. G.

[1984] On a generalized Carleson inequality, *Studia Math.* **78**, 245-251.

Duren, P. L.

[1970] "Theory of H^p Spaces," Academic Press, New York.

Duren, P. L., Romberg, B. W., and Shields, A. L.

[1969] Linear functionals on H^p spaces with $0 < p < 1$, *J. Reine Angew. Math.* **238**, 32-60.

Edwards, R. E.

[1967] "Fourier Series: a Modern Introduction," Vols. I, II, Holt, Rinehart and Winston, New York.

Fabes, E. B., Jodeit Jr., M., and Lewis, J. E.

[1977] Double layer potentials for domains with corners and edges, *Indiana Univ. Math. J.* **26**, 95-114.

Fabes, E. B., Jodeit Jr., M., and Rivière, N. M.

[1978] Potential techniques for boundary value problems on C^1-domains, *Acta Math.* **141**, 165-186.

Fabes, E. B., Johnson, R. L., and Neri, U.

[1976] Spaces of harmonic functions representable by Poisson integrals of functions in *BMO* and $L_{p,\lambda}$, *Indiana Univ. Math. J.* **25**, 159-170.

Fabes, E. B., and Kenig, C.

[1981] On the Hardy H^1 space of a C^1 domain, *Ark. Mat.* **19**, 1-22.

Fabes, E. B., Kenig, C., and Serapioni, R.

[1982] The local regularity of solutions of degenerate elliptic equations, *Comm. Partial Differential Equations* **7**, 77–116.

Fefferman, C.

[1970] Inequalities for strongly singular convolution operators, *Acta Math.* **124**, 9–36.

[1971] Characterizations of bounded mean oscillation, *Bull. Amer. Math. Soc.* **77**, 587–588.

[1973] Pointwise convergence of Fourier series, *Ann. of Math.* **98**, 551–571.

Fefferman, C., and Muckenhoupt, B.

[1974] Two non-equivalent conditions for weight functions, *Proc. Amer. Math. Soc.* **45**, 99–104.

Fefferman, C., and Stein, E. M.

[1971] Some maximal inequalities, *Amer. J. Math.* **93**, 107–115.

[1972] H^p spaces of several variables, *Acta Math.* **129**, 137–193.

Fefferman, R.

[1979] A note on singular integrals, *Proc. Amer. Math. Soc.* **74**, 266–270.

Figa-Talamanca, A.

[1965] Translation invariant operators in L^p, *Duke Math. J.* **32**, 495–502.

Figa-Talamanca, A., and Gaudry, G. I.

[1967] Density and representation theorems for multipliers of type (p, q), *J. Austral. Math. Soc.* **7**, 1–6.

García-Cuerva, J.

[1983] An extrapolation theorem in the theory of A_p weights, *Proc. Amer. Math. Soc.* **87**, 422–426.

Garnett, J. B.

[1981] "Bounded Analytic Functions," Academic Press, New York.

Garnett, J. B., and Jones, P. W.

[1978] The distance in *BMO* to L^∞, *Ann. of Math.* **108**, 373–393.

[1982] *BMO* from dyadic *BMO*, *Pacific Math. J.* **99**, 351–372.

Gatto, A. E., and Gutiérrez, C. E.

[1983] On weighted norm inequalities for the maximal function, *Studia Math.* **76**, 59–62.

Gaudry, G. I.

[1966] Quasimeasures and operators commuting with convolutions, *Pacific J. Math.* **18**, 461–476.

Gehring, F. W.

[1973] The L^p-integrability of the partial derivatives of a quasiconformal mapping, *Acta Math.* **130**, 265–277.

Gilbarg, D., and Trudinger, N.

[1977] "Elliptic Partial Differential Equations of Second Order," Springer-Verlag, Berlin and New York.

Goldberg, D.

[1979] A local version of real Hardy spaces, *Duke Math. J.* **46**, 27–42.

Harboure, E.

[1984] Two weighted Sobolev and Poincaré inequalities and some applications, preprint.

Harboure, E., Macías, R. A., and Segovia, C.

[1984a] Boundedness of fractional operators on L^p spaces with different weights, *Trans. Amer. Math. Soc.* **285**, 629–647.

[1984b] Extrapolation results for classes of weights, preprint.

Hardy, G. H., and Littlewood, J. E.

[1930] A maximal theorem with function-theoretic applications, *Acta Math.* **54**, 81–116.

Hayes, C. A.

[1976] Derivation of the integrals of L^q-functions, *Pacific J. Math.* **64**, 173–180.

Hayman, W. K., and Kennedy, P. L.
[1976] "Subharmonic Functions," Academic Press, New York.
Hedberg, L.
[1972] On certain convolution inequalities, *Proc. Amer. Math. Soc.* **36**, 505-510.
Heinig, H. P.
[1976] Weighted maximal inequalities for l^r-valued functions, *Canad. Math. Bull.* **19**, 445-453.
Hörmander, L.
[1960] Estimates for translation invariant operators in L^p spaces, *Acta Math.* **104**, 93-140.
Hruščev, S. V.
[1984] A description of weights satisfying the A_∞ condition of Muckenhoupt, *Proc. Amer. Math. Soc.* **90**, 253-257.
Hunt, R. A.
[1968] "On the convergence of Fourier series, Orthogonal Expansions and their Continuous Analogues" (*Proc. Conf., Edwardsville, Ill., 1967*), Southern Illinois Univ. Press, Carbondale, pp. 235-255.
[1972] An estimate for the conjugate functions, *Studia Math.* **44**, 371-377.
Hunt, R. A., Kurtz, D. S. and Neugebauer, C. F.
[1983] A note on the equivalence of A_p and Sawyer's condition for equal weights, *Conf. in Harmonic Analysis in Honor of Antoni Zygmund*, Chicago, Ill., 1981, Vol. I, II, 156-158. Wadsworth Math. Ser., Belmont.
Hunt, R. A., Muckenhoupt, B., and Wheeden, R. L.
[1973] Weighted norm inequalities for the conjugate function and Hilbert transform, *Trans. Amer. Math. Soc.* **176**, 227-251.
Hunt, R. A., and Young, W.-S.
[1974] A weighted norm inequality for Fourier series, *Bull. Amer. Math. Soc.* **80**, 274-277.
Janson, S.
[1976] On functions with conditions on the mean oscillation, *Ark. Mat.* **14**, 189-196.
[1978] Mean oscillation and commutators of singular integral operators, *Ark. Mat.* 263-270.
Jawerth, B.
[1984] Weighted inequalities for maximal operators: linearization, localization and factorization, preprint.
Jawerth, B., and Torchinsky, A.
[1985] Local sharp maximal functions, *J. of App. Theory* **43**, 231-270.
Jodeit Jr., M.
[1971] A note on Fourier multipliers, *Proc. Amer. Math. Soc.* **27**, 423-424.
Jodeit Jr., M., and Torchinsky, A.
[1971] Inequalities for Fourier transforms, *Studia Math.* **37**, 245-276.
John, F.
[1974] Quasi-isometric mappings, *Sem. Istituto Nazionali di Alta Matematica, 1962-1963*, pp. 462-473.
John, F., and Nirenberg, L.
[1961] On functions of bounded mean oscillation, *Comm. Pure Appl. Math.* **14**, 415-426.
Jones, P. W.
[1980a] Factorization of A_p weights, *Ann. of Math.* **111**, 511-530.
[1980b] Carleson measures and the Fefferman-Stein decomposition of $BMO(R^n)$, *Ann. of Math.* **111**, 197-208.
Journé, J.-L.
[1983] "Calderón-Zygmund Operators, Pseudo-Differential Operators and the Cauchy Integral of Calderón," *Lecture Notes in Math.* **994**, Springer-Verlag, Berlin and New York.

Jurkat, W. B., and Sampson, G.
[1979] The L^p mapping problem for well-behaved convolutions, *Studia Math.* **65**, 227–238.
Kahane, J.-P.
[1970] "Series de Fourier Absolutement Convergent," Springer-Verlag, Berlin and New York.
Katznelson, Y.
[1968] "An Introduction to Harmonic Analysis," Wiley, New York.
Kenig, C.
[1980] Weighted H^p spaces on Lipschitz domains, *Amer. J. Math.* **102**, 129–163.
Kerman, R. A., and Sawyer, E. T.
[1985] Weighted norm inequalities for potentials with applications to Schrödinger operators, the Fourier transform and Carleson measures, preprint.
Kerman, R. A., and Torchinsky, A.
[1982] Integral inequalities with weights for the Hardy maximal function, *Studia Math.* **71**, 277–284.
Kinderlehrer, D., and Stampacchia, G.
[1980] "An Introduction to Variational Inequalities and their Applications," Academic Press, New York.
Kolmogorov, A.
[1925] Sur les fonctions harmoniques conjugueés et les sèries de Fourier, *Fund. Math.* **7**, 23–28.
Koosis, P.
[1980] "Lectures on H^p Spaces," *London Math. Soc. Lecture Note Series* **40**. Cambridge Univ. Press, London and New York.
Krantz, S. G.
[1982] Fractional integration on Hardy spaces, *Studia Math.* **73**, 87–94.
Krikeles, B. C.
[1983] Weighted L^p estimates for the Cauchy integral operator, *Michigan Math. J.* **30**, 231–244.
Kurtz, D. S.
[1980] Littlewood-Paley and multiplier theorems on weighted L^p spaces, *Trans. Amer. Math. Soc.* **259**, 235–254.
Larsen, R.
[1971] "An Introduction to the Theory of Multipliers," Springer-Verlag, Berlin and New York.
Latter, R. H.
[1978] A decomposition of $H^p(R^n)$ in terms of atoms, *Studia Math.* **62**, 92–101.
Löfstrom, J.
[1983] A non-existence theorem for translation invariant operators on weighted L_p-spaces, *Math. Scand.* **53**, 88–96.
Madych, W. R.
[1974] On Littlewood-Paley functions, *Studia Math.* **50**, 43–63.
Marcinkiewicz, J., and Zygmund, A.
[1939] Quelques inégalités pour les opérations linéaries, *Fund. Math.* **32**, 115–121.
Merryfield, K.
[1985] On the area integral, Carleson measures and H^p in the polydisc, *Indiana Univ. Math. J.* **34**, 663–686.
Meyer, Y.
[1979] Produits de Riesz généralisés, *Sém. d'Analyse Harmonique, 1978–1979, Publ. Math., Orsay*, 38–48.

Meyers, N. G.

[1964] Mean oscillation over cubes and Hölder continuity, *Proc. Amer. Math. Soc.* **15**, 717–721.

Miyachi, A.

[1980] On some Fourier multipliers for $H^p(R^n)$, *J. Fac. Sci. Univ. Tokyo* **27**, 157–179.

Miyachi, A., and Yabuta, K.

[1984] On good λ-inequalities, *Bull. Fac. Sci., Ibaraki Univ., Math.*, No. 16, 1–11.

Moser, J.

[1971] On a pointwise estimate for parabolic differential equations, *Comm. Pure Appl. Math.* **24**, 727–740.

Muckenhoupt, B.

[1972] Weighted norm inequalities for the Hardy maximal function, *Trans. Amer. Math. Soc.* **165**, 207–226.

[1974] The equivalence of two conditions for weight functions, *Studia Math.* **49**, 101–106.

[1983] On inequalities of Carleson and Hunt, *Conf. in Harmonic Analysis in Honor of Antoni Zygmund, Chicago, Ill., 1981*, Vol. I, II.

Muckenhoupt, B., and Wheeden, R. L.

[1974] Weighted norm inequalities for fractional integrals, *Trans. Amer. Math. Soc.* **192**, 261–274.

Namazi, J.

[1984] A singular integral, Thesis, Indiana University.

Neri, U.

[1977] Some properties of functions with bounded mean oscillation, *Studia Math.* **61**, 63–75.

Neugebauer, C. J.

[1983] Inserting A_p weights, *Proc. Amer. Math. Soc.* **87**, 644–648.

O'Neil, R.

[1966] Les fonctions conjugées et les intégrales fractionnaires de la classe $L(\log L)^s$, *C.R. Acad. Sci.* **263**, 463–466.

[1968] Integral transforms and tensor products on Orlicz spaces and $L(p, q)$ spaces, *J. D'Anal. Math.* **21**, 4–276.

Ortiz, J. A., and Torchinsky, A.

[1977] On a mean value inequality, *Indiana Math. J.* **26**, 555–566.

Oswald, P.

[1982] Fourier Series and conjugate functions in the classes $\phi(L)$, *Anal. Math.* **8**, 287–303.

Peetre, J.

[1966] On convolution operators leaving $L^{p,\lambda}$ spaces invariant, *Ann. Mat. Pura Appl.* **72**, 295–304.

[1969] On the theory of $L_{p,\lambda}$ spaces, *J. of Funct. Analysis* **4**, 71–87.

[1975] On the trace of potentials, *Ann. Scuola Norm. Sup. Pisa Serie IV*, Vol. II, No. 1, 33–43.

Power, S. C.

[1980] Vanishing Carleson measures, *Bull. London Math. Soc.* **12**, 207–210.

Quek, T. S., and Yap, Y. H.

[1983] Sharpness of Young's inequality for convolution, *Math. Scand.* **53**, 221–237.

Riesz, F.

[1932] Sur un théorème de maximum de M. M. Hardy et Littlewood, *J. London Math. Soc.* **7**, 10–13.

Riesz, M.

[1927] Sur les fonctions conjugees, *Math. Z.* **27**, 218–244.

Rivière, N. M.

[1971] Singular integrals and multiplier operators, *Ark. Mat.* **9**, 243–278.

Rosenblum, M.
[1962] Summability of Fourier series in $L^p(d\mu)$, *Trans. Amer. Math. Soc.* **105**, 32–42.
Rubio de Francia, J. L.
[1980] Vector valued inequalities for operators in L^p spaces, *Bull. London Math. Soc.* **12**, 211–215.
[1981] Boundedness of maximal functions and singular integrals in weighted L^p spaces, *Proc. Amer. Math. Soc.* **83**, 673–679.
[1982] Factorization and extrapolation of weights, *Bull. Amer. Math. Soc.* **7**, 393–395.
[1984] Factorization theory and A_p weights, *Amer. J. Math.* **106**, 533–547.
Rubio de Francia, J. L., Ruiz, F. J., and Torrea, J. L.
[1985] Calderón-Zygmund theory for operator-valued kernels, preprint.
Rudin, W.
[1958] Representation of functions by convolutions, *J. Math. Mech.* **7**, 103–116.
Ruiz, F. J.
[1985] A unified approach to Carleson measures and A_p weights, *Pacific Math.* **117**, 397–404.
Ruiz, F. J., and Torrea, J. L.
[1985a] A unified approach to Carleson measures and A_p weights, II, *Pacific J. Math.* **120**, 189–197.
[1985b] Weighted and vector valued inequalities for potential operators, preprint.
Sagher, Y.
[1977] On the Fejér-F. Riesz inequality in L^p, *Studia Math.* **61**, 269–278.
Sarason, D.
[1979] "Function Theory on the Unit Circle," Virginia Poly. Inst. and State Univ., Blacksburg, Virginia.
Sawyer, E. T.
[1982] A characterization of a two weight norm inequality for maximal operators, *Studia Math.* **75**, 1–11.
Sawyer, S. A.
[1966] Maximal inequalities of weak type, *Ann. of Math.* **84**, 157–174.
Semmes, S.
[1983] Another characterization of H^p, $0 < p < \infty$, with an application to interpolation, *Proc. Conf. on Harmonic Analysis, Cortona, Lecture Notes in Math.* **992**, Springer-Verlag, Berlin and New York.
Shi, X.
[1985] Some remarks on singular integrals, preprint.
Spanne, S.
[1965] Some function spaces defined using the mean oscillation over cubes, *Ann. Scuola Norm. Sup., Pisa* **19**, 593–608.
Stampacchia, G.
[1964] $L^{(p,\lambda)}$ spaces and interpolation, *Comm. Pure Appl. Math.* **17**, 293–306.
Stegenga, D. A.
[1976] Bounded Toeplitz operators on H^1 and applications on the duality between H^1 and the functions of bounded mean oscillation, *Amer. J. Math.* **98**, 573–589.
Stein, E. M.
[1969] Note on the class $L \log L$, *Studia Math.* **31**, 305–310.
[1970] "Singular Integrals and Differentiability Properties of Functions," Princeton Univ. Press, Princeton, New Jersey.
Stein, E. M., and Weiss, G.
[1960] On the theory of harmonic functions of several variables, *Acta Math.* **103**, 26–62.

Stein, E. M., and Weiss, N. J.
[1969] On the convergence of Poisson integrals, *Trans. Amer. Math. Soc.* **140**, 34–54.
Strömberg, J.-O.
[1979a] Bounded mean oscillation with Orlicz norms and duality of Hardy spaces, *Indiana Univ. Math. J.* **28**, 511–544.
[1979b] Non-equivalence between two kinds of conditions on weight functions, *Proc. Symp. Pure Math.*, Vol. 35, Part I, *Amer. Math. Soc.*, Providence, R.I., pp. 141–148.
Strömberg, J.-O., and Torchinsky, A.
[1980] Weights, sharp maximal functions and Hardy spaces, *Bull. Amer. Math. Soc.* **3**, 1053–1056.
Taibleson, M. H., and Weiss, G.
[1980] "The Molecular Characterization of Certain Hardy Spaces," *Astérisque* **77**, Société Math. de France, Paris, 67–149.
Torchinsky, A.
[1976] Interpolation of operations and Orlicz classes, *Studia Math.* **59**, 177–207.
Torrea Hernández, J. L.
[1984] "Integrales Singulares Vectoriales," INMABB-Conicet, Univ. Nac. del Sur, Bahia Blanca, Argentina.
Triebel, H.
[1983] "Theory of Function Spaces," Birkhauser Verlag, Basel, Boston, and Stuttgart.
Uchiyama, A.
[1978] On the compactness of operators of Hankel types, *Tôhoku Math. J.* **30**, 163–171.
[1982] A constructive proof of the Fefferman-Stein decomposition of $BMO(R^n)$, *Acta Math.* **148**, 215–241.
Uchiyama, A. and Wilson, J. M.
[1983] Approximate identities and $H^1(R)$, *Proc. Amer. Math. Soc.* **88**, 53–58.
Varopoulous, N.
[1977] *BMO* functions and the $\bar{\partial}$-equation, *Pacific J. Math.* **71**, 221–273.
Verchota, G.
[1984] Layer potentials and regularity for the Dirichlet problem for Laplace's equation in Lipschitz domains, *J. Funct. Anal.* **59**, 572–611.
Welland, G. V.
[1975] Weighted norm inequalities for fractional integrals, *Proc. Amer. Math. Soc.* **51**, 143–148.
Wheeden, R. L.
[1976] A boundary value characterization of weighted H^1, *L'Enseign. Math.* **22**, 121–134.
Wilson, J. M.
[1985] On the atomic decomposition for Hardy spaces, *Pacific J. Math.* **116**, 201–207.
Yabuta, K.
[1985] Generalizations of Calderón-Zygmund operators, *Studia Math.* **82**, 17–31.
Young, R. M.
[1980] "An Introduction to Nonharmonic Fourier Series," Academic Press, New York.
Young, W.-S.
[1982] Weighted norm inequalities for the Hardy-Littlewood maximal function, *Proc. Amer. Math. Soc.* **85**, 24–26.
Zafran, M.
[1975] Multiplier transformations of weak-type, *Ann. of Math.* **101**, 34–44.
Zó, F.
[1976] A note on the approximation of the identity, *Studia Math.* **55**, 111–122.
[1978] A note on approximation to the identity with iterated kernels, *Anal. Math.* **4**, 153–158.

Zygmund, A.
[1929] Sur les fonctions conjuguées, *Fund. Math.* **13**, 284–303.
[1956] On a theorem of Marcinkiewicz concerning interpolation of operators, *J. Math. Pures Appl.* **35**, 223–248.
[1968] "Trigonometric Series," 2nd Ed., Cambridge Univ. Press, London and New York.

Index

D

E

I

J

K

L

Zygmund class $L \ln L$, 92, 102

www.ingramcontent.com/pod-product-compliance
Lightning Source LLC
Chambersburg PA
CBHW031621210326
41599CB00021B/3254